Optical Data Processing

WILEY SERIES IN PURE AND APPLIED OPTICS

Advisory Editor
STANLEY S. BALLARD University of Florida

Lasers, BELA A. LENGYEL
Ultraviolet Radiation, second edition, LEWIS R. KOLLER
Introduction to Laser Physics, BELA A. LENGYEL
Laser Receivers, MONTE ROSS
The Middle Ultraviolet: its Science and Technology, A. E. S. GREEN, *Editor*
Optical Lasers in Electronics, EARL L. STEELE
Applied Optics, *A Guide to Optical System Design/Volume 1*, LEO LEVI
Laser Parameter Measurements Handbook, HARRY G. HEARD
Gas Lasers, ARNOLD L. BLOOM
Advanced Optical Techniques, A. C. S. VAN HEEL, *Editor*
Infrared System Engineering, R. D. HUDSON
Laser Communication Systems, WILLIAM K. PRATT
Optical Data Processing, A. R. SHULMAN

Optical Data Processing

ARNOLD ROY SHULMAN

The Advanced Development Division
Goddard Space Flight Center

JOHN WILEY & SONS, INC.

NEW YORK · LONDON · SYDNEY · TORONTO

10 9 8 7 6 5 4 3 2

Library of Congress Catalog Card Number: 70–82970
SBN 471 78980 1.

Printed in the United States of America

To my mother
SELMA SHULMAN
and in memory of my father
ALEXANDER SHULMAN
for encouraging me to pursue this work

Preface

Technology is advancing so rapidly that it is extremely difficult for scientists and engineers to keep pace with developments in their own fields. It is reasonable therefore to expect that developments outside their specialty may go unnoticed or unappreciated. One area in which considerable advancement has recently occurred is optical data processing. It is the purpose of this book to make available the general technology of the subject to those not normally engaged in work in this field. This book therefore serves the general scientific community by providing a technical background in and working knowledge of many of the disciplines related to optical data processing. One of its most important aspects is the attempt to unify the apparently divergent disciplines of this field.

Optical Data Processing is based on my experience as an electrical engineer in acquiring a working knowledge in this subject. It presents the necessary introductory information in a way that will help others who are interested in understanding its basic principles. Most available literature on optical data processing is at a rather advanced level and does not contain adequate introductory material. To fill this gap I have attempted to present an introductory treatment of coherent optics with sufficient emphasis on background details to satisfy most engineers.

The choice of material and method of presentation is based on my own experience and reflects the type of information I should have found useful in my transition to this field. The object is to give the reader a feel for the various principles and disciplines and to provide a basic understanding of practical techniques to stimulate his interest in pursuing more advanced treatments of individual topics. For such further study the material in this book constitutes a basic survey of the areas of knowledge required.

In attempting a complete treatment of background subjects I may be accused of verbosity because of my excessive attention to detail. However, I feel that unless the reader has been working with the principles involved the details are important in a book of this kind. Specifically, mathematical examples are worked out to avoid placing the burden of derivation on the less mathematically oriented reader and to provide him with a guide to mathematical analysis and practical techniques to be used in the solution of complex problems.

I should like to thank Mr. Gerald J. Grebowsky for his generous help in the preparation of this book. To a good measure I have relied on his advice and opinions. In addition I should like to point out that I have modified and revised material that appeared in a document written by Mr. Grebowsky for NASA and converted it to Chapter 4. I feel that any good points this book may have were enhanced by Mr. Grebowsky's suggestions. I should also like to thank Mr. Jacques Breton who was able to take many of my "way-out" ideas and bring them to earth by showing possible techniques for their implementation.

In addition, nearly all of the photographs in this book were obtained from NASA, Goddard Space Flight Center, Greenbelt, Maryland 20771, where several groups are presently engaged in work in the field of coherent optics.

Arnold R. Shulman

Bowie, Maryland
March, 1969

Contents

Introduction   1

1.   Geometric Optics   5
2.   Characteristics of Light   60
3.   Fourier Transforms   95
4.   The Fourier Transform by Diffraction of Light   156
5.   Optical Spectrum Analysis   237
6.   Characteristics of Photographic Film   302
7.   Optical Filtering and Correlation   324
8.   Analysis of Optical Data Processing Systems   358
9.   Zone Plates   387
10. Holography   453
11. Holographic Techniques   502
12. Properties and Techniques for Photographic Reproductions   549

Appendices   585
References   702
Index   707

Introduction

Through the years techniques developed in one field are eventually applied to other fields. One field in which there has been a growing interest in recent years is the field of optical data processing. Optical data processing presents challenges to almost every phase of engineering. Optical data processing is a field, however, which has universal appeal. It permits selection of a narrow area of work for specialization or permits traversing not only the bounds of one speciality to another but from one engineering field to another. Photography enthusiasts will find the optical data processing field extremely appealing. This field offers the amateur as well as the professional photographer an opportunity to explore every facet of photography. Certain areas of photography rely heavily on chemistry, which offers important areas of work in the optical data processing field for chemists. Various areas of optical data processing also offer challenges to physicists, materials engineers, and mechanical engineers. Electrical engineers in particular will find optical data processing closely related to electronics particularly in the area of mathematical analysis.

The mathematical techniques used in optical data processing are identical in a great many respects to those used in electronics. One problem that may be encountered is that unless recent courses have been taken in certain specific topics, the mathematical techniques may have been forgotten. One of the purposes of this book is to reintroduce not only the mathematics, but also the optical techniques from a relatively simple level (Freshman college level) and build up the necessary background to permit analysis of the complex phenomena and provide the necessary foundation for continued study at higher levels.

For the most part optical systems are commonly thought of only in terms of focused imaging as found in cameras, microscopes, and telescopes. This book will show that using basic lens systems it is possible to implement most of the mathematical operations found in modern com-

1

munication theory. Optical data processing permits with a relatively simple configuration of lenses and apertures the implementation of mathematical functions such as multiplication, Fourier transformation, correlation, and convolution. Through the Fourier transform operation, the frequency distribution of a signal is represented by a light amplitude distribution in a plane, filtering of a signal optically is a simple operation since it requires only the blocking of the light at points corresponding to unwanted frequencies. Optical filtering techniques require only simple configurations of standard optical components, while the electronic counterparts performing the same type of operation would result in complex hardware.

The signals in an optical system are in the form of light amplitude distributions in planes perpendicular to the optical axis. An input signal must therefore be in the form of optical transmission variations in a plane — for example, a photographic transparency. A signal can vary from point to point in a plane and thus depends on two coordinates. This two-dimensional characteristic of optical signals represents an advantage over electronic systems that are limited to only one coordinate (usually time). Two-dimensional signals such as pictures or printed matter (pages) can be processed as a whole without scanning as in electronic systems (which is required to reduce them to a single dimension). Thus an entire block or page of information can be processed simultaneously rather than bit by bit.

The two-dimensional capability of optical systems can be used to provide simultaneous multichannel operation without significant increase in the optical systems complexity. In the multichannel optical system the second dimension is used to separate independent channels. Multiple signal channels are usually arranged in the form of bars or stripes with each input signal variation appearing as optical transmission variations along the length of a bar. Desired operations are performed only with respect to the dimensions along which the individual signals vary. All of the channels can be processed at the same time without scanning or time sharing. The number of independent channels is limited only by the resolution of the optical system. This multichannel capability therefore can lead to considerable simplification of equipment when many signals are available and require simultaneous processing.

The already attractive capabilities of optical systems can be further enhanced by applying holographic techniques. Holograms are photographic recordings of interference patterns which are capable of producing three-dimensional images. The three-dimensional aspects of the images produced by holograms are quite startling and have received a great deal of publicity. One of the more important uses for holograms, however, has been little publicized. The hologram can be a basic component in optical data processing systems. A hologram in effect records both the amplitude and phase variations of a light wave. Through the use of holograms, signals can be processed that have complex amplitude and phase variations. Complex signal representations are especially useful in pattern recognition and complex filtering applications. In addition to enhancing the capabilities of optical processing systems holograms may provide a new means for storing information. The advantages of a hologram storage device would include a high storage density per unit volume (with redundancy) such that sections of the hologram recording can be completely destroyed without losing a significant amount of the stored information. Holography thus not only enhances optical processing capabilities but can also provide a means for storing information with high reliability.

The ultimate potential of optical computing or data processing systems is not yet conceivable. Working systems have been developed for the processing of information in the specialized areas of seismography, radio astronomy, and radar. In addition optical systems have the potential capability of processing communications signals and other complex data signals.

Rapid development of optical systems have been hindered by limitations in obtaining real-time optical inputs to the optical processor. Photographic film is presently being used almost exclusively as an input media and the conversion of signals to this format cannot be accomplished too easily. It is expected that this restriction will be eliminated in the near future by developments being made in other areas—for example, photochromics, electrochromics, cathodochromics, electromagnetics, light modulators, etc. (Some of these newer techniques and their applications are discussed in the book.)

The inherent flexibility and simplicity of optical systems warrant their wider use. This book will describe the necessary basic techniques for starting work in this relatively new field of optical data processing.

This book is intended for individuals with some technical background; it is therefore suitable for use by engineers or physicists as well as technicians. Engineers will find this book descriptive and enlightening in this new field, providing the necessary background and insight to permit work on a practical level. Technicians, even without a mathematical background, will find the book provides them with sufficient knowledge and understanding of the field to work productively. There will be those who might feel that the book should either have been written for technicians or engineers, but not for both. Engineers with this opinion will probably feel that the first chapters are unnecessary because these deal basically with fundamentals. In order to avoid such possible criticism I have attempted to present the fundamentals using illustrations that normally are not found in textbooks but at the same time are of general interest and can be useful in optical processing. In addition I have tried to provide sufficient information at a level that would be useful for engineering assistants and technicians to gain sufficient knowledge of the field so that they would be productive. It is expected that engineers will want this material available for reference but might not necessarily have to read it carefully sentence by sentence. Depending on the individual some may find the review material not necessary for their understanding of the phenomena explained later in the book.

In any event I have not diluted this book to make it have a little something for everyone. This book contains the necessary basic tools to impart confidence and understanding for work in the optical data processing field. To cut down the amount of reading diagrams are used profusely. The mathematical results obtained are also explained in words to assure understanding. I did try to put as much pertinent basic information as I could between two covers, with the hope that sufficient background material is presented to prepare one for reading higher level material relative to the current work going on in this field.

1

Geometrical Optics

Optics is the branch of physics that deals with the nature and properties of light. The properties of light are difficult to describe exactly and therefore descriptions are usually methods for approximation of the various phenomena. Three general methods are used for describing different properties of light. Each of these methods is particularly suited to explaining different types of light phenomena. The three different treatments have been broadly classified as geometrical optics, physical optics, and quantum optics. Geometrical optics is used when the propagation of light is assumed to be in straight lines or rays. Physical optics is used when the interaction of light with light is being considered — that is, the wavelike nature of light is being considered. The variations between the results predicted by geometrical and physical optics is only detectable when carefully controlled experiments are made. This tends to imply that geometrical optics will produce results in good agreement with large-scale experimental results provided minute variations from predicted results can be disregarded. When the fine variations from the results obtained by geometrical optics must be accounted for, the more rigorous methods of physical optics must be used. Quantum optics requires a particle (photon) treatment of light and is applicable when light interacts with matter (i.e., the photoelectric effect). There is no single theory that explains all light phenomena. For any particular problem the theory used is that which applies to the phenomena.

Originally it was thought that light obeyed three simple laws concerning its propagation, reflection, and refraction. Based on these laws it was possible to explain many optical phenomena by geometry. The method of treating optical phenomena by geometry is called *geometrical optics*. Geometrical optics makes use of rays to approximate light phenomena. Although geometrical optics proves to be somewhat lacking in very carefully controlled experiments, the results (to a first approximation) are generally acceptable.

The first law of geometrical optics states that light propagates in straight paths in a homogeneous medium and is called the *law of rectilinear propagation*. Two laws concerning reflection and refraction are described in detail later in this chapter. The three laws are all summed up in Fermat's principle which states that light proceeds from one point to another by an optical path whose optical length is either a maximum or minimum.

Geometrical optics does not explain all the properties of light and physical optics is used to explain the wavelike nature of light. The wavelike properties of light are introduced in Chapter 2 and form the basis for optical data-processing techniques. The quantum nature of light is important when dealing with the interaction of light with matter. Except for certain isolated cases the quantum nature of light is of little interest in optical data processing. The quantum nature of light, however, is extremely important in the construction of devices and materials used in optical data processing—for example, lasers and photochromic materials. In optical data processing as discussed in this book it will just be necessary to be aware of the fact that light does (when considered on a microscopic level) consist of energy particles called *photons*. Since optical data processing in general deals with the macroscopic effects of light, the quantum nature of light is essentially outside the scope of this book and will not be discussed in detail.

REFLECTION AND REFRACTION FROM PLANE SURFACES

Reflection of a Point Source from a Plane Surface

Geometrical optics is primarily the study of the laws of reflection and refraction. Ray diagrams are generally used to show the direction of energy flow. Figure 1.1 indicates how reflection is depicted by a ray diagram.

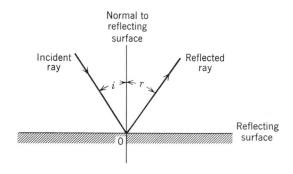

Figure 1.1 Reflection of a ray.

An incident ray is shown to strike a reflecting surface at point O. The angle between the incident ray and the normal to the reflection surface is called the *angle of incidence* ($\angle i$). The incident ray is reflected from point O such that the angle between the normal and the reflected ray (angle of reflection [$\angle r$]) is equal to the angle of incidence. This holds true only in optically homogeneous, isotropic materials (isotropic means that the properties are the same in all directions). The law of reflection states that the angle between the incident ray and the normal to a reflecting surface equals the angle between the normal and the reflected ray — that is, $\angle i = \angle r$.

Image of a Point Source in a Plane Surface

Figure 1.2 shows the image of a point source formed by reflection. (A point source of light can only be approximated since it is really a mathematical abstraction and any real light source will always have finite dimensions.) A light ray emanating from the point source S, striking normal to the reflecting surface, will be reflected back on itself (so as to be reflected back to the source S). If this ray is extended beyond the reflecting surface, a point S' can be found such that $SN = NS'$. The image S' of the point source S is located on a line through S, normal to the reflecting surface, and is as far in the reflecting surface as the object point source S is in front of it. A ray ST, not normal to the reflecting surface, will on striking the surface be redirected along TU such that the angle of incidence ($\angle i$) the ray ST makes with the normal to the reflecting surface is equal to the angle ($\angle r$) the reflected ray TU makes with the normal (the law of reflection). This makes the redirected ray TU appear to an observer to come from the image S'. Another ray SV makes an angle ($\angle i'$) with the normal and is reflected along VW such that $\angle i' = \angle r'$.

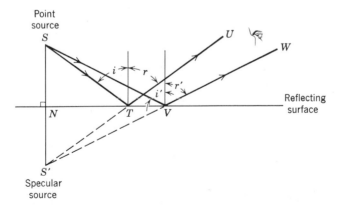

Figure 1.2   Image formed by reflection of a point source from a plane surface.

The reflected rays would appear to an observer to be diverging along straight lines from a light source at S'. Reflections of light that take place in a definite direction (which is determined by the incident ray being in the same plane as the reflected ray and the normal to the reflecting surface) are called *specular reflections*. The point S' is therefore called the *specular image* of S.

Highly polished reflecting surfaces are often referred to as mirrors. The polished surfaces of many metals reflect almost 100 percent of the incident light and therefore are often used as mirrors. Reflections from a rough surface are diffused or scattered as shown in Figure 1.3(a). Reflections from a smooth surface are regular as shown in Figure 1.3(b).

Image of an Object in a Plane Surface

The image of a point source of light formed by reflection in a plane mirror is located the same distance behind the mirror surface as the object is in front of it. For an extended object such as shown in Figure 1.4 each point in the image can be located as previously described for the point source shown in Figure 1.2. Figure 1.4 shows the essential construction for locating the image of an object (only the rays at the extremes of the object were used). The rays from each point on the object will enter the eye in a diverging beam of light; that is, a ray of light that enters the eye enters from a definite direction. The human brain interprets this to mean

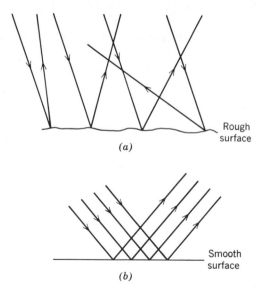

Rough surface

(a)

Smooth surface

(b)

Figure 1.3 *(a)* Diffuse reflections from an irregular surface. *(b)* Regular reflections from a smooth surface.

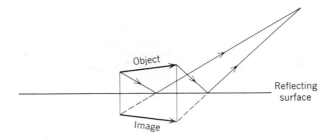

Figure 1.4   Image of an object formed by reflection.

that the object lies along a straight line in the direction from which the light entered the eye. When two (or more) rays are received simultaneously from a given point on an object, the brain assumes that the point lies at the intersection of the lines along which the rays entered the eye. The law of reflection tends to be ignored because of its simplicity; however, it becomes extremely important in practical experimental setups.

Images by Multiple Reflections

Figure 1.5 shows how the images of an object placed between two mirrors at 90-degree angles to each other can be found. Figure 1.5(a) shows the location of image 1 formed by the reflection of the object in the reflecting surface A. Figure 1.5(b) shows image 2 of the object formed in the reflecting surface B. Figure 1.5(c) shows the image 3 formed by the reflection of image 1 in the reflecting surface B. Figure 1.5(d) shows the composite images of Figure 1.5(a), (b), and (c) superimposed on each other. The location of the images of any object in mirrors inclined to each other can be found by following procedures similar to those illustrated for mirrors placed at 90 degrees to each other.

The law of reflection states that the reflection from a plane surface is such that the angle of reflection ($\sphericalangle NOB$ of Figure 1.6(a)) is equal to the angle of incidence ($\sphericalangle AON$) and both angles lie in the same plane. If the reflecting surface on which the ray falls is rotated, the reflected ray is also rotated. The amount that the reflected ray is rotated is twice that through which the mirror has been rotated. This can be seen from Figure 1.6(b). Figure 1.6(a) shows an incident ray striking a mirror at an angle i. This incident ray is reflected at an angle r, where

$$\sphericalangle i = \sphericalangle r \qquad (1-1)$$

Figure 1.6(b) shows the mirror rotated through an angle φ. The position through which the normal has been rotated is therefore also φ. The

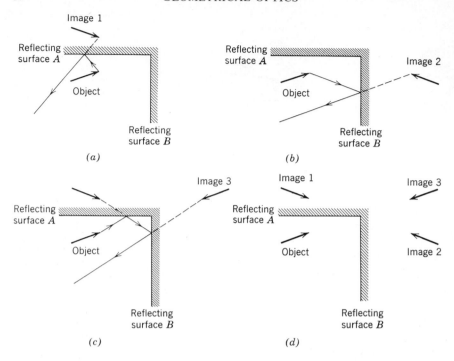

Figure 1.5 Images of an object placed between mirrors 90 degrees. (*a*) Image 1 of object in reflecting surface *A*. (*b*) Image 2 of object in reflecting surface *B*. (*d*) Images of object formed by two reflecting surfaces 90 degrees to each other.

original incident ray, therefore, now strikes the mirror at a new angle of incidence

$$i' = i + \varphi = r' \tag{1-2}$$

The reflected ray has therefore been rotated through twice the angle through which the mirror has been rotated.

Reflection from Inclined Mirrors

Figure 1.7 shows two mirrors inclined at an angle θ to each other. A ray from light source S strikes mirror A and is reflected to mirror B. The reflected ray from mirror B will be deviated from the original direction of the incident ray by an angle equal to twice the angle between the mirrors; that is, a ray reflected from two plane mirrors is deviated an angle twice the angle between the mirrors. This can be shown as follows. In $\triangle AOB$

$$\angle\theta + \angle\varphi + \angle\psi = 180° \quad \text{or} \quad \angle\theta = 180° - \angle\varphi - \angle\psi \tag{1-3}$$

From the law of reflection we have

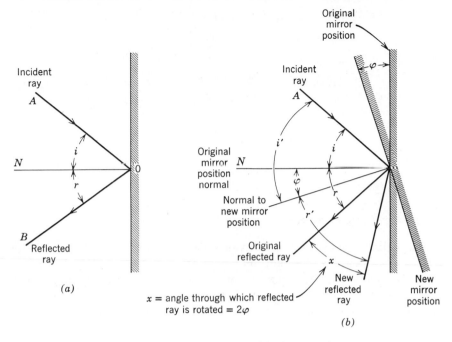

Figure 1.6   Movement of a reflected ray with mirror rotation.

$$\measuredangle \varphi + \measuredangle r = 90° \tag{1-4}$$

and

$$\measuredangle \psi + \measuredangle i' = 90° \tag{1-5}$$

Then substituting (1-4) and (1-5) into (1-3)

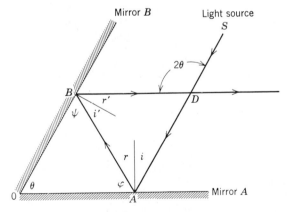

Figure 1.7   Reflection of a ray from two inclined plane mirrors.

$$\measuredangle\theta = 180° - (90° - \measuredangle r) - (90° - \measuredangle i') = 180° - 90° + \measuredangle r - 90° + \measuredangle i'$$
$$\measuredangle\theta = \measuredangle r + \measuredangle i' \qquad (1\text{-}6)$$

In $\triangle DAB$

$$2i' + 2r + \measuredangle BDA = 180°$$
$$\measuredangle BDA = 180° - 2i' - 2r \qquad (1\text{-}7)$$

and

$$\measuredangle ADS = 180°$$

Therefore

$$\measuredangle SDB = 180° - \measuredangle BDA = 180° - 180° + 2i' + 2r = 2i' + 2r$$
$$\measuredangle SDB = 2i' + 2r = 2\measuredangle\theta \qquad (1\text{-}8)$$

INDEX OF REFRACTION

The velocity of light (for all wavelengths) is a constant in vacuum. The velocity of light in different mediums is, however, different for the different mediums. The index of refraction, n, of a medium is the ratio of the velocity of light in a vacuum, c, to the velocity of light (of a specific wavelength) in the medium, v. The index of refraction, n, is defined by

$$n = \frac{c}{v} \qquad (1\text{-}9)$$

where c is a constant 3×10^8 meters/sec, and v is the velocity of light in the medium.

The index of refraction of water is 1.3330. This means that it will take the same time for light to travel through 1 cm of water as it will to travel through 1.3330 cm of air. The refractive index, n, can therefore be taken to indicate numerically the equivalent distance that light could travel in air when it has traveled 1 cm in a substance. Table 1.1 shows the refractive indices of same common substances.

TABLE 1.1

Substance	Refracted Index	Critical Angle with Air ($\sin i_c = 1/n$) (Degrees)
Air	1.00029	90
Flint glass	1.53–1.96	39
Crown glass	1.48–1.62	42
Diamond	2.42	24
Water	1.3330	49
Carbon	1.31	50

REFRACTION

When light passes from one medium to another, two new light beams are formed at the interface of the two mediums. Consider an incident ray making an angle, i, with the normal to an interface between two mediums as shown in Figure 1.8(c). The incident ray will be broken up

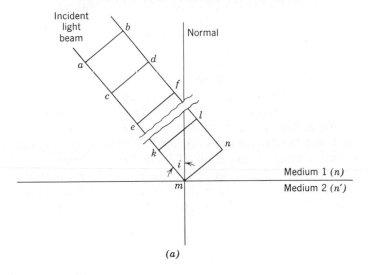

(a)

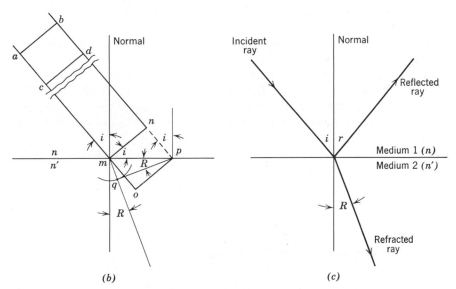

(b)   *(c)*

Figure 1.8.

into two new rays at the interface—a reflected and a refracted ray. The reflected ray travels back into the original medium such that

$$\sphericalangle i = \sphericalangle r \tag{1-1}$$

where $\sphericalangle r$ is the angle the reflected ray makes with the normal. The velocity of the light ray in a medium is given by

$$v = \frac{c}{n} \tag{1-10}$$

where v is the velocity of light, c is a constant 3×10^8 meters/sec, and n is the index of refraction of medium 1.

In Figure 1.8a it will take light a specific time, traveling at a velocity v, to travel from position $\overline{ab}$ to position $\overline{cd}$. In an equal amount of time the light will travel from position $\overline{cd}$ to position $\overline{ef}$ (if the distances $\overline{ac} = \overline{bd} = \overline{ce} = \overline{df}$). When the light reaches $\overline{mn}$, a portion of the light beam has struck the interface surface (at point m). If there were no interface, the light at $\overline{mn}$ would reach position $\overline{op}$ in the next time interval (traveling at a velocity v and where $\overline{mo} = \overline{np} = \overline{ac} = \overline{bd}$, etc.) as shown in Figure 1.8(b). The velocity of the light in the second medium (v') is given by

$$v' = \frac{c}{n'} \tag{1-11}$$

A portion of the light leaving point m will be traveling in the medium 2 at a new velocity (v'). A portion of the light reaching point m will be reflected (not shown in Figure 1.8(b)). The change in velocity of the light in the medium 2 is accompanied by an abrupt change in the direction of travel of the light called *refraction*.

In the time that the light travels from n to p in medium 1 the light does not travel the equal distance from m to o in medium 2. The relation between the distance $\overline{mq}$ that the light actually travels and the distance $\overline{mo}$ that it would have traveled in medium 1 can be found simply by the relation

$$t = \text{distance/velocity.} \tag{1-12}$$

Using the velocity v and v' in an equal time interval t we obtain

$$t = \frac{\overline{mo}}{v} = \frac{\overline{mq}}{v'} \tag{1-13}$$

Substituting for v and v' from (1-10) and (1-11)

$$\frac{\overline{mo}}{c/n} = \frac{\overline{mq}}{c/n'} \tag{1-14}$$

and rearranging terms we obtain

$$\overline{mq} = \frac{n}{n'} \, \overline{mo} \tag{1-15}$$

An arc centered at m of length $\overline{mq}$ corresponds to all the possible directions that the light may have taken in the second medium. Similarly arcs can be drawn for all points on the wavefront mn as they travel into medium 2. The line $\overline{pq}$ tangent to these arcs represents the wavefront in medium 2. The angle between the normal to the surface and the refracted ray $\overline{mq}$ is called the *angle of refraction R*. From the geometry of Figure 1.8(b) the sine of the angle i and R are given as

$$\sin i = \frac{\overline{np}}{\overline{mp}} \tag{1-16}$$

$$\sin R = \frac{\overline{mq}}{\overline{mp}} \tag{1-17}$$

The ratio of the sines gives us the relationship

$$\frac{\sin i}{\sin R} = \frac{\overline{np}}{\overline{mq}} \tag{1-18}$$

In the discussion above $\overline{mo} = \overline{np}$ and therefore

$$\overline{mq} = \frac{n}{n'} \overline{np} \tag{1-19}$$

Substituting for $\overline{mq}$ in the ratio expression we obtain

$$\frac{\sin i}{\sin R} = \frac{n'}{n} \tag{1-20}$$

which is known as *Snell's law*. For two given media the ratio n'/n is constant.

LATERAL DISPLACEMENT OF A RAY BY REFRACTION

Figure 1.9 shows that a ray of light on emerging into the original medium after passing through a parallel plate of glass will be laterally displaced but not deviated; that is, the angle with which the ray emerges from the glass is the same as the angle at which it entered. We can also

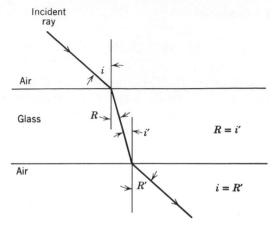

Figure 1.9 Lateral displacement of a ray passing through a parallel sided glass plate.

$$n_{\text{air glass}} = \frac{V_{\text{air}}}{V_{\text{glass}}} = \frac{\sin i}{\sin R} \left.\begin{array}{c} \\ \\ \end{array}\right\} \quad \frac{V_{\text{air}}}{V_{\text{glass}}} \times \frac{V_{\text{glass}}}{V_{\text{air}}} = 1$$

$$n_{\text{glass air}} = \frac{V_{\text{glass}}}{V_{\text{air}}} = \frac{\sin i^1}{\sin R^1}$$

infer from the equation shown in Figure 1.9 that the index of refraction from glass to air must be the reciprocal of the index of refraction from air to glass. Figure 1.10 shows that a ray of light is laterally displaced but not deviated when passing through more than one parallel-sided glass plate and back into the original medium.

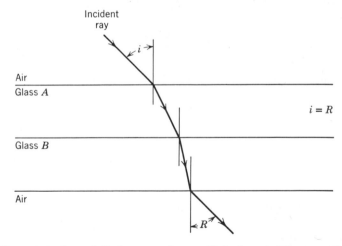

Figure 1.10 Lateral displacement of a ray of light through different media.

$$\frac{V_{\text{air}}}{V_{\text{glass A}}} \times \frac{V_{\text{glass A}}}{V_{\text{glass B}}} \times \frac{V_{\text{glass B}}}{V_{\text{air}}} = 1 \qquad \angle i = \angle R$$

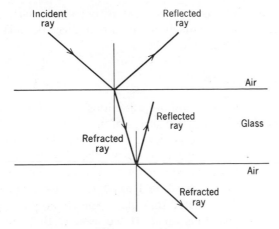

Figure 1.11 Refraction and reflection of a ray of light passing through a parallel-sided glass plate.

Figure 1.9 shows an incident ray of light entering a glass and being refracted when entering and leaving the glass. The reflected rays were omitted in this figure for clarity. Figure 1.11 shows the same incident ray as shown in Figure 1.9, but it also shows that a reflected ray is produced at each interface. The reflected rays of light can give rise to multiple reflections.

REFLECTION FROM AN AIR TO GLASS SURFACE

The amount of light that is reflected from a transparent surface depends on the angle of incidence and the refractive index of the material. For incident light perpendicular to the surface the ratio of the intensity I of the reflected light to the intensity of the incident light I_0 is given by the equation

$$\frac{I}{I_0} = \left[\frac{n_2 - n_1}{n_2 + n_1}\right]^2 \qquad (1\text{-}21)$$

where n_1 and n_2 are the refracted indices of the two materials. If we consider an incident light ray in air striking a glass surface, we can determine the portion of light that enters the glass by using (1-21). If we consider the refracted index of glass to be 1.53 (from Table 1.1), the ratio of the intensities of reflected light to incident light is

$$\frac{I}{I_0} = \left[\frac{n_2 - n_1}{n_2 + n_1}\right]^2 = \left[\frac{1.53 - 1.00}{1.53 + 1.00}\right]^2 = \left[\frac{0.53}{2.53}\right]^2 = (0.21)^2 = 0.044$$

This means that a little over 4 percent of perpendicularly incident light will be reflected back on itself at an air-to-glass boundary (about 96 percent of the light will enter the glass).

MULTIPLE REFLECTIONS

Figure 1.12 shows how multiple reflections of a point source S in front of a thick glass plate can cause a series of multiple images. In Figure 1.12 the effects of refraction are neglected for clarity in the diagram. A ray from the point source S is shown to strike the glass plate at point A and is partially reflected, which will produce a bright spot on the screen at position 1. An observer at position 1 appears to see image S_1 behind the glass plate. We have calculated that glass reflects about 4 percent of the perpendicular light that falls on it. If we assume that the ray from the source S to point A is nearly perpendicular to the glass surface, we can also assume that the reflected ray to position 1 will be about 4 percent of the intensity of the incident beam. This means that about 96 percent of the incident light will continue into the glass to strike the rear surface at point B. At point B 4 percent of the incident light will be reflected toward point C and 96 percent of the incident light to point B will be transmitted through the glass and be lost. A bright spot will be formed on the screen at position 2 and the image S_2 can be seen from that position. Image S_2 (as seen from position 2) will have an intensity of approximately 1/25 the incident light to point B (which is approximately the same as the incident light to point A). At point C 4 percent of the incident light from point B will be reflected toward point D. At point D 4 percent of the incident light is reflected toward point E, which permits an observer at position 3 to see image S_3. At point D 96 percent of the incident light continues through the glass and is lost. A few simple calculations will show that the light is very rapidly attenuated. In fact it is attenuated so rapidly that only a few images can be seen. The light reflected to position 1 is only 4 percent (1/25) the incident light at point A and the light at position 2 is also approximately 4 percent of the incident light as it passes through the glass; 96 percent of the incident light at point A will arrive at point B (4 percent of the light at point A was reflected to position 1). The light reaching point B can therefore be considered to have approximately the same intensity as the incident light reaching point A. The reflected light from point B toward position 2 is therefore slightly less than 1/25 of the incident light to point A. The light to position 1 and position 2 can therefore be considered to be equal. One twentyfifth of the incident light to point C will be reflected toward point D. This means that because of the reflection of light from point C, only 1/625 of the original light will be

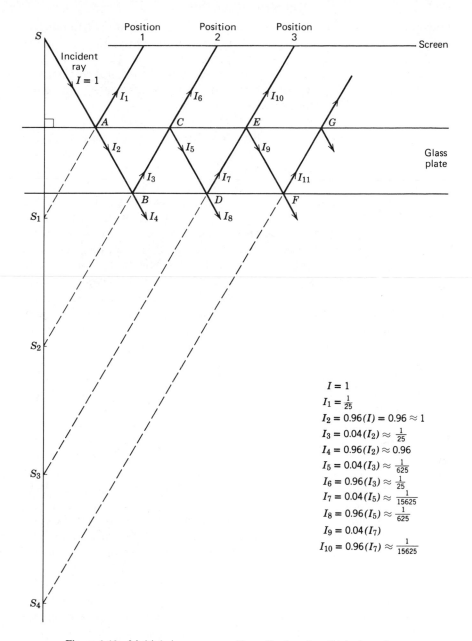

Figure 1.12 Multiple images caused by reflections in a thick glass plate.

19

reflected toward point D. One twentyfifth of the incident light to point D will be directed toward point E. This means that the image produced at position 3 will have an intensity of approximately 1/15625 the intensity of the incident light to point A. We can see that the intensity of each of the images formed by multiple reflections from a plate glass are quite rapidly attenuated after the first two reflections.

Figure 1.13 shows how a series of multiple images are formed by multiple reflections of a point source S in front of a thick glass plate, the back of which has been silvered. In Figure 1.13 the effects of refraction were neglected for clarity in the diagram. The ray being considered is initially reflected to position 1 and produces a bright spot on the screen. The image S_1 can be seen if the eye is placed at position 1. The reflected light to position 1 is 4 percent of the incident light and therefore produces an image S_1 which appears to have an intensity of only 1/25 the source S. The light transmitted through the glass to point B is essentially equal to the incident light. The light reaching position 2 is therefore approximately equal to the incident light since the reflection from the silvered surface at point B will be almost total reflection. This means that an extremely bright image (in comparison to that produced at position 1) will be seen from position 2. The reflected ray from point C will have an intensity of 4 percent the incident ray. The reflected ray from point C striking point D will be totally reflected to point E since the rear surface of the glass is silvered. This means that the intensity of the light reaching position 3 will be approximately equal to the intensity of the light that reached position 1. Four percent of the incident light to point E will be reflected to point F. At point F the intensity of the reflected ray will be approximately equal to the intensity of the incident ray to point F since the rear surface is silvered. The intensity of the light reaching position 4 will therefore be approximately 1/625 the incident light to point A. By similar reasoning the intensity of the image S_5 from position 5 will be 1/15625 the incident light to point A.

The photograph in Figure 1.14 shows images formed in a thick glass plate that has been silvered on the back surface. (Appendix I gives a simple procedure for making such a rear-surfaced mirror.) The images of the high-intensity filament are easy to identify, and five images are clearly visible. There is a less intense image formed on each side of a bright image. The intensity of the less bright images is approximately 1/25 that of the center-reflected bright image. The remaining images are 1/25 the intensity of the previous image (see Figure 1.13). Figure 1.15 shows a diagram for the construction of the first three images shown in Figure 1.14. The refraction effects of the glass have been included in this diagram. Rays are emitted from the light source S in all directions.

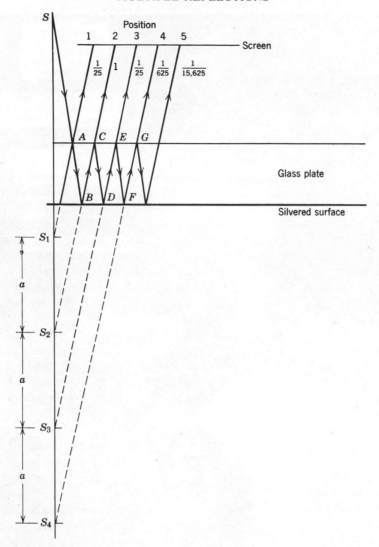

Figure 1.13 Multiple images caused by reflections in a thick glass plate the rear surface of which is silvered.

For simplicity we shall only consider three particular rays of the infinite number emitted. Considering ray 2 we can see that it will be reflected from the rear-surface mirror to the eye forming an apparent image 2 along the line on which the light is reaching the eye. Image 2 has an intensity approximately equal to the incident light. Ray 1 is reflected from

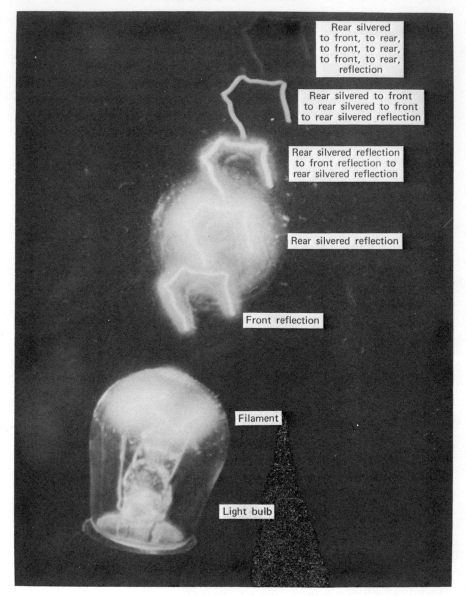

Rear silvered
to front, to rear,
to front, to rear,
to front, to rear,
reflection

Rear silvered to front
to rear silvered to front
to rear silvered reflection

Rear silvered reflection
to front reflection to
rear silvered reflection

Rear silvered reflection

Front reflection

Filament

Light bulb

Figure 1.14   Reflections from a rear surfaced mirror.

the front surface and produces image 1 in the mirror which has an intensity approximately 1/25 of the incident light intensity. Ray 3 is reflected three times. Two of these reflections occur from the rear-silvered surface

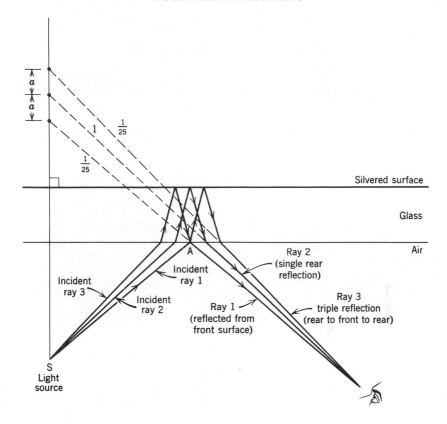

Figure 1.15 Rays from an object producing images by multiple reflections.

without any appreciable loss in light intensity. There is also a single reflection from the front surface (at point A) making the reflected ray approximately 1/25 the incident light intensity. Image 3 produced by ray 3 will therefore be 1/25 the light intensity of the incident light to the glass plate. Notice the difference between Figures 1.13 and 1.15. In Figure 1.13 the reflected light strikes a screen producing different spots of light. The eye placed at any of these spots can see the corresponding image. In Figure 1.15 the eye is at a fixed location and can see all images at the same time.

The various effects described are made use of in the automobile day-nite mirrors. Figure 1.16(a) shows a possible configuration for the construction of such a mirror. During the daylight hours the reflection of the object is from the silvered surface. Nearly all the incident light is therefore reflected to the observer. At night, however, a condition of high

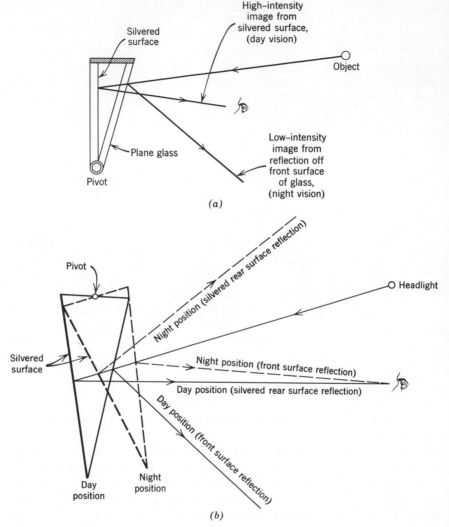

Figure 1.16 (a) Operation of an automobile Day-nite mirror (mirror shown in day position). (b) Prismatic Day-nite mirror.

reflectance is undesirable; if all the incident light from a car's headlights were reflected into the observers eyes, he would not be able to see the dimly lit road. In order for the observers eyes to remain "dark adapted" it is important to reduce the intensity of the light reflected from the mirror. At night therefore the reflection of the object off the front surface of a plane glass reflects to the observer only 1/25 of the incident light (there is also a reflection off the back surface of the front glass which will give a slight double image. The front glass can be made thin to reduce this double image). These reduced intensity reflections are sufficient to make a driver aware of highly illuminated objects (such as headlights) in the mirror but they are observed at approximately the same intensity as the road illumination. By simple rotation of the mirror either the reflection from the silvered surface or from the plane glass surface can be made available to the driver. The front and back surfaces of a glass that are not parallel to each other can be made to act as a day-nite mirror. Such a mirror is called a *prismatic mirror*. A prismatic mirror can be constructed more simply and made more rigid than a mirror made with two glass plates. It also does not have the problem of a double image. The prismatic mirror is silvered on its rear surface, which is used for day vision, and is illustrated in Figure 1.16(b).

It can be seen that extremely important uses are made of mirrors and parallel-sided plane glass plates. In optical data processing and holography these same items are also very important. A simple beam splitter can be made by use of a glass plate. Figure 1.8 shows that the incident ray is broken up into two rays by a glass plate. This effect is often made use of when a ray (or beam) of light is to be broken into two (or more) rays or when it is desired to attenuate a light beam.

KALEIDOSCOPE

Another device that uses the principles just described is the kaleidoscope. Kaleidoscopes have been used for years as toys. The kaleidoscope does not have any specific direct application in optical data processing or holography but knowledge of its operation could be useful when working in these areas. For example, the use of such a device might simplify the construction of certain reference signals or filters for use in optical data processing. The kaleidoscope consists of two flat reflecting surfaces, usually enclosed in a tube as shown in Figure 1.17. Either plane glass plates or mirrors are used, usually inclined at 60 degrees to each other and arranged so that one end is wider than the other. A peep hole is made to lie close to the line of intersection between the two mirrors. By holding the kaleidoscope against a bright background a variety of symmetrical patterns can be seen.

Either front-surface mirrors or plane flat glass can be used as reflectors in the kaleidoscope. Plane glass can be used instead of mirrors because at large angles of incidence (grazing angles) almost total reflection is obtained (compared to 4 percent of the light being reflected when the incident light is perpendicular). When glass is used instead of mirrors, the

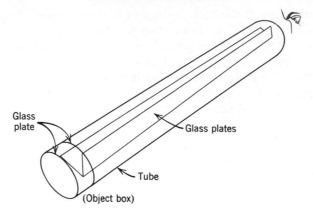

Figure 1.17   Kaleidoscope construction.

backs are blackened to prevent any reflection from them. Ideally the surface of the glasses should be silvered so as to form front-surface mirrors.

The symmetrical patterns obtained are determined by the angle between the mirrors. Symmetrical patterns can be obtained for those angles between the mirrors whose values are 180°/n where n is any integer. Symmetrical patterns can therefore be obtained at angles between the mirrors of 180°, 90°, 60°, 45°, 36°,.... As the angle between the mirrors decreases, the complexity of the pattern increases. Table 1.2 gives the number of fields in each of the symmetrical patterns.

Table 1.2 shows that there are six fields in a symmetrical pattern for a kaleidoscope constructed with the mirrors placed at 60 degrees to each other. Figure 1.18 shows the view obtained from a kaleidoscope

TABLE 1.2

n	Angle between Mirrors 180°/n Degrees	Number of Fields in Symmetrical Pattern
1	180	2
2	90	4
3	60	6
4	45	8
5	36	10

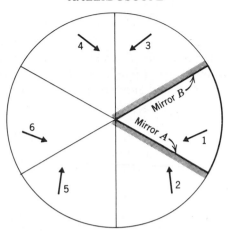

Figure 1.18 Six field kaleidoscope view (mirrors at 60 degrees).

with mirrors at 60 degrees to each other. The object 1 (shown as an arrow) between the mirrors *A* and *B* forms an image 2 in mirror *A* and an image 3 in mirror *B*. Image 3 forms an image 5 of itself in mirror *A*. Image 2 forms an image 4 of itself in mirror *B*. Image 6 is formed by the reflection of image 5 in mirror *B* and the reflection of image 4 in mirror *A*. Actually image 6 is a superposition of two reflections (images 4 and 5) superimposed on each other.

A kaleidoscope usually is made so that small objects can be placed a short distance from the ends of the mirror. Unless the eye can be placed absolutely within the plane of both mirrors, the edges of the mirrors will appear to cut the object and the image will not appear to be perfectly symmetrical. A plane object such as a drawing on a sheet of paper when placed against the far edges of the mirrors, will give a symmetrical pattern. A solid object placed with the solid angle between the mirrors will also form a symmetrical pattern. The major problem of a kaleidoscope is to place the eye in the plane of both mirrors — that is, place the eye so that it is at the intersection of both mirrors. When the eye is placed at the intersection of both mirrors, the pattern will always appear symmetrical, even if the object is beyond the far end of the mirrors. It is therefore necessary to construct the kaleidoscope so that the eye can be placed as close as possible to the line of intersection at the end of the mirrors. In addition it is desirable to keep the objects as close as possible to the far end of the mirrors. Usually it is quite difficult to obtain perfect symmetry but as a toy, unless the observer is specifically looking for perfect symmetry, any unsymmetry will usually be unnoticed.

GRAPHICAL DETERMINATION OF ANGLE OF REFRACTION

Figure 1.19 shows a graphical construction for determining the angle of refraction knowing the angle of incidence and the indices of refraction of the medium. $\overline{ab}$ represents a plane surface separating the mediums A and B. We shall consider that medium A represents glass A of Figure 1.10, medium B represents glass B, and $\overline{ab}$ is the separation between them. An incident ray $\overline{ST}$ makes an angle i with the normal to the surface $\overline{ab}$ in medium A, whose index of refraction is n. The graphical construction consists of initially drawing a line parallel to $\overline{ST}$ ($\overline{st}$). With s as a center two arcs are struck with their radius proportional to n and n', respectively, where n' is the index of refraction of glass B. At the intersection of the arc n and st (i.e., at point t) a line is drawn parallel to the normal $\overline{NN'}$ (i.e., line $\overline{tu}$ is parallel to line $\overline{NN'}$). This normal intersects the arc n' at point u. The line $\overline{su}$ is then drawn. Through point T the refracted ray $\overline{TU}$ is drawn parallel to $\overline{su}$. The angle $\overline{TU}$ makes with the normal is correct based on the respective index of refraction of the two materials and the angle of incidence of the incident ray.

TOTAL REFLECTION

Figure 1.20 shows rays diverging outward from a point source S located in a medium whose index of refraction is n. The rays are shown striking the boundary between it and a second medium whose index of refraction is n' (where n is greater than n'). Ray A is normal to the boundary surface and proceeds through it without deviation (4 percent of the incident light is reflected back on itself). Ray B strikes the surface at

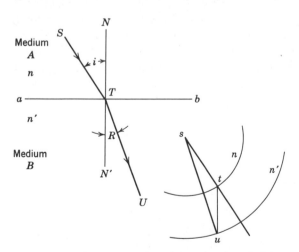

Figure 1.19   Graphical construction of the refraction of a light ray.

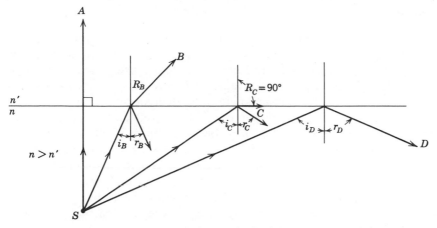

Figure 1.20 Total reflection.

an angle of incidence i_B and is partially reflected and partially refracted at the surface. The angle of reflection r_B is equal to the angle of incidence i_B by the law of reflection. The diffracted portion of ray B bends away from the normal making an angle R_B with the normal because the index of refraction in the second medium is less than the index of refraction in the first medium; that is, $n > n'$. The ray C is shown to hit the boundary surface at an angle of incidence i_C such that the refracted ray bends away from the normal exactly 90 degrees ($R_C = 90°$, where i_C is called the *critical angle*). The angle i_C is called the critical angle because any further increase in the angle of incidence will prevent the ray from being refracted into the second medium. Ray D shows this. When the angle of incidence is made greater than the critical angle i_C, the ray is totally internally reflected at the surface back into the original medium. The critical angle can be found from Snell's law and Figure 1.20. When $R_C = 90°$, Snell's law 1.20 becomes

$$\sin i_C = \frac{n'}{n} \tag{1-22}$$

From this equation we can see that total internal reflections can occur only when the incident ray (as shown in this case) is in the medium of greater index of refraction; that is, $n > n'$. Table 1.1 shows the critical angle for various materials with air. For example, the critical angle for crown glass can be found by taking its index of refraction ($n = 1.50$) and substituting in the equation

$$\sin i_C = \frac{1}{1.50} = 0.67$$

$$i_C = 42°$$

PRISMS

Table 1.1 gives the critical angle for crown glass in air as approximately 42 degrees. It is possible therefore to construct a prism with angles 45–45–90 degrees as shown in Figure 1.21 and cause an incident ray to be totally reflected from one surface. In Figure 1.21 the incident ray is striking the front surface at an angle of 90 degrees and as we have seen previously approximately 4 percent will be reflected back on itself. Total reflection will not take place because the incident light is in the medium of smaller index of refraction. At the second surface, however, total reflection will take place because the light is in the medium of greater index of refraction and is striking the surface at an angle greater than the critical angle (42 degrees). As can be seen in the figure the incident ray is 90 degrees to the reflected ray; that is, the incident ray has been deviated 90 degrees.

Figure 1.22 again shows a 45–45–90 degree prism being used different-ly from the method shown in Figure 1.21. In Figure 1.22 the incident ray is made to reflect the light totally twice before emerging from the prism. The incident ray strikes the reflecting surfaces again at 45 degrees (which is an angle greater than the critical angle of 42 degrees) and is therefore totally reflected as shown. Figure 1.23 shows three possible paths of light rays through a totally reflecting 45–45–90 degree prism. Figure 1.23(a) shows the prism acting as a plane mirror, reflecting the rays in such a way as to deviate them 90 degrees. In Figure 1.23(b) the prism acts so as to divert the rays 180 degrees causing inverted images. In Figure 1.23(c) the prism acts to divert the rays so that images will be inverted.

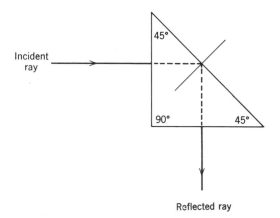

Figure 1.21 Totally reflecting prism (single reflection).

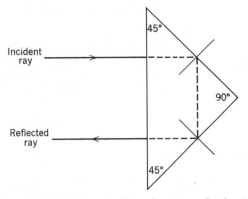

Figure 1.22 Totally reflecting prism (two reflections).

DISPERSION

We have seen that the index of refraction indicates the distance a light wave can travel in air in a given time as related to the distance that it can travel in the substance in the same time. Table 1.1 shows that the index of refraction varies for different substances. The index of refraction of a particular substance also varies with the frequency of the light. If we consider this effect in Snell's law,

$$\frac{\sin i}{\sin R} = \frac{n'}{n} \tag{1-20}$$

we can determine the effect of this variation in refractive index due to frequency. We can write

$$\sin R = \frac{n}{n'} \sin i \tag{1-23}$$

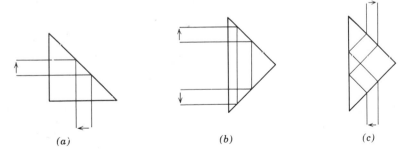

Figure 1.23 Three possible paths of light rays through a totally reflecting prism.

Assuming medium 1 to be vacuum and to have an index of refraction of

$$n = 1$$

for all frequencies

$$\sin R = \frac{n}{n'} \sin i = \frac{1}{n'} \sin i \qquad (1\text{-}24)$$

For a given angle of incidence, i, this shows that the angle of refraction, R, is dependent on the value of n'. n' varies with frequency and therefore the angle of refraction, R, varies with frequency.

Light from most sources is a mixture of different frequencies. A mixture of many different frequency light waves in the visible portion of the spectrum can be called *white light*. When white light is made incident on a prism, the light will be dispersed by the prism into a spectrum; that is, the prism refracts the various frequencies of light different amounts (indicated by equation 1-24) causing the components of the light wave to be separated. Breaking up of the white light into its components is called *dispersion*. Because the index of refraction of glass is lower for red light than for violet, the diffraction angle R for red light will be greater than that for violet light. Figure 1.24 shows relative deviation angles of different frequencies of light caused by the differences in index of refraction of glass to the different frequencies of light.

Luminous Fountain

In order for total reflection to take place the light (at the interface) must originate in the medium of greater optical density. In the case of air and

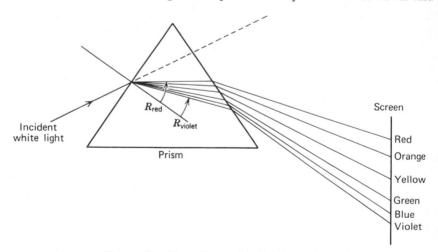

Figure 1.24 Dispersion of white light by a prism.

water total reflection can take place only when the source of light is inside the water. Figure 1.25 shows a technique used in water fountain design to make the stream appear luminous. A light source behind a glass plate projects light into a nozzle from which a stream of water is flowing. The light entering the nozzle continues in the water stream and is internally reflected from the inner surfaces of the water stream. Because the surface of the water stream is not regular, there is enough light scattered and diffusely reflected so as to make the water stream appear luminous.

LIGHT GUIDES

It is possible to make light travel a path along a solid rod that has been twisted and bent into a complex curved shape (but usually not bent over a radius more than the diameter of the rod itself). In the luminous fountain the light tended to follow (by total internal reflections) the path of the water stream. By having a solid rod of glass or plastic it is possible to have the light essentially follow the shape of the rod by total internal reflections. Transparent rods as shown in Figure 1.26 are called light guides. If the surfaces of the rod are smooth and regular, efficient conduction of the light along the rod is possible through total internal reflections. Total internal reflections will take place when the light rays strike the walls at angles greater than the critical angle. The light on reaching the far end of the rod will emerge from the end because the angle of incidence is below the critical angle; that is the light tends to strike the end nearly normally and therefore i_C is small.

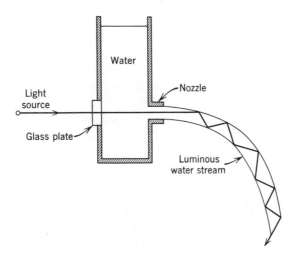

Figure 1.25   Luminous fountain.

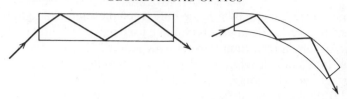

Figure 1.26   Ray path in light guides.

The conditions for total internal reflection exist at any smooth interface between two transparent media having different refractive indices. From this we might assume that a smooth glass rod in air should conduct light efficiently. Actually there usually are enough minute surface defects to reduce the efficiency; that is total internal reflection does not take place. When surface defects are present, the light is absorbed or scattered as it is in the luminous fountains. Similar losses do occur in prisms; however, in prisms these losses are generally not considered serious because only a few reflections take place. In light guides hundreds of reflections may take place so that even small losses become important.

Figure 1.27 shows ray paths in two light guides. Both light guides are bent through the same angle and each of the incident light rays strikes the polished front face surface at the same angle. The number of reflections in the thin rod is considerably more than in the thick rod. Although the distance between the front and end faces of each light guide is the same, the light ray in the thin rod will have traveled a considerably longer optical path than the ray in the thick rod. The intensity of the emerging ray will be reduced because of the absorption of the light by the rod material (losses due to total internal reflections are small if the surface is smooth). The longer the optical path in the rod, the greater the absorption of light and the lower the intensity of the emerging light from the rod. From this it can be deduced that the thinner the rod, the greater the optical path length and therefore the greater will be the light absorption and the lower the light intensity of light from the end face.

Under proper conditions (smooth, polished surface walls) total internal reflections are an exceptionally efficient means of reflection. Total internal reflections can be made so efficient that no significant change in light intensity will occur even after hundreds of total internal reflections. No mirror or metallic reflector is superior in this regard. We have previously seen how a few reflections from a plane glass surface will cause a significant reduction in light intensity. In addition the efficiency of the reflection of light by polished metal surfaces is related to the wavelength of the light being reflected. Table 1.3 gives representative figures for the percent of incident light reflected from various polished metals.

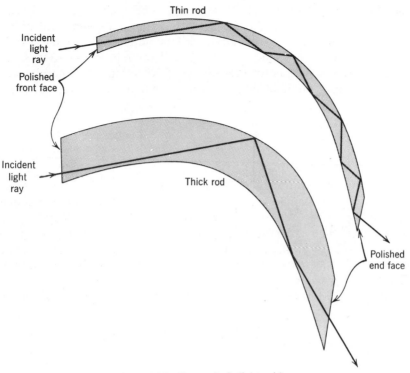

Figure 1.27   Ray paths in light guides.

The significant point brought out by Table 1.3 is that light tends to be reflected better at the longer wavelengths (red end of the visible spectrum) than at short wavelengths (violet end of the visible spectrum). Multi-colored images will therefore become color distorted after multiple-reflections. In the range of visible light successive images after each

TABLE 1.3

Wavelength in microns	Percent Incident Light Reflected			
	Silver	Aluminum	Tin	Steel
*0.5	90	—	—	55
*0.7	94	—	—	58
1.0	97	71	54	63
2.0	98	82	61	77
4.0	98	92	72	88

*Range of visible light.

reflection will first tend to become yellow and then redder and redder because of the slightly higher reflectivity of the red component of light. Even a small difference in reflectivity will be noticeable after a few reflections. The efficiency of the total internal reflection process can be so high that even after hundreds (or thousands) of reflections the emerging color of the light from the light guide will be essentially the same as when it entered the light guide.

OPTICAL TUNNEL

A rectangular glass or plastic bar can be used as a type of kaleidoscope for multiduplication of images. This type of operation has many applications and can be adapted to many more. Consider, for example, that many copies of a printed circuit board are to be made. A transparency is usually prepared and then exposure after exposure is made using what is called a *step and expose camera*. Each individual board must be accurately positioned before exposure and requires costly time for making the individual exposures.

Figure 1.28(a) shows a glass bar that can be used to form multiple images. At one end of the bar a transparency is mounted and illuminated. If the eye is placed at the other end of the bar (at the image point) multiple, independent, nonoverlapping images will be seen. Figure 1.28(b) shows how the multiple images are formed by the bar. Only the images formed directly from the top and bottom surfaces are shown. Image 1 can be viewed directly through the bar. Image 2 is viewed by looking at a single reflection of the object transparency from the bottom surface. Image 2 therefore appears to lie in a line directly below the actual object and along the line of viewing. Other images form in a similar manner. Figure 1.28(c) shows the image fields seen from the image point (only the vertical fields going through the center are numbered). The further out the image being viewed is from the center direct-viewed image, the more reflections are required to form the image and the more accurately the bar must be constructed. If use is to be made of internal surface reflections, the bar must be relatively long with respect to its cross section in order for the angles of incidence to be large and exceed the critical angle. The use of internal reflections is desirable because of the greater reflection efficiency.

An alternate method for achieving the same results is by use of an optical tunnel. An optical tunnel is a long square hole in a block of glass that has been silvered. The operation of the optical tunnel is identical to that of the glass bar. The optical tunnel has the advantage of rigidity over a glass bar. A glass bar tends to be flexible and bend which will

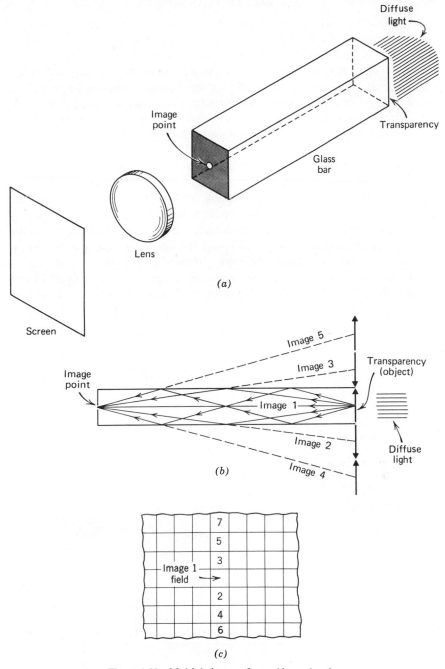

Figure 1.28 Multiple images formed by a glass bar.

distort the images. In addition lenses and objects are more easily positioned at the exact ends of an optical tunnel which make it more convenient
to use than a glass bar. In order for either device to operate properly for
precision applications, extremely accurate polished and aligned sides
are required. For extremely accurate applications the light tunnel is
usually used since it is easier to maintain precision alignment. The light
tunnel is usually constructed from four heavy glass bars, and polished to
the required flatness and annealed to form the optical tunnel, as shown in
Figure 1.29(a).

A possible application for such a multiple-image device is an optical
printer. Let us assume that Figure 1.29(b) consists of 25 identical image
fields formed by either an optical tunnel or bar. The position of each of
these fields is fixed relative to the optical tunnel and its ends (with a given
lens). We can therefore place in any field a transparency that we may wish
to print. By illuminating a specific field we illuminate the transparency
we wish to print and, using the optical tunnel in reverse, the image of
the illuminated section will appear at the opposite end of the tunnel where
it can be printed. Such a device can be useful as a computer or optical
tape printout.

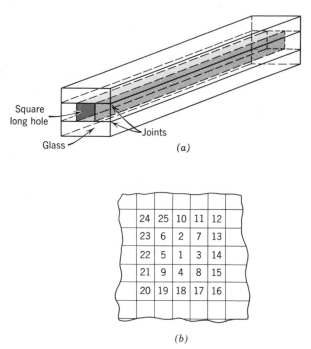

Figure 1.29   (a) Optical tunnel. (b) Images formed by optical tunnel (or bar).

The optical tunnel also has potential uses in optical memory devices. A Cathode Ray Tube (CRT) can be imaged by a lens to one end of an optical tunnel and multiple images of it will be produced by the optical tunnel. These identical images of the CRT can be projected by a lens to many different locations. Let us assume we project 100 images of the CRT. At each of these 100 positions we can have a photocell pickups read the light from the CRT spot arriving at each respective image. The CRT spot can be controlled and positioned to any point on the face of the tube. Each spot location of the CRT face corresponds to a computer memory address. By placing a photographic transparency in front of each photocell light from the CRT spot (for a given location on the CRT faceplate) can produce a coded light output to each photocell. We can therefore obtain a 100-bit readout for each spot position on the CRT. It is reasonable to assume that 100,000-spot locations on the CRT can be achieved without too much difficulty and that for each of these we can produce a 100-bit word. Such a memory would have quick access and be relatively simple to fabricate. In effect this memory uses the light tunnel similar to the way it was used in optical tape printout except that the position of the light source is reversed.

FIBER OPTICS

When a light guide's diameter is made quite small, it can be referred to as a *fiber*. A group of fibers put together is referred to as a *fiber optics bundle*. It was pointed out that for total internal reflections to take place the light must be in the medium of greater optical density and the surface interface must be sharp. Any contamination of the surface will tend to reduce the efficiency of total internal reflections. This is particularly important for small fibers because more reflections take place per unit length than in light guides. Any contamination of the surface of a fiber will tend to reduce its efficiency and thus have a drastic effect on the amount of transmitted light. To avoid possible fiber-surface contamination (oil, moisture, acids, smoke, etc.) it is a usual practice to coat each fiber core with a material of low index of refraction. This coating serves to protect the fiber from contamination (which is ever present and quite difficult to avoid).

The transmission of light by fiber optics is quite similar to the transmission of electromagnetic energy by waveguides. For simplicity and ease in understanding, ray diagrams can be used to explain most of the phenomena but the basic wave nature of the phenomena must be remembered. When light travels down a fiber of small diameter, the light tends to concentrate on the surface of the fiber. When such an uncoated fiber comes in contact with another uncoated fiber, there is leakage across the region of contact. In order to prevent this leakage of light at points of contact the fibers are coated. We see, therefore, that the fiber coating (called *cladding*) serves not only as a vital protective covering but also to prevent light leakage to adjacent fibers. This leakage is sometimes called *optical crosstalk* and is due to the penetration of the electromagnetic field into adjacent fibers. The extent of this penetration is only in the order of the wavelength of the light being used. A nonconducting fiber brought to within this distance will absorb some of the light energy from the conducting fiber. A thick coating on each fiber prevents it from getting near enough to any other fiber to permit the absorption of energy from it. A two-wavelength-deep coating over each fiber core is usually sufficient to prevent light leakage and protects the surfaces.

Coated single strands of fibers placed together in a bundle will act independently of each other. Figure 1.30 shows such a bundle of fiber

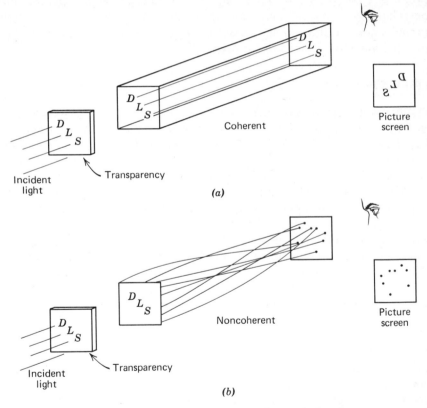

Figure 1.30 Coherent and noncoherent fiber optics bundles.

optics. Each fiber in the bundle transmits light to the other end proportional to the amount of incident light on it. This means that if one end were placed over a film transparency (slide) it would transmit this picture, fiber by fiber to the other end. At the far end the picture would be seen, regardless of the twists or bends the bundle of fibers was subjected to. When the fibers are routed between the input and output face plates in such a way that the image at the input is preserved at the output, the fiber optic bundle is said to be coherent (see Figure 1.30(a)). Figure 1.30(a) shows fiber optics being used to transmit the letters DLS which is being imaged on the input end face. Each fiber of the bundle will pick up and transmit the average amount of light falling on its input cross section. At the far end the letters appear to be reversed. The reason for this is that at the input end of the fiber optics bundle we are looking at the front side of the letters DLS (image correct), while at the output end we are apparently looking at the rear of the image of the letters on the input face, which accounts for the apparent reversal. When the bundles are arranged with no particular attention paid to the relative positions of the fibers between the input and output positions, they are referred to as noncoherent fiber optics bundles. Usually such noncoherent bundles are used for the purpose of carrying light from the input to the output, with no attempt made to preserve an input image form at the output. A noncoherent bundle (shown in Figure 1.30(b)) with deliberate disorientation of the fibers could be used for coding optical images. Only a similar bundle to that used for coding can be used to decode the coded image.

SPHERICAL MIRRORS

Mirrored surfaces can be made to have spherical surfaces. A spherical surface is said to be concave if the portion of the inner surface of a sphere is mirrored. The spherical surface is said to be convex when the surface that is mirrored corresponds to a portion on the outer surface of a sphere. Figure 1.31(a) shows a concave and a convex spherical mirror. Because the surfaces of the mirrors are spherical, we can see that there is a *radius of curvature* to the surface that has an origin O. The origin O is therefore the center of curvature of the mirror. The optical axis of the mirror is a line through the center of the mirror and the center of curvature O.

Let us consider a concave mirror as shown in Figure 1.31(b). A ray AB is shown parallel to the optical axis incident on the spherical surface at B. The line OB is the radius of curvature of the spherical mirror; that is, OB is normal to the spherical surface at B. The angle ABO is therefore the angle of incidence θ that the ray makes with the mirror. Since the angle of reflection must equal the angle of incidence (by

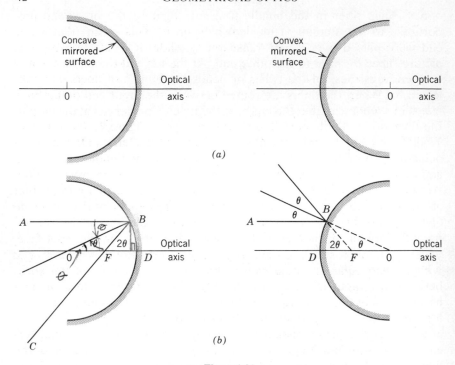

Figure 1.31.

the law of reflection), the angle OBF must also be equal to θ. Since the ray AB is parallel to the optical axis, we can see from plane geometry (parallel lines cut by a transversal) that angle $BFD = 2\theta$. We shall assume that the size of the mirror is small as compared to its radius of curvature; that is, the aperture of the mirror is small. This assumption permits us to consider the spherical mirror to be essentially flat, and the arc BD can be considered to be a line perpendicular to the optical axis. In triangle ODB

$$BD = OD \tan \theta \tag{1-25}$$

and in triangle BDF

$$BD = FD \tan 2\theta \tag{1-26}$$

Then

$$OD \tan \theta = FD \tan 2\theta \tag{1-27}$$

and since the angle θ is small, we can write

$$OD\,(\theta) = FD\,(2\theta) \tag{1-28}$$

$$OD = 2FD \tag{1-29}$$

Any ray parallel to the optical axis will be reflected to pass through point F. For a convex lens the reflected ray will appear to pass through point F. Point F can therefore be called the *focal point* of a spherical mirror. We have thus shown that the focal length, DF, of a spherical mirror is located on the optical axis and is <u>halfway</u> between the center of curvature and the mirrored surface. By similar reasoning the corresponding results can be obtained for a convex mirror. For a concave mirror the ray parallel to the optical axis passes through the focal point F after being reflected. The rays reflected from a convex mirror will appear to be originating at the focus F behind the mirror.

Figure 1.32(a) shows an object placed away from a concave mirror by a distance greater than its radius of curvature. From point A on the object there are an infinite number of rays being emitted (one at every angle). From this infinite number of rays we select and draw the ray AB that is parallel to the optical axis. We know that this ray upon reflection will pass through the focal point F. We now go back to point A on the object and select the ray that goes through O — that is, ray AO or AE

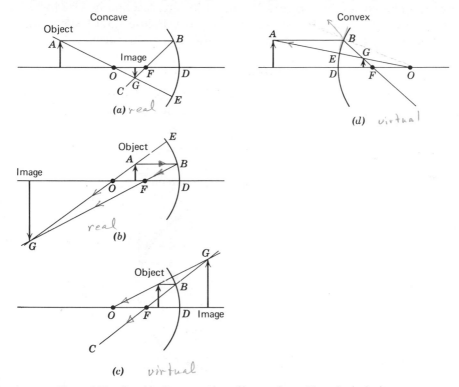

Figure 1.32 Graphical construction of images formed by spherical mirrors.

(*AO* extended). We select this ray (*AE*) because we know that any ray passing through *O* will be normal to the spherical surface and will be reflected back on itself. The intersection of the reflection of these two rays will locate the corresponding image point, *G*.

Similar constructions are shown for different locations of the object with both concave and convex mirrors. When the reflected rays from the mirror actually pass through the image, it is called a *real* image. A *virtual* image is one that the reflected rays from the mirror appear to pass through but actually do not. The real images when viewed by the eye cannot be distinguished from a virtual image. Because the reflected rays actually pass through the real image, it can be projected on a screen, but the virtual image cannot be projected on a screen since the rays of light do not actually pass through the image.

Figure 1.32(b) shows the object between the center of curvature of the lens and its focus. The image is formed outside the center of curvature and is inverted and magnified. If we compare Figure 1.32(a) and (b), we see that we can make the object in one the image in the other and vice versa. The images in Figure 1.32(a) and (b) are real and inverted while in Figure 1.32(c) the image is virtual and erect. Figure 1.32(d) shows that a convex mirror always forms an erect, reduced virtual image.

ANALYTIC LOCATION OF OBJECT AND IMAGE

The location of the image of an object can be found analytically. Figure 1.33 shows Figure 1.32(a) redrawn with the following distances indicated:

$$\overline{OD} = \text{radius of curvature} = r$$
$$\overline{FD} = \text{focal length of mirror} = f$$
$$p = \text{distance of object from mirror}$$
$$q = \text{distance of image from mirror}$$

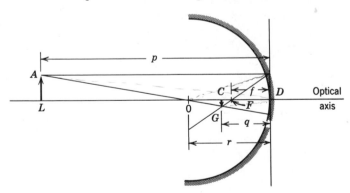

Figure 1.33.

From triangles ALO and OCG we can see that

$$\frac{AL}{GC} = \frac{LO}{OC} = \frac{p-r}{r-q} \qquad \text{(1-30)}$$

From triangles ALD and GCD we can see that

$\angle ACD = \angle CDG$

$$\frac{AL}{LD} = \frac{GC}{CD} \qquad \text{(1-31)}$$

Since $LD = p$ and $CD = q$, then

$$\frac{AL}{p} = \frac{GC}{q} \qquad \text{(1-32)}$$

$$\frac{AL}{GC} = \frac{p}{q} \qquad \text{from 1-30} \qquad \text{(1-33)}$$

then

$$\frac{AL}{GC} = \frac{p}{q} = \frac{p-r}{r-q} \qquad \text{(1-34)}$$

$$p(r-q) = q(p-r)$$
$$pr - pq = pq - qr$$
$$pr + qr = 2pq \qquad \text{(1-35)}$$

multiplying by $1/pqr$

$$\frac{1}{q} + \frac{1}{p} = \frac{2}{r} \qquad f = \frac{r}{2} \qquad \text{(1-36)}$$

but $r = 2f$; then

$$\frac{2}{r} = \frac{1}{f} = \frac{1}{p} + \frac{1}{q} \qquad \text{(1-37)}$$

This equation applies to both concave and convex mirrors. Negative values indicate distances behind the mirror. For convex mirrors r and f are negative and for concave mirrors q may have a negative value (indicating a virtual image behind the mirror).

SPHERICAL MIRROR MAGNIFICATION

The ratio of the image size to object size is the magnification of the spherical mirror. Consider triangles ALD and GCD in Figure 1.33.

$$\frac{AL}{p} = \frac{GC}{q} \qquad \text{from 1-32} \qquad \text{(1-38)}$$

Then the magnification (neglecting signs of p and q) is

$$M = \frac{GC}{AL} = \frac{q}{p} \tag{1-39}$$

Consider the following example:

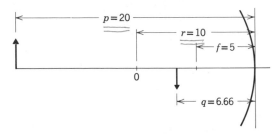

with $r = 10$ cm and $p = 20$ cm. Determine the size of the image if the object is 1 cm high. The magnification is

$$M = \frac{q}{p} \tag{1-39}$$

We find q from

$$\frac{1}{p} + \frac{1}{q} = \frac{1}{f} \qquad , \quad f = \frac{r}{2} \tag{1-37}$$

$$\frac{1}{q} = \frac{1}{f} - \frac{1}{p} = \frac{1}{5} - \frac{1}{20} = 0.2 - 0.05 = 0.15$$

$$q = \frac{1}{0.15} = 6.66 \text{ cm} \qquad M = \frac{q}{p} = \frac{6.66}{20} = 0.333$$

If the object is 1 cm high, the image size is 0.333 cm.

LENSES

Figure 1.34(a) shows how a prism deviates incident monochromatic (single frequency) collimated light (rays in a collimated beam of light are drawn parallel to each other). The prism deviates each ray of the collimated light beam in such a way that the rays still remain parallel to each other. Figure 1.34(b) shows how a prism can be modified to make the deviated rays converge to a point F. In Figure 1.34(a) we can see that ray 3 must be deviated more than ray 1 in order to have the rays converge at a point. In order to achieve convergence the thickness of the prism must be changed to achieve the desired results. By having the thickness

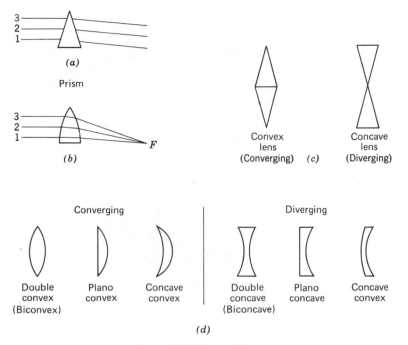

Figure 1.34 Types of lenses.

of the prism change continuously from point to point the desired result can be achieved. Convergence can be achieved by making the face of the prism curved so that it becomes thickened at the bottom and thinner at the top.

Figure 1.34(c) shows how lenses can be considered to be made up of two prisms. Figure 1.34(d) shows the two basic types of lenses — converging (thicker at center than at rim) and diverging (thinner at center than rim) — in some of their varying forms.

Lens Terms

Figure 1.35 illustrates the descriptive terms associated with lenses. We are assuming that the aperture of the lens is small compared to the radius of curvatures and that all rays make very small angles with the optical axis.

Lens Plane. The lens plane is a plane through the points of intersection of the two lens surfaces. The lens axis is one line in this plane usually used for illustration purposes.

Optical Axis. A line through the geometric center of the lens, perpendicular to the plane through the points of intersection of the two lens

GEOMETRICAL OPTICS

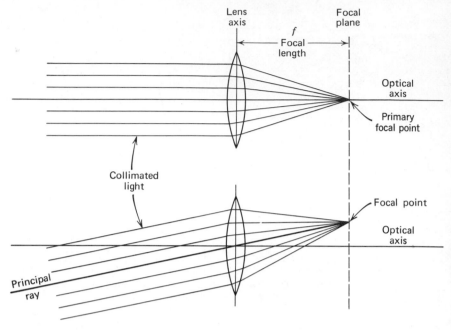

Figure 1.35.

surfaces is the optical axis. The optical axis lies on a line containing the centers of curvature of the lens surfaces.

Primary Focal Point. The primary focal point is the point where a collimated monochromatic light beam (the rays of which are incident to the lens parallel to the optical axis) is brought to a point of convergence (or divergence). The focal point is the same distance on either side of the lens depending only on which side the light is made incident to the lens.

Focal Length. The distance between the lens axis and the primary focal point is the focal length.

Thin Lens. A lens whose thickness is small compared with its focal length is a thin lens. For thin lenses the distance can be approximated by measuring to the lens surfaces.

Focal Plane. A plane through the primary focal point perpendicular to the optical axis is a focal plane.

Principal Ray. The principal ray is a ray in a collimated light beam passing through the intersection of the lens axis and the optical axis. Rays parallel to the principal ray are brought to a focus in the focal plane in line with the principal ray.

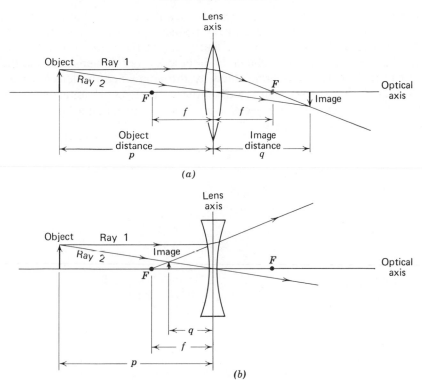

Figure 1.36 (*a*) Convex lens. (*b*) Concave lens.

LENS EQUATIONS

The relationship between the object distance, p, and the image distance, q, in terms of the focal length of a lens is given by

$$\frac{1}{f} = \frac{1}{p} + \frac{1}{q}$$

(1-37)

When the image distance, q, takes on negative values, the image is virtual and on the same side of the lens as the object. The focal length, f, is positive for converging lenses and negative for diverging lenses. We can also see from (1-37) that when the object is at infinity, the image will be a focal length from the lens; that is, $1/p = 1/\infty = 0$. Figure 1.36(a) shows how these facts can be used graphically to verify the equation for a convex lens. We can take any point on the object (in this case we have taken the head of the arrow) and consider only two of the infinite number of rays that can be drawn from it. Ray 1 is chosen to be parallel to the optical axis. This ray must be deviated by the lens so that it will pass through the focal point; that is, rays from an object located at infinity will be parallel to the optical axis and focus at the primary focal point. Ray 2 is selected to pass through the intersection of the lens axis and the optical axis. This ray is undeviated by the lens since the lens surfaces that this ray passes through are essentially parallel. The intersection of these two rays from the object locates the corresponding image point. The magnification is given by

$$M = -\frac{q}{p} \tag{1-39}$$

EXAMPLE. Determine the size of image of Figure 1.36(a) when $f = 10$ cm, $p = 25$ cm, and object is 2 cm high,

$$\frac{1}{f} = \frac{1}{p} + \frac{1}{q}$$

$$\frac{1}{q} = \frac{1}{f} - \frac{1}{p} = 0.1 - 0.04 = 0.06$$

$$q = 16.6 \text{ cm}$$

$$M = -\frac{q}{p} = -\frac{16.6}{25} = -0.666$$

size

$$2 \times -0.666 = -1.333 \text{ cm}$$

where the minus sign indicates that the image is inverted.

Essentially the same construction can be used to locate an image in a concave lens. Figure 1.36(b) shows ray 1 originating at the arrow head of the object and traveling to the lens along a line parallel to the optical axis. This ray is deviated by the concave lens so as to diverge and will appear to have originated at the focal point as shown. Ray 2 (also originating at the arrow head of the object) is undeviated by the lens since the portion of the lens this ray travels through has essentially parallel faces. The image therefore appears to be at the location where these two rays appear to intersect. The light rays actually do not intersect at this point but appear to and therefore a virtual image is formed.

EXAMPLE. Determine the size of the image of Figure 1.36(b) when $f = -10$ cm $p = 25$ cm, and the object size is 2 cm high.

$$\frac{1}{f} = \frac{1}{p} + \frac{1}{q} \tag{1-37}$$

$$\frac{1}{q} = \frac{1}{f} - \frac{1}{p} = -0.1 - 0.04 = -0.14$$

$$q = -7.15 \text{ cm}$$

$$M = -\frac{q}{p} = -\frac{-7.15}{25} = +0.286$$

size

$$(0.286)(2) = +0.572 \text{ cm}$$

where the + sign indicates an upright image.

LENS ABERRATIONS

Single lenses can have different types of defects called *aberrations*; the more important ones are listed below. Generally lens aberrations increase with increasing aperture. Aberrations are usually corrected, at least in part, by combinations of lenses of different curvatures and types of glass.

Spherical Aberrations. Rays of light parallel to the optical axis should be brought to a focus at the principal focal point. Imperfections in the lens surfaces can produce spherical aberrations that cause the rays to cross (focus) to more than one point on the optical axis. Usually with simple spherical surface lenses the marginal rays focus nearer the lens than those passing through the center.

Chromatic Aberrations. The index of refraction of glass varies with the frequency of the light. This causes the focal length of a lens to be less for violet light than for red light and is chromatic aberration.

Curvilinear Distortions. Distortions of an image's edges due to the aperture placed so that the rays pass through the lens at a large angle to the optical axis are curvilinear distortions.

Incident Barrel Distortion. Barrel distortion makes what should be a square image have bowed-out sides. It is produced by a simple lens when the aperture is in front of the lens.

Pin-Cushion Distortion. This type of distortion makes what should be a square image have bowed-in sides. It is produced by a simple lens when the aperture is behind the lens.

Coma. A type of skewed spherical aberration is coma. When this defect is present it makes images of points appear pear-shaped.

THE CAMERA

A pinhole can be used as a lens in certain applications—for example, a pinhole camera. A picture of an object can be taken by placing a small aperture between the object and a photographic film (usually the photographic film is placed 7 to 14 cm from the pinhole). A ray of light from a point on the object goes through the aperture in straight lines as shown in Figure 1.37(a). When a pinhole is used as a lens in a pinhole camera, it should have a diameter of approximately 0.3–0.5 mm. If the diameter is made appreciably smaller than 0.3 mm, diffraction effects (as explained in the next chapter) will tend to make the image fuzzy. If the pinhole is made appreciably larger than 0.5 mm, a diverging beam of light (i.e., more than one ray of light can be drawn from a point on the object to the film) will be produced from each point on the object thus tending to produce a fuzzy image. The size of the pinhole is therefore a compromise between image brightness (large pinhole) and resolution (small pinhole).

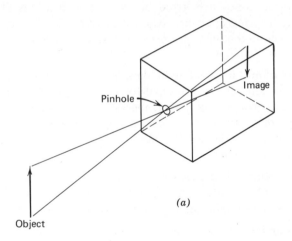

(a)

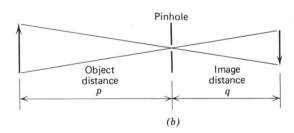

(b)

Figure 1.37.

A simple pinhole can be made in thick aluminum foil. A thin sewing needle (approximately #10) is used to puncture the aluminum foil to make a pinhole approximately 0.5 mm in diameter. Any burrs that may have formed around the pinhole are carefully sanded away with very fine sandpaper or emery cloth. The inside of the pinhole camera should, like other cameras, be painted a flat, dull black to avoid internal reflections.

From the diagram in Figure 1.37(b) we can see that the relation between the object and image size is given by

$$\frac{\text{object size}}{\text{image size}} = \frac{p}{q} \qquad (1\text{-}40)$$

A camera using a lens for imaging will require shorter exposure than a pinhole camera since it gathers more light. The lens must be placed a definite distance from the film to permit a sharp image to be formed. A pinhole forms an image regardless of its relative position to the film; that is, the relative position of the pinhole from the film only determines the image size.

Relative Aperture ($\mathscr{F}$/number)

The amount of light admitted by a pinhole is proportional to its area. If the diameter of the pinhole is doubled, the area of the pinhole is quadrupled. This means that for a given exposure of the film, one quarter of the time will be needed when the diameter of the pinhole is doubled. The relative aperture ($\mathscr{F}$/number) of a pinhole camera can be approximated by dividing the aperture diameter into the pinhole to film distance; for example, a pinhole of diameter 0.4 mm placed 100 mm from the film has a $\mathscr{F}$/number of approximately

$$\mathscr{F}/\text{number} = \frac{\text{focal length}}{\text{diameter}}$$

$$\frac{100 \text{ mm}}{0.4 \text{ mm}} = \frac{f}{250} \qquad (1\text{-}41)$$

The $\mathscr{F}$/number is used to express the aperture diameter as a fraction of the focal length. A $\mathscr{F}$/250 pinhole means that the aperture is stopped down to a diameter 1/250 of its focal length. If the diameter of the pinhole is doubled to $\mathscr{F}$/125, four times as much light will reach the film and the corresponding exposure time will be one quarter the previous value but the image will not be as sharp. One advantage of a pinhole camera is its large depth of field; that is, images of objects very near the pinhole are in focus and sharp as well as more distant objects.

The large depth of field of a small aperture can be appreciated by attempting a simple experiment. Reading fine print by holding it close to the eye is difficult if not impossible. By holding a pinhole very close to the eye we can now focus sharply on the fine print held near the eye. (For this experiment the pinhole can be made by simply putting a pencil point through a piece of paper).

We have seen that the focal length of a lens is the point where a collimated beam of incident light will converge. The location of this point is dependent on the index of refraction of the lens material. The greater the index of refraction of the lens material, the shorter will be the lens focal length. For a plane-convex lens the shortest possible focal distance occurs when $f = r$, where f is the focal length and r is the radius.

Let us now consider the lens shown in Figure 1.38. From the equation for the $\mathscr{F}$/number

$$\mathscr{F}/\text{number} = \frac{\text{focal length}}{\text{diameter}} \tag{1-41}$$

we can see that when the focal length is equal to the radius of the lens, the largest aperture is two times the radius and therefore the smallest possible $\mathscr{F}$/number is

$$\mathscr{F}/\text{number} = \frac{\text{focal length}}{\text{diameter}} = \frac{r}{2r} = \frac{1}{2} \tag{1-41}$$

If we now reduce the area of the lens with an iris by one half, [i.e., since $A = \pi r^2$ if we are to reduce the area by one half, we must divide the radius by $1/\sqrt{2}$ or

$$\frac{A}{2} = \frac{\pi r^2}{2} = \pi \left(\frac{r}{\sqrt{2}}\right)\left(\frac{r}{\sqrt{2}}\right)\Big]$$

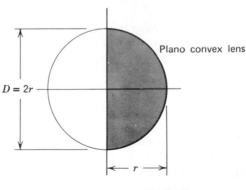

$$\mathscr{F}/\text{number} = \frac{\text{focal length}}{\text{diameter}} = \frac{r}{2r}$$

Figure 1.38 $\mathscr{F}$/Number of a plano-convex lens.

we must double the exposure time. The $\mathscr{F}$/number corresponding to one half the area of the smallest possible $\mathscr{F}$/number is therefore

$$\mathscr{F}/\text{number} = \frac{\text{focal length}}{\text{diameter}} = \frac{r}{2(r/\sqrt{2})} = \frac{\sqrt{2}}{2} = 0.707$$

The following sequence lists increasing $\mathscr{F}$/numbers, each requiring double the exposure time of the previous listed $\mathscr{F}$/number starting with the smallest possible $\mathscr{F}$/number; that is, each successive $\mathscr{F}$/number has a reduced diameter of $1/\sqrt{2}$ or the effective lens surface is reduced by one half.

0.500, 0.707, 1.000, 1.414, 2.000, 2.828, 4.000, 5.656, 8.000, 11.312, 16.000, 22.624, 32.000, 45.248, 64.000, 90.496, 128.000 . . .

Lenses for photographic work are usually marked with the $\mathscr{F}$/number using only one decimal place. The $\mathscr{F}$/number indicates the amount of light energy that a lens can transmit; that is, two lenses with the same $\mathscr{F}$/number will transmit the same amount of light and give equivalent exposures regardless of focal length.

LENS COATING

In order to reduce (or prevent) multiple reflections from lens surfaces, coatings are usually used on them. Multiple surface reflections can cause flare spots or extraneous images to be formed. An uncoated lens (with a given $\mathscr{F}$/number) in a camera transmits as much light to the film as a coated lens (with the same $\mathscr{F}$/number) but bright spots in an object will, because of multiple reflections, be scattered in the image. This tends to make the images of shadows brighter (density greater in the negative) and bright points less bright (density less dense in the negative) which will reduce the image contrast. Coated lenses reduce the amount of scattered light due to multiple reflections and thus increase image contrast.

Field of View

If one stands back from a window and looks out, the field of view is limited by the size of the window. As the observer moves closer to the window, the angular field of view is widened. Moving further from the window narrows the field of view. We can define the angular field of view by the angle θ, as shown in Figure 1.39(a) for a pinhole camera. Field of view is important in selecting camera lenses and in viewing holograms.

The image size in a camera is determined by the focal length of the lens used. The field of view of a lens depends on the design of the lens and not its focal length. Generally a normal lens is considered to be one

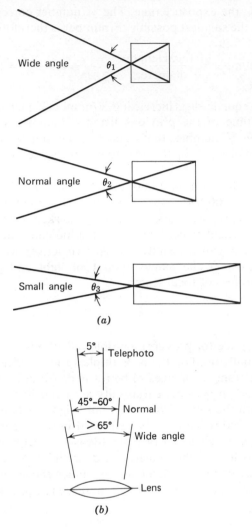

Figure 1.39   Illustration of field of view. (*a*) Field of view for pinhole cameras with different pinhole to film distances. (*b*) Angular view of various types of lenses.

whose focal length equals the diagonal dimension of the film to be exposed. For portrait work a lens of focal length one and one-half to twice the diagonal of the film is usually used. This produces a picture with pleasing perspective but requires a larger object distance than the normal lens. Lenses with focal lengths shorter than the diagonal (wide-angle lenses) permit shorter object distances to be used.

DEPTH OF FOCUS

The focal plane of a lens is the plane in which the images of distant objects are the sharpest. Sharp images can be obtained (in planes parallel to the focal plane) a small distance on either side of the focal plane. The depth of focus is the distance through which the focal plane can be moved without a detectable change in image quality. If we consider a point on an object far from the lens, it will image as a point in the focal plane. As the plane on which the image is being formed is moved out of the focal plane of the lens, the image point will become enlarged.

In order to view the imaged point with proper perspective it should be viewed from a distance equal to the lens focal length f times the magnification of the picture M; that is,

$$\text{Viewing distance} = \text{focal length} \times \text{magnification} = f \times M$$

The angular resolution limit of the eye is about 1/3000 as illustrated in Figure 1.40; that is, two points separated by 1/3000 their distance from the eye will be just barely resolvable. Two points on an object will be indistinguishable if they subtend an angle of less than 1 in 3000.

The circle of confusion is the size of a point image. We can therefore defocus a lens image of a point until it is less than or equal to $f/3000$; that is, the depth of focus is the distance along the optical axis that we can move the image plane with no apparent change in size of the circle of confusion. Only after the image of the point becomes greater than $f/3000$ will it start to appear blurred from a normal viewing distance. For normal photography ($f/1000$ is adequate) the resolution of the lens does not have to be as high as for microfilming, but probably the highest resolution requirements for lenses are imposed in optical data processing where the viewing distances can be very small. The larger the lens aperture (lower $\mathscr{F}$/numbers), the shorter is the depth of focus.

DEPTH OF FIELD

The depth of focus is the distance the image plane can be moved without an apparent change in image quality. On the object side of the lens there is a region within which all object points will be imaged sharply by the lens (in the image plane). The region within which all object points

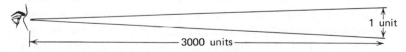

Figure 1.40   Resolution of the eye.

will image sharply is called the *depth of field*. The depth of field of a lens is dependent only on the diameter of the aperture and not its focal length.

HYPERFOCAL DISTANCE

The hyperfocal distance is that distance in front of a lens beyond which all objects are essentially in focus when the lens is focused at infinity; that is, with the lens focused at infinity the distance beyond which the focus of a point source of light does not exceed the circle of confusion. When a lens is focused on the hyperfocal distance, all objects from infinity to one half the hyperfocal distance will appear in sharp focus. Stopping down a lens to smaller apertures reduces the hyperfocal distance. The greatest depth of field for any specific lens aperture is obtained by focusing on the hyperfocal distance. It is important to remember that the hyperfocal distance is based on an assumed tolerable circle of confusion.

SCHLIEREN PHOTOGRAPHY

Schlieren photography is a photographic technique that is particularly suited to recording variations in the index of refraction of the medium through which a light beam is passed. The index of refraction of a gas (or liquid) changes with temperature. As the temperature of a gas is increased, it tends to become less dense and its index of refraction decreases. The nonuniformity of the index of refraction of air can be observed by noticing the waviness of objects seen through the heated air rising from a hot radiator. It is the differences in the optical density of the air that cause the objects to appear to move.

Figure 1.41(a) shows a simple Schlieren system. A light source, S, radiates light to a lens, L_1, which is a focal distance away. The light leaving lens L_1 is collimated and is incident on lens L_2. Lens L_2 brings this collimated light to a focus (a focal length from the lens). Lens L_3, which is a focal length away from the focal point of lens L_2, collimates the light to lens L_4. Lens L_4 brings this collimated light to a focus (a focal length from the lens). The light leaving the focal point of lens L_4 is imaged on a screen. A knife-edge is placed at the focal points of lens L_2 and lens L_4 so as to block some of the light. A change in the index of refraction of the air between lens L_3 and L_4 will cause some of the light that would have reached the screen to be refracted so as to strike (and be blocked) by knife-edge K_2. This will cause a reduced image intensity corresponding to those areas where the light is blocked. In the areas where the light that would have been blocked by knife-edge K_2 is now refracted so as to strike the screen, the image will have increased intensity.

When white light is used to obtain the Schlieren effect, color changes can be observed since air has a different index of refraction for different frequencies of light.

Figure 1.41(b) shows the configuration of Figure 1.41(a) in a more practical form. Cylindrical lenses are used to produce line focuses rather than point focuses. The lenses L_1 and L_2 form a sharp line image from the light source on knife-edge K_1. Knife-edge K_1 is placed about halfway into the line image. The same procedure is followed for knife-edge K_2. The knife-edges should be positioned so as to be exactly aligned with the line images. This is particularly true of knife-edge K_2. When knife-

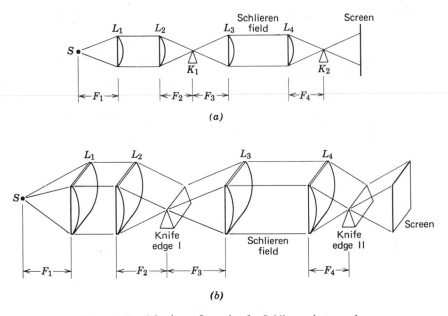

Figure 1.41   A basic configuration for Schlieren photography.

edge K_2 is properly aligned, it will produce uniform darkening of the image on the screen as it is gradually moved into the light beam.

The area between lens L_3 and lens L_4 is referred to as the Schlieren field. Temperature differences as small as 10°F in an air stream can easily be detected. By decreasing the sensitivity (positioning knife-edges) temperatures of more than 5000°F can be recorded. It is therefore possible to take pictures of the heat rising from a hand placed in the Schlieren field by detecting thermal changes transmitted to a gas or liquid. For certain applications this technique can be considered as a possible technique for obtaining a real time input for optical data processing.

2

Physical Optics Fundamentals

CHARACTERISTICS OF LIGHT

Light is defined as electromagnetic radiation capable of inducing visual sensations in the eye. Like any electromagnetic radiation, light can be represented by two mutually perpendicular vibrations (electric and magnetic). In modern theory it is an accepted fact that light energy is emitted (and absorbed) by atoms in discrete units or quanta as postulated by Einstein. This leads to the paradox that light has characteristics that exhibit both particlelike and wavelike behavior. When light interacts with particles, it can be treated as quanta; and when light interacts with light, it can be treated as waves. Because optical data processing primarily involves the interaction of light with light, only the wave properties need be considered here. The quantum characteristics of light must be considered when dealing with the details of optical sources (primarily lasers) and photographic emulsions. Although these areas are important, they will only be touched on in Appendix 3 to explain laser operation as a light source for optical data processing.

Originally it was thought that light traveled in straight lines. This premise formed the basis of geometrical optics. In geometrical optics the location of an image can be found by plane geometry since it is assumed that light travels in a straight line in a homogeneous medium. When rays of light are deviated at the surfaces separating two different media, they are deviated through definite angles again permitting the techniques of plane geometry to be applied. The phenomena of interference and diffraction cannot be satisfactorily explained using the techniques of geometrical optics. In order to describe interference and diffraction the wavelike nature (phase and amplitude) of light must be considered. Actually geometrical optics is an approximation valid for certain optical phenomena whereas physical optics is a more exact method of treating these phenomena.

60

LIGHT BEAMS AND SOURCES

Electromagnetic radiation such as light passing through a point can be represented by a wave. Consider a thin transparent plane placed in the path of a parallel light beam. The light passing through each point of the transparent plane can be represented by a wave. Because there are an infinite number of points on the surface of the plane, an infinite number of waves are required to represent the light beam. Each of these waves is traveling in the same direction as the beam. The amplitude of each wave corresponds to the amplitude of light at the point through which the wave passes. In describing light it is usually sufficient to consider only a few sample waves that are representative of those in the beam. The waves shown in Figure 2.1 are not continuous. Because light energy is emitted in discrete units, the waves will be of finite length and are therefore discontinuous. Normally such wave trains last for approximately 10^{-8} sec. Surfaces of equal phase, called *wavefronts*, are formed when points of corresponding phase are connected together. The dotted lines in Figure 2.1 shows corresponding points of equal phase (maximums) on different wave trains connected together to form parallel, equally spaced wavefronts. The idea of wavefronts is important in optical data processing. In Figure 2.1 the wave trains are all shown to be in phase with one another. This is not normally the case; usually there is no fixed relation between one atom emitting light and another atom emitting light. There is therefore no fixed relationship between one wave train and another. The wave trains in Figure 2.1 are all shown to be of the same frequency

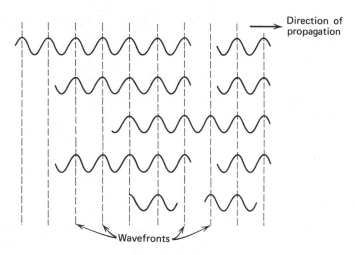

Figure 2.1   Wave trains in parallel light beam.

and phase (but because they each have discontinuities they will produce a beam of light that will contain other frequencies in addition to the frequency of each wave train). In addition we have shown the amplitude of each wave train to be the same. This is usually not the case. Most light sources, such as tungsten lamp, emit light of many frequencies (polychromatic), that is, wave trains of varying lengths, amplitudes, and phases. Figure 2.2 shows sample wave trains of a polychromatic parallel light beam that is more representative of the usual case (although for simplicity we still show the wave-train amplitudes to be the same). Some light sources, such as mercury vapor lamps, do emit light at certain discrete frequencies. In these cases emission at the characteristic frequency is predominant and by filtering can be made to provide a nearly monochromatic light source.

Figure 2.3 shows sample wave trains that are of the same frequency, phase, and amplitude and that are also continuous. Figure 2.3 can be used to represent a light beam from a laser. In a laser the emission of light from individual atoms is controlled because the emission field from an excited atom stimulates other atoms to emit light of the same wavelength and with such phase that the emission field increases. This process occurs repeatedly and the light radiated from the laser is essentially

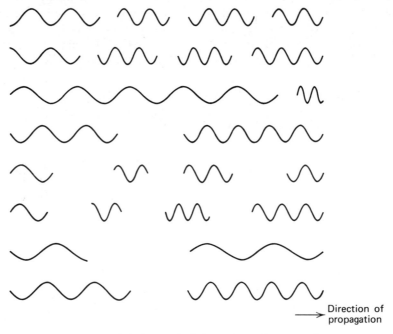

Direction of
propagation

Figure 2.2 Polychromatic light beam sample wave trains.

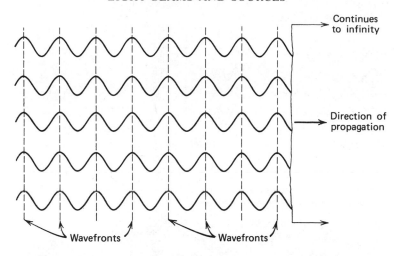

Figure 2.3   Monochromatic coherent light beam sample wave trains.

monochromatic, continuous, and to a high degree uniform in phase. Appendix 3 gives a more detailed discussion of laser operation.

The sample waves shown in Figures 2.1, 2.2, and 2.3 were drawn with equal amplitudes for convenience only. Amplitude variations are present in practical cases. However, instantaneous values of amplitude cannot be measured directly and the variations of light amplitude with respect to time are neglected in optical data-processing applications.

Figure 2.3 also shows constant-phase wavefronts, (as previously stated wavefronts are surfaces representing constant phase). Usually the wavefronts are chosen to correspond to the points of maximum amplitude as shown in Figure 2.3. The light at all points on a wavefront is in phase. In the case of a light beam from a laser the wavefronts are approximately plane wavefronts as shown in Figure 2.3. (A point source produces spherical wavefronts.) A wavefront moves in the direction of the light waves it represents. For example, if the light waves are propagating to the right in Figure 2.3, the wavefronts also move to the right.

Figure 2.4 shows the difference between wave trains of a collimated and noncollimated light beam. Collimated light is composed of wave trains propagating along parallel lines. Figure 2.4(a) shows wave trains that produce parallel, evenly spaced wavefronts. Such a representation can quite accurately describe a laser beam. Figure 2.4(b) shows wave trains that produce wavefronts that are parallel but nonuniformly (randomly) spaced. Figure 2.4(c) shows wave trains for which it is not

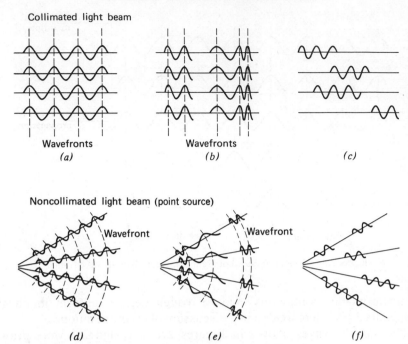

Figure 2.4   Wave trains in a collimated and noncollimated light beam.

possible to draw propagating wavefronts (surfaces of equal phase) since the wave trains themselves are random. Figure 2.4(d) shows wave trains radiating out from a point light source producing spherical wavefronts that are evenly spaced. Figure 2.4(e) shows a point source that produces wavefronts that are not uniformly spaced but the wavefronts still form spheres of equal phase. Figure 2.4(f) shows wave trains in a noncollimated (divergent) light beam. The wave trains in this beam are random so that it is not possible to draw propagating wavefronts (surfaces of equal phase).

INTERFERENCE

Light waves are not impeded by the presence of other light waves, but when crossing each other they produce a combined effect (at the point of crossing) called *interference*. The principle of superposition can be used to find the interference effect; that is, the amplitude at any point, at any instant, can be found by adding the instantaneous amplitudes that would be produced at the point by the individual waves if each were present alone. Beyond the point of intersection the light waves continue in the original directions of propagation as though there had been no intersection. Interference between two light waves does not involve loss of energy. The fact that the amplitude is reduced (or destroyed) whenever the waves meet in opposite phase is counterbalanced by an increased amplitude in regions of reinforcement; for example, two equal amplitude waves meeting exactly out of phase will cancel (produce a resultant of zero) but the amplitude will be twice that of either wave where they meet in phase. In effect the energy of the two waves taken separately is equal to the energy of the two waves taken together.

The additive properties of two waves at a point of intersection are shown in Figure 2.5. Here we assume that the waves are of the same amplitude and frequency. It is seen that the result of adding two waves at a point is dependent on the phase difference between the two waves. The

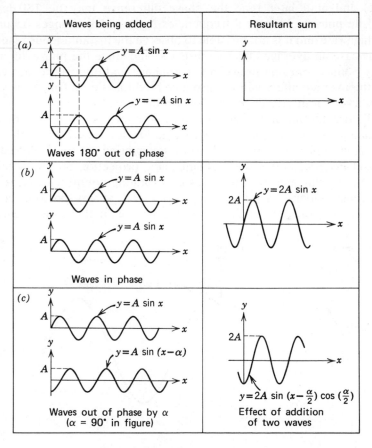

Figure 2.5 Addition of two waves.

resultant sum is a minimum (zero for equal amplitudes) when the phase difference is 180 degrees (opposite phase); maximum when the phase difference is zero (in phase); and takes values between the maximum and minimum for phase differences between 0 and 180 degrees. From this discussion it should be noted that the frequencies of two interfering waves must be the same in order to produce a stationary interference pattern. When the frequencies are the same, the phase of each of the waves varies at the same rate and the relative phase difference will remain constant. The constant phase difference of equal frequency waves implies that an in-phase point will always be an in-phase point, if the frequencies are different, the phase of each of the waves varies at a different rate. The difference between the rate of change of the phase will not produce a constant phase difference. Thus at one instant the phase difference may be zero (in-phase) and at a later time the phase difference may be 180 degrees (opposite phase). At optical frequencies these phase changes occur at a very high rate and it is not possible to observe the instantaneous resultant. In this case an average affect is observed and since the phase difference at any point is varying between the in-phase and opposite phase condition, the average affect will not produce bright and dark fringes as would be observed for interferences of waves of the same frequency.

In Figure 2.5 the sinusoidal function $y = A \sin x$ is shown as a function of x and y. Another common representation for functions of this type used rotating phasors. Figure 2.6 shows a rotating phasor A plotted on a real and imaginary axis with its sinusoidal projection on the imaginary axis. We can imagine that the phasor is rotating counterclockwise with an angular velocity of ω. Stopping the phasor at any position permits the projection on the imaginary axis to be obtained. In the Figure 2.6

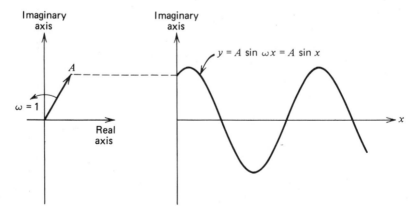

Figure 2.6   Projection of a rotating phasor on the imaginary axis.

we can see that the projection of the phasor on the imaginary axis is the function value for that position of the phasor. For the function

$$y = A \sin x$$

we can consider that ω is a constant equal to one ($\omega = 1$) and therefore

$$y = A \sin x = A \sin \omega x.$$

Consider two phasors rotating at the same angular frequency as shown in Figure 2.7(a). It can be seen that the angle θ between the phasors must

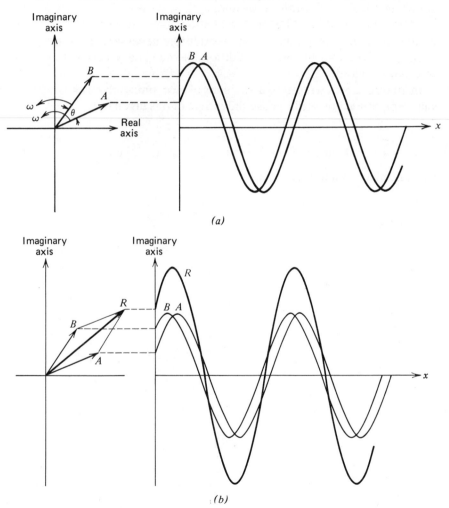

(a)

(b)

Figure 2.7 (a) Projection of two rotating phasors on imaginary axis. (b) Addition of phasors.

remain constant if they rotate at the same angular velocity. Thus the sum of the two phasors at any instant can be found by the vector addition shown in Figure 2.7(b). This vector sum is the amplitude of the resultant sine waves that could be obtained by adding the value of the two sine waves at the corresponding instant. Thus by using phasors the amplitude of two sine waves can be found by geometric methods.

For the investigation of the quantitative properties of wave addition the use of phasor (vector) notation is convenient. Figure 2.8(a) shows the familiar method for addition of phasors applied to the three examples shown in Figure 2.5. In these examples the two waves were assumed to have equal amplitudes. This vector addition gives the desired quantitative results without the point by point addition necessary in Figure 2.7. Figure 2.8(b) shows the phasor addition for the general case with phasors of equal amplitude.

In Figure 2.5(c) we see that the sum of two sinusoidal waves with the same amplitude but with a phase difference of α is given by

$$A \sin x + A \sin (x - \alpha) = 2A \sin \left(x - \frac{\alpha}{2}\right) \cos \frac{\alpha}{2} \qquad (2\text{-}1)$$

since using trigonometric identity

$$\sin (X \pm Y) = \sin X \cos Y \pm \cos X \sin Y$$

we can let

$$X = \left(x - \frac{\alpha}{2}\right) \qquad Y = \frac{\alpha}{2}$$

then

$$A \sin x = A \sin \left[\left(x - \frac{\alpha}{2}\right) + \frac{\alpha}{2}\right]$$

$$= A \left[\sin \left(x - \frac{\alpha}{2}\right) \cos \frac{\alpha}{2} + \cos \left(x - \frac{\alpha}{2}\right) \sin \frac{\alpha}{2}\right]$$

and

$$A \sin (x - \alpha) = A \sin \left[\left(x - \frac{\alpha}{2}\right) - \frac{\alpha}{2}\right]$$

$$= A \left[\sin \left(x - \frac{\alpha}{2}\right) \cos \frac{\alpha}{2} - \cos \left(x - \frac{\alpha}{2}\right) \sin \frac{\alpha}{2}\right]$$

Therefore

$$A \sin x + A \sin (x - \alpha) = 2A \sin \left(x - \frac{\alpha}{2}\right) \cos \frac{\alpha}{2} \qquad (2\text{-}1)$$

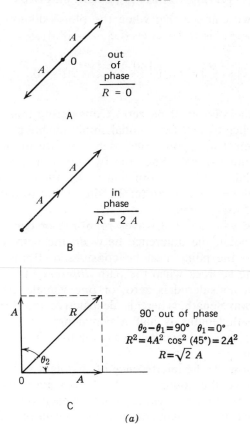

(a)

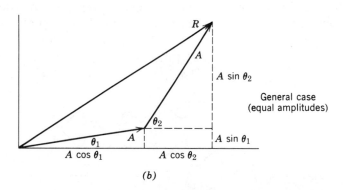

(b)

Figure 2.8 Phasor addition of complex amplitudes. General case for equal amplitudes.
$R^2 = [A \cos\theta_1 + A \cos\theta_2]^2 + [A \sin\theta_1 + A \sin\theta_2]^2$; $R^2 = A^2\{[\cos\theta_1 + \cos\theta_2]^2 + [\sin\theta_1 + \sin\theta_2]^2\}$;
$R^2 = 2A^2[1 + \cos(\theta_2 - \theta_1)]$; $R^2 = 4A^2 \cos^2(\theta_2 - \theta_1)/2$.

From (2-1) we can see that when the phase difference between the two sinusoidal waves is 180 degrees (i.e., $\alpha = 180°$),

$$\cos\left(\frac{\alpha}{2}\right) = \cos\left(\frac{180°}{2}\right) = \cos 90° = 0$$

and therefore the sum must be zero. Considering this another way we can state that when two waves of equal amplitude but phased 180 degrees to each other meet at a point, they will interfere in such a way as to produce a minimum (zero). When the two equal amplitude waves meet at a point in phase with each other ($\alpha = 0°$), then $\cos \alpha/2 = 1$ in (2-1) and the sum will be a maximum (of $2A \sin x$); that is, the two waves will interfere to produce a maximum.

To determine whether two waves reinforce or cancel (destructively interfere) at a point, the difference between the path lengths from the light sources to the point must be considered. Reinforcement occurs for two in-phase sources when the path *difference* from the sources to the point being considered is zero, or one wavelength, or an integer multiple of a wavelength. There is destructive interference when the *difference* in path lengths is half a wavelength, or one and one-half wavelengths, or two and one-half wavelengths, or any odd multiple of half wavelengths.

In order to *observe* the interference of two light waves at a point it is necessary to have the sources of the light waves *always* maintain a constant phase relationship with respect to each other. If we assume, for example, that the two light sources start out in phase, it is necessary for these light sources to maintain the same respective phase relationship if the interference between the two light waves is to be observable. Let us consider an example to make this point clear. Figure 2.9 shows

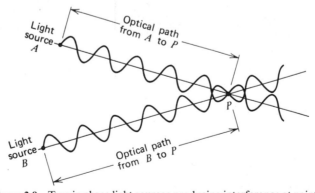

Figure 2.9   Two in phase light sources producing interference at point P.

two in-phase light sources A and B. Light from source A is assumed to reach point P in phase with respect to the phase of the light at A because the optical path length AP is a multiple of the wavelength of the light; that is, $\overline{AP} = n\lambda$, where n is a whole number. Light from source B is assumed to reach point P out of phase (with the light arriving from source A) because the optical path length $\overline{BP}$ is an odd multiple of $\lambda/2$; that is, $\overline{BP} = m\lambda/2$, where m is an odd number. Under these conditions the light from source A will cancel the light from source B at point P. Notice in Figure 2.9 that although the two light waves will cancel at point P, they will continue unaffected past point P. A dark spot would be present at point P under the conditions stated. If the relative phases of the two light sources changed with respect to each other, point P would no longer be a point of cancellation; for example, if light from source A were made 180 degrees out of phase with the light being emitted from source B, the light reaching point P would reinforce because the difference in path lengths $\overline{AP}$ and BP is considered to be a multiple of $\lambda/2$ as stated. If the phases of the sources A and B drifted with respect to each other, point P would be a point of cancellation (minimum) only at the instant when the relative phases of the two sources are in phase. At the instant when the two sources are 180 degrees out of phase point P will be a point of maximum reinforcement. Point P will assume values between these two values as the relative phase between the two sources changes from 0 to 180 degrees.

The light from most sources is produced by atoms that are independent of each other and therefore the light produced usually has frequent haphazard phase changes. It is to be expected, therefore, that two separate light sources will not be capable of producing *observable* interference effects. In order to produce observable interference effects it is necessary to have two sources emitting light waves that always maintain the same relative phases to each other. It is therefore usual to start with only one light source and split the light from it to produce two phase-related light beams. In this way any haphazard changes of phase of the original light source will be imparted to the two light beams, thus maintaining a constant relative phase between them. Two such light beams are said to be mutually coherent.

Experiments Showing Interference of Light

The difficulties in obtaining two independent light sources that can produce light in phase with each other have been pointed out. In addition the fact that light waves have very short wavelengths makes it extremely difficult to measure path-length differences in the order of a half wavelength. These difficulties can be overcome in various ways.

Fresnel used two mirrors, A and B, to reflect light from a source S to a screen as shown in Figure 2.10(a). The obstruction is used to prevent light from the source S from reaching the screen directly. Light from source S can only reach the screen by reflection from mirrors A and B. Mirror A reflects the light so that it appears to be coming from S_1 and mirror B reflects the light so that it appears to be coming from S_2. The two apparent sources S_1 and S_2 radiate light in phase (because they are both images of the same source). The mirrors A and B are adjusted to be nearly parallel so that the images S_1 and S_2 are very close together. The interference pattern on the screen is a series of narrow parallel light and dark bands. The center light band is equally distant from S_1 and S_2, while the first dark band on each side of it is a half wavelength nearer to one source than the other (the second dark band is one and one-half wavelengths nearer, the third is two and one-half wavelengths nearer, etc. while the bright bands path differences on each side are, respectively, one, two, three, etc, wavelengths).

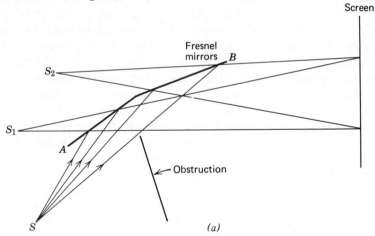

(a)

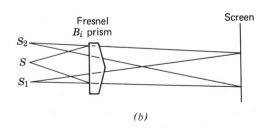

(b)

Figure 2.10   Methods for demonstrating interference.

An alternate method used by Fresnel to demonstrate interference was by means of the Fresnel biprism shown in Figure 2.10(b). A plate of glass called a *biprism* (a double bevel on one side) is used to refract light from a slit in different directions through each half. The light from slit S after it passes through the biprism appears to be coming from two sources, slits S_1 and S_2. If the biprism is made so that the images S_1 and S_2 are very close together, interference bands will be seen.

The interference patterns can be formed with the Fresnel mirrors or a biprism using a white light source. When white light is used, the bands will appear colored. The wavelengths for different colors are not the same and therefore the dark bands for red light are not in the same place as the dark bands for other wavelengths (other colors). This accounts for the multicolored interference pattern produced by the Fresnel mirrors. The same fact accounts for the multicolored interference pattern when the Fresnel biprism is used; but in addition one must remember that the prism will bend different colors different amounts, causing the positions of the images S_1 and S_2 to be slightly different for different colors.

COHERENT LIGHT BEAMS

Two light beams combined to form interference patterns are said to be *mutually coherent* beams. Light beams from two different sources (except possibly lasers) cannot be combined to produce interference patterns because the light wave trains emitted are independent and of random phases. A monochromatic light beam can be split to produce two mutually coherent light beams. This is possible since any discontinuities in one beam will also appear in the other beam.

The term *coherence* implies phase correlation between points in a light beam. There are two types of coherence: namely, spatial and temporal coherence. Figure 2.11 shows a comparison between spatial and temporal coherence. The sources (point sources) used in Figure 2.11 produce spherical wavefronts.

Spatial Coherence

Spatial coherence is a measure of the phase correlation between two points on the same wavefront. Spatial coherence can be measured by passing a wavefront through two slits as shown in Figure 2.11(a). If the light coming through the slits forms an interference pattern, the light source is said to display spatial coherence. The contrast of the fringe pattern is a measure of the spatial coherence. A 100 percent spatially coherent light beam will produce fringe patterns that vary in intensity from zero to maximum intensity (100 percent contrast). A noncoherent

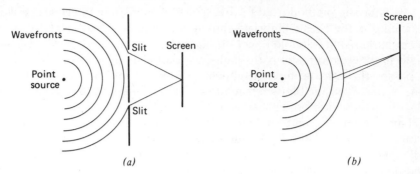

Figure 2.11 Comparison between spatial and temporal coherence. (*a*) Spatial coherence: two points on a wavefront are examined to see if they can be made to produce fringe patterns. (*b*) Temporal coherence: two points on different wavefronts are examined to see if they can be made to produce fringe patterns.

beam will not form a fringe pattern. Therefore the contrast of a fringe pattern of noncoherent beam of light is zero (actually no fringe pattern).

Because electromagnetic radiation is a continuous phenomenon (no abrupt discontinuities), there is always some spatial coherence between two closely spaced points in a light beam. Hence a monochromatic light beam passed through a pinhole will be nearly 100 percent spatially coherent. For this reason early studies of interference patterns were made by passing a monochromatic light beam through a pinhole to obtain spatially coherent light.

Laser light beams have wave trains that are all in phase. When any two points on a wavefront in a laser beam are tested for spatial coherence, very high contrast patterns are produced. These high contrast patterns indicate the expected high degree of spatial coherence in a laser beam.

Temporal Coherence

As shown in Figure 2.11(b) the degree of similarity between wavefronts separated in time is a measure of temporal coherence. The experiment to determine temporal coherence is usually based on the interference of two successive wavefronts of light with each other.

The term *coherence length* is often used in work with lasers. Coherence length refers to the maximum length separating two different wavefronts that can be made to produce fringe patterns. Beyond a certain path distance no interference fringes will be observed; that is, the wave trains being combined are no longer derived from the same wave train and will have no correlation to each other. Thus path-length differences less than the coherence length are required to observe interference effects. Appendix 4 shows one technique that can be used to measure temporal coherence.

Figure 2.12 is similar to Figure 2.4, showing the difference between wave trains of a collimated and noncollimated light beam. Collimated light from a monochromatic source is composed of parallel, evenly spaced wavefronts as shown in Figure 2.12(a). The wavefronts in this light beam will be spatially and temporally coherent. When the wavefronts are parallel, but nonuniformly (randomly) spaced, as shown in Figure 2.12(b), the light beam will be spatially coherent and temporally noncoherent. Figure 2.12(c) shows a spatially and temporally incoherent light beam. For such wave trains it is not possible to draw wavefronts because points of equal phase on different wave trains do not form surfaces of uniform phases. Figure 2.12(d) shows spherical wavefronts produced from a monochromatic point light source. The wavefronts are evenly spaced producing spatially and temporally coherent light. Figure 2.12(e) shows a point source that produces light which is spatially coherent, but temporally incoherent, because the wave fronts produced are not uniformly spaced; that is, the wavefronts form spheres of equal phase but not equally spaced. Figure 2.12(f) shows wave trains in a noncoherent light beam. The wave trains in this beam are random so that it is not possible to draw propagating wavefronts. From this we can deduce that a

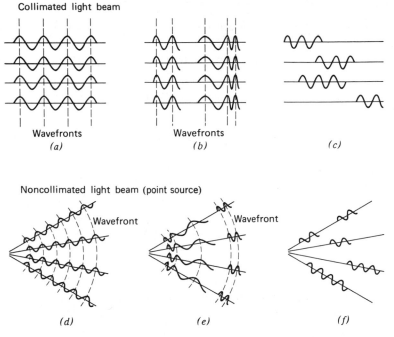

Figure 2.12 Wavetrains in a collimated and noncollimated light beam.

monochromatic light source (such as a laser light source) would be required to produce the uniformly spaced wavefronts shown in Figures 2.12(a) and (d). Figures 2.12(c) and (f) represent the wave trains from a polychromatic light source such as a tungsten lamp. The basic concepts illustrated in Figure 2.12 will be useful in understanding optical data processing that will be explained later.

INTENSITY AND AMPLITUDE

The brightness of light in a beam of light can be indicated by the intensity that is proportional to the square of the amplitude. Mathematically this is expressed as

$$I = KA^2 \tag{2-2}$$

For our purposes the constant K can be assumed to be equal to one. This allows the equation to be written as

$$I = A^2 \tag{2-3}$$

In the case of several waves at a point the combined intensity is equal to the square of the resultant amplitude. For example, the general case shown in Figure 2.8(b) gives the results of adding two waves of equal amplitude and frequency. The resultant amplitude is given by

$$R = 2A \cos\left(\frac{\theta_2 - \theta_1}{2}\right) \tag{2-4}$$

while the resultant intensity is given by

$$I = R^2 = 4A^2 \cos^2\left(\frac{\theta_2 - \theta_1}{2}\right) \tag{2-5}$$

For two waves in phase (i.e., $\theta_1 = \theta_2$) this gives

$$I = 2^2 A^2 = 4A^2 \tag{2-6}$$

Repeated use of vector addition for n waves of equal amplitude, A, and the same frequency gives

$$I = n^2 A^2 \qquad n \text{ waves in phase} \tag{2-7}$$

For the general case in which the n waves (equal amplitude and frequency) have fixed phase relations between each other the resultant intensity will have some value in the range from zero to n^2A^2. That is,

$$0 \leqslant I \leqslant n^2A^2 \qquad n \text{ waves with constant phase relations.}$$

The upper limit n^2A^2 is seen to be the special case in which all n waves are in phase. In all other cases where the n waves have fixed phase differences between each other (but are not all in phase), the resultant intensity will be less than n^2A^2.

The dependence of intensity on the phase relation of the combining waves is due to the $\cos^2[(\theta_2 - \theta_1)/2]$ term of (2-5). If two waves have a random phase relation, this term must be averaged over the range 0–360 degrees. For two random waves the intensity is found to be $I = 2A^2$. Repeating for n random waves gives

$$I = nA^2 \qquad n \text{ waves with random phase relations.} \qquad (2\text{-}8)$$

As an example of the intensity relations given above, let us consider two horns that produce sound of the same frequency and the same amplitude. The resultant sound intensity produced by these two horns together would normally be

$$I = nA^2 = 2A^2$$

The resultant intensity is given by the random case above because two horns normally produce sound waves that have no fixed phase relation with each other. If extra care is taken in the design of the two horns, it might be possible to construct them so that the sound produced by each was exactly the same in amplitude and in phase. The resultant sound intensity produced by these two special horns would be

$$I = n^2A^2 = 4A^2$$

In each case a single horn produces the same amplitude of sound. It is significant to note that two horns of the usual type produce sound twice as intense ($I = 2A^2$) as a single horn, and two special horns, which produce sound in phase with each other, will produce sound four times as intense ($I = 4A^2$) as a single horn. This implies that although there is no difference in the input power the special horns can be heard at a greater distance than that of the usual horns.

In optical data processing it is important to distinguish between the amplitude of the light and the intensity of the light. All light detectors (for example, the eye, photocells, photographic film, etc.) respond to the intensity of the incident light. The time response of all light detectors is too slow to follow the instantaneous variations of the amplitude of the light. Detectors respond to the intensity of the incident light that is analogous to saying the light detectors are responding to the "light power." In electrical engineering we can measure the amplitude of a voltage with a voltmeter and we can measure the power absorbed by a circuit with a wattmeter. The comparable device to the voltmeter does not exist for optics and therefore the amplitude of light cannot be measured directly. The intensity (comparable to power) can be measured optically. It is therefore possible to rewrite (2-3) as

$$A = \sqrt{I}$$

From this we can find the amplitude of the light incident on a light detector by taking the square root of each measured value of intensity; however, we cannot determine anything regarding the phase of the light from its intensity.

DIFFRACTION

When waves meet obstacles, they usually are able to bend around them. This bending of waves around obstacles is called *diffraction*. We are all aware of the fact that sound waves are capable of bending around buildings. Likewise we are aware that water waves can bend around piers. The bending of light waves was not apparent at first. Careful experiments have proved that light also is capable of bending around corners (indicating its wavelike nature); because of the short wavelengths of light, this bending is difficult to observe.

In order to illustrate the phenomena of diffraction let us consider some commonly observed examples using sound. Figure 2.13 shows two intersecting streets, a marching band on one and an observer on the other. Between the marching band and the observer is a building blocking the sound from the band from directly reaching the observer. Most of us have been in a similar circumstance, but not many of us have been aware of the principles of diffraction taking place. Usually the observer is impressed with the change in quality of the music as the band comes into view. When the band is far around the corner, only the longer wavelength notes are heard (probably only the drum beat). As the band moves closer to the corner the higher pitched (short wavelengths) instruments begin to be heard. Before the band is in view the base instruments sound much louder than the higher pitched instruments, but as the band comes

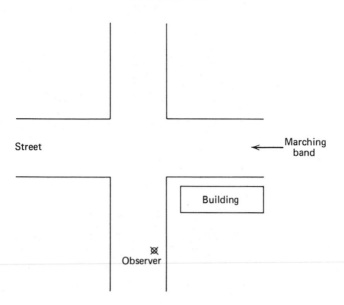

Figure 2.13 Diffraction of sound around a building.

into view there is a somewhat sudden increase in the intensity of the sound from the higher pitched instruments. In fact the effect is usually so pronounced that casual observers believe that only the drummer was playing before the band reached the corner and that the band began to play just as it reached the corner. The effect is caused by the fact that the longer wavelengths of sound bend around corners much more readily than do the shorter wavelengths. Diffraction effects also occur for other types of waves, for example, water waves and electromagnetic waves.

When light waves pass through an aperture or are in part blocked by an obstacle, they spread beyond the limit of the geometric shadow of the obstacle by the phenomenon known as *diffraction*. Huygen's principle states that each point on a wavefront can be considered as a new source of a wave. With this in mind the diffraction phenomena can easily be explained. Figure 2.14 shows a portion of the light from a monochromatic point source being blocked by a baffle in which there is a small aperture. By Huygen's principle the same aperture can be considered as a new light source.

On the light side of the baffle the light spreads out from the pinhole past its geometric shadow by diffraction as if the pinhole were a light source.

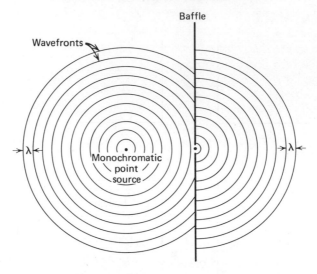

Figure 2.14   Diffraction of light at a small aperture.

Experiment Showing Diffraction of Light

Thomas Young performed an experiment similar to that shown schematically in Figure 2.15. In his experiment Young first passed sunlight through pinhole A. From our discussion of spatial coherence it is apparent that the light radiated from pinhole A will have a high degree of spatial coherence. By Huygen's principle wavelets spread out from pinhole A and reach pinholes B and C at the same instant, (pinholes B and C are equally distant from pinhole A).

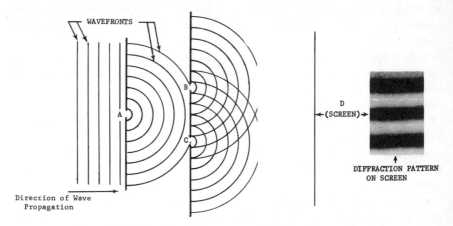

Figure 2.15   Experimental arrangement for Young's experiment.

In the previous discussion of spatial coherence it was stated that a coherent light beam can be split into two beams that can be made to interfere with each other. In effect, this is done by the pinholes B and C. Thus, by using pinholes in the arrangement shown in Figure 2.15 Young was able to produce two mutually (spatially) coherent light sources for his investigation of interference patterns.

When discussing Young's experiment it is convenient to replace the pinholes by narrow slits and use a monochromatic source of light. The narrow slits will produce cylindrical wavefronts instead of the spherical wavefront produced by the pinholes. Figure 2.15 can also be used to represent the cylindrical wavefronts emerging from the slits. Either the spherical or cylindrical wavefronts will produce an interference pattern of alternate bright and dark bands (formed perpendicular to the line joining the pinholes, or parallel to the slits). Because the light leaving the two slits (B and C) has a constant phase relationship, the light reaching the screen D will interfere, forming the symmetrical pattern of bright and dark bands as shown in Figure 2.15. Two respective wavefronts leaving slits B and C will radiate out towards screen D. Because both wavefronts start out phase related (and all succeeding wavefronts start out with the same phase relationship), certain points on screen D will always be points of constructive interference, while other points will always be points of destructive interference. If either slit B or C is covered, the dark and bright bands disappear and the screen becomes illuminated by a single broad band. Some common diffraction and interference patterns are discussed in Appendix 7 and Appendix 8.

Diffraction from Multiple Slits

The basic operation of an optical processor is easily understood after considering a light beam with plane wavefronts that is diffracted by equally spaced slits. When a plane wavefront from a monochromatic coherent light source passes through a series of slits (Figure 2.16), each slit will act as a new source of a cylindrical wave in accordance with Huygen's principle. The successive wavefronts spread out on the right side of the slits. The wavefronts from adjacent slits interfere constructively and destructively with each other as shown in Figure 2.15. Appendix 6 describes how Moiré patterns are formed and can be used to determine the interference patterns produced by two coherent point light sources.

Zero-Order Wavefront Formation

Figure 2.16 shows the construction of zero-order wavefronts produced by slits in a plane. When these slits are illuminated by plane wavefronts

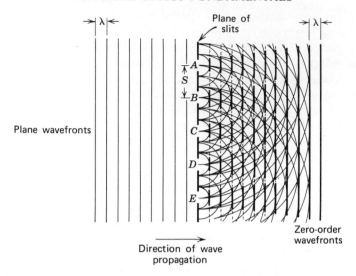

Figure 2.16　Formation of zero-order wavefronts.

each slit will radiate in phase cylindrical wavefronts. In Figure 2.16 it can be seen that as the cylindrical wavefronts travel out (in the direction perpendicular to the slit plane) they become more and more plane as shown by the darkened portions of the dotted lines representing the wavefronts. The *zero-order wavefronts* are formed by planes parallel to the plane of the slits and tangent to the cylindrical wavefronts produced by the slits. The zero-order wavefronts travel in the same direction and have the same spacing (one wavelength, λ) as the original plane wavefronts. The respective phases of the zero-order wavefront will (sufficiently far from the plane of the slits) be the same as they would be if the slits were not present.

First-Order Diffraction Wavefront Formation

Figure 2.17 shows the formation of first-order diffraction wavefronts. The cylindrical wavefronts also merge at large distances from the slits into plane waves traveling at an angle upward or downward from the horizontal as shown. The first-order up wavefronts are formed by planes tangent to the mth wavefront of slit A, $(m+1)$ wavefront of slit B, $(m+2)$ wavefront of slit B, $(m+2)$ wavefront of slit C, etc. Similarly the first-order down wavefronts are formed by planes tangent to the mth wavefront of slit A, $(m-1)$ wavefront of slit B, $(m-2)$ wavefront of slit C, etc. By diffraction theory and trigonometry it can be shown that the sine of the angle of the direction of propagation is related to the wavelength

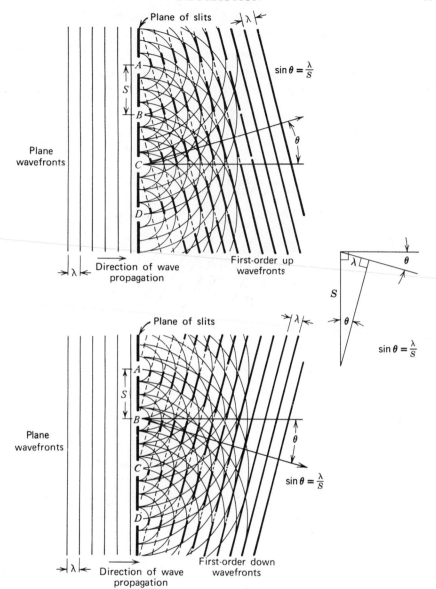

Figure 2.17   Formation of first-order wavefronts.

of the light and the spacing between the slits S in accordance with the formula

$$\sin \theta = \frac{\lambda}{S}$$

PHYSICAL OPTICS FUNDAMENTALS

where λ is the wavelength of light, S is the slit spacing, and θ is the angle of wavefront propagation from the horizontal.

Second-Order Diffraction Wavefront Formation

Figure 2.18 shows the formation of the second-order diffraction wavefronts. Second-order wavefronts are formed by planes tangent to the mth

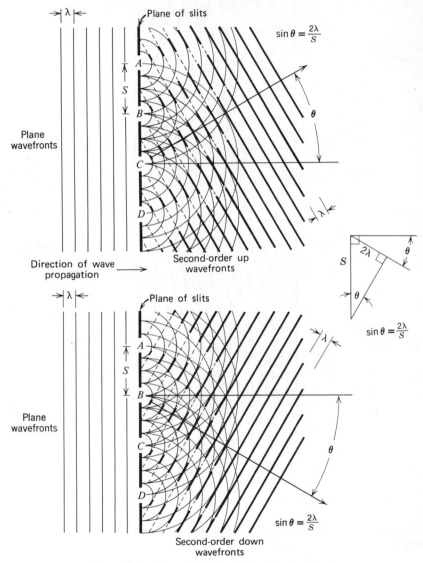

Figure 2.18 Formation of second-order wavefronts.

wavefront of slit A, ($m \pm 2$) wavefront of slit B, ($m \pm 4$) wavefront of slit C, etc. The cylindrical wavefronts merge at large distances from the slits into plane waves as shown. The angle from the horizontal at which the second-order wavefronts travel is given by

$$\sin \theta = \frac{2\lambda}{S}$$

where λ is the wavelength of light, S is the slit spacing, and θ is the angle of wavefront propagation from horizontal.

nth-Order Diffraction Wavefront Formation

From the above discussion it can be seen that there are third-order diffraction wavefronts, fourth-order, and so on, to n-order diffraction wavefronts, each existing simultaneously. The angles from the horizontal at which the n-order diffraction wavefront is propagating is given by the formula

$$\sin \theta = \frac{n\lambda}{S}$$

where n is the order of the diffraction wavefront.

It has been shown how plane waves are diffracted by multiple slits to form the various order wavefronts. These diffraction wavefronts are mutually coherent (originate from the same source) and can interfere with each other. Let us consider a screen parallel to the zero-order wavefronts. Each order of diffraction wavefronts would illuminate the screen uniformly if it were the only order present. When two or more wavefronts (of different orders) are presented, however, they will not have the same relative phase with respect to each other at every point on the screen. Because the relative phase relation determines how the wavefronts combine, a variation in light intensity will occur on the screen. In the case of slits this interference pattern will consist of light and dark bands. In effect the same pattern could be produced by using separate collimated light sources for each order of diffraction wavefronts. These separate sources would have to be mutually coherent and incident at the same angle as the corresponding order of diffraction wavefronts. Because this interference pattern is formed by wavefronts created by diffraction, it is called a *diffraction pattern*.

We have seen that because each slit is illuminated by a plane wavefront, identical wavefronts will be emitted from each slit. The phase and form of each wavefront emanating from any slit will be identical to any other. As these wavefronts travel away from the plane of the slits we see that plane wavefronts begin to be formed, traveling at various angles to the plane of the slits. When the slits are formed in photographic film, varying thicknesses of emulsion cause phase changes in the light leaving the slits. These phase changes will change the diffraction pattern formed by the slits. Consider the two cases shown in Figure 2.19(a). In each case the varying thickness of the emulsion will cause phase changes in the light transmitted. At times these phase changes may be deliberately introduced (phase holograms) while at other times they are to be avoided. Undesirable phase changes are usually eliminated by placing the photographic emulsion between optical flats with an oil of matching index of refraction. The oil fills the voids between the optical flats and the film to make the photographic emulsion appear to be of constant thickness as shown in Figure 2.19(b). Appendix 7 includes the diffraction patterns of various types of apertures. Appendix 5 shows how plane-oriented crystals produce diffraction of X-rays (Bragg effect).

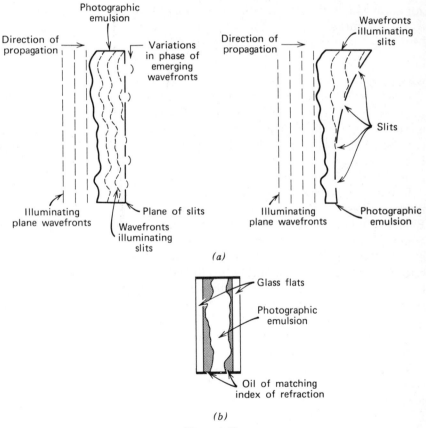

Figure 2.19.

FOCUSING DIFFRACTION PATTERN (SPECTRUM ANALYSIS)

The different orders of diffraction wavefronts caused by the diffraction of light through slits can be focused by a lens. It was shown previously that parallel incident rays to a lens will be focused to a point in the back focal plane of the lens. When the incident rays are parallel to the optical axis, the focal point A will be located in the focal plane of the lens and on the optical axis. When the incident collimated beam is not parallel to the optical axis, the focal point B will be off the optical axis but still in the back focal plane of the lens as shown in Figure 2.20. Using this principle it is possible to focus the different orders of wavefronts (since each order is essentially a collimated light beam inclined to the optical axis) with a lens to points in the focal plane determined by the incident angle ϕ.

It was shown that orders of diffraction wavefronts will be set up when plane wavefronts illuminate a set of slits as shown in Figures 2.16, 2.17, and 2.18. The directions of propagation of the zero-order wavefronts and of the first-order up and down wavefronts are shown in Figure 2.21. These directions are indicated by rays and can be compared to Figures 2.16, and 2.17. The various orders of diffraction wavefronts produced by the slits are brought to focus in the manner illustrated in Figure 2.20 for the first-order up. The zero-order diffraction wavefronts can be represented by parallel horizontal rays and would therefore be brought to a focus at the back focal point A of the lens. This focusing of zero-order wavefronts is illustrated in Figure 2.21. Figure 2.21 therefore combines in one diagram the focusing of various orders of diffracted wavefronts (zero, first-order up, first-order down). It is to be noted that when the diffraction pattern is imaged in the back focal plane of a lens, the imaged

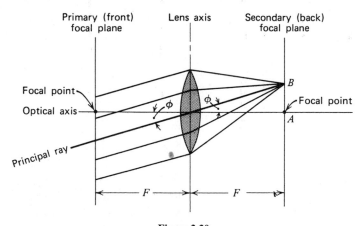

Figure 2.20.

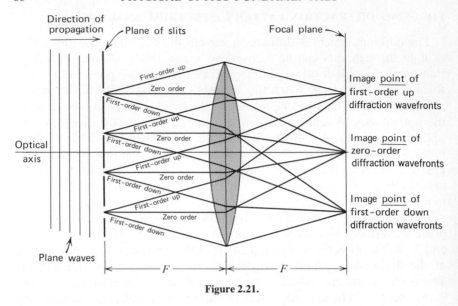

Figure 2.21.

pattern consists of *points* of light on a line *perpendicular* to the direction of the slits.

The principle shown in Figure 2.21 can be interpreted as a form of frequency analysis (spectrum analysis). Let us consider only the first-order diffraction wavefronts. The distance of the first-order image points from the optical axis is inversely proportional to the slit spacing. If the space between slits is considered as one period (or one spatial wavelength), the distance of the first-order points from the optical axis can be considered proportional to the spatial frequency of the slits. This interpretation will be discussed later in more detail and, in fact, will be used as the basic principle in many optical data processing methods.

DIFFRACTION GRATING

We have discussed how a prism can refract a beam of white light in such a way as to form a spectrum. The spreading out of a light beam into its component colors (spectrum) is known as *dispersion*. A diffraction grating produces a much better spectrum than prisms. There are two basic types of diffraction grating—a transmission type and a reflection type. A transmission grating can be made by ruling a large number of parallel, equally spaced scratches on glass. The scratches will tend to block the light and the spaces between the scratches will transmit light. In this way the spaces between scratches become, in effect, equally spaced slits. A reflection grating can be made by ruling fine parallel lines on a polished metal plate. The space between the ruled lines is capable of reflecting light while the ruled lines will not reflect light. The surfaces between ruled lines in effect form (reflecting) slits.

From Figures 2.17 and 2.18 we have seen that the various order wavefronts are diffracted at an angle θ that is related to the wavelength of the light. The angle θ is the angle between the zero-order and the diffracted-order wavefront. Color is determined by the frequency of the light or by its associated wavelength, therefore when white light is used, each color (wavelength) will be deflected at an angle θ related to its wavelength.

The diffraction grating operation is easily explained by Huygen's principle. Each slit of a diffraction grating can be considered as a light source. Plane wavefronts incident on the grating will therefore make all these light sources in phase. Figures 2.16, 2.17, and 2.18 can therefore be applied to analysis of diffraction gratings.

POLARIZED LIGHT

Interference and diffraction can occur with any type of wave. Transverse waves such as light waves can display effects of polarization that cannot be duplicated in longitudinal waves such as sound waves. It has already been stated that light is transmitted like any other electromagnetic radiation. The only restriction on electromagnetic radiation is that the electric and magnetic vibrations must be perpendicular to each other. In Figure 2.22 we can compare a linearly (vertically) polarized light wave (electric field oscillates only in a vertical direction) with light waves of ordinary light (a mixture of waves in all possible transverse directions). Ordinary light, has electric field vibrations that are random. When the direction of the electric field vibrations are not random but can be specified somewhat regularly the light is said to be *polarized* with respect to the electric field vibrations.

In Figure 2.22 the diagrams on the left represent the envelope of the electric field vectors at an instant of time. A line drawn from a point on the envelope perpendicular to the Z axis represents the electric field at the point of intersection on the Z axis. The distance from the Z axis to the envelope (length of line perpendicular to Z axis) represents the amplitude of the electric field and the direction of the line represents the direction of the electric field. The diagrams on the right demonstrate the particular characteristics of the types of polarization illustrated — that is, the electric field characteristics in a plane perpendicular to the Z axis at the point on the Z axis being considered.

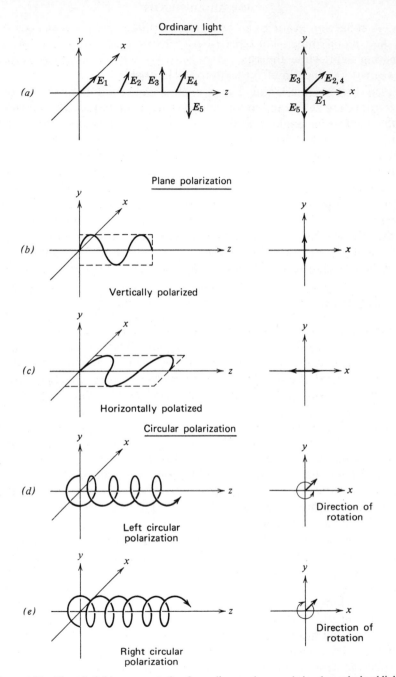

Figure 2.22 Electric field representation for ordinary, plane, and circular polarized light.

90

The first diagram in Figure 2.22 shows that ordinary light waves are random. At any instant of time the field vectors may be in any direction at any point along the direction of propagation. The second diagram shows a vertically plane polarized wave (also called *linear polarized*). In this wave the electric field vibrations are confined to the YZ plane. The diagram on the left illustrates that the electric field varies in amplitude with respect to Z. The diagram on the right indicates that the amplitudes are restricted to up and down variations in the YZ plane.

The third diagram in Figure 2.22 shows a horizontally plane polarized wave. This wave has characteristics similar to those of the vertically polarized waves except that the vibrations are restricted to the XZ plane. It is to be noted that as shown in these two examples of plane polarization a plane polarized wave varies in amplitude only. The plane determined by the Z axis (direction of propagation) and the direction of the electric field is called the *plane of polarization*.

The fourth and fifth diagrams in Figure 2.22 illustrate the two types of circular polarization. The envelopes can be traced out by rotating a line of fixed length about the Z axis while simultaneously moving along the Z axis. Because the rotation can be in either of two directions, we can distinguish between right- and left-circular polarization. The diagrams on the right indicate that at any given point on the Z axis the electric field is constant in amplitude but its direction changes as the vector rotates about the Z axis. The tip of the vector sweeps out a circle as it is rotated. This characteristic of the electric field gives rise to the descriptive title of circular polarization.

It is noted that the two types of polarization shown in Figure 2.22 are the special cases of amplitude variation only or direction variation only. If both the amplitude and direction change, elliptical curves will be swept out by the electric field vector at a point on the Z axis. This is called *elliptical polarization*. It should be noted that plane and circular polarization are special cases of elliptical polarization.

There are many methods of producing polarized light but probably the most important in optical processing is the laser. The basic operation of a laser is explained in Appendix 3. The laser produces essentially a highly collimated, monochromatic, coherent, and *polarized* light beam. The laser produces plane polarized light which is collimated monochromatic and coherent.

A light beam is said to be *polarized* when the electric field vibrations at all points in the beam have the same polarization. When the individual waves in the light beam do not have the same polarization, the light beam is said to be *unpolarized*. Figure 2.23 shows sample representations of the waves in an unpolarized light beam.

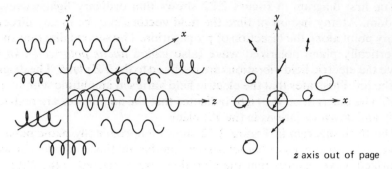

Figure 2.23   Unpolarized light beam.

It is possible to mix polarized light with unpolarized light and have a resultant light beam partially polarized. The degree of polarization will be dependent on the proportion of polarized wave trains to unpolarized wave trains in a light beam. Degree of polarization in percent is given by

$$\text{percent polarization} = \frac{I_{max} - I_{min}}{I_{max} + I_{min}} \times 100 \text{ percent}$$

where I_{max} and I_{min} are maximum and minimum intensities of light passed by a rotating polarizer.

Polarizing by Reflection

Ordinary light can be polarized by reflection. When ordinary light is reflected (at other than normal incidences) from a plane dielectric surface, the electric field of the reflected light is partially polarized perpendicular to the plane of incidence (the plane of incidence is the plane containing the incident ray and the normal to the surface).

Let us consider an incident light beam with components as shown in Figure 2.24. A larger fraction of the component perpendicular to the plane of incidence is reflected than of the component in the plane of incidence. At a specific angle, called the *polarizing angle* only the vertical component will be reflected producing a vertically plane polarized reflected beam. The intensity of this reflected beam will be small since only 15 percent of the vertical component is reflected (85 percent is transmitted as a component of the refracted beam). The component in the plane of incidence is not reflected and appears as the horizontal component of the refracted beam.

Sir David Brewster found the relation between the index of refraction of the reflecting surface medium and the polarizing angle. (Sir Brewster

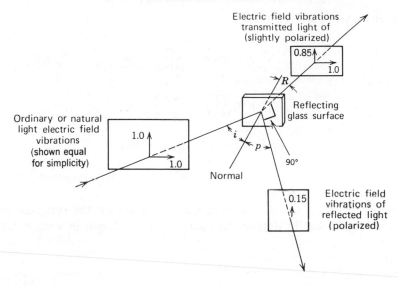

Figure 2.24 Polarization produced when ordinary light is incident to a glass surface at the Brewster angle.

also invented the kaleidoscope described in Chapter 1). Brewster's law of polarization is given by

$$\frac{n}{n'} = \tan p$$

where n' is the index of a refraction material in which light is traveling before reflection, n is the index of refraction of reflecting material, and p is the polarizing angle. For glass and air as shown in Figure 2.24

$$n = \tan p$$

By the law of reflection we know that the angle of incidence, i, is equal to the angle of reflection r, that is,

$$\angle i = \angle r.$$

When the polarization is a maximum

$$\angle i = \angle r = \angle p.$$

We have previously found the index of refraction related the angle of incidence, i, with the angle of refraction, R, as follows:

$$n = \frac{\sin i}{\sin R}$$

From Brewster's law we have

$$n = \tan p = \frac{\sin p}{\cos p}$$

therefore,

$$n = \frac{\sin i}{\sin R} = \frac{\sin p}{\cos p}$$

and

$$\sin R = \cos p = \sin(p + 90°)$$

or

$$R = p + 90°$$

From this we can see that at maximum polarization the reflected and refracted rays are 90 degrees to each other as shown in Figure 2.24.

3

Fourier Transforms

In this chapter the basic mathematics necessary to understand and explain the optical processes is presented. In general a somewhat naive approach is used so as to permit ease in understanding. Those who are not familiar with the subject, or have forgotten it because of lack of use, will find this chapter (plus the referenced appendices) particularly helpful. This chapter (plus its appendices) although not a complete treatment of the mathematics does give the mathematical background necessary to convert from other scientific areas to optics.

FOURIER SERIES

Introduction

Most nonsinusoidal periodic waves can be expressed in terms of sinewave components of different frequencies. One advantage of this type of analysis is that each sine-wave component can be handled according to the laws governing sine-wave calculations. The results of analysis for each of the component sine waves can be combined to form the final analysis. Usually only a few terms are necessary because as a rule higher frequency terms have relatively small effect. An equation expressing the components of a periodic wave is known as a Fourier series and can be expressed as

$$y = f(x) = A_0 + A_1 \cos x + B_1 \sin x + A_2 \cos 2x + B_2 \sin 2x$$
$$+ A_3 \cos 3x + B_3 \sin 3x + \cdots + A_n \cos nx + B_n \sin nx$$

$$(3\text{-}1)$$

The meaning of (3-1) becomes clear when we consider it as a way of approximating (and sometimes expressing exactly, depending on the function) a periodic function with different frequency sine (and/or cosine) waves. As a simple illustration let us consider the periodic wave form shown in the graph of Figure 3.1.

From a cursory look it would appear quite difficult to find sine waves that would add together in such a way as to approximate this wave form. Fourier analysis is a method for easily determining the amplitudes, frequencies, and relative phases of sine waves that will add together to make up a desired wave form.

Let us consider the addition of the two sine waves shown in the graphs of Figure 3.2. The two sine waves shown in Figures 3.2(a) and 3.2(b) when added together (shown in Figure 3.2(c)) approximate the waveform of Figure 3.1. To be more specific the sum of the functions shown in Figures 3.2(a) and 3.2(b) can be written

$$y = A \sin x + A \sin 2x$$

and can be used to approximate the original square wave of Figure 3.1. Fourier analysis is a method for determining the component sine waves (in this case Figures 3.2(a) and 3.2(b) of a given periodic waveform (in this case Figure 3.1) and permitting the periodic waveform to be approximated to any desired degree of accuracy.

The Fourier series expressed in (3-1) can be written in the simplified form

$$y = f(x) = A_0 + \sum_{n=1}^{n=\infty} \left\{ A_n \cos nx + B_n \sin nx \right\} \tag{3-2}$$

Fourier analysis consists of determining the coefficient $A_0, A_1, B_1, A_2, B_2, A_3, B_3, \ldots A_n, B_n$ in (3-1) or its equivalent equation (3-2). These coefficients are referred to as the *Fourier coefficients*. Specific terms in the Fourier series expressed in (3-1) refer to specific items in the final

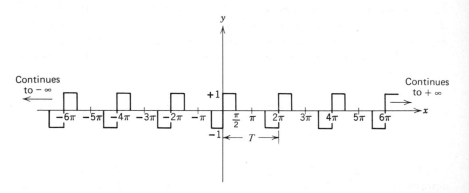

Figure 3.1 Graph of a periodic square wave.

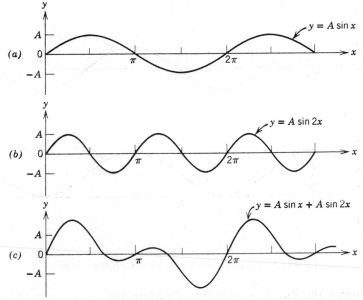

Figure 3.2 Sum of two waves approximating the square wave form of Fig. 3.1.

Fourier analysis. The terms in (3-1) are shown grouped to indicate the specific items in the final analysis that they represent,

$$y = f(x) = \underbrace{A_0}_{\substack{\text{DC or steady} \\ \text{state value,} \\ \text{average} \\ \text{value}}} + \underbrace{A_1 \cos x + B_1 \sin x}_{\substack{\text{fundamental or} \\ \text{1st harmonic}}} + \underbrace{A_2 \cos 2x + B_2 \sin 2x}_{\text{2nd harmonic}}$$

$$+ \underbrace{A_3 \cos 3x + B_3 \sin 3x}_{\text{3rd harmonic}} + \cdots + \underbrace{A_n \cos nx + B_n \sin nx}_{n\text{th harmonic}} \quad (3\text{-}1)$$

Fourier Series DC Term

The term A_0 in (3-2) refers to the average value (which is sometimes called the DC or steady state value). The graph shown in Figure 3.3(a) illustrates a sinusoidal waveform that has an average value of zero; that is, in one cycle of the waveform the area above the x axis is equal to the area below the x axis and therefore the average value is zero ($y = A_0 = 0$).

In the graph of Figure 3.3(b) the average value of the sinusoidal waveform is shown to be C_0; that is, in one cycle the area under the curve above $y = C_0$ is equal to the area under the curve below $y = C_0$, giving an average value of $A_0 = C_0$. In Figure 3.1 we can see that the average

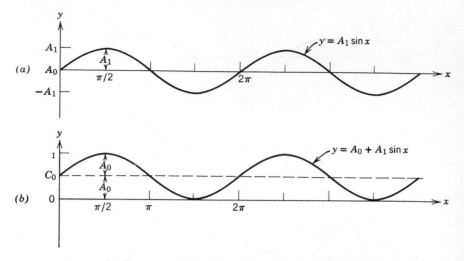

Figure 3.3   Graph illustrating A_0 term of a fourier series.

value under the curve is also zero because the area under the curve above the x axis is equal to the area under the curve below the x axis.

Fourier Series First Harmonic

The fundamental or first harmonic is the sinusoidal waveform with the same period as the given periodic waveform. Referring to Figure 3.1 we can see that the period of the periodic waveform is

$$T = 2\pi$$

where T is the period of waveform; that is, the smallest interval along the x axis that contains one cycle of the waveform has a length of 2π. The frequency corresponding to this period is given by

$$f = \frac{1}{T} = \frac{1}{2\pi}$$

where f is the frequency. Figure 3.2(a) shows a sinusoidal waveform with a period

$$T = 2\pi$$

which is the same as the period of the periodic waveform of Figure 3.1.

The phase and amplitude of the sinusoid of Figure 3.2(a) must be properly selected if it is to represent the first harmonic of the periodic waveform of Figure 3.1. In order to see how the phase and amplitude of the first harmonic is determined so that it will best match the desired periodic waveform we refer to (3-1). From (3-1) we can see that the first harmonic is given by two terms; that is,

$$\text{first harmonic} = A_1 \cos x + B_1 \sin x$$

The first harmonic therefore represents the sum of the two sinusoidal waves of the same frequency. The addition of two sinusoidal waveforms of the same frequency results in a third sinusoid of the same frequency. The amplitude and phase of the third sinusoidal wave is related to the amplitudes and phases of the two original sinusoidal waves. Graphically as shown in Figure 3.4 the third sinusoid can be found by adding individual ordinates of the two original sinusoids to form the third sinusoid.

The point by point addition of two sinusoids to form a third sinusoid representing the sum can be a tedious and awkward procedure. A simpler procedure can be used by noting that the peak amplitude A_1 of the cosine waveform is displaced 90 degrees to the peak amplitude B_1 of

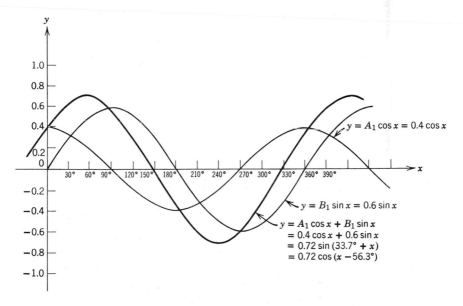

Figure 3.4 Sum of two sinusoids. $C_1 = \sqrt{A_1{}^2 + B_1{}^2} = \sqrt{0.4^2 + 0.6^2} = 0.72$; $\tan\theta = \dfrac{A_1}{B_1} = \dfrac{0.4}{0.6} = 0.67$; $\theta = 33.7$. First harmonic $= C_1 \sin(\theta + x) = 0.72 \sin(33.7 + x)$.

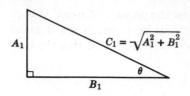

Figure 3.5 Right triangle showing relationship between the coefficients A_1 and B_1.

the sine waveform. We can therefore relate the coefficients A_1 and B_1 to a right triangle as shown in Figure 3.5. If the legs of the right triangle are A_1 and B_1 respectively, then the hypotenuse C_1 is given by

$$C_1 = \sqrt{A_1^2 + B_1^2} \qquad (3\text{-}3)$$

The amplitude of the resultant sinusoid is equal to C_1. This can be shown as follows:

$$\text{first harmonic} = A_1 \cos x + B_1 \sin x$$

$$= \sqrt{A_1^2 + B_1^2} \left\{ \frac{A_1}{\sqrt{A_1^2 + B_1^2}} \cos x + \frac{B_1}{\sqrt{A_1^2 + B_1^2}} \sin x \right\} \qquad (3\text{-}4)$$

From Figure 3.5 we see that

$$\sin \theta = \frac{A_1}{\sqrt{A_1^2 + B_1^2}} \qquad \text{and} \qquad \cos \theta = \frac{B_1}{\sqrt{A_1^2 + B_1^2}}$$

then

$$\text{first harmonic} = \sqrt{A_1^2 + B_1^2} \left\{ \sin \theta \cos x + \cos \theta \sin x \right\} \qquad (3\text{-}5)$$

Using the trigonometric identity

$$\sin (\alpha + \beta) = \sin \alpha \cos \beta + \cos \alpha \sin \beta$$

we can rewrite (3-5) as

$$\text{first harmonic} = \sqrt{A_1^2 + B_1^2} \left\{ \sin (\theta + x) \right\} \qquad (3\text{-}6)$$

We see therefore that we have converted (3-4) to (3-6). This can be interpreted to mean that any sine and cosine wave of the same frequency (3-4), when added together, will result in another sinusoid of the same frequency (3-6). The angle θ is a constant because both sine waves have the same frequency ($f = 1/2\pi$) and can be thought of as a phase angle which from Figure 3.5 can be defined by

$$\tan \theta = \frac{A_1}{B_1} \qquad (3\text{-}7)$$

We can therefore see that the magnitude of the cosine and sine wave (A_1 and B_1 terms) determine the phase as well as the amplitude of the resultant sinusoidal wave. We can see from Figure 3.4 that the result of adding a cosine wave ($A_1 \cos x = 0.4 \cos x$) and a sine wave ($B_1 \sin x = 0.6 \sin x$) is a sinusoidal waveform of amplitude

$$C_1 = 0.72$$

The resultant sinusoidal waveform can be represented as either a sine or cosine wave. When expressed as a sine wave it can be written

$$C_1 \sin (\theta + x) = 0.72 \sin (33.7° + x)$$

or expressed as a cosine wave it can be written

$$C_1 \cos (x - 90° + \theta) = 0.72 \cos (x - 56.3°)$$

Fourier Series Second-nth Harmonic

We have seen that the first harmonic of a periodic waveform is the sinusoid with the same period as the periodic waveform and with amplitude and phase that best approximate the periodic waveform being analyzed. The second harmonic is that sinusoid, of twice the frequency of the first harmonic, with amplitude and phase such that when added to the first harmonic best approximates (or matches) the periodic waveform. Using this same type of reasoning the nth harmonic is that sinusoid of frequency n times the fundamental and of amplitude and phase which when added to the other harmonics best approximates the periodic waveform being analyzed. In a Fourier series (as shown in equation 3-1) only integral multiples of the fundamental will be found; that is, the frequencies of the sine and cosine waves can only have frequencies

$$f, 2f, 3f, 4f, \cdots , nf$$

where n is an integer and f is the frequency.

We can state this another way by saying that in a Fourier series there *cannot* be a harmonic at a fractional multiple of the fundamental frequency (since only integral multiples of the fundamental are allowed); for example, harmonics at 1.2 or 2.6 times the fundamental frequency cannot exist.

Appendix 9 is a somewhat rigorous derivation of the Fourier series. This appendix provides the necessary mathematical background (in depth so that it can be used as a math refresher course) to permit the reader to understand the mathematical descriptions that will follow.

Referring back to (1-3),

$$y = f(x) = A_0 + A_1 \cos x + B_1 \sin x + \cdots + A_n \cos nx + B_n \sin nx \quad (3\text{-}1)$$

the value of each term in (1-3) can be easily determined but the procedures are somewhat lengthy. The results of the derivation are summarized below and their use is illustrated:

$$A_0 = \frac{1}{2\pi} \int_0^{2\pi} f(x)\, dx \tag{3-8}$$

$$A_1 = \frac{1}{\pi} \int_0^{2\pi} f(x) \cos x\, dx \tag{3-9}$$

$$B_1 = \frac{1}{\pi} \int_0^{2\pi} f(x) \sin x\, dx \tag{3-10}$$

$$A_2 = \frac{1}{\pi} \int_0^{2\pi} f(x) \cos 2x\, dx \tag{3-11}$$

$$B_2 = \frac{1}{\pi} \int_0^{2\pi} f(x) \sin 2x\, dx \tag{3-12}$$

$$A_n = \frac{1}{\pi} \int_0^{2\pi} f(x) \cos nx\, dx \tag{3-13}$$

$$B_n = \frac{1}{\pi} \int_0^{2\pi} f(x) \sin nx\, dx \tag{3-14}$$

As an illustrative example of how the Fourier series representing a periodic waveform is determined, let us consider a periodic square waveform as shown in Figure 3.1. We desire to determine the coefficients of (3-1) as follows:

$$y = f(x) = A_0 + A_1 \cos x + B_1 \sin x + A_2 \cos 2x + B_2 \sin 2x$$
$$+ A_3 \cos 3x + B_3 \sin 3x + A_4 \cos 4x + B_4 \sin 4x + \cdots \tag{3-15}$$

that is, the function $f(x)$ representing the wave of Figure 3.1 is to be represented by (3-15). From Figure 3.1 and the appropriate equations we obtain

A_0 from (3-8):

$$A_0 = \frac{1}{2\pi} \int_0^{2\pi} f(x)\ dx = \frac{1}{2\pi} \left\{ \int_0^{\pi/2} dx - \int_{3\pi/2}^{2\pi} dx \right\}$$

$$= \frac{1}{2\pi} \left\{ x \Big|_0^{\pi/2} - x \Big|_{3\pi/2}^{2\pi} \right\}$$

$$= \frac{1}{2\pi} \left\{ \frac{\pi}{2} - \left(2\pi - \frac{3\pi}{2} \right) \right\} = 0$$

A_1 from (3-9):

$$A_1 = \frac{1}{\pi} \int_0^{2\pi} f(x) \cos x\ dx = \frac{1}{\pi} \left\{ \int_0^{\pi/2} \cos x\ dx - \int_{3\pi/2}^{2\pi} \cos x\ dx \right\}$$

$$= \frac{1}{\pi} \left\{ \sin x \Big|_0^{\pi/2} - \sin x \Big|_{3\pi/2}^{2\pi} \right\} = \frac{1}{\pi} \left\{ +1 - 1 \right\} = 0$$

B_1 from (3-10):

$$B_1 = \frac{1}{\pi} \int_0^{2\pi} f(x) \sin x\ dx = \frac{1}{\pi} \left\{ \int_0^{\pi/2} \sin x\ dx - \int_{3\pi/2}^{2\pi} \sin x\ dx \right\}$$

$$= \frac{1}{\pi} \left\{ -\cos x \Big|_0^{\pi/2} + \cos x \Big|_{3\pi/2}^{2\pi} \right\}$$

$$= \frac{1}{\pi} \left\{ +1 + 1 \right\} = \frac{2}{\pi}$$

A_2 from (3-11):

$$A_2 = \frac{1}{\pi} \int_0^{2\pi} f(x) \cos 2x\ dx = \frac{1}{\pi} \left\{ \int_0^{\pi/2} \cos 2x\ dx - \int_{3\pi/2}^{2\pi} \cos 2x\ dx \right\}$$

$$= \frac{1}{\pi} \left\{ \frac{1}{2} \sin 2x \Big|_0^{\pi/2} - \frac{1}{2} \sin 2x \Big|_{3\pi/2}^{2\pi} \right\} = (0 - 0) - (0 - 0) = 0$$

B_2 from (3-12):

$$B_2 = \frac{1}{\pi} \int_0^{2\pi} f(x) \sin 2x\ dx = \frac{1}{\pi} \left\{ \int_0^{\pi/2} \sin 2x\ dx - \int_{3\pi/2}^{2\pi} \sin 2x\ dx \right\}$$

$$= \frac{1}{\pi} \left\{ -\frac{1}{2} \cos 2x \Big|_0^{\pi/2} + \frac{1}{2} \cos 2x \Big|_{3\pi/2}^{2\pi} \right\} = \frac{1}{\pi} \left\{ \left(\frac{1}{2} + \frac{1}{2} \right) + \left(\frac{1}{2} + \frac{1}{2} \right) \right\} = \frac{2}{\pi}$$

A_3 from (3-13):

$$A_3 = \frac{1}{\pi} \int_0^{2\pi} f(x) \cos 3x \, dx = \frac{1}{\pi} \left\{ \int_0^{\pi/2} \cos 3x \, dx - \int_{3\pi/2}^{2\pi} \cos 3x \, dx \right\}$$

$$= \frac{1}{\pi} \left\{ \frac{1}{3} \sin 3x \Big|_0^{\pi/2} - \frac{1}{3} \sin 3x \Big|_{3\pi/2}^{\pi} \right\}$$

$$= \frac{1}{\pi} \left\{ -\frac{1}{3} + \frac{1}{3} \right\} = 0$$

B_3 from (3-14):

$$B_3 = \frac{1}{\pi} \int_0^{2\pi} f(x) \sin 3x \, dx = \frac{1}{\pi} \left\{ \int_0^{\pi/2} \sin 3x \, dx - \int_{3\pi/2}^{2\pi} \sin 3x \, dx \right\}$$

$$= \frac{1}{\pi} \left\{ -\frac{1}{3} \cos 3x \Big|_0^{\pi/2} + \frac{1}{3} \cos 3x \Big|_{3\pi/2}^{2\pi} \right\}$$

$$= \frac{1}{\pi} \left\{ \left(0 + \frac{1}{3} \right) + \left(\frac{1}{3} - 0 \right) \right\} = \frac{1}{\pi} \left(\frac{2}{3} \right) = \frac{2}{3\pi}$$

A_4 from (3-13):

$$A_4 = \frac{1}{\pi} \int_0^{2\pi} f(x) \cos 4x \, dx = \frac{1}{\pi} \left\{ \int_0^{\pi/2} \cos 4x \, dx - \int_{3\pi/2}^{2\pi} \cos 4x \, dx \right\}$$

$$= \frac{1}{\pi} \left\{ \frac{1}{4} \sin 4x \Big|_0^{\pi/2} - \frac{1}{4} \sin 4x \Big|_{3\pi/2}^{2\pi} \right\} = \frac{1}{\pi} \left\{ (0-0) - (0-0) \right\} = 0$$

B_4 from (3-14):

$$B_4 = \frac{1}{\pi} \int_0^{2\pi} f(x) \sin 4x \, dx = \frac{1}{\pi} \left\{ \int_0^{\pi/2} \sin 4x \, dx - \int_{3\pi/2}^{2\pi} \sin 4x \, dx \right\}$$

$$= \frac{1}{\pi} \left\{ -\frac{1}{4} \cos 4x \Big|_0^{\pi/2} + \frac{1}{4} \cos 4x \Big|_{3\pi/2}^{2\pi} \right\}$$

$$= \frac{1}{\pi} \left\{ \left(-\frac{1}{4} + \frac{1}{4} \right) + \left(\frac{1}{4} - \frac{1}{4} \right) \right\} = 0$$

We can now write (3-15) as

$$y = f(x) = A_0 + A_1 \cos x + B_1 \sin x + A_2 \cos 2x + B_2 \sin 2x + A_3 \cos$$
$$3x + B_3 \sin 3x + A_4 \cos 4x + B_4 \sin 4x + \cdots \qquad (3\text{-}15)$$

$$= 0 + 0 + \frac{2}{\pi} \sin x + 0 + \frac{2}{\pi} \sin 2x + 0 + \frac{2}{3\pi} \sin 3x + 0 + 0 + \cdots$$

$$= \frac{2}{\pi} \sin x + \frac{2}{\pi} \sin 2x + \frac{2}{3\pi} \sin 3x + \cdots \qquad (3\text{-}16)$$

From (3-16) we can (with some difficulty) deduce the general Fourier series for this waveform to be

$$y = f(x) = \frac{2}{\pi} \sum_{n=1}^{\infty} \frac{1}{n} \left[1 - \cos \frac{n\pi}{2} \right] \left[\sin nx \right]$$

From this we can deduce that when n is odd ($\cos (n\pi/2) = 0$) the coefficient will be $2/\pi n$; when n is a multiple of 4 ($\cos (n\pi/2) = +1$) the coefficient will be zero. For all other even n ($\cos n\pi/2 = -1$) the coefficient will be $4/\pi n$.

From Figure A9.1 in Appendix 9 we could have determined immediately that the waveform in Figure 3.1 was an odd function. Odd functions contain only sine terms and therefore in determining the coefficient for (3-15) we would not have to evaluate the A_n terms (since odd functions contain no cosine terms). Equation (3-16) shows that only sine terms exist in the Fourier series. For the function shown in Figure 3.1 there is no position where the y axis can be placed with respect to the waveform that will make it appear an even function—that is, so that only cosine terms appear in the Fourier series. The axis could, however, be placed so that both sine and cosine terms appear in the Fourier series. In such a case we could write any harmonic as a single sinusoid by combining the sine and cosine terms (as shown in Figure 3.4) to produce a single sine (or cosine) wave that is phase shifted with respect to the axis. The fact that odd functions produce only sine terms (with no phase shift) or even functions produce only cosine terms (with no phase shift) should not be confused with functions that are neither odd nor even and produce sine or cosine terms with a phase shift; for example, the Fourier series can be written in either of the following two forms

$$f(t) = A_0 + C_1 \cos (x + \theta_1) + C_2 \cos (2x + \theta_2) + C_3 \cos (3x + \theta_3) + \cdots$$
$$+ C_n \cos (nx + \theta_n)$$

or

$$f(t) = A_0 + C_1 \sin (x + \varphi_1) + C_2 \sin (2x + \varphi_2) + C_3 \sin (3x + \varphi_3) + \cdots$$
$$+ C_n \sin (nx + \varphi_n)$$

where $\varphi_n + 90° = \theta_n$.

FOURIER INTEGRALS

When Fourier analysis is applied to a function whose period is essentially infinite, a Fourier integral will result. A function whose period is infinite is a way of saying that the function does not repeat itself. When the period of a function gets larger, the frequency of the fundamental gets smaller, as illustrated in Figure 3.6.

Using this logic it can be seen that as the period of repetition of a function approaches infinity, the frequency of the fundamental approaches zero (DC). We have seen that in the Fourier series each harmonic is an integer multiple of the fundamental. This means that as the fundamental frequency approaches zero, the separation between the harmonics will also get smaller, approaching zero. In effect, for a nonrepetitive function the separation between harmonics is zero. This means that when the fundamental frequency approaches zero *all* frequencies will be represented. This is the basic difference between the Fourier series and the Fourier integral; that is, all frequencies may be represented by a Fourier integral while only whole-number multiples of the fundamental frequency may be represented by a Fourier series. Figure 3.7 illustrates the differences between a Fourier series and a Fourier integral.

The Fourier series applies to periodic functions (with certain restrictions but a detailed discussion of these restrictions is outside the scope of this book). In the derivation of the Fourier series we took a periodic function of position that is, (3-1),

$$y = f(x)$$

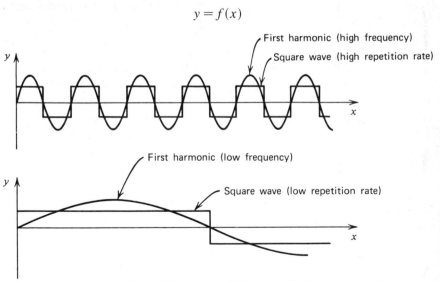

Figure 3.6　Effect of changing repetition rate on the first harmonic.

	Repetition rate	Example of repetition rate	Frequency components determined by analysis
Fourier series	Finite (repetitive wave)	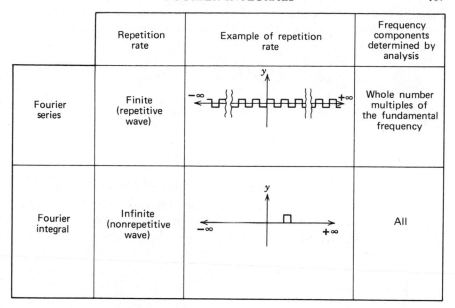	Whole number multiples of the fundamental frequency
Fourier integral	Infinite (nonrepetitive wave)		All

Figure 3.7 Differences between a Fourier series and a Fourier integral.

where the displacement, y, was a function of position, x. We are not restricted to making the variable a function of position only. We can, for example, let the variable be associated with time (instead of position). Equation (3-1) written as a function of position is

$$y = f(x) = A_0 + A_1 \cos x + B_1 \sin x + A_2 \cos 2x + B_2 \sin 2x + \cdots$$
$$+ A_n \cos nx + B_n \sin nx \qquad (3\text{-}1)$$

By letting $x = \omega t$ we can write (3-1) as a function of time t as follows:

$$y = g(t) = A_0 + A_1 \cos \omega t + B_1 \sin \omega t + A_2 \cos 2\omega t + B_2 \sin 2\omega t + \cdots$$
$$+ A_n \cos n\omega t + B_n \sin n\omega t \qquad (3\text{-}17)$$

In (3-17) we introduced a new *constant* ω which is the angular frequency. For a function of time the period T is defined as the time duration for one completed cycle. As appropriate one cycle can be expressed in 2π radians, 360 degrees, or time in seconds. When one cycle is expressed as 2π radians, and the time for one complete cycle is expressed as T seconds (period $= T$), then the angular frequency ω can be expressed in radians per second, or

$$\omega = \text{angular frequency} = \frac{2\pi \text{ radians}}{T \text{ seconds}}$$

Figure 3.8 shows the graph of the function $y = \sin \omega t$. From this graph we can see that with the scale chosen the sine wave has completed its first cycle (one period, T) when

$$t = \frac{2\pi}{\omega}$$

where ω is the constant whose value is given by

$$\omega = \frac{2\pi}{T} \quad \text{and} \quad T = \frac{2\pi}{\omega}$$

Appendix 10 discusses the effects of changing axes for the sine-wave function shown in Figure 3.8. In addition Appendix 10 discusses the results of changing the limits and variables when integrating.

A_0 can be found from (3-8) to be

$$A_0 = \frac{1}{2\pi} \int_0^{2\pi} f(x) \, dx \tag{3-8}$$

where 2π is one period of the periodic function.

We can let $x = \omega t$ and then

$$dx = \omega \, dt$$

$$f(x) = f(\omega t) = g(t)$$

Substituting these values in (3-8) we have

$$A_0 = \frac{1}{2\pi} \int g(t) \omega \, dt \tag{3-18}$$

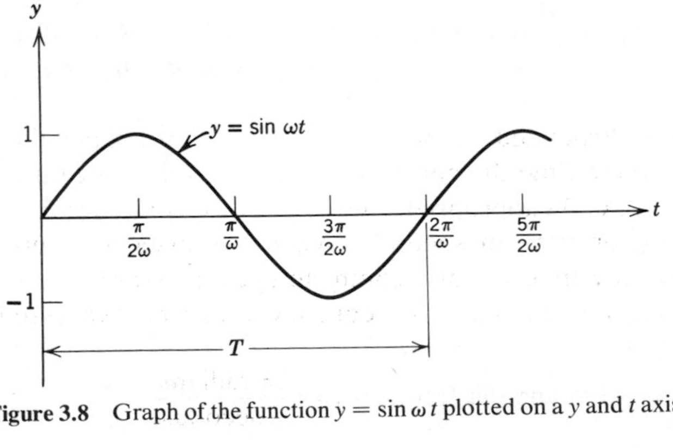

Figure 3.8   Graph of the function $y = \sin \omega t$ plotted on a y and t axis.

The limits of integration for (3-18) are determined by noting that because we let $x = \omega t$, then when $x = 2\pi$, $t = 2\pi/\omega$ and when $x = 0$, $t = 0$.

We can therefore put in the limits for the integration in (3-18) as follows:

$$A_0 = \frac{1}{2\pi} \int_0^{2\pi/\omega} g(t)\omega \, dt = \frac{\omega}{2\pi} \int_0^{2\pi/\omega} g(t) \, dt \tag{3-19}$$

When $\omega = 2\pi/T$, then $T = 2\pi/\omega$ and we can substitute these values in (3-19) giving

$$A_0 = \frac{\omega}{2\pi} \int_0^{2\pi/\omega} g(t) \, dt = \frac{1}{T} \int_0^T g(t) \, dt \tag{3-20}$$

Changing the limits of integration will not change the value of the integral as long as the limits represent one complete cycle of the periodic function. Therefore (3-20) can be written as

$$A_0 = \frac{1}{T} \int_0^T g(t) \, dt = \frac{1}{T} \int_{-T/2}^{T/2} g(t) \, dt \tag{3-21}$$

Correspondingly in (3-13)

$$A_n = \frac{1}{\pi} \int_0^{2\pi} f(x) \cos nx \, dx \tag{3-13}$$

We can let

$$x = \omega t$$
$$dx = \omega \, dt$$
$$f(x) = f(\omega t) = g(t)$$

$$A_n = \frac{1}{\pi} \int_0^{2\pi} f(x) \cos nx \, dx \tag{3-13}$$

$$= \frac{1}{\pi} \int_0^{2\pi/\omega} g(t) [\cos (n\omega t)] \omega \, dt \tag{3-22}$$

$$= \frac{\omega}{\pi} \int_0^{2\pi/\omega} g(t) \cos (n\omega t) \, dt \tag{3-23}$$

The limits of integration in (3-23) were obtained by noting that because $x = \omega t$, then when $x = 2\pi$, $t = 2\pi/\omega$ and when $x = 0$, $t = 0$. We can change the limits of integration in (3-23) without changing its value as long as the integration limits represent one complete cycle. Therefore noting that $\omega = 2\pi/T$ we can write

$$A_n = \frac{\omega}{\pi} \int_0^{2\pi/\omega} g(t) \cos(n\omega t)\, dt \tag{3-23}$$

$$= \frac{2}{T} \int_0^T g(t) \cos(n\omega t)\, dt \tag{3-24}$$

$$= \frac{2}{T} \int_{-T/2}^{T/2} g(t) \cos(n\omega t)\, dt \tag{3-25}$$

Similarly (3-14) can be written

$$B_n = \frac{1}{\pi} \int_0^{2\pi} f(x) \sin(nx)\, dx \tag{3-14}$$

$$= \frac{2}{T} \int_0^T g(t) \sin(n\omega t)\, dt \tag{3-26}$$

$$= \frac{2}{T} \int_{-T/2}^{T/2} g(t) \sin(n\omega t)\, dt \tag{3-27}$$

By letting $x = \omega t$ (3-2) can be written

$$y = f(x) = A_0 + \sum_{n=1}^{\infty} A_n \cos(nx) + \sum_{n=1}^{\infty} B_n \sin(nx) \tag{3-2}$$

$$= g(t) = A_0 + \sum_{n=1}^{\infty} A_n \cos(n\omega t) + \sum_{n=1}^{\infty} B_n \sin(n\omega t) \tag{3-28}$$

Substituting (3-21), (3-25), and (3-27) into (3-28) we get

$$g(t) = A_0 + \sum_{n=1}^{\infty} A_n \cos(n\omega t) + \sum_{n=1}^{\infty} B_n \sin(n\omega t) \tag{3-28}$$

$$= \frac{1}{T} \int_{-T/2}^{T/2} g(t)\, dt + \sum_{n=1}^{\infty} \left[\frac{2}{T} \int_{-T/2}^{T/2} g(t) \cos(n\omega t)\, dt \right] \cos(n\omega t)$$

$$+ \sum_{n=1}^{\infty} \left[\frac{2}{T} \int_{-T/2}^{T/2} g(t) \sin(n\omega t)\, dt \right] \sin(n\omega t) \tag{3-29}$$

The t in the integrals of (3-29) is replaced by the limits of integration to obtain the final results and any letter can therefore be used as the variable of integration; that is, in evaluating the integral the final result is in terms of the limits of the integration. For example, consider the following two integrals:

$$\int_{-T/2}^{T/2} dt = \left[t \right]_{-T/2}^{T/2} = \frac{T}{2} + \frac{T}{2} = T$$

and

$$\int_{-T/2}^{T/2} d\tau = \left[\tau \right]_{-T/2}^{T/2} = \frac{T}{2} + \frac{T}{2} = T$$

Notice that the integration variables, when evaluated at the limits of integration, disappear in the final result.

Because t under the integral sign in (3-29) drops out when evaluated at the limits of integration, we can let

$$t = \tau$$

then (3-29) can be written

$$y = g(t) = \frac{1}{T} \int_{-T/2}^{T/2} g(\tau) \, d\tau + \sum_{n=1}^{\infty} \left[\frac{2}{T} \int_{-T/2}^{T/2} g(\tau) \cos (n\omega\tau) \, d\tau \right] \cos (n\omega t)$$

$$+ \sum_{n=1}^{\infty} \left[\frac{2}{T} \int_{-T/2}^{T/2} g(\tau) \sin (n\omega\tau) \, d\tau \right] \sin n\omega t \qquad (3\text{-}30)$$

We can now move the $\cos (n\omega t)$ and $\sin (n\omega t)$ under their respective integral signs in the summation terms because they are constant with respect to τ.

$$y = g(t) = \frac{1}{T} \int_{-T/2}^{T/2} g(\tau) \, d\tau + \sum_{n=1}^{\infty} \frac{2}{T} \int_{-T/2}^{T/2} g(\tau) \cos (n\omega\tau) \cos (n\omega t) \, d\tau$$

$$+ \sum_{n=1}^{\infty} \frac{2}{T} \int_{-T/2}^{T/2} g(\tau) \sin (n\omega\tau) \sin (n\omega t) \, d\tau \qquad (3\text{-}31)$$

Combining terms gives

$$y = g(t) = \frac{1}{T} \int_{-T/2}^{T/2} g(\tau) \, d\tau + \sum_{n=1}^{\infty} \frac{2}{T} \int_{-T/2}^{T/2} g(\tau) [\cos (n\omega\tau) \cos (n\omega t)$$

$$+ \sin (n\omega\tau) \sin (n\omega t)] \, d\tau \qquad (3\text{-}32)$$

Using the trigonometric relationship

$$\cos(\alpha - \beta) = \cos\alpha\cos\beta + \sin\alpha\sin\beta \tag{3-33}$$

We can write (3-32) as

$$y = g(t) = \frac{1}{T}\int_{-T/2}^{T/2} g(\tau)\,d\tau + \sum_{n=1}^{\infty} \frac{2}{T}\int_{-T/2}^{T/2} g(\tau)\cos[n\omega(\tau - t)]\,d\tau \tag{3-34}$$

In (3-34) because $\cos 0° = 1$, we can take the first term (the A_0 term) and express it as follows:

$$\frac{1}{T}\int_{-T/2}^{T/2} g(\tau)\,d\tau = \frac{1}{T}\int_{-T/2}^{T/2} g(\tau)\cos[n\omega(\tau - t)]_{n=0}\,d\tau \tag{3-35}$$

Substituting the value of $n = 0$ in (3-35) makes $\cos n\omega(\tau - t) = 1$, making both sides identical. The second term in (3-34) can be expressed as follows:

$$\sum_{n=1}^{\infty} \frac{2}{T}\int_{-T/2}^{T/2} g(\tau)\cos[n\omega(\tau - t)]\,d\tau = \sum_{n=1}^{\infty} \frac{1}{T}\int_{-T/2}^{T/2} g(\tau)\cos[n\omega(\tau - t)]\,d\tau$$

$$+ \sum_{n=1}^{\infty} \frac{1}{T}\int_{-T/2}^{T/2} g(\tau)\cos[n\omega(\tau - t)]\,d\tau \tag{3-36}$$

Because

$$\cos\alpha = \cos(-\alpha) \tag{3-37}$$

equation (3-36) can be written as

$$\sum_{n=1}^{\infty} \frac{2}{T}\int_{-T/2}^{T/2} g(\tau)\cos[n\omega(\tau - t)]\,d\tau = \sum_{n=1}^{\infty} \frac{1}{T}\int_{-T/2}^{T/2} g(\tau)\cos[n\omega(\tau - t)]\,d\tau$$

$$+ \sum_{n=-1}^{-\infty} \frac{1}{T}\int_{-T/2}^{T/2} g(\tau)\cos[n\omega(\tau - t)]\,d\tau \tag{3-38}$$

The first term on the right of the equals sign sums for all positive integers of n — corresponding to $\cos\alpha$ in (3-37). The second term on the right side of the equals sign sums all negative integers of n — corresponding to $\cos(-\alpha)$ in (3-37). Equation (3-34) can now be rewritten substituting values of (3-35) and (3-38) giving

$$y = g(t) = \frac{1}{T} \int_{-T/2}^{T/2} g(\tau)\, d\tau + \sum_{n=1}^{\infty} \frac{2}{T} \int_{-T/2}^{T/2} g(\tau) \cos\left[n\omega\left(\tau - t\right)\right] d\tau \qquad (3\text{-}34)$$

$$= \left[\frac{1}{T} \int_{-T/2}^{T/2} g(\tau) \cos\left[n\omega\left(\tau - t\right)\right] d\tau\right]_{n=0}$$

$$+ \sum_{n=1}^{\infty} \frac{1}{T} \int_{-T/2}^{T/2} g(\tau) \cos\left[n\omega\left(\tau - t\right)\right] d\tau$$

$$+ \sum_{n=-1}^{-\infty} \frac{1}{T} \int_{-T/2}^{T/2} g(\tau) \cos\left[n\omega\left(t - t\right)\right] d\tau \qquad (3\text{-}39)$$

From Figure 3-9 we can see that all integer values of n (from $-\infty$ to $+\infty$) are permitted by (3-39). Equation (3-39) can now be written as

$$y = g(t) = \sum_{-\infty}^{\infty} \frac{1}{T} \int_{-T/2}^{T/2} g(\tau) \cos\left[n\omega\left(\tau - t\right)\right] d\tau \qquad (3\text{-}40)$$

Letting T approach ∞ (i.e., $T \rightarrow \infty$) in the equation

$$\frac{1}{T} = \frac{\omega}{2\pi}$$

we find that as T gets larger, ω gets smaller. Eventually ω gets infinitesimally small so that

$$\omega \rightarrow d\omega \qquad \text{and} \qquad \frac{1}{T} = \frac{\omega}{2\pi} \rightarrow \frac{d\omega}{2\pi}$$

That is,

$$\lim_{T \rightarrow \infty} \frac{1}{T} = \frac{d\omega}{2\pi} \qquad (3\text{-}41)$$

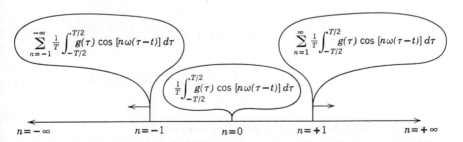

Integer values of n

Figure 3.9   Integer values of n and the corresponding term in equation (3.39).

Equation (3-40) represents a summation of the amplitudes of the component frequencies $n\omega$, where ω is the fundamental frequency. When the period T becomes larger, the fundamental frequency ω becomes smaller (causing the spectrum to become denser). Equations (3-40) and (3-41) indicate that as the period T becomes larger, the amplitudes of the individual components become smaller. We can see that in the limit as $T \to \infty$, the magnitude of each component will be infinitesimally small and there will be a continuous range of frequency components. Because the magnitude of a component becomes infinitesimal, we can no longer speak of a discrete component. Physical interpretation will now require the consideration of a finite interval of components and the contribution of such an interval. This interpretation leads to a spectral density or contribution per unit frequency interval.

It is also important to notice that as $T \to \infty$ in (3-40), the increments of ω will get smaller. We have seen that in the Fourier series for a repetitive waveform only integer multiples of the fundamental frequency ω will be found; that is, $n\omega = 0, \omega, 2\omega, 3\omega, \ldots, n\omega$. As $\omega \to d\omega$ the frequencies found will be $nd\omega = 0, d\omega, 2d\omega, 3d\omega, \ldots, nd\omega$. Because $d\omega$ is infinitesimally small, actually ω will take on all values (as compared to only integer multiples of the fundamental when T is small) and $n\omega \to \omega$. Because the increments are infinitesimally small, the summation in (3-40) can be replaced by an integral over ω; that is, using (3-41) we obtain

$$\frac{1}{T} \sum_{n=-\infty}^{\infty} \to \frac{1}{2\pi} \int_{-\infty}^{\infty} d\omega \qquad (3-42)$$

Substituting (3-41) and (3-42) in (3-40) and applying the limit $T \to \infty$, and $n\omega \to \omega$ we get

$$y = g(t) = \sum_{-\infty}^{\infty} \frac{1}{T} \int_{-T/2}^{T/2} g(\tau) \cos\left[n\omega(\tau-t)\right] d\tau \qquad (3-40)$$

$$= \int_{-\infty}^{\infty} \frac{d\omega}{2\pi} \int_{-\infty}^{\infty} g(\tau) \cos\left[\omega(\tau-t)\right] d\tau \qquad (3-43)$$

$$y = g(t) = \frac{1}{2\pi} \int_{-\infty}^{\infty} d\omega \int_{-\infty}^{\infty} g(\tau) \cos\left[\omega(\tau-t)\right] d\tau \qquad (3-44)$$

Equation (3-44) is referred to as *Fourier's integral*.

The representation of the Fourier integral (3-44) is somewhat awkward to work with. For this reason the Fourier integral and even the Fourier series are most often represented with complex notation. Complex notation simplifies the manipulation of the equations and makes the Fourier transform simpler to understand.

Complex Notation

The use of complex notation simplifies working with the Fourier series and Fourier integrals. Complex notation basically involves the representation of cosines and sines in the form

$$\cos n\omega t = \frac{e^{jn\omega t} + e^{-jn\omega t}}{2} \tag{3-45}$$

and

$$\sin n\omega t = \frac{e^{jn\omega t} - e^{-jn\omega t}}{2j} \tag{3-46}$$

where $j = \sqrt{-1}$.
Equation 3-28,

$$y = g(t) = A_0 + \sum_{n=1}^{\infty} A_n \cos (n\omega t) + \sum_{n=1}^{\infty} B_n \sin (n\omega t) \tag{3-28}$$

can be written

$$y = g(t) = A_0 + \sum_{n=1}^{\infty} A_n \left\{ \frac{e^{jn\omega t} + e^{-jn\omega t}}{2} \right\} + \sum_{n=1}^{\infty} B_n \left\{ \frac{e^{jn\omega t} - e^{-jn\omega t}}{2j} \right\} \tag{3-47}$$

$$= A_0 + \sum_{n=1}^{\infty} \frac{A_n}{2} (e^{jn\omega t} + e^{-jn\omega t}) + \sum_{n=1}^{\infty} \frac{B_n}{2j} (e^{jn\omega t} - e^{-jn\omega t}) \tag{3-48}$$

$$= A_0 + \sum_{n=1}^{\infty} \frac{A_n}{2} (e^{jn\omega t} + e^{-jn\omega t}) + \sum_{n=1}^{\infty} \frac{B_n}{2j} \frac{j}{j} (e^{jn\omega t} - e^{-jn\omega t}) \tag{3-49}$$

$$= A_0 + \sum_{n=1}^{\infty} \frac{A_n}{2} (e^{jn\omega t} + e^{-jn\omega t}) - \sum_{n=1}^{\infty} \frac{jB_n}{2} (e^{jn\omega t} - e^{-jn\omega t}) \tag{3-50}$$

$$= A_0 + \sum_{n=1}^{\infty} \left(\frac{A_n}{2} e^{jn\omega t} + \frac{A_n}{2} e^{-jn\omega t} - \frac{jB_n}{2} e^{jn\omega t} + \frac{jB_n}{2} e^{-jn\omega t} \right) \tag{3-51}$$

$$= A_0 + \sum_{n=1}^{\infty} \left\{ e^{jn\omega t} \left(\frac{A_n}{2} - \frac{jB_n}{2} \right) + e^{-jn\omega t} \left(\frac{A_n}{2} + \frac{jB_n}{2} \right) \right\} \tag{3-52}$$

$$= A_0 + \sum_{n=1}^{\infty} \left\{ e^{jn\omega t} \left(\frac{A_n - jB_n}{2} \right) + e^{-jn\omega t} \left(\frac{A_n + jB_n}{2} \right) \right\} \tag{3-53}$$

$$y = g(t) = A_0 + \sum_{n=1}^{\infty} \{C_n e^{jn\omega t} + D_n e^{-jn\omega t}\} \qquad (3\text{-}54)$$

where

$$C_n = \frac{A_n - jB_n}{2} \qquad (3\text{-}55)$$

and

$$D_n = \frac{A_n + jB_n}{2} \qquad (3\text{-}56)$$

In (3-25) and (3-27)

$$A_n = \frac{2}{T} \int_{-T/2}^{T/2} g(t) \cos(n\omega t)\, dt \qquad (3\text{-}25)$$

$$B_n = \frac{2}{T} \int_{-T/2}^{T/2} g(t) \sin(n\omega t)\, dt \qquad (3\text{-}27)$$

we have previously seen how the t in the integrals can be replaced by any letter. Equations (3-25) and (3-27) can therefore be written as

$$A_n = \frac{2}{T} \int_{-T/2}^{T/2} g(\tau) \cos(n\omega\tau)\, d\tau \qquad (3\text{-}57)$$

$$B_n = \frac{2}{T} \int_{-T/2}^{T/2} g(\tau) \sin(n\omega\tau)\, d\tau \qquad (3\text{-}58)$$

Substituting these values in (3-55) we have

$$C_n = \frac{A_n - jB_n}{2} = \frac{2}{2T} \int_{-T/2}^{T/2} g(\tau) \cos(n\omega\tau)\, d\tau - \frac{2j}{2T} \int_{-T/2}^{T/2} g(\tau) \sin(n\omega\tau)\, d\tau \qquad (3\text{-}59)$$

$$= \frac{1}{T} \int_{-T/2}^{T/2} g(\tau) [\cos(n\omega\tau) - j \sin(n\omega\tau)]\, d\tau \qquad (3\text{-}60)$$

Euler's equation states that an imaginary exponential can be expressed as

$$e^{\pm j\theta} = \cos\theta \pm j \sin\theta \qquad (3\text{-}61)$$

then (3-60) can be written as

$$C_n = \frac{1}{T} \int_{-T/2}^{T/2} g(\tau) [e^{-jn\omega\tau}]\, d\tau \qquad (3\text{-}62)$$

The value of D_n (3-56) is found in a similar manner,

$$D_n = \frac{A_n + jB_n}{2} = \frac{2}{2T} \int_{-T/2}^{T/2} g(\tau) \cos (n\omega\tau) \, d\tau + \frac{2j}{2T} \int_{-T/2}^{T/2} g(\tau) \sin (n\omega\tau) \, d\tau \tag{3-63}$$

$$= \frac{1}{T} \int_{-T/2}^{T/2} g(\tau) [\cos (n\omega\tau) + j \sin (n\omega\tau)] \, d\tau \tag{3-64}$$

Using (3-61) we can write (3-64) as

$$D_n = \frac{1}{T} \int_{-T/2}^{T/2} g(\tau) [e^{jn\omega\tau}] \, d\tau \tag{3-65}$$

Substituting (3-62) and (3-65) in (3-54) we get

$$y = g(t) = A_0 + \sum_{n=1}^{\infty} C_n e^{jn\omega t} + \sum_{n=1}^{\infty} D_n e^{-jn\omega t} \tag{3-54A}$$

$$= A_0 + \sum_{n=1}^{\infty} \left[\frac{1}{T} \int_{-T/2}^{T/2} g(\tau) e^{-jn\omega\tau} d\tau \right] e^{jn\omega t}$$

$$+ \sum_{n=1}^{\infty} \left[\frac{1}{T} \int_{-T/2}^{T/2} g(\tau) e^{jn\omega\tau} d\tau \right] e^{-jn\omega t} \tag{3-66}$$

The equation for A_0 was given in (3-21). Changing the variable for A_0 in (3-21) from t to τ, (3-66) can be written as

$$y = g(t) = \frac{1}{T} \int_{-T/2}^{T/2} g(\tau) \, d\tau + \sum_{n=1}^{\infty} \left[\frac{1}{T} \int_{-T/2}^{T/2} g(\tau) e^{-jn\omega\tau} d\tau \right] e^{jn\omega t}$$

$$+ \sum_{n=1}^{\infty} \left[\frac{1}{T} \int_{-T/2}^{T/2} g(\tau) e^{jn\omega\tau} d\tau \right] e^{-jn\omega t} \tag{3-67}$$

Equation (3-67) has a succession of terms,

$$e^{-jn\omega\tau} \times e^{jn\omega t} \qquad \text{and} \qquad e^{jn\omega\tau} \times e^{-jn\omega t}$$

both of which are summed from $n = 1$ to $n = \infty$, that is,

$$\sum_{n=1}^{\infty} e^{-jn\omega\tau} e^{jn\omega t} + \sum_{n=1}^{\infty} e^{jn\omega\tau} e^{-jn\omega t} \tag{3-68}$$

If we look at the second summation, we see that changing the signs of the exponents and changing the summation from $n = -1$ to $n = -\infty$ will give equivalent results, that is,

$$\sum_{n=1}^{\infty} e^{jn\omega\tau} e^{-jn\omega t} = \sum_{n=-1}^{-\infty} e^{-jn\omega\tau} e^{jn\omega t}$$

Therefore (3-68) can be written

$$\sum_{n=1}^{\infty} e^{-jn\omega\tau} e^{jn\omega t} + \sum_{n=1}^{\infty} e^{jn\omega\tau} e^{-jn\omega t} = \sum_{n=1}^{\infty} e^{-jn\omega\tau} e^{jn\omega t} + \sum_{n=-1}^{-\infty} e^{-jn\omega\tau} e^{jn\omega t}$$

$$= \left[\sum_{n=1}^{\infty} + \sum_{n=-1}^{-\infty} \right] [e^{-jn\omega\tau} e^{jn\omega t}] \qquad (3\text{-}69)$$

Equation (3-67) can now be written as

$$y = g(t) = \frac{1}{T} \int_{-T/2}^{T/2} g(\tau)\, d\tau + \sum_{n=1}^{\infty} \left[\frac{1}{T} \int_{-T/2}^{T/2} g(\tau)\, e^{-jn\omega\tau}\, d\tau \right] e^{jn\omega t}$$

$$+ \sum_{n=1}^{\infty} \left[\frac{1}{T} \int_{-T/2}^{T/2} g(\tau) e^{jn\omega\tau}\, d\tau \right] e^{-jn\omega t} \qquad (3\text{-}67)$$

$$= \frac{1}{T} \int_{-T/2}^{T/2} g(\tau)\, d\tau + \left[\sum_{n=1}^{\infty} + \sum_{n=-1}^{-\infty} \right] \left[\frac{1}{T} \int_{-T/2}^{T/2} g(\tau) e^{-jn\omega\tau}\, d\tau \right] e^{jn\omega t} \quad (3\text{-}70)$$

If we look at the second term (3-70), we can see that it is summed for all integers except zero. When $n = 0$, it reduces to the first term of (3-70), that is,

$$\left\{ \left[\frac{1}{T} \int_{-T/2}^{T/2} g(\tau) e^{-jn\omega\tau}\, d\tau \right] e^{jn\omega t} \right\}_{n=0} = \frac{1}{T} \int_{-T/2}^{T/2} g(\tau)\, d\tau$$

We can therefore write (3-70) as

$$y = g(t) = \frac{1}{T} \int_{-T/2}^{T/2} g(\tau)\, d\tau + \left[\sum_{n=1}^{\infty} + \sum_{n=-1}^{-\infty} \right] \left[\frac{1}{T} \int_{-T/2}^{T/2} g(\tau) e^{-jn\omega\tau}\, d\tau \right] e^{jn\omega t}$$

$$(3\text{-}70)$$

$$= \sum_{-\infty}^{\infty} \left[\frac{1}{T} \int_{-T/2}^{T/2} g(\tau) e^{-jn\omega\tau}\, d\tau \right] e^{jn\omega t} \qquad (3\text{-}71)$$

which is the summation of all negative and positive integers including $n = 0$. Equation (3-71) is another way of expressing the Fourier series. The terms inside the brackets represent the amplitudes and phases of the Fourier components when a harmonic analysis is made of the function $g(t)$.

Letting $T \to \infty$ in the equation $1/T = \omega/2\pi$ we found from equations (3-40) and (3-41) T gets larger, ω gets smaller and eventually ω gets infinitesimally small so that $\omega \to d\omega$ and also $n\omega \to \omega$ then

$$\frac{1}{T} = \frac{d\omega}{2\pi}$$

and the summation

$$\frac{1}{T} \sum_{n=-\infty}^{\infty} \to \frac{1}{2\pi} \int_{-\infty}^{\infty} d\omega$$

The effect of letting $T \to \infty$ permits us to obtain the Fourier integral in complex form from

$$y = g(t) = \sum_{-\infty}^{\infty} \left[\frac{1}{T} \int_{-T/2}^{T/2} g(\tau) e^{-jn\omega\tau} \, d\tau \right] e^{jn\omega t} \qquad (3\text{-}71)$$

as follows:

$$y = g(t) = \int_{-\infty}^{\infty} \frac{d\omega}{2\pi} \left[\int_{-\infty}^{\infty} g(\tau) e^{-j\omega\tau} \, d\tau \right] e^{j\omega t} \qquad (3\text{-}72)$$

$$y = g(t) = \frac{1}{2\pi} \int_{-\infty}^{\infty} e^{j\omega t} \left[\int_{-\infty}^{\infty} g(\tau) e^{-j\omega\tau} \, d\tau \right] d\omega \qquad (3\text{-}73)$$

The terms in the brackets represent the amplitudes and phases of the Fourier harmonics of the Fourier series. Equation (3-73) is the Fourier integral in complex form; that is, it is the complex form of (3-44). We can convert (3-73) to (3-44) by rewriting equation (3-73) as

$$y = g(t) = \frac{1}{2\pi} \int_{-\infty}^{\infty} d\omega \left[\int_{-\infty}^{\infty} g(\tau) e^{-j\omega(\tau-t)} d\tau \right] \qquad (3\text{-}74)$$

Using Euler's equation (3-61) we can write (3-74) as

$$y = g(t) = \frac{1}{2\pi} \int_{-\infty}^{\infty} d\omega \left[\int_{-\infty}^{\infty} g(\tau) \{ \cos \omega (\tau - t) - j \sin \omega (\tau - t) \} \, d\tau \right]$$

$$= \frac{1}{2\pi} \int_{-\infty}^{\infty} d\omega \int_{-\infty}^{\infty} g(\tau) \cos \omega (\tau - t) \, d\tau$$

$$- \frac{j}{2\pi} \int_{-\infty}^{\infty} d\omega \int_{-\infty}^{\infty} g(\tau) \sin \omega (\tau - t) \, d\tau \qquad (3\text{-}75)$$

By noting that sine is an odd function it can be shown that the second integral in (3-75) is zero. Since $\sin A = -\sin(-A)$ when integrating this integral from $-\infty$ to $+\infty$, with respect to ω, every term due to a positive ω will be cancelled by the term due to a negative ω. Then (3-75) reduces to

$$y = g(t) = \frac{1}{2\pi} \int_{-\infty}^{\infty} d\omega \int_{-\infty}^{\infty} g(\tau) \cos \omega(\tau - t)\, d\tau \qquad (3\text{-}76)$$

Referring back to (3-44) it is seen that (3-44) and (3-76) are identical in form. This shows that (3-44) and (3-73) are different forms of the same expression; that is, both (3-44) and (3-73) give the Fourier transform relation. However, the exponential form of (3-73) is the commonly used one because exponentials are easier to work with.

THE FOURIER TRANSFORM

The Fourier transform is used to resolve a given function into its exponential components. The function $F(\omega)$ is the Fourier transform of the function $f(t)$ and represents the amplitude and phase of the various frequency components making up the function $f(t)$. The function $F(\omega)$ is a function of frequency and is called the *frequency-domain representation* of the time-domain representation of the function). In optics work we are usually interested in functions $f(x)$ in the space domain and the Fourier transform of $f(x)$. A time-domain representation of a function specifies the amplitude of the function at each instant of time, while a space-domain representation of the same function would specify the amplitude as a function of position. A frequency-domain representation of a function specifies the relative amplitudes and phases of the frequency components of the function. Each of these representations (i.e., time, space, or frequency domain) uniquely specify the function.

The function $F(\omega)$ is generally complex and is represented by a magnitude $|F(\omega)|$ and a phase $\theta(\omega)$, that is,

$$F(\omega) = |F(\omega)| e^{j\theta(\omega)}$$

A graph of the function $F(\omega)$ must therefore contain information concerning its magnitude and phase. When $F(\omega)$ is either real or imaginary, only one graph is necessary because phase variations occur only for complex $F(\omega)$.

Let us now just consider the time- and frequency-domain relationship, understanding that the same relationships will exist between the space and frequency domains; that is, changing the meaning of the variable from time to space is all that is required. The Fourier integral in complex form is given by (3-73) and can be rewritten with $f(t)$ in place of $g(t)$.

$$y = f(t) = \frac{1}{2\pi} \int_{-\infty}^{\infty} e^{j\omega t} \left[\int_{-\infty}^{\infty} f(\tau) e^{-j\omega \tau} \, d\tau \right] d\omega \qquad (3\text{-}73\text{A})$$

We can let

$$F(\omega) = \frac{1}{\sqrt{2\pi}} \int_{-\infty}^{\infty} f(\tau) e^{-j\omega \tau} \, d\tau \qquad (3\text{-}77)$$

and then

$$\sqrt{2\pi} F(\omega) = \int_{-\infty}^{\infty} f(\tau) e^{-j\omega \tau} \, d\tau \qquad (3\text{-}78)$$

which is the portion of (3-73A) in the brackets. Therefore (3-73A) can be written

$$y = f(t) = \frac{1}{2\pi} \int_{-\infty}^{\infty} e^{j\omega t} \left[\int_{-\infty}^{\infty} f(\tau) e^{-j\omega \tau} \, d\tau \right] d\omega \qquad (3\text{-}73\text{A})$$

$$= \frac{1}{2\pi} \int_{-\infty}^{\infty} e^{j\omega t} \left[\sqrt{2\pi} F(\omega) \right] d\omega$$

$$= \frac{\sqrt{2\pi}}{2\pi} \int_{-\infty}^{\infty} F(\omega) e^{j\omega t} \, d\omega$$

$$= \frac{1}{\sqrt{2\pi}} \int_{-\infty}^{\infty} F(\omega) e^{j\omega t} \, d\omega \qquad (3\text{-}79)$$

If we examine $F(\omega)$ more closely, we see that (3-77),

$$F(\omega) = \frac{1}{\sqrt{2\pi}} \int_{-\infty}^{\infty} f(\tau) e^{-j\omega \tau} \, d\tau \qquad (3\text{-}77)$$

was written in terms of τ. We know that since $F(\omega)$ is shown as a definite integral, we can change the variable under the definite integral without changing the value of $F(\omega)$. We can therefore write (3-77) in terms of t as follows:

$$F(\omega) = \frac{1}{\sqrt{2\pi}} \int_{-\infty}^{\infty} f(\tau) e^{-j\omega \tau} \, d\tau \qquad (3\text{-}77)$$

$$F(\omega) = \frac{1}{\sqrt{2\pi}} \int_{-\infty}^{\infty} f(t) e^{-j\omega t} \, dt \qquad (3\text{-}80)$$

Therefore in (3-80) $F(\omega)$ is given in terms of t (τ in (3-77) was used in place of t to avoid confusion when integrating). There is then a symmetrical arrangement between (3-79) and (3-80) that can easily be seen when these equations are written below each other as follows:

$$f(t) = \frac{1}{\sqrt{2\pi}} \int_{-\infty}^{\infty} F(\omega) e^{j\omega t} \, d\omega \qquad (3\text{-}79)$$

$$F(\omega) = \frac{1}{\sqrt{2\pi}} \int_{-\infty}^{\infty} f(t) e^{-j\omega t} \, dt \qquad (3\text{-}80)$$

In (3-80) $F(\omega)$ is known as the Fourier transform of $f(t)$. Equation (3-79) gives the function of t, $[f(t)]$, in terms of $F(\omega)$ and is known as the *inverse Fourier transform*. When $f(t)$ is given as a function of time, the Fourier transform gives the amplitudes and phases of the infinite number of frequencies making up $f(t)$.

Instead of defining $F(\omega)$ as in (3-80) we could have defined $F(\omega)$ as follows:

$$F(\omega) = \int_{-\infty}^{\infty} f(t) e^{-j\omega t} \, dt \qquad (3\text{-}81)$$

then substituting (3-81) in the equation (3-73A) gives

$$f(t) = \frac{1}{2\pi} \int_{-\infty}^{\infty} F(\omega) e^{j\omega t} \, d\omega \qquad (3\text{-}82)$$

Equations (3-81) and (3-82) also define the Fourier transform and the inverse Fourier transform of $f(t)$, respectively.

We might also have defined $F(\omega)$ as

$$F(\omega) = \frac{1}{2\pi} \int_{-\infty}^{\infty} f(t) e^{-j\omega t} \, dt \qquad (3\text{-}83)$$

then substituting (3-83) into (3-73A) gives

$$f(t) = \int_{-\infty}^{\infty} F(\omega) e^{j\omega t} \, d\omega \qquad (3\text{-}84)$$

Again (3-83) and (3-84) also define the Fourier transform and the inverse Fourier transform of $f(t)$, respectively. The appearance of the factor $1/2\pi$ (or $1/\sqrt{2\pi}$) is determined by the way $F(\omega)$ was defined. The pairs of equations (3-79) and (3-80), (3-81) and (3-82), or (3-83) and (3-84) are all Fourier transforms and inverse Fourier transforms of each other, respectively, and each of the forms appear in different literature. Equation (3-73A) requires a factor of $1/2\pi$ but it does not specify where this term must appear. Because $1/2\pi$ is a constant, the combinations discussed here (or any other factor combination) are all suitable, for example, in (3-82).

$$f(t) = \frac{1}{2\pi} \int_{-\infty}^{\infty} F(\omega) e^{j\omega t} \, d\omega \tag{3-82}$$

we can substitute

$$\omega = 2\pi \mathscr{F} \qquad \text{and} \qquad d\omega = 2\pi d\mathscr{F}$$

then

$$f(t) = \int_{-\infty}^{\infty} F(\omega) e^{j2\pi \mathscr{F} t} \, d\mathscr{F} \tag{3-85}$$

Then from (3-81)

$$F(\omega) = \int_{-\infty}^{\infty} f(t) e^{-j2\pi \mathscr{F} t} \, dt \tag{3-86}$$

Equations (3-85) and (3-86) are the Fourier transform relations between $f(t)$ and $F(\omega)$.

When the function $f(t)$ is an even function, the Fourier transform as given by (3-80) reduces to

$$F(\omega) = \frac{1}{\sqrt{2\pi}} \int_{-\infty}^{\infty} f(t) \cos(\omega t) \, dt \tag{3-87}$$

Equation (3-87) is known as the *Fourier cosine transform*. When $f(t)$ is an odd function, the Fourier transform reduces to

$$F(\omega) = \frac{1}{\sqrt{2\pi}} \int_{-\infty}^{\infty} f(t) \sin \omega t \, dt \tag{3-88}$$

and is known as the *Fourier sine transform*. This can be shown by making use of the fact

$$\cos(\alpha - \beta) = \cos \alpha \cos \beta + \sin \alpha \sin \beta \tag{3-33}$$

and writing (3-44) as follows:

$$f(t) = \frac{1}{2\pi} \int_{-\infty}^{\infty} d\omega \int_{-\infty}^{\infty} f(\tau) \cos[\omega(\tau - t)] \, d\tau \tag{3-44A}$$

$$= \frac{1}{2\pi} \int_{-\infty}^{\infty} d\omega \int_{-\infty}^{\infty} f(\tau) \cos(\omega \tau) \cos(\omega t) \, d\tau$$

$$+ \frac{1}{2\pi} \int_{-\infty}^{\infty} d\omega \int_{-\infty}^{\infty} f(\tau) \sin(\omega \tau) \sin(\omega t) \, d\tau$$

$$= \frac{1}{2\pi} \int_{-\infty}^{\infty} \cos(\omega t) \, d\omega \int_{-\infty}^{\infty} f(\tau) \cos(\omega \tau) \, d\tau$$

$$+ \frac{1}{2\pi} \int_{-\infty}^{\infty} \sin(\omega t) \, d\omega \int_{-\infty}^{\infty} f(\tau) \sin(\omega \tau) \, d\tau \tag{3-89}$$

If $f(t)$ is an even function, the integral (τ replacing t) term in (3-89)

$$\int_{-\infty}^{\infty} f(\tau) \sin \omega\tau \, d\tau = 0$$

since the product $[f(\tau) \sin \omega\tau]$ is an odd function, positive and negative values of integration variables produce terms that cancel. Thus when $f(t)$ is even, equation (3-89) becomes

$$f(t) = \frac{1}{\sqrt{2\pi}} \int_{-\infty}^{\infty} \cos(\omega t) \, d\omega \frac{1}{\sqrt{2\pi}} \int_{-\infty}^{\infty} f(\tau) \cos(\omega\tau) \, d\tau \qquad (3\text{-}90)$$

where

$$F(\omega) = \frac{1}{\sqrt{2\pi}} \int_{-\infty}^{\infty} f(\tau) \cos(\omega\tau) \, d\tau = \frac{1}{\sqrt{2\pi}} \int_{-\infty}^{\infty} f(t) \cos(\omega t) \, dt \qquad (3\text{-}91)$$

Equation (3-91) is the Fourier cosine transform of $f(t)$ and (3-90) states that the inverse Fourier cosine transform of $F(\omega)$, is $f(t)$.

Similarly when $f(t)$ is an odd function,

$$\int_{-\infty}^{\infty} f(\tau) \cos \omega\tau \, d\tau = 0 \qquad (3\text{-}89)$$

in (3-89) and the Fourier sine transform of $f(t)$ is

$$F(\omega) = \frac{1}{\sqrt{2\pi}} \int_{-\infty}^{\infty} f(\tau) \sin(\omega\tau) \, d\tau = \frac{1}{\sqrt{2\pi}} \int_{-\infty}^{\infty} f(t) \sin(\omega t) \, dt \qquad (3\text{-}92)$$

and the inverse sine transform gives

$$f(t) = \frac{1}{2\pi} \int_{-\infty}^{\infty} \sin \omega t \, d\omega \int_{-\infty}^{\infty} f(\tau) \sin(\omega\tau) \, d\tau = \frac{1}{\sqrt{2\pi}} \int_{-\infty}^{\infty} \sin \omega t F(\omega) \, d\omega \qquad (3\text{-}93)$$

Example of a Fourier Transform

The Fourier transform is an extremely important operation in optical data processing. It will be shown how coherent optical processors make use of the Fourier transform relationship that exists between the light amplitude distribution at the front and back focal planes of a lens. In optical data processors space-domain functions are used as input signals (that is, optical signals are usually spatial variations whereas electronic signals are usually time variations) and the Fourier transform operation is optically easy to implement. In order to ensure an appreciation of the Fourier transform we shall examine in some detail the Fourier transform of some of the common functions.

Most people intuitively recognize the Fourier transform of certain functions, although they may not be aware of it. Probably the most familiar Fourier transform is that shown in Figure 3.10. Figure 3.10(a) shows the plot of a d-c signal. A d-c signal is a function $f(t)$ that is constant, and therefore contains only one frequency component at $\omega = 0$. When we make a plot of the Fourier transform we are plotting the frequencies present in the function. Since a constant function is a DC signal, one and only one frequency component will be present at $\omega = 0$ as shown in Figure 3.10(b). The relative amplitude of the $\omega = 0$ component in Figure 3.10(b) will be determined after we first consider the Fourier transform of a rectangular pulse and the meaning of the unit impulse function.

The Fourier Transform of a Rectangular Pulse

We have seen that the Fourier transform can be given by any one of the equations (3-80), (3-81), (3-83), or (3-86). The choice of the equation used only determines the inverse transform equation that must be used. If we decide that we will find the Fourier transform of a rectangular pulse by (3-83), then we must use (3-84) to obtain the inverse Fourier transform. Let us use (3-83) to find the Fourier transform of the rectangular pulse shown in Figure 3.11.

$$F(\omega) = \frac{1}{2\pi} \int_{-\infty}^{\infty} f(t)\, e^{-j\omega t}\, dt \qquad (3\text{-}83)$$

The rectangular pulse (Figure 3.11A) is defined by

$$f(t) = \begin{cases} A & t \leq |T/2| \\ 0 & t > |T/2| \end{cases}$$

and therefore in (3-83)

$$F(\omega) = \frac{1}{2\pi} \int_{-\infty}^{\infty} f(t)\, e^{-j\omega t}\, dt \qquad (3\text{-}83)$$

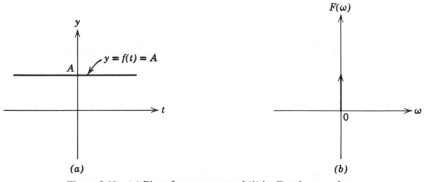

Figure 3.10 (a) Plot of a constant and (b) its Fourier transform.

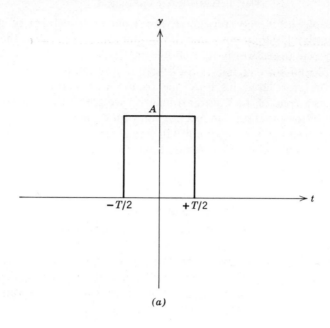

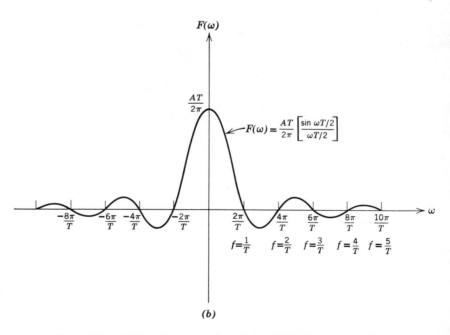

Figure 3.11 (a) Plot of a rectangular pulse and (b) its Fourier transform.

we can change the limits of integration from $(-\infty$ to $+\infty)$ to $(-T/2$ to $+T/2)$. We also note that the value of the function $f(t)$ between the limits $-T/2$ to $+T/2$ is a constant A.

We can now solve for $F(\omega)$ as follows:

$$F(\omega) = \frac{1}{2\pi} \int_{-T/2}^{T/2} A e^{-j\omega t} \, dt$$

$$= -\frac{A}{2\pi j\omega} \left[e^{-j\omega t} \right]_{-T/2}^{T/2}$$

$$= \frac{A}{\pi \omega} \left[\frac{e^{j\omega(T/2)} - e^{-j\omega(T/2)}}{2j} \right] = \frac{A}{\pi \omega} \sin \frac{\omega T}{2}$$

$$= \frac{AT}{2\pi} \left[\frac{\sin \omega T/2}{\omega T/2} \right] \tag{3-94}$$

From this we can see that $F(\omega)$ is a real function and therefore can be represented graphically by a single curve as shown in Figure 3.11(b). Appendix 2A is a table of values for the function $(\sin x)/x$.

If we had used equation (3-81) instead of (3-83) to find the Fourier transform of the rectangular pulse, we would have found that the final result would have been

$$F(\omega) = AT \left[\frac{\sin \omega T/2}{\omega T/2} \right] \tag{3-95}$$

The difference between equation (3-94) and (3-95) is the factor $1/2\pi$. It is therefore necessary to use the proper matching equations to find the Fourier transform and the inverse Fourier transform.

The Unit Impulse Function

Let us consider a unit step function as shown in Figure 3.12. The derivative (dy/dt) is zero for all values of t except at $t = 0$. The unit step

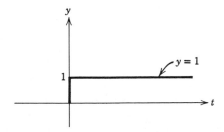

Figure 3.12 A unit step function.

function is discontinuous at $t = 0$ and the derivative does not exist at this point. The derivative of a unit step function, therefore, does not exist but by assuming a function that can approximate the unit step function, the derivative of the unit step function can be found. Let us assume a function $u(t)$ approximates the unit step function and then let us examine what this function can be.

If we consider $u(t)$ to be a function as shown in Figure 3.13, we can see that as the distance a gets smaller, the function $u(t)$ approaches a unit step function. The best approximation to the unit step function shown in Figure 3.12 is that of Figure 3.13(c). A unit step function $u(t)$ is shown in Figure 3.14(a). The slope of the line AB in Figure 3.14(a) is $1/a$, while the slope of the remainder of the function $u(t)$ is equal to zero. This is the derivative of the function and is shown plotted in Figure 3.14(b).

From Figure 3.14(b) we can see that the derivative of the unit step-function approximation is a rectangular pulse of height $1/a$ and width a. We can see that if a is varied, the area $(1/a \times a)$ of the derivative of the

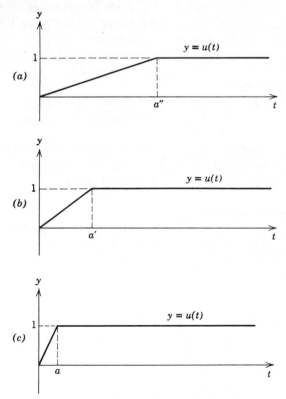

Figure 3.13 Unit step function approximation.

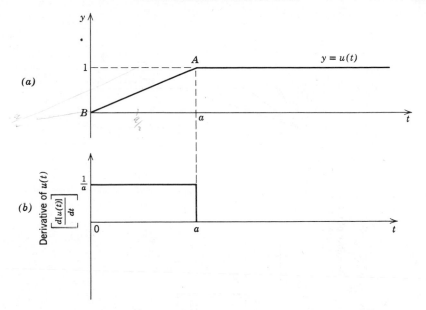

Figure 3.14 Approximation of a unit step function and its derivative.

step-function approximation will remain constant. Replotting the unit step-function approximation of Figure 3.13 on one graph as shown in Figure 3.15 we can see the effect of varying the distance a on the derivative of the step-function approximation.

From Figure 3.15 we can see that the area of each rectangle will be a constant $(1/a \times a = 1)$ equal to one. When the distance a approaches zero, the height of the rectangle goes toward infinity. The width of the rectangle therefore goes towards zero but the area of the rectangle will still be one. A unit impulse function is defined as the derivative of a unit step function. We know that the derivative of a unit step function does not exist but we can say that the limit of the derivative of the step-function approximation $u(t)$, as the distance a goes to zero, is a unit impulse function (denoted by $\delta(t)$). Specifically, we have defined the following:

$$\delta(t) = \lim_{a \to 0} \frac{d}{dt}[u(t)] \tag{3-96}$$

It was shown in Figure 3.15 that the function $d[u(t)]/dt$ is a rectangle of height $1/a$ and width a. We can therefore rewrite (3-96) as

$$\delta(t) = \lim_{a \to 0} \frac{1}{a}[u(t) - u(t-a)] \tag{3-97}$$

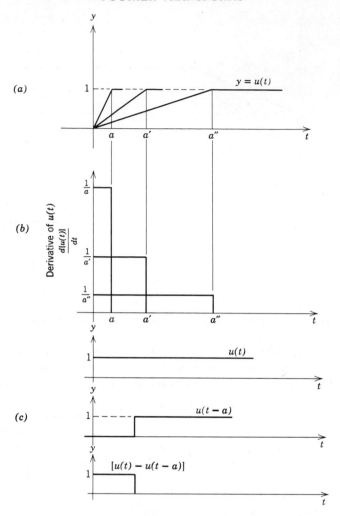

Figure 3.15 Effect of varying the distance (*a*).

where the function in the brackets is defined as shown in Figure 3.15(c). Then as the distance *a* gets smaller, the factor $1/a$ gets bigger and the factor

$$[u(t) - u(t-a)] \rightarrow [u(t) - u(t)] = 0$$

We can see therefore that in the limit as the distance *a* approaches zero, the function $\delta(t)$ will still be represented by a rectangle of unit area but will have infinite height and zero width. We can also see that the δ function (unit impulse function) will exist only when $t = 0$ and it will be zero for all other values of t and may be defined by the following relations:

$$\left.\begin{aligned} \delta(t) &= 0 \qquad \text{for } t \neq 0 \\ \int_{-\infty}^{\infty} \delta(t)\, dt &= 1 \end{aligned}\right\} \tag{3-98}$$

We shall make use of the unit impulse function in the solution to certain optical problems even though such distributions actually do not exist. The use of the unit impulse function does simplify certain derivations although physically it cannot exist. In actual work because of the finite resolution of measuring equipment, a unit step function cannot be distinguished from a pulse with a ramp of small but finite duration, which further justifies its use.

The unit impulse function represented by (3-97) can be taken to represent an impulse-function limiting process without regard to the actual shape of the pulse. In the example we have considered we have assumed that the impulse function was a limiting form of a rectangle (rectangular pulse). The function represented by (3-98) can be used to represent limits of other pulse forms; for example, the function $\delta(t)$ can be used to represent the limit of a triangular pulse, exponential pulse, Gaussian pulse, etc. Appendix 11 discusses other limiting definitions for a δ function. Appendix 12 shows in detail how the Fourier transform of two-different functions can be determined. These illustrative examples show the mathematical techniques for determining the Fourier transform of a function $\cos \omega_0 t$ over a finite interval and the Fourier transform of a Gaussian function. Appendix 13 is a table showing common examples of functions $f(t)$ and their Fourier transforms. Appendix 13 also indicates how the scales of the functions are related and how they might be changed to meet specific problems

The Fourier Transform of a Constant

Let us now return to the problem of finding the Fourier transform of a constant given by $f(t) = A$. The Fourier transform of a rectangular pulse of height A and width T was given in (3-95) as

$$F(\omega) = AT \left[\frac{\sin \omega T/2}{\omega T/2} \right] \tag{3-95}$$

If we let $T \rightarrow \infty$, the rectangular pulse tends to be a constant function A as shown in Figure 3.10(a). The Fourier transform of a constant A can therefore be considered the Fourier transform of the rectangular pulse as $T \to \infty$. Therefore

$$F[A] = \lim_{T \to \infty} AT \left[\frac{\sin \omega T/2}{\omega T/2} \right] \tag{3-99}$$

$$= 2\pi A \lim_{T \to \infty} \frac{T}{2\pi} \left[\frac{\sin \omega T/2}{\omega T/2} \right] \tag{3-100}$$

The limit of this function is an impulse function $\delta(\omega)$ as demonstrated in Appendix 11. Thus the transform of a constant A can be written as

$$F(A) = 2\pi A \delta(\omega) \tag{3-101}$$

for example, when $A = 1$

$$F(1) = 2\pi \delta(\omega). \tag{3-102}$$

From this we see that when $f(t)$ is a constant, it will contain only one frequency component $(\omega = 0)$. The amplitude at $\omega = 0$ is indeterminate because the amplitude of $\delta(\omega)$ approaches infinity. This difficulty leads to the spectral density interpretation which will be discussed later.

Power Spectrum

In electrical engineering power in watts is equal to the square of the voltage divided by the resistance (that is, $P = E^2/R$, $P \not\propto E^2$) or power is proportional to the square of the voltage. In optics the intensity of the light (power) is proportional to the square of the light amplitude (that is, $I \not\propto A^2$). The intensity spectrum (power spectrum) can be obtained for periodic functions by squaring the absolute values of the Fourier transform and can be written as follows:

$$I(\omega) = \text{power spectrum (as a function of } \omega) = |F(\omega)|^2, \tag{3-103}$$

This means that the power spectrum is the magnitude of the Fourier transform squared. Another way of stating this is that the power spectrum can be obtained by multiplying the Fourier transform by its complex conjugate; that is, since $F(\omega)$ is complex, it can be written as

$$F(\omega) = R(\omega) + jS(\omega) \tag{3-104}$$

the magnitude of $F(\omega)$ squared is found as follows:

$$|F(\omega)|^2 = [R(\omega) + jS(\omega)][R(\omega) - jS(\omega)]$$

$$= R(\omega)^2 + jR(\omega) S(\omega) - jR(\omega) S(\omega) + S(\omega)^2$$

$$= R(\omega)^2 + S(\omega)^2 \tag{3-105}$$

This can also be written as

$$\text{power spectrum (as a function of } (\omega)) = F(\omega)F^*(\omega) \quad (3\text{-}106)$$

where $F^*(\omega)$ means the complex conjugate of $F(\omega)$, or if

$$F(\omega) = R(\omega) + jS(\omega) \quad (3\text{-}107)$$

then

$$F^*(\omega) = R(\omega) - jS(\omega) \quad (3\text{-}108)$$

that is, the signs of the j terms are changed to obtain the complex conjugate.

A given function $f(t)$ contains both amplitude and phase information. The Fourier transform $F(\omega)$ of $f(t)$ also contains both amplitude and phase information of each and all the spectrum components. It is therefore possible by knowing $F(\omega)$ to obtain $f(t)$ and likewise if $f(t)$ is known, it is possible to obtain $F(\omega)$.

$$f(t) \quad \xleftarrow{\text{Fourier transform}} \quad F(\omega) \quad (3\text{-}109)$$
$$\text{(a function of } t) \qquad \xrightarrow{\hspace{2cm}} \qquad \text{(amplitude spectrum)}$$

When the power spectrum $|F(\omega)|^2$ is obtained from the amplitude spectrum $F(\omega)$, the phase information is lost. This means that given an amplitude spectrum $F(\omega)$ the power spectrum $|F(\omega)|^2$ can be determined, but given a power spectrum $|F(\omega)|^2$ the amplitude spectrum cannot be found. For a given $F(\omega)$ there is one, and only one, unique power spectrum $|F(\omega)|^2$ that can be derived from it. For a given power spectra $|F(\omega)|^2$ there is no one uniquely defined amplitude spectrum that can be formed from it; that is, because the power spectrum has no phase information, many amplitude spectra $F(\omega)$ can be deduced containing different phases that might form $|F(\omega)|^2$.

An aperiodic function can be described by its energy density spectrum whereas a periodic function is described by its power spectrum. The energy density spectrum relates how the signal energy is distributed over the frequency spectrum. An energy density spectrum can be considered to have units of (amplitude)²/cycle/second and can be plotted as shown in Figure 3.16. The total area under the energy density spectrum curve is the total energy contained in the signal.

The power spectrum has units of (amplitude)² as distinguished from the energy density spectrum, which has units of (amplitude)²/cycle/second. The power spectrum is used to describe periodic signals. Since periodic signals are made up of a finite number of descrete frequency components,

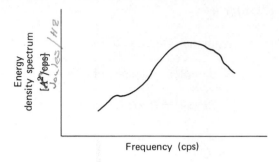

Figure 3.16   Energy density spectrum.

the amplitude squared of each component can be physically represented on a graph. The energy density spectrum is used to represent nonperiodic waves because nonperiodic waves contain an infinite number of frequency components. If, for example, we assume a nonperiodic wave form to be a pulse of finite width, we can easily see that the power it contains is finite. Likewise if this finite pulse contains an infinite number of frequency components, each of these frequency components must have an infinitesimally small amplitude (but is proportional to $F(\omega)$ the frequency spectrum). The energy density spectrum is therefore a measure of how the signal energy is distributed in frequency. Since the energy density spectrum contains no phase information, it will not uniquely specify a signal nor will it uniquely specify the amplitude variation of a signal with time. For example, two signals can have the same energy density spectrum if they contain the same frequency components with the same amplitudes; however, if the phases of the two signals are different, their wave shapes can be drastically different.

Random functions, which are outside the scope of this book, are described by the power density spectrum which is defined similarly to the energy density function.

Matched Filter

Let us consider the system shown in Figure 3.17. An input waveform of known amplitude and phase spectra is fed into a system that has a known amplitude, frequency, and phase response. The output amplitude spectrum is found by multiplying the input amplitude spectrum by the system amplitude frequency response. The output phase is obtained by adding

Figure 3.17   Representation of a system.

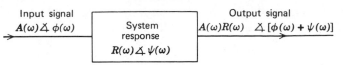

Figure 3.18   Representation of signals and system response.

the input phase spectrum to the system phase response as shown in Figure 3.18. Let us suppose that we can make the system amplitude frequency response equal to the input amplitude spectra and, further, we can make the phase frequency response of the system equal but opposite to the input phase spectrum. Under such conditions the amplitude spectrum of the output will be equal to the input power spectrum and the output phase spectrum will be zero. The system is called a *matched filter* for the particular input waveform when

$$\angle \phi(\omega) + \angle \psi(\omega) = 0 \qquad \text{and} \qquad R(\omega) = A(\omega)$$

that is, the output phase spectrum is zero, and the input power spectrum (input power spectrum is $|A(\omega)|^2$) equals the output amplitude spectrum. In order for the output amplitude spectrum (that is, $|A(\omega)||R(\omega)|$) to equal the input power spectrum $|A(\omega)|^2$ we must have the condition that $|R(\omega)| = |A(\omega)|$. A matched filter for an input signal $A(\omega) \angle \phi(\omega)$ is a system having an amplitude frequency response equal to $A(\omega)$ and a phase frequency response equal to $-\angle \phi(\omega)$. Figure 3.19 shows a representation of the matched filter system and the signal output for this matched filter system.

CORRELATION

Autocorrelation

The process of correlation is one of many different techniques for measuring the similarity between two quantities. The autocorrelation function is a plot of the similarity between a waveform and a displaced version of itself, plotted as a function of the displacement. Let us consider the autocorrelation function of a function $f_1(t)$ with a displaced version of itself, $f_1(t+\tau)$, where τ is the displacement. To compare these two

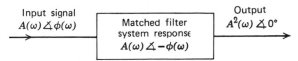

Figure 3.19   Representation of signals for a matched filter.

functions we can multiply the function $f_1(t)$ by the displaced version of itself, $f_1(t+\tau)$, and integrate for all values of t that is,

$$\int f_1(t) f_1(t+\tau)\, dt$$

This could be done graphically to obtain an approximate answer; for example, each ordinate on one curve could be multiplied by the corresponding ordinate on the displaced curve for a given displacement, τ. The sum (approximate integral) of all the products for this given displacement, τ, might represent the correspondence between the two curves for that value of τ. The process can be repeated for other values of τ until a plot is obtained for the correspondence versus different displacements of τ. Mathematically the process can be performed more exact by multiplying the function $f_1(t)$ by its displaced version $f_1(t+\tau)$ and integrating for all values of t, that is,

$$\int_{-\infty}^{\infty} f_1(t) f_1(t+\tau)\, dt$$

This integral from $-\infty$ to $+\infty$ will give an infinite result for periodic functions that exists from $-\infty$ to $+\infty$. The area under the curve, however, for any one period will be the same, therefore we can restrict our integration to one period to obtain

$$\int_{-T_1/2}^{T_1/2} f_1(t) f_1(t+\tau)\, dt \qquad (3\text{-}110)$$

where T_1 is the period of the function. There is one detail that is not considered by (3-110). Let us consider the two periodic waveforms shown in Figure 3.20. The sinusoid $y = A \sin \omega t$ (waveform 1) will bound a larger area than one cycle of the higher frequency sinusoid ($y = A \sin 2\omega t$, waveform 2). In order to normalize these two curves with respect to frequency (that is, have the bounded areas for one period of each waveform be related only to the respective amplitudes of the other waveform during one period) it is necessary to divide each waveform by its respective period. Following this type of logic it is necessary to divide (3-110) by the period of the function so that the correlation will be independent of the frequency of the wave and only dependent on the relative amplitude (time average of product). Equation (3-110) now becomes the auto-correlation function of $f_1(t)$ given by

$$\varphi_{11}(\tau) = \frac{1}{T_1} \int_{-T_1/2}^{T_1/2} f_1(t) f_1(t+\tau)\, dt \qquad (3\text{-}111)$$

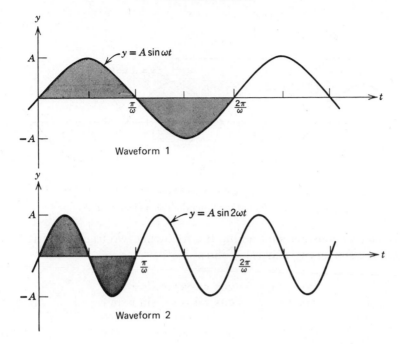

Figure 3.20   Plot of the periodic functions. $y = A \sin \omega t$ and $y = A \sin 2\omega t$.

where T_1 is the period of the waveform, $f_1(t)$ is a periodic function of t, τ is the displacement of the function, and the subscript $_{11}$ means the function is correlated with itself. If we did not normalize (by dividing by $1/T_1$), changing the frequency of waveform $f_1(t)$ would change the value of the autocorrelation function even though the waveshape stayed the same; and then a change in frequency would only appear to be a change of scale. By normalizing (dividing by $1/T_1$) we are in effect relating the shaded areas of each curve in Figure 3.20 to the shaded areas of the other curve.

Let us now take a sinusoidal function

$$y = A \sin (\omega t + \theta) \qquad (3\text{-}112)$$

as shown in Figure 3.21 and determine its autocorrelation function. From (3-111) we have

$$\varphi_{11}(\tau) = \frac{1}{T_1} \int_{-T_1/2}^{T_1/2} f_1(t) f_1(t + \tau)\, dt \qquad (3\text{-}111)$$

$$\phi_{11}(\tau) = \frac{\omega}{2\pi} \int_{-\pi/\omega}^{\pi/\omega} A \sin (\omega t + \theta)\, A \sin (\omega t + \theta + \omega \tau)\, dt \qquad (3\text{-}113)$$

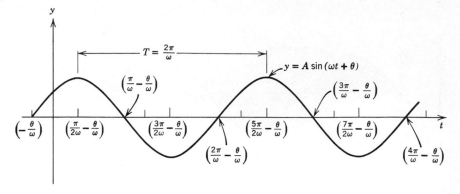

Figure 3.21   Plot of the function. $y = A \sin(\omega t + \theta)$.

Now let us consider subtracting the following two trigonometric functions of the sum of two angles as follows:

$$\cos(x+y) = \cos x \cos y - \sin x \sin y$$
$$\cos(x-y) = \cos x \cos y + \sin x \sin y$$

$$\cos(x-y) - \cos(x+y) = 2 \sin x \sin y \qquad (3\text{-}114)$$

Then

$$\sin x \sin y = \tfrac{1}{2}[\cos(x-y) - \cos(x+y)] \qquad (3\text{-}115)$$

If we let

$$x = \omega t + \theta + \omega \tau \qquad \text{and} \qquad y = \omega t + \theta$$

then (3-115) can be written

$$\sin(\omega t + \theta + \omega \tau)\sin(\omega t + \theta) = \tfrac{1}{2}[\cos\{(\omega t + \theta + \omega \tau) - (\omega t + \theta)\}$$
$$- \cos\{(\omega t + \theta + \omega \tau) + (\omega t + \theta)\}]$$
$$= \tfrac{1}{2}\cos\omega\tau - \tfrac{1}{2}\cos(2\omega t + 2\theta + \omega\tau) \qquad (3\text{-}116)$$

Then in (3-113) we can write

$$A\sin(\omega t + \theta + \omega\tau)\,A\sin(\omega t + \theta) = A^2/2[\cos\omega\tau - \cos(2\omega t + 2\theta + \omega\tau)]$$
$$(3\text{-}117)$$

Now substituting (3-117) into (3-113) we get

$$\varphi_{11}(\tau) = \frac{\omega}{2\pi}\int_{-\pi/\omega}^{\pi/\omega}\frac{A^2}{2}[\cos\omega\tau - \cos(2\omega t + 2\theta + \omega\tau)]\,dt$$

$$= \frac{A^2\omega}{4\pi}\int_{-\pi/\omega}^{\pi/\omega}\cos\omega\tau\,dt - \frac{A^2\omega}{4\pi}\int_{-\pi/\omega}^{\pi/\omega}\cos(2\omega t + 2\theta + \omega\tau)\,dt$$
$$(3\text{-}118)$$

$$= \frac{A^2}{4\pi} \cos \omega\tau \int_{-\pi/\omega}^{\pi/\omega} dt - \frac{A^2\omega}{4\pi} \underbrace{\int_{-\pi/\omega}^{\pi/\omega} \cos (2\omega t + 2\theta + \omega\tau) \, dt}_{\text{integral of cosine over one cycle is zero}}$$

integral of cosine over one cycle is zero (3-119)

Therefore,

$$\varphi_{11}(\tau) = \frac{A^2\omega}{4\pi}[\cos \omega\tau] \left[\frac{\pi}{\omega} + \frac{\pi}{\omega}\right] = \frac{A^2\omega 2\pi}{4\pi\omega} \cos (\omega\tau)$$

$$\varphi_{11}(\tau) = \frac{A^2}{2} \cos \omega\tau \qquad (3\text{-}120)$$

From this we can conclude that the autocorrelation function of any sinusoidal function is a cosine function of the same frequency as the original sinusoid; that is, the angular frequency ω of the original function

$$y = A \sin (\omega t + \theta) \qquad (3\text{-}112)$$

is the same as that of the autocorrelation function

$$\varphi_{11}(\tau) = \frac{A^2}{2} \cos \omega\tau \qquad (3\text{-}120)$$

Figure 3.22 shows a comparison between a sinusoidal function and its

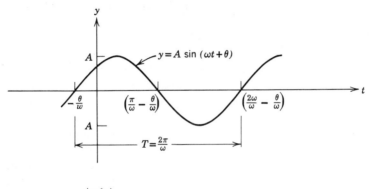

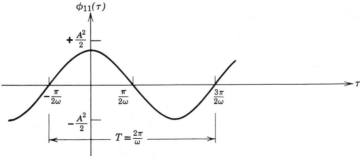

Figure 3.22 Comparison between a sinusoid and its autocorrelation function.

autocorrelation function and graphically shows the period of the two to be the same.

Let us now consider autocorrelation from a somewhat different point of view to assure better understanding of the process. Let us select an example of an autocorrelation function that we can synthesize by combining Fourier components. We have already seen how the Fourier series gives the amplitudes and phases of the sinusoidal waves making a periodic waveform. Let us consider the periodic square waveform shown in Figure 3.23(a). This square wave can be approximated by its first and third harmonics as shown. Let us determine and synthesize the autocorrelation function of the square wave by using its first and third harmonics. From (3-120) we can see that the autocorrelation of a sine wave is equal to one half the square of the amplitude of the sine wave. We can therefore take one half of the square of each of the harmonics to determine the autocorrelation function for each. Figure 3.23(b) shows the first harmonic and one half the first harmonic with the amplitude squared. Figure 3.23(c) shows the third harmonic and the third harmonic with one half the amplitude squared. Each of these components is replotted in Figure 3.23(d) as cosine functions in phase with each other at the origin. The amplitude of these cosine functions is added to obtain the autocorrelation function shown in Figure 3.23(d).

The above synthesizing of the autocorrelation function was accomplished by finding the autocorrelation function of each of the harmonics of the function (these will be cosine waves with frequencies corresponding to the frequency of the respective harmonic and with amplitudes equal to one half the amplitude squared of the harmonic), that is, $A^2/2$. Adding all the autocorrelation functions of the components (harmonics) of the function together synthesizes the autocorrelation of the original function; that is, in Figure 3.23(d) we see the autocorrelation function of each harmonic plotted and their sum is the synthesized autocorrelation of the approximated square wave. We can see that the autocorrelation function must always be an even function (a cosine function) and the phase of any harmonic component is not considered; that is, harmonics are plotted as cosine functions of zero phase. Because the phase of the individual harmonics is not considered, signals of different form can have the same autocorrelation function.

Crosscorrelation

We have seen that the autocorrelation function can be plotted as a graph that will indicate the similarity between a periodic waveform and a displaced version of itself. It is also possible to determine the similarity between two periodic waveforms that are not identical. The crosscorrela-

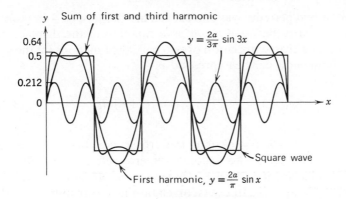

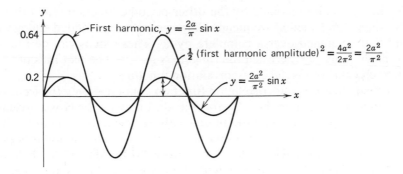

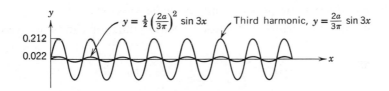

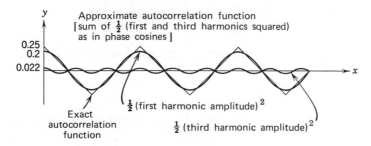

Figure 3.23

tion function of two periodic waveforms can also be plotted as a graph to indicate the similarity between them as a function of the displacement between them. The crosscorrelation of two periodic functions (of the same fundamental frequency) can be expressed as

$$\varphi_{12}(\tau) = \frac{1}{T_1} \int_{-T_1/2}^{T_1/2} f_1(t) f_2(t+\tau) \, dt \qquad (3\text{-}121)$$

where the subscript $_{12}$ indicates that the crosscorrelation involves two functions, $f_1(t)$ and $f_2(t)$. The second numeral refers to the fact that $f_2(t)$ has the displacement τ. Also τ is independent of t and can take any value between $-\infty$ and $+\infty$. In crosscorrelation just as in autocorrelation one of the functions $[f_2(t)]$ is displaced by an amount τ. The displaced function is then multiplied by the other periodic function $[f_1(t)]$ that has the same fundamental frequency. The products are added (integrated) and then averaged over a complete cycle. These steps can be repeated for each value of τ between the range $-\infty$ to $+\infty$. The plot thus obtained for each value of τ is the crosscorrelation function.

We saw that the autocorrelation function contained only the frequency components present in the waveform itself. Likewise the crosscorrelation function of two waveforms contains only frequencies common to both waveforms. This makes it possible to use the crosscorrelation process as a filter. When we crosscorrelate a waveform with a sine wave, the output will contain only the component with the frequency of the sine wave.

Correlation functions for periodic and aperiodic waveforms.

For a *periodic waveform* the autocorrelation function can be written as

$$\varphi_{11}(\tau) = \frac{1}{T_1} \int_{-T_1/2}^{T_1/2} f_1(t) f_1(t+\tau) \, dt \qquad (3\text{-}111)$$

and the crosscorrelation of two periodic functions can be written as

$$\varphi_{12}(\tau) = \frac{1}{T_1} \int_{-T_1/2}^{T_1/2} f_1(t) f_2(t+\tau) \, dt \qquad (3\text{-}121)$$

For *aperiodic waveforms* the autocorrelation function can be written as

$$\varphi_{11}(\tau) = \int_{-\infty}^{\infty} f_1(t) f_1(t+\tau) \, dt \qquad (3\text{-}122)$$

and the crosscorrelation function of two aperiodic functions can be written as

$$\varphi_{12}(\tau) = \int_{-\infty}^{\infty} f_1(t) f_2(t+\tau) \, dt \qquad (3\text{-}123)$$

The formulas for the aperiodic case are the mathematical formulas defined as correlation integrals. When applied to truly periodic functions which exist over the entire range of a variable from $-\infty$ to $+\infty$ the integrals become infinite. Because the integral is the same over any period of the periodic functions, the limits of integration are reduced to include only one cycle to obtain a finite integral. Division by the length of the period is introduced to normalize and eliminate the effects of frequency as previously described. The correlation functions defined in this way permit similar operations to be performed on two different classes of signals.

Power Spectrum and Autocorrelation

We found previously that the power spectrum of periodic functions could be obtained by multiplying the Fourier transform by its complex conjugate. Assuming that a function $f_1(t)$ has a complex amplitude spectrum, $F(\omega)$, then its power spectrum is obtained from

$$I(\omega) = |F(\omega)|^2. \qquad (3\text{-}103)$$

The spectrum of the autocorrelation function of a periodic signal is proportional to the square of the magnitude of the transform of the signal as shown by the discussion of Figure 3.23. Because the square of the transform is the power spectrum, the autocorrelation function of a periodic waveform contains the same information as the power spectrum. The power spectrum contains no phase information and neither does the autocorrelation function. We can see the effect of this by considering two different periodic functions. Let us assume both functions have the same harmonic content; that is, the respective amplitudes of each of the harmonics in the two waveforms are the same. Even though the respective phases of each of the harmonics may be different, the autocorrelation function of the two waveforms will be the same (and so will the power spectrum of each be the same).

We have seen how the Fourier series can be used to determine the components (phases and amplitudes) of a periodic waveform. The Fourier series can also be used to synthesize a periodic waveform by combining Fourier components of known amplitudes and phases. The autocorrelation function of a periodic waveform can be synthesized by combining the Fourier series components with their amplitudes squared and making the phases of each of the components coincide (that is, with reference to the origin, each of the component frequencies will be in phase (cosines) as shown in Figure 3.23).

We can see that in the matched filter (Figure 3.19) the output fulfills the same conditions of (3-103); that is, the amplitude spectrum of the output of a matched filter is proportional to the autocorrelation of the input waveform. The autocorrelation function has a maximum for zero displacement (τ). For periodic functions no other value of the autocorrelation function can exceed the maximum at zero displacement. For aperiodic functions no other value can exceed or equal this maximum at zero displacement.

Convolution

The expression for the convolution of periodic functions $f_1(t)$ and $f_2(t)$ is given by

$$\rho_{12}(\tau) = \frac{1}{T_1} \int_{-T_1/2}^{T_1/2} f_1(t) f_2(\tau - t) \, dt \tag{3-124}$$

which appears quite similar to crosscorrelation. The difference between the crosscorrelation and convolution is the appearance of a negative sign for t in $f_2(\tau - t)$. Convolution therefore involves an additional step that is not involved in crosscorrelation. This step represents a folding-back operation of $f_2(t + \tau)$. The meaning of this can easily be seen from Figure 3.24.

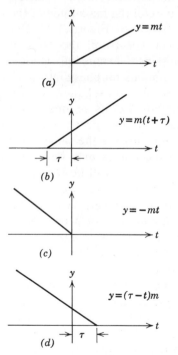

Figure 3.24 Graphs showing the addition of a constant to one variable.

We can see from Figure 3.24(a) that the line $y = mt$ goes up to the right. When we add the value τ to t in the function $y = mt$, it moves the curve to the left as shown in Figure 3.24(b). The line $y = -mt$ goes up to the left and is shown in Figure 3.24(c). Adding τ moves the line to the right as shown in Figure 3.24(d). If we just look at Figures 3.24(b) and (d), we see that they appear to be reversed representations of each other. If we consider the y axis as a mirror, then Figures 3.24(b) and (d) are mirror images of each other. The folding operation is shown in Figure 3.25 for a function represented on a transparency. Figure 3.25(a) shows the function $y = mt$. Figure 25(b) shows the displaced function $y = m(t+\tau)$. Figure 3.25(c) and (d) show the transparency being turned over and Figure 3.25(e) shows it completely turned over or the folding operation completed. Figure 3.25(f) shows the original line and the displaced folded line superimposed. Convolution is a measure of the similarity between a function $f_1(t)$ and a function f_2 that is folded and moved a distance τ with respect to $f_1(t)$ and is represented by $f_2(\tau-t)$.

def⁰.

The process of convolution (and also crosscorrelation) of two periodic functions of the same fundamental frequency retains the harmonics common to both the original given functions. The spectrum of convolution is the product of the spectra of the individual functions, but the

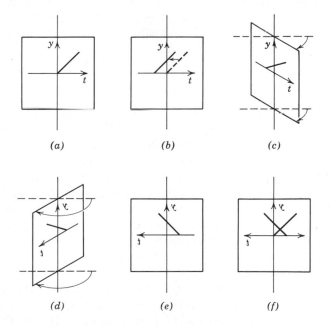

(a)   (b)   (c)

(d)   (e)   (f)

Figure 3.25   The folding back operation.

spectrum of crosscorrelation is the product of the complex conjugate of the undisplaced function with the spectrum of the displaced function, that is,

$$P_{12} = \text{convolution spectrum} = F_1(\omega)\, F_2(\omega) \qquad (3\text{-}125)$$

$$\Phi_{12} = \text{crosscorrelation spectrum} = F_1^*(\omega)\, F_2(\omega) \qquad (3\text{-}126)$$

The main difference between convolution and crosscorrelation is the folding-back process involved with one of the functions during convolution. If we crosscorrelate two functions, one of which has been folded back, we are actually doing the process of convolution.

Examination of the equations for crosscorrelation and convolution for periodic functions permits us to point out certain distinctions between them,

$$\text{crosscorrelation } \varphi_{12}(\tau) = \frac{1}{T_1} \int_{-T_1/2}^{T_1/2} f_1(t)\, f_2(t+\tau)\, dt \qquad (3\text{-}121)$$

$$\text{convolution } \rho_{12}(\tau) = \frac{1}{T_1} \int_{-T_1/2}^{T_1/2} f_1(t)\, f_2(\tau-t)\, dt \qquad (3\text{-}124)$$

The crosscorrelation of two functions could give different results depending on which of the functions is displaced. This is why in determining the crosscorrelation spectrum the complex conjugate of the undisplaced function appears, that is,

$$\Phi_{12}(\omega) = F_1^*(\omega)\, F_2(\omega) \qquad (3\text{-}127)$$

In convolution the displacement can be applied to either function provided it is the function that is folded and the spectrum is always $F_1(\omega)\, F_2(\omega)$.

Figure 3.26 shows a graphical comparison between convolution and correlation for aperiodic functions. In this example it is assumed that a function $f_1(t)$ is to be convolved and correlated with the function of $f_2(t)$. It is noted that in both convolution and correlation the function $f_2(t)$ is displaced to the left by τ producing $f_2(t+\tau)$ as shown in the displacement diagrams. The next step in the convolution process is the folding or reflection about $t = 0$ for the function $f_2(t+\tau)$ to produce $f_2(\tau-t)$. There is no corresponding step in the correlation process. This accounts for the difference between the two operations. The function $f_1(t)$ is then multiplied by $f_2(\tau-t)$ for convolution and by $f_2(t+\tau)$ for correlation. The area under the respective multiplication curves show the respective value of the convolution and correlation function for a given τ. In general the area under this curve is given by an integral of the products from $-\infty$ to $+\infty$.

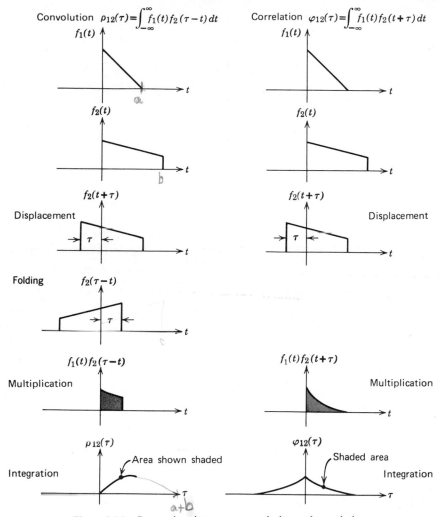

Convolution $\rho_{12}(\tau) = \int_{-\infty}^{\infty} f_1(t) f_2(\tau - t)\, dt$

Correlation $\varphi_{12}(\tau) = \int_{-\infty}^{\infty} f_1(t) f_2(t + \tau)\, dt$

$f_1(t)$

$f_1(t)$

$f_2(t)$

$f_2(t)$

Displacement

$f_2(t+\tau)$

$f_2(t+\tau)$

Displacement

Folding $f_2(\tau - t)$

Multiplication $f_1(t) f_2(\tau - t)$

$f_1(t) f_2(t + \tau)$ Multiplication

Integration $\rho_{12}(\tau)$

Area shown shaded

$\varphi_{12}(\tau)$ Shaded area

Integration

Figure 3.26 Comparison between convolution and correlation.

The result of this integration is a function of τ. The integration graphs indicate the integral (area under multiplication curve) plotted for each value of τ giving the respective convolution and correlation functions.

The convolution function can be obtained optically by a similar method to that used to obtain the correlation function. The additional operation of folding required in convolution can be accomplished by a 180-degree rotation of a film. Thus by a 180-degree rotation of the film to represent the folding operation the convolution function can be

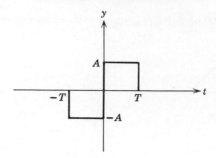

Figure 3.27 Double square waveshape.

obtained using the same basic optical configuration as for crosscorrelation. Convolution is an extremely important operation in the analysis of linear systems; for example, the output of a linear system can be determined by convolving the given input with the impulse response of the system.

Convolution Functions and Fourier Transforms

The Fourier transform of the double square wave pulse of Figure 3.27 can be determined in a straightforward manner using (3-81)

$$F(\omega) = \int_{-\infty}^{\infty} f(t) e^{-j\omega t} \, dt \qquad (3\text{-}81)$$

This can be shown as follows:
The square waveshape function can be written as

$$y = f(t) = -A$$

when $-T \leqslant t \leqslant 0$ and

$$y = f(t) = A$$

when $0 \leqslant t \leqslant T$ at all other values $y = 0$. Then

$$F(\omega) = \int_{-T}^{0} (-A) e^{-j\omega t} \, dt + \int_{0}^{T} (A) e^{-j\omega t} \, dt \qquad (3\text{-}128)$$

$$= \left[\frac{+A}{j\omega} e^{-j\omega t} \right]_{-T}^{0} + \left[\frac{-A}{j\omega} e^{-j\omega t} \right]_{0}^{T}$$

$$= \left[\frac{A}{j\omega} e^{-j\omega 0} - \frac{A}{j\omega} e^{j\omega T} \right] + \left[\frac{-A}{j\omega} e^{-j\omega T} - \frac{-A}{j\omega} e^{-j\omega 0} \right]$$

$$= \frac{A}{j\omega} - \frac{A}{j\omega} e^{j\omega t} - \frac{A}{j\omega} e^{-j\omega t} + \frac{A}{j\omega}$$

$$= \frac{2A}{j\omega} - \frac{A}{j\omega} \left[e^{j\omega T} + e^{-j\omega T} \right]$$

$$= \frac{2A}{j\omega} - \frac{2A}{j\omega} \left[\frac{e^{j\omega T} + e^{-j\omega T}}{2} \right] = \frac{2A}{j\omega} \left[1 - \cos \omega T \right]$$

$$= \frac{4A}{j\omega} \left[\tfrac{1}{2} - \tfrac{1}{2} \cos \omega T \right] = - \frac{4AjT}{\omega} \frac{\sin^2 (\omega T/2)}{T}$$

$$= -2AjT \frac{\sin^2 (\omega T/2)}{\omega T/2} \tag{3-129}$$

$$= -AjT^2\omega \left[\frac{\sin (\omega T/2)}{\omega T/2} \right]^2 \tag{3-130}$$

Appendix 2B is a table of values for the function $[(\sin x)/x]^2$ Figure 3.28 shows the double square waveshape and its Fourier transform as de-termined by (3-130).

Another way of obtaining the same result shown in Figure 3.28 is to consider the Fourier transform of a rectangular pulse (Figure 3.29(a)) and the two pulse spikes (Figure 3.29(b)) reproduced from the table of Fourier transforms given in Appendix 13. When the rectangular pulse (Figure 3.30(a)) is convolved with the two pulse spikes (Figure 3.30(b)), the result is the double square waveshape of Figure 3.30(c). The Fourier transform of the rectangular pulse of Figure 3.29(a) and the Fourier trans-form of the two pulse spikes of Figure 3.29(b) are shown in Figure 3.31(a) and (b). Figure 3.31(c) is the product of these two Fourier transforms. The convolution theorem states that the product of two Fourier trans-forms (spectra) is equal to the Fourier transform of the two functions convolved with each other. Figure 3.32 shows a graphical representation of the convolution theorem. This theorem is extremely useful in optical data processing since the product of two Fourier transforms (spectra) is relatively easy to obtain optically.

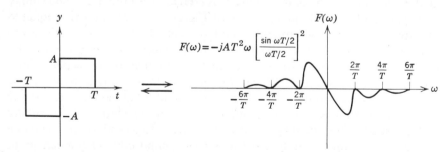

$$F(\omega) = -jAT^2\omega \left[\frac{\sin \omega T/2}{\omega T/2} \right]^2$$

Figure 3.28 Double square wave and its Fourier transform.

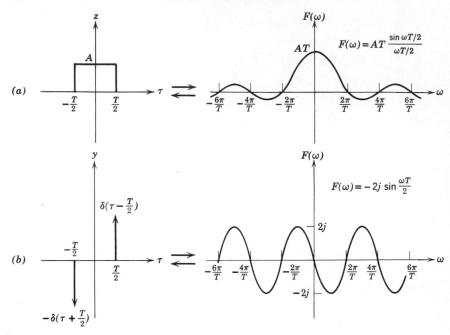

Figure 3.29 Two waveshapes and their Fourier transforms.

We have seen that the convolution of two functions has a spectrum that is equivalent to multiplying the respective spectra of the two functions as shown in Figure 3.32. We could also have obtained the spectrum of the double square waveshape of Figure 3.27 by making use of the fact that the spectrum of the sum of two functions is equivalent to the sum of the respective spectra of the two functions. Figure 3.33 shows the procedure graphically. The double square waveshape of Figure 3.27 can be considered to be the sum of two rectangular pulses; that is, the sum of pulses in Figures 3.33(a) and (b) equals the waveform of Figure 3.33(c) and is the same as the double pulse waveshape of Figure 3.27. The sum of the spectra of the individual pulses of Figures 3.33(a) and (b) is equal to the spectrum of the two functions added together. Figure 3.33(a) shows a rectangular pulse and its spectrum. In Appendix 14 it is shown that a shift in the time domain is equivalent to multiplication by a phase shift in the frequency domain. We can therefore redraw the rectangular pulse of Figure 3.30(a) shifted in time by an amount $T/2$ (Figure 3.33(a)) and its corresponding spectrum multiplied by a phase shift $e^{-j(\omega T/2)}$. Figure 3.33(b) shows the rectangular pulse of Figure 3.30(a) shifted in time by an amount $-T/2$ and made to have a negative amplitude. The spectrum of the negative rectangular pulse shown in Figure 3.33(b) is therefore the negative of that of Figure 3.33(a) and multiplied by a phase shift of

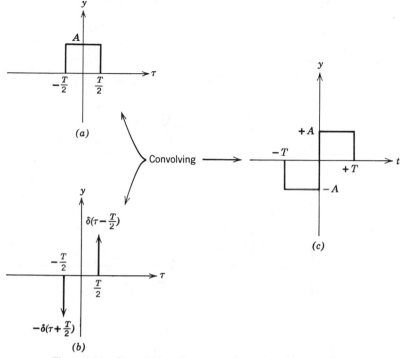

Figure 3.30   Convolution of a rectangular pulse with two spikes.

$e^{j(\omega T/2)}$. Figure 3.33(c) shows that the sum of the two spectra of Figures 3.33(a) and (b) added together result in

$$F(\omega) = F_A(\omega) + F_B(\omega)$$

$$F(\omega) = -AT \left[\frac{\sin(\omega T/2)}{\omega T/2}\right] \left[e^{j(\omega T/2)} - e^{-j(\omega T/2)}\right]$$

$$F(\omega) = -AT \left[\frac{\sin(\omega T/2)}{\omega T/2}\right] 2j \left[\frac{e^{j(\omega T/2)} - e^{-j(\omega T/2)}}{2j}\right]$$

$$F(\omega) = -AT2j \left[\frac{\sin(\omega T/2)}{\omega T/2}\right] \left[\sin\frac{\omega T}{2}\right]$$

$$F(\omega) = -AT2j \left[\frac{\sin(\omega T/2)}{\omega T/2}\right] \frac{\omega T}{2} \left[\frac{\sin(\omega T/2)}{\omega T/2}\right]$$

$$= -AT^2 j\omega \left[\frac{\sin(\omega T/2)}{\omega T/2}\right]^2 \qquad\qquad (3\text{-}130)$$

which is the same result (3-130) obtained by the convolution process previously described.

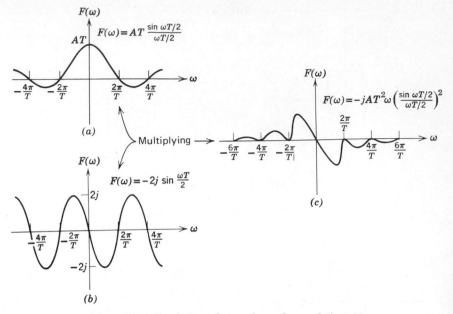

Figure 3.31 Fourier transforms of waveforms of Fig. 3.30.

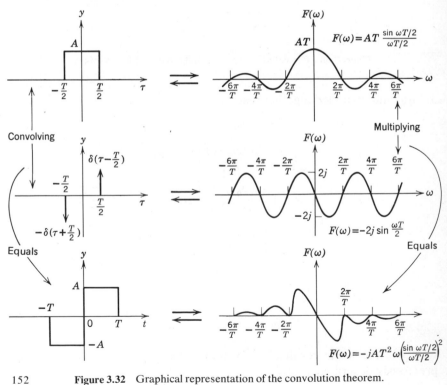

152   Figure 3.32 Graphical representation of the convolution theorem.

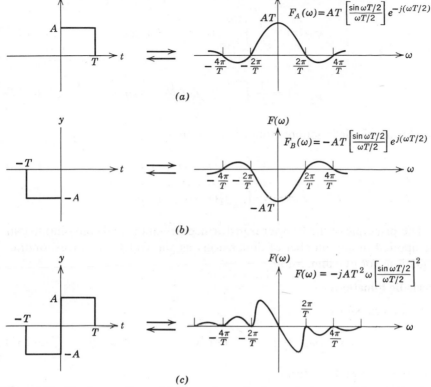

Figure 3.33   Fourier transform of the sum of two signals equals the sum of the two Fourier transforms.

TWO-DIMENSIONAL FOURIER TRANSFORM, CORRELATION, AND CONVOLUTION

The Fourier transform in one dimension was given by

$$F(\omega) = \int_{-\infty}^{\infty} f(t) e^{-j\omega t} \, dt \tag{3-81}$$

and the inverse Fourier transform in one dimension was given by

$$f(t) = \frac{1}{2\pi} \int_{-\infty}^{\infty} F(\omega) e^{j\omega t} \, d\omega \tag{3-82}$$

We have also seen that correlation and convolution are quite similar to each other. In one dimension the crosscorrelation and convolution equations for periodic functions were given by

$$\varphi_{12}(\tau) = \frac{1}{T_1} \int_{-T_1/2}^{T_1/2} f_1(t) f_2(t+\tau) \, dt \qquad (3\text{-}121)$$

$$\rho_{12}(\tau) = \frac{1}{T_1} \int_{-T_1/2}^{T_1/2} f_1(t) f_2(\tau-t) \, dt \qquad (3\text{-}124)$$

and for aperiodic functions from Figure 3.26

$$\varphi_{12}(\tau) = \int_{-\infty}^{\infty} f_1(t) f_2(t+\tau) \, dt \qquad (3\text{-}123)$$

$$\rho_{12}(\tau) = \int_{-\infty}^{\infty} f_1(t) f_2(\tau-t) \, dt \qquad (3\text{-}131)$$

The principle of the Fourier transform, correlation, and convolution can be applied in any number of dimensions as shown by the corresponding equations for two dimensions.

Periodic Functions

Fourier Transform

$$F(\omega_x, \omega_y) = \int_{-\infty}^{\infty} \int_{-\infty}^{\infty} f(x, y) \, e^{-j(\omega_x x + \omega_y y)} \, dx \, dy \qquad (3\text{-}132)$$

Inverse Fourier Transform

$$f(x, y) = \frac{1}{(2\pi)^2} \int_{-\infty}^{\infty} \int_{-\infty}^{\infty} F(\omega_x, \omega_y) \, e^{j(\omega_x x + \omega_y y)} \, d\omega_x \, d\omega_y \qquad (3\text{-}133)$$

Crosscorrelation

$$\varphi_{12}(\tau_x, \tau_y) = \frac{1}{T_x, T_y} \int_{-T_x/2}^{T_x/2} \int_{-T_y/2}^{T_y/2} f_1(x, y) f_2(x+\tau_x, y+\tau_y) \, dx \, dy \qquad (3\text{-}134)$$

Convolution

$$\rho_{12}(\tau_x, \tau_y) = \frac{1}{T_x, T_y} \int_{-T_x/2}^{T_x/2} \int_{-T_y/2}^{T_y/2} f_1(x, y) f_2(\tau_x-x, \tau_y-y) \, dx \, dy \qquad (3\text{-}135)$$

Aperiodic Functions

Fourier transform

$$F(\omega_x, \omega_y) = \int_{-\infty}^{\infty} \int_{-\infty}^{\infty} f(x, y) \, e^{-j(\omega_x x + \omega_y y)} \, dx \, dy \qquad (3\text{-}132)$$

Inverse Fourier Transform

$$f(x, y) = \frac{1}{(2\pi)^2} \int_{-\infty}^{\infty} \int_{-\infty}^{\infty} F(\omega_x, \omega_y) e^{j(\omega_x x + \omega_y y)} d\omega_x \, d\omega_y \quad (3\text{-}133)$$

Crosscorrelation

$$\varphi_{12}(\tau_x, \tau_y) = \int_{-\infty}^{\infty} \int_{-\infty}^{\infty} f_1(x, y) f_2(x + \tau_x, x + \tau_y) \, dx \, dy \quad (3\text{-}136)$$

Convolution

$$\rho_{12}(\tau_x, \tau_y) = \int_{-\infty}^{\infty} \int_{-\infty}^{\infty} f_1(x, y) f_2(\tau_x - x, \tau_y - y) \, dx \, dy$$

$$(3\text{-}137)$$

$$\mathcal{F}(f_1(x)) = F_1$$

$$\mathcal{F}(f_2(x)) = F_2$$

$$\mathcal{F}(f_1(x) f_2(x)) = F_1 * F_2$$

$$\mathcal{F}^{-1}(F_1 F_2) = f_1(x) * f_2(x)$$

4

The Fourier Transform by Diffraction of Light

This chapter presents a mathematical analysis of the approximations required to obtain the Fourier transform representation of an ideal lens. The physical significance of approximations and the variations from ideal results produced by neglecting terms in the mathematical formulation are demonstrated. Approximations are considered in terms of output signals in an optical spectrum analyzer, optical imaging system, and an optical correlator system.

We have implied that the light amplitude distribution in the back focal plane of a lens represents a spectrum of the light amplitude distribution in the front focal plane. We will now show that there is, in fact, a Fourier transform relationship between the front focal plane and the back focal plane light amplitude distribution of a lens. Indeed the light amplitude distribution in the back focal plane of a lens is proportional to the two-dimensional Fourier transform of the light amplitude distribution of a two-dimensional object (transmission function) in the front focal plane of the lens. We will now analyze aperture limitations that are required to obtain the Fourier transform representation of a focused diffraction pattern.

The optical Fourier transform is primarily based on the phenomena of light diffraction. The lens in effect focuses the diffraction pattern. The focused diffraction pattern then turns out to be a good approximation of the Fourier transform of the object causing the diffraction. We will first consider an ideal lens ignoring lens aberrations and diffraction effects at the edge of the lens. We know these effects will modify our results but we shall assume that we can restrict the range of variables to a region in which ideal assumptions are valid within experimental accuracies.

INITIAL ASSUMPTIONS

Consider Figure 4.1 which shows a plane Z that is perpendicular to planes F, F', L, Z'. The line formed by the intersection of planes Z and Z' is perpendicular to planes F, F', and L. In plane L we place a lens (as shown in Figure 4.2) so that its optical axis is the line formed by the intersection of planes Z and Z'. We also designate the optical axis as the z axis. We have seen that rays parallel to the optical axis will be focused to a point in the focal plane (on the optical axis). Figure 4.2(a) considers only the rays in the Z plane (to permit simplified drawings to be used) emitted from plane F parallel to the optical axis. Figure 4.2(a) shows the rays in a perspective drawing while Figure 4.2(b) shows the plane of the paper to be the Z plane.

We shall initially assume we are dealing with an ideal lens, the properties of which can be summarized as follows:

1. The lens (in plane L, Figure 4.2) is so thin that it can be represented by the plane L (perpendicular to the optical axis) at which all refraction is assumed to take place.

2. Principal rays will not be deviated; that is, principal rays are those passing through the intersection of the optical axis of the lens and the lens plane L (which is point 0).

3. The distance between plane L and plane F' in Figure 4.2 is one focal length, f, so that all parallel incident rays will be focused to the same point in the back focal plane as struck by the associated principal ray.

4. A plane P perpendicular to the incident parallel rays to a lens will cut these incident parallel rays in such a manner that the optical path length of any of the parallel rays from the plane P to the common focal point in the plane F' will be the same.

Geometric Considerations for Optical Path Length

In Figure 4.2 the principal ray and the optical axis of the lens coincided. The optical axis is the line formed by the intersection of planes Z and Z'.

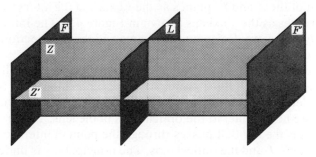

Figure 4.1 Designation of planes.

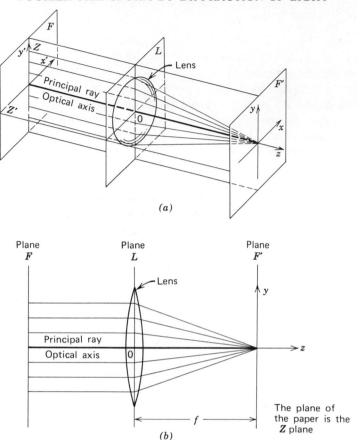

Figure 4.2 Representation of the focusing action of a lens in plane Z, (collimated light parallel to optical axis of the lens).

We designate the optical axis as the z axis. We further designate the intersection of the Z and F' planes as the y axis and the intersection of the Z' and F' planes as the x axis as shown in Figure 4.2. The intersection of the Z and F planes we designate the y' axis and the intersection of the Z' and F planes we designate the x' axis. Since planes F and F' are parallel the x and x' axes are parallel and the y and y' axes are parallel. Let us now consider the effect of the principal ray not coincident with the optical axis of the lens. Figure 4.3 shows parallel rays incident to the lens but these rays are not parallel to the optical axis of the lens. The principal ray of these rays is the ray that passes through the point of intersection of the plane of the lens, L and the optical axis. The principal ray is undeviated by the lens. The rays parallel to the principal ray will be brought to a focus in

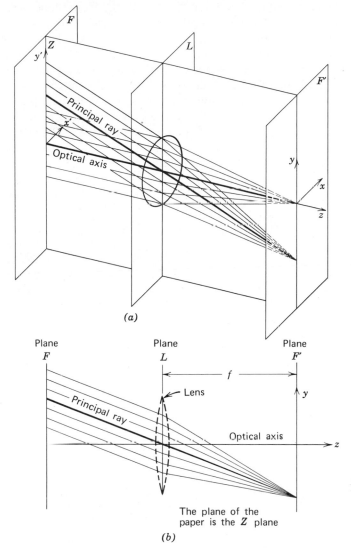

(a)

(b)

Figure 4.3 Focusing action of a lens in plane Z for collimated light not parallel to optical axis.

plane F' (which is a focal length from plane L) but not on the optical axis of the lens.

The optical path length of any ray parallel to the principal ray will be the same from a plane perpendicular to the principal ray to the focal point.

In Figure 4.2 the optical path length of each ray from plane F to plane F' is the same; that is, all path lengths (of rays parallel to the principal ray) from a plane perpendicular to the principal ray to the focal point are equal. The optical path lengths from plane F in Figure 4.3 to plane F' will *not* be the same (for the rays not parallel to the optical axis) since plane F is not perpendicular to the principal ray. Figure 4.4 shows a plane P which is perpendicular to the principal ray and which intersects plane F at the point of intersection of plane F with the optical axis. For simplicity in the diagrams only the rays in the Z plane will be shown. Plane P intersects plane F along the x' axis and Figure 4.4(b) shows the intersection of the P plane with the Z plane. The principal ray makes an angle θ with the optical axis of the lens and therefore (since plane P is perpendicular to the principal ray and plane F is perpendicular to the optical axis) planes F and P are also inclined to each other at an angle θ.

We have made the intersection of the Z and F' planes the y axis, with

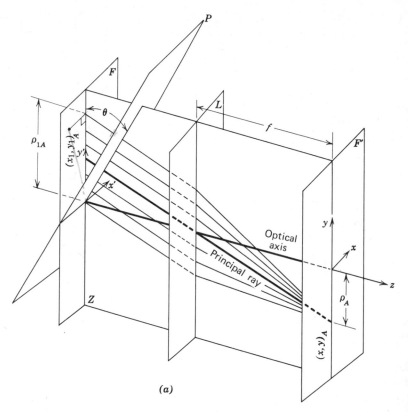

(a)

Figure 4.4 (a) Intersecting optical axis at plane F and perpendicular to principal ray.

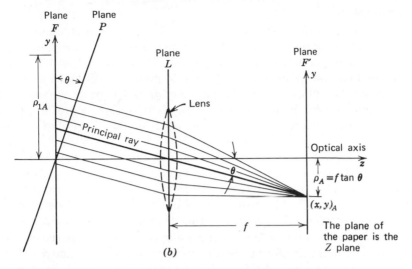

Figure 4.4 (*b*) The plane of the paper is the Z plane.

the origin of the *y* axis at the intersection of the optical axis (*z* axis) with the *F'* plane. For any principal ray there will be a point (*x*, *y*) in the *F'* plane at which all rays parallel to this principal ray will be focused. Let us consider a focal point that is on the negative *y* axis in plane *F'* as shown in Figure 4.4(a) which we shall designate (*x*, *y*)$_A$. The principal ray for the point (*x*, *y*)$_A$ passes through the point of intersection of the optical axis with the *L* plane. The plane *P* is drawn perpendicular to the principal ray. Any ray leaving plane *P* parallel to this principal ray will travel the same optical path distance to the focal point (*x*, *y*)$_A$ in plane *F'* as the principal ray. Again, for simplicity in drawing, Figure 4.4(a) shows only rays in the Z plane parallel to the principal ray. All of these rays will travel the same optical path distance from plane *P* to *F'*.

Definition of ρ

In the *F'* plane the point (*x*, *y*)$_A$ is a distance ρ_A from the optical axis as shown in Figure 4.4(a). Because the principal axis and the optical axis determine the Z plane, the distance (ρ_A) that the point (*x*, *y*)$_A$ is from the optical axis is always measured along the line of intersection between the Z and *F'* planes. The distance between the *L* and *F'* planes is a focal length, *f*; that is, the length of the optical axis between planes *L* and *F'* is *f*. From this we can see that

$$\rho = f \tan \theta. \tag{4-1}$$

Definition of ρ_1

Let us now consider any point $(x_1, y_1)_A$ in plane F and determine its contribution to the point $(x, y)_A$ in the F' plane. From Figure 4.4(a) we see that the Z plane contains the principal ray. If in the F plane we drop a perpendicular from point $(x_1, y_1)_A$ to the Z plane, the distance (in the F plane) from the optical axis to the intersection of the perpendicular with the Z plane we designate ρ_{1A}; that is, ρ_{1A} is the projection of the point $(x_1, y_1)_A$ on the Z plane. In Figure 4.4(a) ρ_{1A} is on the y' axis and ρ_A is on the y axis.

We can consider another point $(x, y)_B$ in the F' plane as shown in Figure 4.5. In the F' plane the point $(x, y)_B$ is a distance ρ_B from the optical axis. The principal ray to point $(x, y)_B$ and the optical axis determine a plane Z'. The projection of point $(x_1, y_1)_A$ in plane F to the intersection of plane Z' determines the distance ρ_{1B}; that is, ρ_{1B} is the distance from the optical axis to the intersection of the perpendicular from $(x_1, y_1)_A$ to the line formed by the intersection of the Z' and F planes.

From Figures 4.4(a) and 4.5 we can see that the distance ρ_m in plane F'

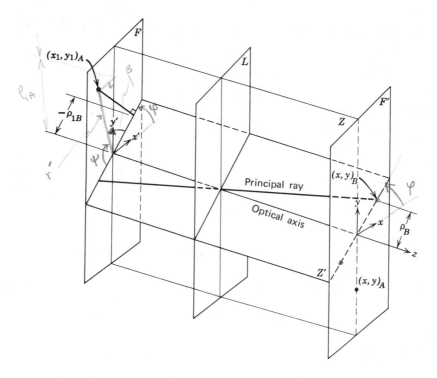

Figure 4.5 Location of Z' plane determined by locations of point in plane F'.

is the distance of any point $(x, y)_m$ from the optical axis. ρ_{1m} in plane F is the projected distance of any point $(x_1, y_1)_m$ from the optical axis on the line formed by intersection of the F plane and a Z plane. The distance ρ_m in the F' plane represents a different quantity than ρ_{1m} in the F plane. In plane F' the distance ρ_m always represents the distance from the optical axis to any point $(x, y)_m$. In plane F distance ρ_{1m} is the projection of any point $(x_1, y_1)_m$ in plane F on the line formed by the intersection of plane F with the plane through the optical axis and a point $(x, y)_m$ in plane F'. The value of ρ_{1m}, therefore, depends on the point $(x, y)_m$ being considered in plane F'. The value of ρ_{1m} will be a function of $x_1, y_1, x,$ and y, but ρ_m will depend only on x and y. The distance from the optical axis to the point being considered in plane F' will always be a positive quantity since it represents a distance. ρ_{1m} represents a coordinate of the intersection with the F plane of a ray parallel to the principal ray for example, in Figure 4.4(a) the intersection of the perpendicular from the point being considered $(x_1, y_1)_A$ to the intersection of planes F and Z determines ρ_{1A}. The plane P being perpendicular to the principal ray represents a plane from which rays parallel to the principal ray will all travel equal path distances to the focal point $(x, y)_A$. From plane P we therefore have to add a path length to reach plane F where the ray from $(x_1, y_1)_A$ (parallel to the principal ray) is being radiated. For the point $(x, y)_A$ in plane F' the sign and value of ρ_{1m} will depend on the location of the point (x_1, y_1) being considered in plane F. We can specify a general rule for the sign of ρ_{1m} by requiring that for any point $(x, y)_m$ in plane F' for which we have to add path lengths to rays parallel to the principal ray (when going from plane P to the point in plane F) ρ_{1m} will be positive. Thus ρ_{1m} is positive when the points $(x_1, y_1)_m$ and $(x, y)_m$ are on opposite sides of the optical axis (and ρ_{1m} is negative when they are on the same side of the optical axis as shown in Figure 4.6).

Figure 4.6 shows a plane P' added to the planes shown in Figure 4.5. The plane P' is drawn perpendicular to the principal ray from point $(x, y)_B$ and passes through the point of intersection of the optical axis with the F plane. Because the point $(x, y)_B$ is shown to have a positive y coordinate value (in plane F'), the value of ρ_{1B} is negative (since it also has a positive y' coordinate value). From Figure 4.6 we can verify this by seeing that going from plane P' (along the ray that is parallel to the principal ray and passes through ρ_{1B}) to arrive at plane F we must subtract the path length from plane P' to F; that is, the value of ρ_{1B} should be negative. Figure 4.7 shows how the value of ρ_{1m} is dependent on the location of the point in plane F' being considered (the P and P' planes shown in Figures 4.4 and 4.6, respectively, have been omitted for clarity in the drawings).

see
Fig. 4-4
4-24
Pg 167

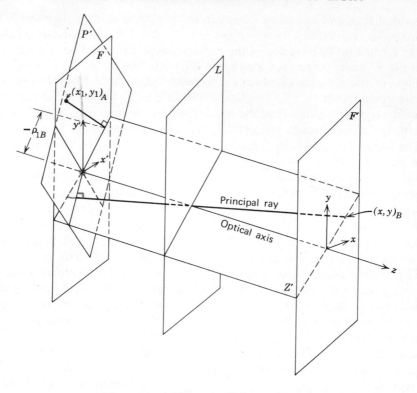

Figure 4.6 Addition of a P' plane to Fig. 4.5.

Relationships Involving ρ and ρ_1

Let us now examine the geometric relationship of the various lines in planes F and F'. From Figure 4.5 we can draw the lines in the F' plane as shown in Figure 4.8(a). From Figure 4.8(a) we can write the following relationships:

$$\cos \varphi = \frac{x}{\rho_B} \tag{4-2}$$

$$\sin \varphi = \frac{y}{\rho_B} \tag{4-3}$$

$$\rho_B = (x^2 + y^2)^{1/2} \tag{4-4}$$

or more generally for any ρ in plane F'

$$\cos \varphi = \frac{x}{\rho} \tag{4-5}$$

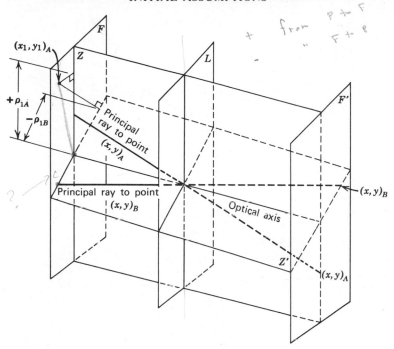

Figure 4.7 Dependent of ρ_1 on location of point in plane F'.

$$\sin \varphi = \frac{y}{\rho} \tag{4-6}$$

$$\rho = (x^2 + y^2)^{1/2}. \tag{4-7}$$

From Figure 4.5 we can also draw the line in the F plane as shown in Figure 4.8(b). The intersection of the Z' and F planes forms the ρ axis centered on the optical axis. The angle φ is the angle the ρ_1 axis makes with the x axis. The point $(x_1, y_1)_A$ is a distance r' from the optical axis. Then

$$r' = (x_1^2 + y_1^2)^{1/2}. \tag{4-8}$$

If we assume that the line r' is at an angle α' with respect to the x' axis, then

$$\varphi = \alpha' - \beta'. \tag{4-9}$$

From Figure 4.8(b) we see that

$$\alpha' = 180° - \psi \tag{4-10}$$

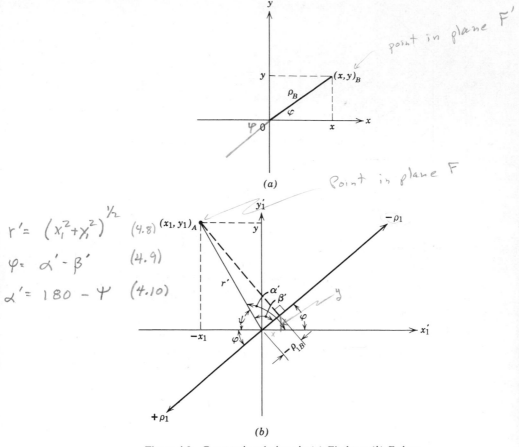

$$r' = \left(x_1^2 + y_1^2\right)^{1/2} \quad (4.8)$$

$$\varphi = \alpha' - \beta' \quad (4.9)$$

$$\alpha' = 180 - \psi \quad (4.10)$$

point in plane F'

Point in plane F

(a)

(b)

Figure 4.8 Geometric relations in (a) F' plane, (b) F plane.

then from the trigonometric relationship

$$\cos \psi = -\cos (180° - \psi) \quad (4\text{-}11)$$

$$\cos \psi = -\cos (\alpha') \quad (4\text{-}12)$$

and since

$$\cos \psi = -\frac{x_1}{r'} \quad (4\text{-}13)$$

then

$$\cos \psi = -\cos \alpha' = -\frac{x_1}{r'} \quad (4.14)$$

$$\cos \alpha' = \frac{x_1}{r'} \quad (4\text{-}15)$$

and

$$\sin \psi = \frac{y_1}{r'} \qquad (4\text{-}16)$$

$$\sin \alpha' = \frac{y_1}{r'}. \qquad (4\text{-}17)$$

From Figure 4.8(b) we see that ρ_{1B} is negative and therefore

$$-\rho_{1B} = r' \cos \beta'. \qquad (4\text{-}18)$$

The value $-\rho_{1B}$ is the projection of point $(x_1, y_1)_{A_1}$ on the intersection of the Z' and F planes that permits finding the optical path from any point $(x_1, y_1)_A$ in plane F to point $(x, y)_B$ in plane F'.

$$\rho_{1B} = -r' \cos (\alpha' - \varphi) \qquad (4\text{-}19)$$

Using the trigonometric relationship

$$\cos (x - y) = \cos x \cos y + \sin x \sin y \qquad (4\text{-}20)$$

(4-19) can be written

$$\rho_{1B} = -r' (\cos \alpha' \cos \varphi + \sin \alpha' \sin \varphi) \qquad (4\text{-}21)$$

$$\rho_{1B} = -r' \left(+\frac{x_1}{r'} \cos \varphi + \frac{y_1}{r'} \sin \varphi \right) \qquad (4\text{-}22)$$

$$= -x_1 \cos \varphi - y_1 \sin \varphi \qquad (4\text{-}23)$$

Substituting

$$\cos \varphi = \frac{x}{\rho_B} \qquad \text{from } 4.8\,(a) \qquad (4\text{-}2)$$

and

$$\sin \varphi = \frac{y}{\rho_B} \qquad (4\text{-}3)$$

and

$$\rho_B = (x^2 + y^2)^{1/2} \qquad (4\text{-}4)$$

in (4-23) we have

$$\rho_{1B} = \frac{-x_1 x}{\rho_B} - \frac{y_1 y}{\rho_B} = \frac{-xx_1 - yy_1}{(x^2 + y^2)^{1/2}} \qquad (4\text{-}24)$$

from which we can see that ρ_B is dependent on x and y, and ρ_{1B} is dependent on x, y, x_1, and y_1 as our previous discussion indicated. We can rewrite (4-24) for any ρ_{1m} as (dropping subscript m)

pg 163

$$\rho_1 = \frac{-xx_1 - yy_1}{(x^2 + y^2)^{1/2}}. \qquad (4\text{-}25)$$

Determination of Optical Path Length from Any Point in Plane F to a Point in Plane F'

We shall now use Figure 4.9 to determine an expression for the optical path length from any point $(x_1, y_1)_A$ in plane F to point $(x, y)_B$ in plane F'. The principal ray of Figure 4.6 is shown as line $\overline{CO}$ in Figure 4.9. Plane P' is perpendicular to this principal ray and all others parallel to it (such as $\overline{DE}$). The optical path length r for the principal ray is given by the geometric length

$$r = l_1 + l_2 + l_3 \tag{4-26}$$

From Figure 4.9

$$l_1 = (f^2 + \rho_B^2)^{1/2}. \tag{4-27}$$

By similar triangles l_2 is found to be

$$\frac{l_2}{f} = \frac{d}{l_1} \quad \Longrightarrow \quad \frac{l_2}{d} = \frac{f}{l_1} \tag{4-28}$$

$$l_2 = \frac{df}{l_1}. \tag{4-29}$$

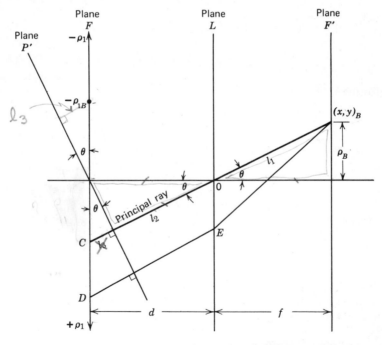

Figure 4.9 Geometry for optical pathlength for Fig. 4.6. The plane of the paper is the Z' plane.

The optical path length for the principal ray from plane P' to point $(x, y)_B$ in plane F' is $l_1 + l_2$ from 4.27

$$l_1 + l_2 = l_1 + \frac{df}{l_1} = \frac{l_1^2 + df}{l_1} = \frac{f^2 + \rho_B^2 + df}{(f^2 + \rho_B^2)^{1/2}}. \tag{4-30}$$

We have assumed that the optical path length for all rays parallel to the principal ray will be equal from a plane perpendicular to the principal ray to the intersection of the principal ray with the focal plane F'. The path length $(l_1 + l_2)$ is therefore the same for every ray parallel to the principal ray. The optical path length r for the principal ray from plane F to plane F' can be written as

$$r = l_1 + l_2 + l_3 = \frac{f^2 + \rho_B^2 + df}{(f^2 + \rho_B^2)^{1/2}} + l_3. \tag{4-31}$$

From Figure 4.9 the length l_3 can be expressed

$$l_3 = \rho_1 \sin \theta \tag{4-32}$$

We can also see from Figure 4.9 that

$$\sin \theta = \frac{\rho_B}{l_1} = \frac{\rho_B}{(f^2 + \rho_B^2)^{1/2}} \tag{4-33}$$

then

$$l_3 = \frac{\rho_1 \rho_B}{(f^2 + \rho_B^2)^{1/2}} \tag{4-34}$$

and since

$$r = l_1 + l_2 + l_3 \tag{4-26}$$

then

$$r = \frac{f^2 + \rho_B^2 + df}{(f^2 + \rho_B^2)^{1/2}} + \frac{\rho_1 \rho_B}{(f^2 + \rho_B^2)^{1/2}} \tag{4-35}$$

$$r = \frac{f^2 + \rho_B^2 + df + \rho_1 \rho_B}{(f^2 + \rho_B^2)^{1/2}} \tag{4-36}$$

Substituting the values of ρ_1 and ρ_B from (4-4) and (4-25) in (4-36), we obtain

$$r = \frac{f^2 + x^2 + y^2 + df + \left[\dfrac{-xx_1 - yy_1}{(x^2 + y^2)^{1/2}}\right][(x^2 + y^2)^{1/2}]}{(f^2 + x^2 + y^2)^{1/2}} \tag{4-37}$$

$$r = \frac{f^2 + x^2 + y^2 + df - xx_1 - yy_1}{(f^2 + x^2 + y^2)^{1/2}} \tag{4-38}$$

Equation (4-38) gives the optical path length from any point $(x_1, y_1)_A$ in plane F (a distance d in front of the plane of the lens L) to a point $(x, y)_B$ in the back focal plane F'. Using more general notation (4-38) gives the optical path length from a point (x_1, y_1) in plane F to a point (x, y) in plane F'. To simplify further discussion we will write (4-38) as

$$r = R(x, y) - \alpha x_1 - \beta y_1 \tag{4-39}$$

where

$$R(x, y) = \frac{f^2 + x^2 + y^2 + df}{(f^2 + x^2 + y^2)^{1/2}} \tag{4-40}$$

$$\alpha = \frac{x}{(f^2 + x^2 + y^2)^{1/2}} \tag{4-41}$$

$$\beta = \frac{y}{(f^2 + x^2 + y^2)^{1/2}} \tag{4-42}$$

SUMMARY OF FOCAL PROPERTIES OF AN IDEAL LENS

We have thus far considered a point $(x_1, y_1)_A$ in plane F to radiate light in all directions. We have considered only the portion of the radiated light propagating in directions at angle θ with respect to the normal n to plane F. The light rays representing these directions are shown in Figure 4.10 to form a cone, the vertex at point $(x_1, y_1)_A$, and making an angle θ with the normal to that point. For each of these rays a parallel principal ray can be drawn (through the point of intersection of the optical axis with plane L). Each principal ray that is parallel to an element of the cone formed around the normal to plane F at point $(x_1, y_1)_A$ will make an angle θ with respect to the optical axis. Each ray is therefore focused to a point on a ring of radius

$$\rho = f \tan \theta \tag{4-1}$$

where the corresponding principal ray intersects the back focal plane F'. The optical path length from point $(x_1, y_1)_A$ to each point on the ring is given by (4-39). This equation holds for any point A in plane F. We can state general focal properties for an ideal lens as follows:

1. Light radiated from all points in plane F in directions making an angle θ with respect to the normal (to plane F) is focused to a ring of radius

$$\rho = f \tan \theta \tag{4-1}$$

in the back focal plane F'.

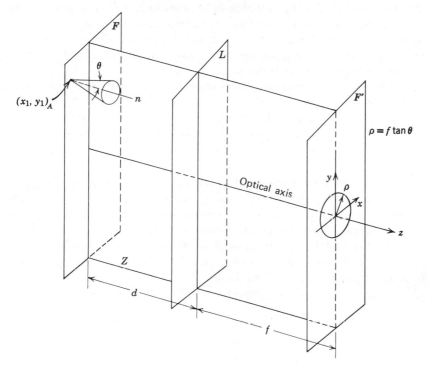

Figure 4.10 Focusing of a cone of light.

2. The optical path length r from any point $(x_1, y_1)_A$ in plane F to any point $(x, y)_B$ in the back focal plane F' is given by (4-39).

A relation between ρ_B and θ can also be obtained from Figure 4.9 and equations (4-4) and (4-27), that is,

$$\cos \theta = \frac{f}{l_1} = \frac{f}{(f^2 + \rho_B^2)^{1/2}} \tag{4-43}$$

$$\rho_B = (x^2 + y^2)^{1/2} \tag{4-4}$$

$$l_1 = (f^2 + \rho_B^2)^{1/2} \tag{4-27}$$

or

$$\cos \theta = \frac{f}{(f^2 + x^2 + y^2)^{1/2}}. \tag{4-44}$$

DIFFRACTION FORMULA

Since we will consider only light distributions on plane surfaces we can use the Rayleigh-Sommerfeld diffraction formula. In rectangular coordinates this diffraction formula has the form

$$A'(x, y, z) = \frac{-1}{2\pi} \int\int A(x_1, y_1) \frac{e^{jkr}}{r} \left[jk - \frac{1}{r} \right] \cos \theta \, dx_1 \, dy_1 \tag{4-45}$$

Equation (4-45) gives the complex light amplitude $A'(x, y, z)$ at any point in space due to a monochromatic coherent light distribution $A(x_1, y_1)$, where (x_1, y_1) denotes any point in plane F [$(x_1, y_1)_A$ denotes point A in plane F]. We can relate the terms in the diffraction formula (4-45) by referring to Figure 4.11 as follows:

1. $A(x_1, y_1)$ is the complex amplitude of monochromatic light for all points in plane F, that is, $z = 0$.

2. $A'(x, y, z)$ is the complex amplitude of light produced by $A(x_1, y_1)$ at a point $(x, y, z)_{A'}$ in space — that is, for z not in plane F, $z > 0$.

3. r is the distance from a point $(x_1, y_1)_A$ in plane F to the point $(x, y, z)_{A'}$.

4. θ is the angle between r and n, where r is directed from $(x_1, y_1)_A$ to $(x, y, z)_{A'}$ and n is the normal to the plane F at point $(x_1, y_1)_A$ in the direction of the positive z axis. The term $\cos \theta$ is usually referred to as the obliquity factor; that is, the obliquity factor is the cosine of the angle that a particular ray makes with the normal to the plane F. The obliquity factor accounts for the fact that the amplitude of light emitted from a point does not propagate equally in all directions.

5. $k = 2\pi/\lambda$, where λ is the wavelength of the monochromatic light.

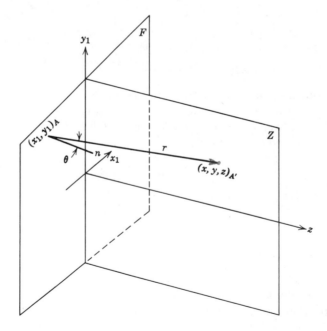

Figure 4.11 Relationship between points $(x, y, z)_A$ and points $(x_1, y_1)_A$ in plane F as indicated in diffraction formula (4-45).

Each of three vector components of the electromagnetic field representing the light amplitude distribution must be determined by the diffraction formula (4-45). The components of the electromagnetic field determine the magnitude and direction of the vector representing the electromagnetic field. When we are interested in only the light amplitude distribution and we can assume polarization effects can be neglected, it is permissible to assume that the light amplitude distribution is a scalar; that is, whenever polarization effects can be neglected, we can consider only the amplitude (not the direction) of an electric field. At any point (x, y, z) in the light amplitude distribution $A(x, y, z)$ we can detect (measure) only the intensity and we will define this measured quantity as the square of the absolute amplitude at that point (x, y, z). We are therefore neglecting any constant associated with the fact that the intensity is proportional to the amplitude squared, that is,

$$\text{intensity} = C_1 \, (\text{amplitude})^2 \qquad (4\text{-}46)$$

where C_1 is the constant of proportionality which we will assume to be equal to one $(C_1 = 1)$ for convenience.

Simplification of Diffraction Formula $-$ Drop $\frac{1}{r}$

We can simplify the diffraction formula (4-45) by considering the relative magnitudes of the terms inside the brackets, that is,

$$\left[jk - \frac{1}{r} \right] = j \frac{2\pi}{\lambda} - \frac{1}{r}. \qquad (4\text{-}47)$$

For wavelengths as long as 100 microns (far infrared) the term

$$\left[\frac{2\pi}{\lambda} = \frac{6.28}{10^2 \times 10^{-4}} = 628 \text{ cm} \right]$$

is a relatively large number, while for path length r greater than 1 cm the term $1/r$ is less than 1. For visible light (0.4 to 0.7 microns) $\lambda \ll 100$ microns and k is of the order

$$k = \frac{2\pi}{\lambda} = \frac{6.28}{0.628 \times 10^{-4}} = 10^5.$$

In most optics cases we will deal with path lengths r greater than 1 cm, and as shown the term $1/r$ will be negligible and can be dropped without any appreciable effect on accuracy. The diffraction formula (4-45) can now be written

$$A'(x, y, z) = \frac{-1}{2\pi} \int \int A(x_1, y_1) \frac{e^{jkr}}{r} [jk] \cos \theta \, dx_1 \, dy_1 \qquad (4\text{-}48)$$

The constant factor jk can be taken outside the integral giving

$$A'(x, y, z) = \frac{-j}{\lambda} \int \int A(x_1, y_1) \frac{e^{jkr}}{r} \cos \theta \, dx_1 \, dy_1 \qquad (4\text{-}49)$$

Obliquity Factor

We stated previously that the obliquity factor $\cos \theta$ is a weighting factor that accounts for the difference in the amount of light radiated in different directions. Because $\cos \theta$ has a maximum value of one at θ equal to zero, this factor has a maximum value of one for light contributions propagated normal to the signal plane F and drops off as the angle increases with respect to the surface normal. Referring to Figure 4.11 if we assume the light from a point (x_1, y_1) contributing to the light at the point $(x, y, z)_{A'}$ travels the straight line r, this line is the light ray at an angle θ to the normal n. We previously showed that through the focal property of an ideal lens this angle is a constant for all light contributing to a point (x, y) in the back focal plane and that x, y, and θ are related by the expression

$$\cos \theta = \frac{f}{(f^2 + x^2 + y^2)^{1/2}}. \qquad (4\text{-}44)$$

Specifically an ideal lens focuses light of constant obliquity factor into a ring of radius

$$\rho = (x^2 + y^2)^{1/2}. \qquad (4\text{-}7)$$

For a given θ, Figure 4.12(a) (Figure 4.12(a-2) with Z plane projection of Figure 4.12(a-1)) shows the significance of this focal effect. In Figure 4.12(a) the points A and B are sample points in the F plane and the points C and D are sample points in a parallel plane F' (which is a distance $z = d + f$ from plane F). Considering just point C we can see that paths AC and BC have obliquity factors of $\cos \theta_1$ and $\cos \theta_2$, respectively. From this example we can see that for a point (for example, C, or D) the obliquity factor depends on the contributing point (x_1, y_1) in plane F.

Likewise if we consider the point A in plane F, we can see that the paths AC and AD have obliquity factors of $\cos \theta_1$ and $\cos \theta_3$, respectively. This indicates that the obliquity factor also depends on the location of the point (x, y, z). The obliquity factor is included in the diffraction formula (4-45)

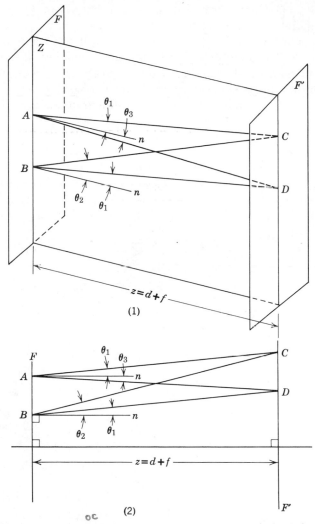

Figure 4.12 (*a*) Unfussed diffraction. Plane of the paper is the *Z* plane.

which we have accepted from the work of Rayleigh-Sommerfeld. The derivation of the obliquity factor is included in the derivation of the diffraction formula (4-45) and will therefore be given here without proof for

$$z = d + f \tag{4-50}$$

as shown in Figure 4.12(a)

$$\cos \theta = \frac{d+f}{[(x-x_1)^2 + (y-y_1)^2 + (d+f)^2]^{1/2}} \tag{4-51}$$

Equation (4-51) includes the coordinates of both the point (x_1, y_1) in plane F and the point (x, y, z) at which the diffracted light amplitude is to be found. The expression for the general case will have a z in place of $(d+f)$ in (4-51). In the diffraction formula the obliquity factor appears under the integral sign because x_1 and y_1 are the variables of integration and these terms appear in the obliquity factor as given by (4-51).

Figure 4.12(b-1) shows a focused diffraction pattern. In Figure

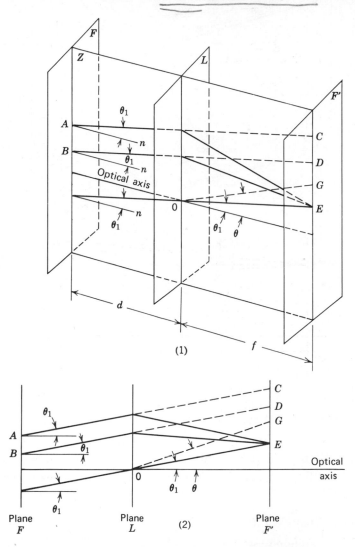

Figure 4.12   (b) Focussed diffraction. Plane of the paper is the Z plane.

4.12(b-2) only the rays AC and BD of Figure 4.12(a) are considered. The rays AC and BD are parallel as indicated by the fact that θ_1 is the same for both. The dotted portion of these rays indicates the path the light followed in Figure 4.12(a). The lens refracts these rays so that these paths are changed to those shown being focused to point E. When we can consider a point such as E, we find that the obliquity factor $\cos \theta_1$ is the same for points A and B and therefore independent of the coordinates (x_1, y_1) of the point in plane F. We recall that to contribute to any point (for example, point G) in plane F' a ray must be parallel to the principal ray (that is, for point G the ray must be parallel to the line OG). Rays parallel to OG will have an obliquity factor $\cos \theta$ that will be different from $\cos \theta_1$ for point E. Thus we can see that the obliquity factor depends on the location of the point (x, y) in the back focal plane F' of the lens.

The obliquity factor for a focused diffraction pattern is given by (4-44)

$$\cos \theta = \frac{f}{(f^2 + x^2 + y^2)^{1/2}} \tag{4-44}$$

and does not depend on the coordinates x_1 and y_1.

The obliquity factor for a focused diffraction pattern therefore is independent of the integration variables x_1 and y_1 and this factor can be taken outside the integral. We can now write the diffraction formula (4-49) as

$$A'(x, y, z) = \frac{-jf}{\lambda (f^2 + x^2 + y^2)^{1/2}} \int \int A(x_1, y_1) \frac{e^{jkr}}{r} \, dx_1 \, dy_1 \tag{4-52}$$

Because we are considering the complex light amplitude at a point (x, y) only in the back focal plane F' of the lens (i.e., constant z), we can let $A'(x, y)$ represent this complex light amplitude at the point (x, y). Then

$$A'(x, y) = \frac{-jf}{\lambda (f^2 + x^2 + y^2)^{1/2}} \int \int A(x_1, y_1) \frac{e^{jkr}}{r} \, dx_1 \, dy_1. \tag{4-53}$$

Diffraction Formula Phase Term

We will now discuss the remaining term which we have not considered (i.e., the e^{jkr}/r term) in equation 4-45 we defined r as the distance from the contributing point in the F plane to the point of interest (in plane F'). In Figure 4.11 the distance r is measured along the straight line from $(x_1, y_1)_A$ to $(x, y, z)_{A'}$. In Figure 4.12(b) the light traveling from A to E does not follow a straight line because of the refraction at the lens plane L. We can assume that the effects due to the length of the refracted path are the same as traveling an equivalent distance in a straight line. The r in the diffraction formula can therefore be interpreted as the optical path length that we determined previously; for example, in Figure 4.9 the optical path length from point C in plane F to point $(x, y)_B$ in plane F' (the principal ray) is given by $r = l_1 + l_2 + l_3$. For a ray parallel to this principal ray such as DE the optical path length from plane P' to plane F' is equal to $l_1 + l_2$ but the total optical path length from plane F to F' (i.e., from point D to $(x, y)_B$) will be given by equation 4-38

$$r = \frac{f^2 + x^2 + y^2 + df - xx_1 - yy_1}{(f^2 + x^2 + y^2)^{1/2}} \tag{4-38}$$

and will not be the same as for the principal ray.   *∴ phase shift.*

Therefore e^{jkr} represents the change in phase over an optical path length r, and $1/r$ is an attenuation factor that decreases the amplitude contribution as the optical path length increases.

Substituting the optical path length r as given by equations (4-38) and (4-39) in (4-53) *pg 170*

$$A'(x, y) = \frac{-jf}{\lambda (f^2 + x^2 + y^2)^{1/2}} \int \int A(x_1, y_1) \frac{e^{jkr}}{r} dx_1\, dy_1 \tag{4-53}$$

$$A'(x, y) = \frac{-jf}{\lambda (f^2 + x^2 + y^2)^{1/2}} \int \int A(x_1, y_1) \frac{e^{jk(R(x,y) - \alpha x_1 - \beta y_1)}}{\dfrac{f^2 + x^2 + y^2 + df - xx_1 - yy_1}{(f^2 + x^2 + y^2)^{1/2}}} dx_1\, dy_1 \tag{4-54}$$

$$A'(x, y) = \frac{-jf}{\lambda} \int \int A(x_1, y_1) \frac{e^{jk[R(x,y) - \alpha x_1 - \beta y_1]}}{f^2 + x^2 + y^2 + df - xx_1 - yy_1} dx_1\, dy_1 \tag{4-55}$$

$$A'(x, y) = \frac{-jf}{\lambda} e^{jkR(x,y)} \int \int A(x_1, y_1) \frac{e^{-jk(\alpha x_1 + \beta y_1)}}{f^2 + x^2 + y^2 + df - xx_1 - yy_1} dx_1\, dy_1 \tag{4-56}$$

Factoring $(f^2 + df)$ from the denominator of (4-56) gives

$$A'(x, y) = \frac{-jf}{\lambda (f^2 + df)} e^{jkR(x,y)} \int \int A(x_1, y_1) \frac{e^{-jk(\alpha x_1 + \beta y_1)}}{1 + \dfrac{x^2 + y^2 - xx_1 - yy_1}{f^2 + df}} dx_1\, dy_1 \tag{4-57}$$

Equation (4-57) can be rewritten

$$A'(x, y) = \frac{-j}{\lambda(f+d)} e^{jkR(x,y)} \int \int A(x_1, y_1) \frac{\exp\left[-j2\pi\left(\frac{\alpha}{\lambda} x_1 + \frac{\beta}{\lambda} y_1\right)\right]}{1 + \frac{x^2 + y^2 - xx_1 - yy_1}{(f^2 + df)}} dx_1 \, dy_1 \tag{4-58}$$

where we have introduced the notation exp (a) for e^a.

To simplify futher discussion we will introduce two new variables u and v defined as

$$u = \frac{\alpha}{\lambda} = \frac{x}{\lambda(f^2 + x^2 + y^2)^{1/2}} \tag{4-59}$$

$$v = \frac{\beta}{\lambda} = \frac{y}{\lambda(f^2 + x^2 + y^2)^{1/2}}. \tag{4-60}$$

where α and β are given by equations (4-41) and (4-42). pg 170

Substituting (4-59) and (4-60) into (4-58) gives

$$A'(x, y) = \frac{-j}{\lambda(f+d)} e^{jkR(x,y)} \int \int A(x_1, y_1) \frac{e^{-j2\pi(ux_1+vy_1)}}{1 + \frac{x^2 + y^2 - xx_1 - yy_1}{(f^2 + df)}} dx_1 \, dy_1 \tag{4-61}$$

$$A'(x, y) = \frac{-j}{\lambda(f+d)} e^{jkR(x,y)} \int \int A(x_1, y_1) \frac{e^{-j2\pi(ux_1+vy_1)}}{1 + \frac{x(x-x_1) + y(y-y_1)}{f(f+d)}} dx_1 \, dy_1 \tag{4-62}$$

Equation (4-62) gives the focused diffraction pattern through a lens.

THE FOURIER TRANSFORM APPROXIMATION

The denominator of (4-62) has the form

Integrand

$$\frac{1}{1 + (M/N)}$$

4-63 (2)

where $M = x(x-x_1) + y(y-y_1)$ and $N = f(f+d)$. We can add M/N and subtract M/N from the numerator of the fraction (4-63) without changing its value; that is

$$\frac{1}{1 + (M/N)} = \frac{1 + (M/N) - (M/N)}{1 + (M/N)} \tag{4-63}$$

then

$$\frac{1 + (M/N) - (M/N)}{1 + (M/N)} = 1 - \frac{M/N}{1 + (M/N)} = 1 - \frac{1}{1 + (N/M)} \tag{4-64}$$

The algebraic identity shown in expressions (4-63) and (4-64) indicates that

$$\frac{1}{1+(M/N)} = 1 - \frac{1}{1+(N/M)} \qquad (4\text{-}65)$$

We can now write (4-62) as

$$A'(x,y) = \frac{-j}{\lambda(f+d)} e^{jkR(x,y)} \int\int A(x_1,y_1)\left[1 - \frac{1}{1+\dfrac{f(f+d)}{x(x-x_1)+y(y-y_1)}}\right]$$

N = f(f+d)

M = x(x-x₁) + y(y-y₁)

$$\times e^{-j2\pi(ux_1+vy_1)}\,dx_1\,dy_1 \qquad (4\text{-}66)$$

Rewriting (4-66) with an integral for each term in the bracket gives

Amplitude

A₀↑²

$$A'(x,y) = \frac{-j}{\lambda(f+d)} e^{jkR(x,y)} \int\int A(x_1,y_1)\,e^{-j2\pi[ux_1+vy_1]}\,dx_1\,dy_1$$

$$+ \frac{j}{\lambda(f+d)} e^{jkR(x,y)} \int\int \frac{A(x_1,y_1)}{1+\dfrac{f(f+d)}{x(x-x_1)+y(y-y_1)}}$$

$$\times e^{-j2\pi[ux_1+vy_1]}\,dx_1\,dy_1 \qquad (4\text{-}67)$$

∴ ie x≈x₁, & y≈y₁.

By restricting the maximum values (aperture limits) of x_1, y_1 (aperture in F plane) and x, y (aperture in F' plane) the second integral of (4-67) can be made negligible compared to the first integral; that is, the denominator of the integral of the second integral can be made large. A detailed discussion of this approximation will follow later but here we shall just assume that it is possible to make the second integral negligible.

Assuming we can neglect the second integral of (4-67) we can write the diffraction formula as

u & v are spatial freq.'s

$$A'(x,y) = \frac{-j}{\lambda(f+d)} e^{jkR(x,y)} \int\int A(x_1,y_1)\,e^{-j2\pi(ux_1+vy_1)}\,dx_1\,dy_1 \qquad (4\text{-}68)$$

Equation (4-68) can be written as

$$A'(x,y) = \left[\frac{-je^{jkR(x,y)}}{\lambda(f+d)}\right] F(u,v) \qquad (4\text{-}69)$$

where $F(u,v)$ is the two-dimensional Fourier transform of $A'(x,y)$; that is,

$$F(u,v) = \int\int A(x_1,y_1)\,e^{-j2\pi(ux_1+vy_1)}\,dx_1\,dy_1 \qquad (4\text{-}70)$$

If we were to measure the light distribution given by (4-69), we would measure the intensity that is the square of the magnitude of the complex amplitude $A'(x, y)$. This intensity is given as

$$I(x, y) = A'(x, y)A'^*(x, y) = \left[\frac{-j}{\lambda(f+d)} e^{jkR(x,y)} F(u, v)\right]$$

$$\times \left[\frac{+j}{\lambda(f+d)} e^{-jkR(x,y)} F^*(u, v)\right] \quad (4\text{-}71)$$

$$I(x, y) = \frac{1}{(\lambda(f+d))^2} F(u, v) F^*(u, v)$$

$$I(x, y) = \frac{|F(u, v)|^2}{(\lambda(f+d))^2} \quad (4\text{-}73)$$

The intensity in the back focal plane of a lens (within the limits to be determined for the approximations made) is given by the square of the magnitude of the Fourier transform of the light amplitude in plane F.

If the aperture in the back focal plane restricts the maximum value of x and y so as to limit the phase variations (to where they can be considered essentially constant) due to the exponential term in the brackets in (4-69), we can write

$$A'(x, y) = KF(u, v) \quad (4\text{-}74)$$

where K is a complex constant given by

$$K = \frac{-je^{jkR(x,y)}}{\lambda(f+d)} \qquad Phase \\ Appx$$

Within the range of x and y for which $e^{jkR(x,y)}$ can be assumed constant (i.e., negligible phase variation) the amplitude distribution in the back focal plane F' is proportional to the Fourier transform of the light amplitude in the plane F as given by (4-74). We shall show later that the restriction on x and y is tighter for this phase limitation than the restriction required for x and y to eliminate the second integral in (4-67). In terms of spectrum analysis in the back focal plane of the lens, the phase limitation is not important because only intensity can be sensed (measured). The Fourier transform relationship between the amplitude distributions in the front and back focal planes of a lens is useful in analyzing cascaded lens systems. Each lens in a cascaded lens system can perform a Fourier transform and therefore a pair of lenses can be made to perform a double Fourier transform operation; that is, the second lens can take the Fourier transform of the Fourier transform of the first lens. This will be discussed in greater detail later in this chapter.

FOURIER COMPONENT (u, v) AND FOCAL PLANE COORDINATES (x, y)

The amplitude distribution $A'(x, y)$ in terms of the Fourier transform of $A(x_1, y_1)$ was given by (4-69) as *pg 180.*

$$A'(x, y) = \left[\frac{-je^{jkR(x, y)}}{\lambda(f+d)}\right] F(u, v) \qquad (4\text{-}69)$$

where

$$F(u, v) = \int\int A(x_1, y_1)\, e^{-j2\pi[ux_1 + vy_1]}\, dx_1\, dy_1 \qquad (4\text{-}70)$$

The Fourier transform coordinates u and v were defined by

$$u = \frac{x}{\lambda(f^2 + x^2 + y^2)^{1/2}} \qquad (4\text{-}59)$$

pg 179

$$v = \frac{y}{\lambda(f^2 + x^2 + y^2)^{1/2}} \qquad (4\text{-}60)$$

We can substitute for u and v in (4-69) to obtain an expression for $A'(x, y)$ in terms of x and y; however, the fact that u and v are each dependent on both x and y makes this substitution somewhat complicated, (4-70) becomes

$$F(u, v) = F\left(\frac{x}{\lambda(f^2 + x^2 + y^2)^{1/2}}, \frac{y}{\lambda(f^2 + x^2 + y^2)^{1/2}}\right)$$

$$= \int\int A(x_1, y_1)\, e^{-j2\pi(ux_1 + vy_1)}\, dx_1\, dy_1$$

$$= \int\int A(x_1, y_1)\, \exp\left[-j2\pi\left(\frac{xx_1}{\lambda(f^2 + x^2 + y^2)^{1/2}}\right.\right.$$

$$\left.\left. + \frac{yy_1}{\lambda(f^2 + x^2 + y^2)^{1/2}}\right)\right] dx_1\, dy_1 \qquad (4\text{-}72)$$

If u were directly proportional to x and v directly proportional to y, the light amplitude at a particular value of x would be related to a particular value of u and a particular value of y would be related to a particular value of v. Unfortunately we can see from (4-59) and (4-60) that u is not only related to x but also depends on y, and likewise v is related to both x and y. We will show later the importance of a linear relationship between the

transform coordinates u and v and the spatial coordinates x and y.

In order to obtain a linear relation between u and x let us consider the series expansion for (4-59); that is,

$$u = \frac{x}{\lambda(f^2 + x^2 + y^2)^{1/2}} \tag{4-59}$$

factoring f^2 we get

$$u = \frac{x}{\lambda f}\left[\frac{1}{\left(1 + \frac{x^2 + y^2}{f^2}\right)^{1/2}}\right] \tag{4-75}$$

The series expansion for $(1 \pm \alpha)^{-n}$ is given by

$$\frac{1}{(1 \pm \alpha)^n} = 1 \mp n\alpha + \frac{n(n+1)\alpha^2}{2!} \mp \frac{n(n+1)(n+2)\alpha^3}{3!} + \cdots \tag{4-76}$$

where $\alpha^2 < 1$ and if we let

$$\alpha = \left(\frac{x^2 + y^2}{f^2}\right) \qquad \text{and} \qquad n = \tfrac{1}{2} \tag{4-77}$$

then in (4-75) the bracket term can be approximated by using (4-76) with values given by (4-77) as follows:

$$\left[\frac{1}{\left(1 + \frac{x^2 + y^2}{f^2}\right)^{1/2}}\right] = 1 - \left[\frac{x^2 + y^2}{2f^2}\right] + \frac{3}{8}\left[\frac{x^2 + y^2}{f^2}\right]^2 - \cdots$$

if $[(x^2 + y^2)/f^2] < 1$. Therefore (4-75) can be written

$$u = \frac{x}{\lambda f}\left[1 - \frac{x^2 + y^2}{2f^2} + \frac{3}{8}\left[\frac{x^2 + y^2}{f^2}\right]^2 - \cdots\right] \tag{4-75A}$$

If we restrict our analysis to an area of the focal plane such that

$$\left[\frac{x^2 + y^2}{f^2}\right] \ll 1$$

then all terms in the brackets of (4-75A) except the first can be neglected giving

$$u = \frac{x}{\lambda f} \tag{4-78}$$

Similarly we can see the approximation for (4-60)

$$v = \frac{y}{\lambda(f^2 + x^2 + y^2)^{1/2}}\qquad(4\text{-}60)$$

is given by

$$v = \frac{y}{\lambda f}\qquad(4\text{-}79)$$

Therefore within a restricted area in the back focal plane of the lens the Fourier transform equation (4-72) can be written

$$F(u, v) = F\left(\frac{x}{\lambda f}, \frac{y}{\lambda f}\right)$$

$$= \int\int A(x_1, y_1)\exp\left[-j2\pi\left(\frac{xx_1}{\lambda f} + \frac{yy_1}{\lambda f}\right)\right]dx_1\,dy_1\qquad(4\text{-}80)$$

From (4-80) and (4-70) we can see that the frequencies are given by coefficients of x_1 and y_1 in the exponential term. The coordinates x and y in the back focal plane are scaled representations of the frequencies u and v, respectively; that is,

$$u = \frac{x}{\lambda f}\qquad(4\text{-}78)$$

and

$$v = \frac{y}{\lambda f}\qquad(4\text{-}79)$$

The light contributions corresponding to a spatial frequency u in the x_1 direction appear at the coordinate $x = \lambda fu$ in the back focal plane of the lens. Similarly the light contributions corresponding to a spatial frequency v in the y_1 direction appear at the coordinate $y = \lambda fv$ in the back focal plane of the lens.

ACCURACY OF LINEAR FREQUENCY APPROXIMATION

The actual restrictions to be imposed on x and y for the above approximation will depend on how accurate the Fourier frequency value must be in a particular application. The error in the approximate frequency of (4-78) and (4-79) as a fraction of the exact value as given by (4-59) and (4-60) is

$$E_f = \frac{x/\lambda f}{x/[\lambda(f^2 + x^2 + y^2)^{1/2}]} - 1\qquad(4\text{-}81)$$

where E_f is the fractional error in frequency.

$$E_f = \frac{(f^2 + x^2 + y^2)^{1/2}}{f} - 1 \tag{4-82}$$

$$= \frac{f[1 + (x^2 + y^2)/f^2]^{1/2}}{f} - 1 \tag{4-83}$$

$$E_f = \left(1 + \frac{x^2 + y^2}{f^2}\right)^{1/2} - 1 \tag{4-84}$$

To change from the frequency fractional error to percent error in frequency we must multiply by 100 percent.

If we let

$$\rho^2 = x^2 + y^2 \tag{4-85}$$

(where ρ is the radius of a circle in the x, y plane) and express ρ in terms of multiples of the focal length f as given by

$$\rho = af \tag{4-86}$$

then

$$\rho^2 = x^2 + y^2 = a^2 f^2 \tag{4-87}$$

and (4-84) can be written

$$E_f = \left(1 + \frac{a^2 f^2}{f^2}\right)^{1/2} - 1 = (1 + a^2)^{1/2} - 1 \tag{4-88}$$

The curve of Figure 4.13 gives the percent error ($100E_f$) of the linear

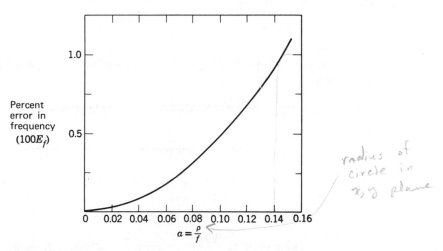

Percent error in frequency ($100E_f$)

1.0

0.5

0   0.02 0.04 0.06 0.08 0.10 0.12 0.14 0.16

$$a = \frac{\rho}{f}$$

Figure 4.13   Percent error in linear frequency approximations.

approximation of frequency as a function of a. For a less than 0.14 the error will be less than 1 percent. Thus the linear approximation

$$u = \frac{x}{\lambda f} \quad \rightarrow \text{focal length} \tag{4-78}$$

and

$$v = \frac{y}{\lambda f} \tag{4-79}$$

are accurate within 1 percent for values of x and y satisfying the restriction

$$(x^2 + y^2)^{1/2} = \rho \leqq 0.14f \tag{4-89}$$

For accuracies better than 1 percent smaller values of a must be imposed as given by (4-88) and Figure 4.13.

Because our approximation requires that we limit our considerations to the area in plane F' within a circle of radius $\rho_{max} = 0.14f$ (for accuracy within 1 percent), the maximum value of $x^2 + y^2$ is specified by

$$(x^2 + y^2)_{max} = \rho_{max}^2 = 0.02f^2 \tag{4-90}$$

Squaring the approximate expressions for u and v, (4-78) and (4-79), and adding we obtain

$$u^2 + v^2 = \frac{x^2}{\lambda^2 f^2} + \frac{y^2}{\lambda^2 f^2} = \frac{x^2 + y^2}{\lambda^2 f^2} \tag{4-91}$$

Applying the restrictions on $x^2 + y^2$ to (4-91) we obtain

$$u^2 + v^2 \leqq \frac{0.02}{\lambda^2} \tag{4-92}$$

The maximum allowed value of u occurs when $v = 0$ and the maximum v occurs when $u = 0$ or

$$u_{max} = \frac{0.14}{\lambda} \quad \text{for } v = 0 \quad \text{(that is, } y = 0) \tag{4-93}$$

$$v_{max} = \frac{0.14}{\lambda} \quad \text{for } u = 0 \quad \text{(that is, } x = 0) \tag{4-94}$$

As an example let us consider green light of wavelength $\lambda = 5461 \times 10^{-8}$ cm. In this case the simplified expressions for u and v are accurate within 1 percent for frequencies in the range given by (4-92)

$$(u^2 + v^2)^{1/2} \leqq \frac{0.14}{\lambda} = \frac{0.14 \times 10^8}{5461} \approx 2560 \text{ cycles/cm}$$

On the x axis ($y = 0$, $v = 0$) the maximum frequency will be

$$u_{max} = 2650 \text{ cycles/cm}$$

On the y axis ($x = 0$, $u = 0$) the maximum frequency is likewise

$$v_{max} = 2650 \text{ cycles/cm}$$

In practice the limitations of the usual available techniques for controlling the input light distribution $A(x_1, y_1)$ (for example, photographic film) restrict the maximum input spatial frequencies to values below the 2560 cycles/cm restriction we have imposed above; that is, we usually will be looking for frequencies in the F' plane only as high as those in the F plane. The linear approximations of the frequency components u and v

$$u = \frac{x}{\lambda f} \tag{4-78}$$

$$v = \frac{y}{\lambda f} \tag{4-79}$$

and equations

$$A'(x,y) = \left[\frac{-je^{jkR(x,y)}}{\lambda(f+d)}\right] F(u,v) \tag{4-69}$$

when

$$F(u,v) = \int \int A(x_1, y_1) e^{-j2\pi(ux_1 + vy_1)} \, dx_1 \, dy_1 \tag{4-70}$$

and

$$I(x,y) = \frac{|F(u,v)|^2}{[\lambda(f+d)]^2} \tag{4-73}$$

are applicable for practical systems and can be expressed

$$A'(x,y) = [K]F\left(\frac{x}{\lambda f}, \frac{y}{\lambda f}\right) \tag{4-95}$$

where

$$K = \frac{-je^{jkR(x,y)}}{\lambda(f+d)}$$

$$I(x,y) = \frac{|F(x/\lambda f, y/\lambda f)|^2}{\{\lambda(f+d)\}^2} \tag{4-96}$$

The accuracy of u and v in (4-95) and (4-96) as given by (4-78) and (4-79) is determined by (4-88) as shown in Figure 4.13 for values

$$a = \frac{\rho}{f} \tag{4-97}$$

multiples of focal length, f.

pg 185

Therefore for a given focal length f the restrictions on the maximum value of ρ determine the accuracy of the linear approximation introduced here; that is, the frequencies u and v have been made proportional to x and y, respectively. These equations also include the approximation assumed earlier in neglecting terms other than the $F(u, v)$ term; that is, in (4-67) we can use only the first integral without introducing any appreciable error. In the next section we will consider that approximation and determine if the restriction

$$x^2 + y^2 \leq 0.02f^2 \tag{4-98}$$

is also a sufficient restriction to assure the validity of neglecting terms other than $F(u, v)$. As mentioned on pg 181.

LIMITS IN DIFFRACTION FORMULA TO APPROXIMATE THE FOURIER TRANSFORM

We shall now return to examine in detail the focused diffraction formula given by (4-66)

$$A'(x, y) = \frac{-j}{\lambda(f+d)} e^{jkR(x,y)} \int \int A(x_1, y_1) \left[1 - \frac{1}{1 + \dfrac{f(f+d)}{x(x-x_1) + y(y-y_1)}} \right]$$

$$\times e^{-j2\pi[ux_1 + vy_1]} dx_1 \, dy_1 \tag{4-66}$$

and consider the limitations required to obtain the Fourier transform approximation given by (4-69) and (4-70)

$$A'(x, y) = \left[\frac{-je^{jkR(x,y)}}{\lambda(f+d)} \right] F(u, v) \tag{4-69}$$

where

$$F(u, v) = \int \int A(x_1, y_1) e^{-j2\pi(ux_1 + vy_1)} dx_1 \, dy_1 \tag{4-70}$$

To obtain the form of a Fourier transform of $A(x_1, y_1)$ the bracketed term in (4-66) must be approximated by a constant. We can consider this term equal to one if we restrict the range of x, y in plane F' and x_1 and y_1 in plane F to satisfy the inequality

$$\left[\frac{1}{1 + \dfrac{f(f+d)}{x(x-x_1) + y(y-y_1)}} \right] \ll 1 \tag{4-99}$$

By referring back to (4-66) we can see that the approximation (4-99) corresponds to making the second integral in (4-67) negligible to the first integral, which has the form of a Fourier transform.

$$A'(x,y) = \frac{-j}{\lambda(f+d)} \, e^{jkR(x,y)} \int\int A(x_1,y_1) e^{-j2\pi(ux_1+vy_1)} \, dx_1 \, dy_1$$

$$+ \frac{j}{\lambda(f+d)} \, e^{jkR(x,y)}$$

$$\times \int\int \frac{A(x_1,y_1)}{1 + \dfrac{f(f+d)}{x(x-x_1)+y(y-y_1)}} \, e^{-j2\pi(ux_1+vy_1)} \, dx_1 \, dy_1 \qquad (4\text{-}67)$$

The complete term inside the brackets of (4-66) is effectively a weighting factor that varies the contribution from each point (x_1,y_1) in plane F to the point (x,y) in plane F'. The diffraction formula (4-45) shows that both the path length and obliquity factor will determine the amount of light reaching a given point in plane F' from a point in plane F. The variable term in the brackets under the integral of (4-66) contains the terms corresponding to path length and obliquity factor from a point in the input plane F to a specific point in plane F'. It is usually assumed that these effects are negligible so that the inequality (4-99) is satisfied. In the following analysis we will attempt to present a more detailed quantitative discussion of this approximation.

When the inequality (4-99) is satisfied, this term in (4-66) can be neglected and (4-69) can be used to approximate the light amplitude contribution from each point (x_1,y_1) in plane F to a point (x,y) in plane F'. The contribution $dA(x,y)$ at point (x,y) in plane F' from an infinitesimally small region $dx_1 \, dy_1$ about point (x_1,y_1) is given exactly by

$$dA'(x,y) = K\left[1 - \frac{1}{1 + \dfrac{f(f+d)}{x(x-x_1)+y(y-y_1)}}\right] A(x_1,y_1) \, e^{-j2\pi(ux_1+vy_1)} dx_1 \, dy_1$$

$$(4.100)$$

where

$$K = \frac{-je^{jkR(x,y)}}{\lambda(f+d)}$$

The relation between (4-66) and (4-100) is that when considering a small region about a point (x_1,y_1) (in plane F and only in that region), we can determine its contribution to any point (x,y) in plane F' by using (4-100).

When we wish to find the contribution of all points (x_1, y_1) in plane F to any point (x, y) in plane F', we must use (4-66), which sums up (integrates) the contribution of all points (x_1, y_1) in plane F to a point (x, y) in plane F'. We can apply the approximation of neglecting the variable term in the brackets of (4-100), which gives

$$dA'(x, y) = KA(x_1, y_1) e^{-j2\pi(ux_1 + vy_1)} dx_1 dy_1 \qquad (4\text{-}101)$$

The total light amplitude $A'(x, y)$ at any point (x, y) in plane F' is therefore obtained by integrating over the range x_1 and y_1. The integration of (4-101) will yield an approximation for the total light amplitude $A'(x, y)$ at least as accurate as the worst individual case of (4-101). This can be seen by assuming a case of three 100-ohm resistors marked with an accuracy of ± 1, ± 5, and ± 10 percent. If these three 100-ohm resistors are wired in series, the total maximum resistance can be no more than

maximum 1 percent value for 100 ohms   101
maximum 5 percent value for 100 ohms   105
maximum 10 percent value for 100 ohms   110
maximum value for the three resistors in series   316 ohm

The value, 316 ohm, represents approximately a 5.3 percent increase over the nominal value of 300 ohms. This illustrates that when adding measurements taken with different accuracies, the total will always have an accuracy better than the measurement with the worst accuracy. In the example with the three resistors, 5.3 percent for the total of the three resistors is less than 10 percent, which is the value for the worst resistor. We can now see that by integrating (adding individual contributions from different points in plane F to a point in plane F') the greatest possible error will be given by the maximum value of the neglected term— the variable in the brackets of (4-66).

Accuracy of Amplitude Approximation

We shall now consider the maximum amplitude error introduced by neglecting the variable term to obtain (4-101). Denoting the error in amplitude by the fraction E_A given by the ratio of the neglected term to the exact factor within the brackets of (4-66), we obtain

$$E_A = \frac{\text{neglected term}}{\text{exact term}} = \frac{\left[1 + \dfrac{f(f+d)}{x(x-x_1) + y(y-y_1)}\right]^{-1}}{1 - \left[1 + \dfrac{f(f+d)}{x(x-x_1) + y(y-y_1)}\right]^{-1}} \qquad (4\text{-}102)$$

$$= \frac{1}{1+\left[\dfrac{f(f+d)}{x(x-x_1)+y(y-y_1)}\right]-1} \qquad (4\text{-}103)$$

$$E_A = \frac{x(x-x_1)+y(y-y_1)}{f(f+d)} \qquad (4\text{-}104)$$

$$E_A = \frac{x^2+y^2-xx_1-yy_1}{f(f+d)} \quad see \qquad (4\text{-}105)$$

457
pg 178

To simplify our discussion we can express x, y, x_1 and y_1 in terms of polar coordinates ρ, φ, r', and α' as done in the early portion of this chapter. The relations between these coordinates can be seen by referring to Figures 4.8(a) and 4.8(b) and are given by the following equations

$$\rho = (x^2+y^2)^{1/2} \qquad (4\text{-}106)$$

$$x = \rho \cos \varphi \qquad (4\text{-}107)$$

$$y = \rho \sin \varphi \qquad (4\text{-}108)$$

$$r' = (x_1^2+y_1^2)^{1/2} \qquad (4\text{-}8) \quad pg \; \begin{array}{l}165\\166\end{array}$$

$$x_1 = r' \cos \alpha' \qquad (4\text{-}109)$$

$$y_1 = r' \sin \alpha' \qquad (4\text{-}110)$$

Substituting in (4-105) we obtain

$$E_A = \frac{x^2+y^2-xx_1-yy_1}{f(f+d)}$$

$$= \frac{\rho^2 \cos^2 \varphi + \rho^2 \sin^2 \varphi - \rho r' \cos \varphi \cos \alpha' \quad \rho r' \sin \varphi \sin \alpha'}{f(f+d)} \qquad (4\text{-}111)$$

$$E_A = \frac{\rho^2 - \rho r'(\cos \varphi \cos \alpha' + \sin \varphi \sin \alpha')}{f(f+d)} \qquad (4\text{-}112)$$

Using the trigonometric identity

$$\cos (x-y) = \cos x \cos y + \sin x \sin y \qquad (4\text{-}20)$$

we can write (4-112) as

$$E_A = \frac{\rho^2 - \rho r' \cos (\varphi - \alpha')}{f(f+d)} \qquad (4\text{-}113)$$

The cosine term in (4-113) can take values between minus one and plus one. Since we are interested in the maximum error, we will consider the case where $\cos(\varphi - \alpha') = -1$; that is, when $\cos(\varphi - \alpha') = -1$, the term $\rho r'$ is added to the ρ^2 term to make E_A a maximum.

Equation (4-113) can be written for

$$\cos(\varphi - \alpha') = -1 \tag{4-114}$$

as

$$E_{A_{MAX}} = \frac{\rho^2 + \rho r'}{f(f+d)} \tag{4-115}$$

Figure 4.14 shows the relative positions of points (x_1, y_1) and (x, y) for the case

$$\cos(\varphi - \alpha') = -1 \tag{4-114}$$

The maximum error defined by (4-115) applies to the light contribution from points (x_1, y_1) located on the line of intersection OT, between planes F and Z' as shown in Figure 4.14.

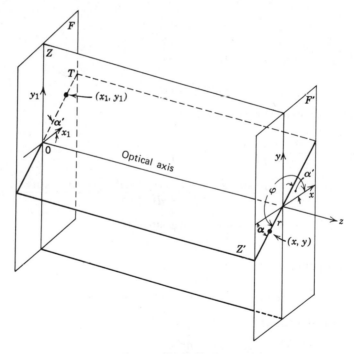

Figure 4.14   Diagram of conditions for maximum E_A.

The plane Z' is a plane containing the optical axis and the point (x, y) in the back focal plane F'. The points (x_1, y_1) are further restricted to the portion of the line of intersection of F and Z' that is on the side of the optical axis opposite to the point (x, y). From the geometry of Figure 4.14 it is clear that the angle φ is equal to $(\pi + \alpha')$. Thus $(\varphi - \alpha')$ is equal to π and

$$\cos(\varphi - \alpha') = \cos \pi = -1 \qquad (4\text{-}114)$$

as required for the maximum E_A given by (4-115). For any point in plane F that does not fall on the line OT the cosine term will be greater than -1 and the value of E_A will be less than that given by (4-115).

Examination of (4-115) shows that the error E_A increases as ρ and r' increase. To determine the maximum value of E_A as a function of ρ and r' we need only to specify the maximum values of ρ and r'. Conversely if we are interested in restricting the values of E_A to be less than or equal to a specified value, the maximum values of ρ and r' must satisfy (4-115) for that particular value of E_A. Reducing either the input or output aperture will reduce the value of E_A. From (4-115) we can see that reducing the radius of the input aperture r' will have less of an effect than a similar reduction in the radius of the aperture ρ in the Fourier transform plane F' (since ρ is a squared term).

LENS APERTURE RESTRICTION

In order to analyze the relation between E_A, ρ, and r' we must consider the interdependence of the maximum values of ρ and r' due to the limitations of a finite lens aperture. We will assume that diffraction effects at the edge of the lens aperture are negligible. Figure 4.15 shows the extreme rays that can pass through a lens aperture of radius R_L to reach the points at the distance ρ_{max} from the optical axis. Any ray parallel to, but above, the upper extreme incident ray to the lens or parallel to, but below, the lower extreme incident ray to the lens will be outside the lens aperture and will not pass through the lens. Therefore, any signal point outside the rays defined by r'_{max} in Figure 4.15 cannot contribute to both of the points $+\rho_{max}$ and $-\rho_{max}$; for example, if the input signal area from plane F extends upward beyond the r'_{max} limit, the additional input signal interval cannot contribute to the spectral point at $+\rho_{max}$ because the necessary light path will fall outside of the lens aperture. Under these circumstances the amplitude at the spectral point at $+\rho_{max}$ will not correspond to the entire signal but only the interval below the $+r'_{max}$ limit and the light intensity at the spectral point will be less than that predicted for the entire signal. From this example it is apparent that the

r'_{max} limit given by Figure 4.15 defines the maximum signal interval over which every point contributes to the spectral points within the range $\pm \rho_{max}$.

The dashed lines in Figure 4.15 represent the extreme rays to a spectral point at a distance ρ that is less than ρ_{max}. The extreme rays for such a case define a maximum signal interval longer than that obtained for ρ_{max}. This means that the signal interval defined by r'_{max} increases as the spectral range of interest defined by ρ_{max} decreases. We can see that for a given maximum frequency (i.e., ρ_{max}) the maximum value of the signal interval r'_{max} is limited by the lens aperture.

We can derive an expression defining the relation between ρ_{max} and r'_{max} by referring to Figure 4.16. Figure 4.16 represents the upper extreme incident ray to the lens and the principal ray contributing to the spectral point at $+\rho_{max}$. Because the extreme ray must be parallel to the principal ray, they must both make an angle θ with the optical axis and we can apply the principles of similar triangles to obtain

$$\frac{\rho_{max}}{f} = \frac{r'_{max}}{a} = \frac{R_L}{a+d} \qquad (4\text{-}116)$$

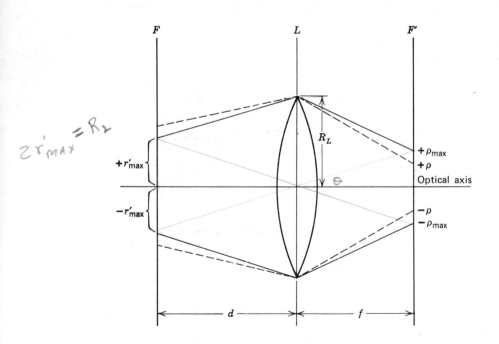

Figure 4.15 Lens aperature limitations on ρ_{max} and r'_{max}.

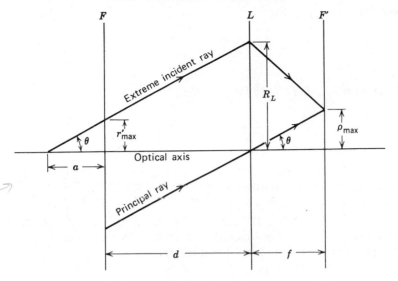

Figure 4.16 Geometry for the relation between ρ_{max} and r'_{max}.

From these relations we obtain two equations for a

$$a = \frac{r'_{max}}{[\rho_{max}/f]} \qquad (4\text{-}117)$$

and

$$a = \frac{R_L}{[\rho_{max}/f]} - d \qquad (4\text{-}118)$$

Because the right-hand side of these equations must be equal, we obtain the result

$$r'_{max} = R_L - d\left[\frac{\rho_{max}}{f}\right] \qquad (4\text{-}119)$$

A lens is usually specified by its $\mathscr{F}$ stop which is defined as

$$\mathscr{F} = \frac{f}{2R_L} \qquad (4\text{-}120)$$

Dividing both sides of (4-119) by f (the focal length of the lens) we obtain

$$\frac{r'_{max}}{f} = \frac{R_L}{f} - \frac{d}{f}\frac{\rho_{max}}{f} \qquad (4\text{-}121)$$

From (4-120) we find that

$$\frac{R_L}{f} = \frac{1}{2\mathscr{F}}$$

and substituting into (4-121) we can write

$$\frac{r'_{\max}}{f} = \frac{1}{2\mathscr{F}} - \frac{d}{f}\frac{\rho_{\max}}{f} \qquad (4\text{-}122)$$

It is obvious that from (4-122) as well as Figure 4.15 that $r'_{\max}$ cannot be greater than the lens aperture radius R_L. Equation (4-122) defines the maximum allowed signal aperture radius $r'_{\max}$ due to the limitations of the lens aperture. In practice the size of the signal aperture is specified by physical considerations or desired size format. We can rearrange terms in (4-122) to define the maximum spectral term $\rho_{\max}$ as

$$\frac{\rho_{\max}}{f} = \frac{1/2\mathscr{F} - r'_{\max}/f}{d/f} \qquad (4\text{-}123)$$

Equation (4-123) defines the maximum allowable ρ for a given $r'_{\max}$ as determined by the restriction of a lens aperture. By rearranging terms in (4-115) we can obtain a second expression specifying the limitations on $\rho_{\max}$ required for an allowed error E_A

$$\frac{\rho_{\max}}{f} = \frac{1}{2}\left[\frac{r'_{\max}}{f}\right]\left[\left(1 + \frac{4E_A(1 + (d/f))}{(r'_{\max}/f)^2}\right)^{1/2} - 1\right] \qquad (4\text{-}124)$$

Applications of Aperture and Accuracy Limitations

The application of (4-123) and (4-124) can be demonstrated by considering the case of $[r'_{\max}/f] = \frac{1}{5}$; for example, $f = 100$ mm and $r'_{\max} = 20$ mm. Figure 4.17 is a graph of $[\rho_{\max}/f]$ plotted as a function of d/f for the specified input aperture; that is,

$$r'_{\max} = \frac{f}{5} \qquad (4\text{-}125)$$

The curves labeled $\mathscr{F} = 1.4$ and $\mathscr{F} = 2$ correspond to (4-123) for the specified values of $\mathscr{F}$. The curves labeled $E_A = 0.02$, $E_A = 0.01$, and $E_A = 0.005$ correspond to (4-124) for the specified values of E_A. In plotting (4-123) and (4-124) it was assumed that $[r'_{\max}/f] = \frac{1}{5}$. The $\mathscr{F}$ curves specify the upper limit on ρ/f due to the lens aperture and the E_A

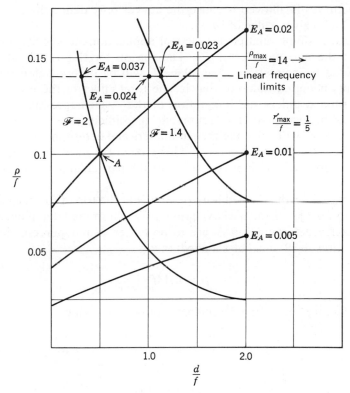

Figure 4.17 Limitation on maximum spectral term $\dfrac{\rho_{max}}{f}$ based on $\dfrac{r'_{max}}{f} = \frac{1}{5}$.

curves specify the upper limit for a given accuracy of the approximation. For a chosen value of d/f (i.e., the input plane F is a specified distance from the lens plane L) the value of $[\rho_{max}/f]$ (i.e., the maximum distance of a focused point from the optical axis) must be below the $\mathscr{F}$ curve (i.e., a selected lens aperture) and the E_A curve (i.e., a desired accuracy).

Let us consider an example of an $\mathscr{F} = 2$ lens (i.e., in the usual terminology this would be written as an $\mathscr{F}/2$ lens) and the maximum error to be allowed is $E_A = 0.02$. The greatest value allowed for $[\rho/f]$ corresponds to point A in Figure 4.17 which is the intersection of the $\mathscr{F} = 2$ curve and the $E_A = 0.02$ curve. The value of $[d/f] = 0.5$ would be selected to obtain the value $[\rho_{max}/f] = 0.10$. For any other value of $[d/f]$ the limit of $[\rho/f]$ would be less than the maximum at point A. For $[d/f]$ less than 0.5 the $E_A = 0.02$ curve specifies a tighter limit on $[\rho/f]$ while for $[d/f]$ greater than 0.5 the $\mathscr{F} = 2$ curve limits $[\rho/f]$. Of course any combination of $[\rho/f]$ and $[d/f]$ corresponding to a point in the area bounded below the two curves is allowed because the curves only define the upper limit on $[\rho/f]$ for a given value of $[d/f]$.

It is not a simple matter in practical applications to specify a desired value of the maximum error E_A. The reason for this is that the error contributed by points (x_1, y_1) in plane F will be different depending on the points being considered; that is, the total effect of the error cannot be determined unless the integration of the exact expression given by (4-67) can be evaluated. To circumvent this difficulty we will consider a more logical way of selecting the parameters and determine the maximum E_A specified by these parameters. This value of E_A will then specify the maximum error of the approximation for the particular set of parameters we select.

In Figure 4.17 the dashed horizontal line at the top of the graph corresponds to $[\rho/f] = 0.14$. We have previously determined that if accuracies in frequency determination E_f are to be better than 1 percent (in plane F'), equation (4-89) must be satisfied; that is,

$$\rho \leqq 0.14f \tag{4-89}$$

We can factor f^2 out of the denominator of (4.115) giving

$$E_A = \frac{\rho^2 + \rho r'}{f(f+d)} \tag{4-115}$$

$$E_A = \frac{(\rho/f)^2 + (\rho/f)(r'/f)}{1 + (d/f)} \tag{4-126}$$

From (4-126) we can see that the error E_A decreases as $[d/f]$ increases. We do not wish to lower the previous limit of

$$E_f = \frac{\rho}{f} = 0.14$$

We can improve (decrease) the error E_A by using the largest possible value of $[d/f]$. Referring to Figure 4.17 we can see that for a $\mathscr{F} = 2$ lens, the maximum value of $[d/f]$ is 0.3 (which is given by the intersection of the dashed horizontal line $[\rho_{max}/f] = 0.14$ and $\mathscr{F} = 2$ curve. As noted on Figure 4.17, $E_A = 0.037$ at this point—obtained from (4-115) or (4-126) as follows:

$$\mathscr{F}/2 \text{ lens}, \qquad \frac{d}{f} = 0.3, \qquad \frac{\rho}{f} = 0.14, \qquad \frac{r'}{f} = \frac{1}{5}$$

then

$$E_A = \frac{(\rho/f)^2 + (\rho/f)(r'/f)}{1 + (d/f)}$$

$$= \frac{(0.14)^2 + 0.14(0.2)}{1.3} \sim 0.037$$

If we consider a lens with an $\mathscr{F} = 1.4$ aperture we see from Figure 4.17 that we can increase $[d/f]$ to a value of 1.1 and reduce the error to $E_A = 0.023$ for the same value of $[\rho/f] = 0.14$. We thus obtain the result that the lens with the lower $\mathscr{F}$ number (larger aperture for a given focal length) will provide greater flexibility; that is, for a given input signal aperture a smaller $\mathscr{F}$ number lens will allow greater distances between the input signal plane and the lens to take advantage of the increased accuracy. For a given distance between the signal input plane and the lens the larger the lens aperture, the higher the input frequencies that can be transformed within a given accuracy. We can also infer that small $\mathscr{F}$ number lenses will (at a given input plane F to lens distance) permit larger input apertures to be used.

In practice the required values of $[r'_{max}/f]$ and $[\rho_{max}/f]$ may be lower than those considered above. In such cases the error E_A would be less than the $E_A = 0.023$ determined here and higher $\mathscr{F}$ number lenses (lenses with smaller apertures for a given focal length) may be used. Here we have considered an extreme case and determined what amounts to an extreme error $E_A = 0.023$ (or 2.3 percent). This extreme value of the error introduced by the Fourier transform approximation can be acceptable for most applications and justifies using the transformation approximation.

In the next sections we shall consider the phase of the focused light distribution in the back focal plane F'. It will be shown that selecting $\lceil d/f \rceil = 1$ has advantages in reducing the phase terms not related to the Fourier transform. In Figure 4.17 the point corresponding to $[d/f] = 1$ and $[\rho/f] = 0.14$ is shown to have an error value $E_A = 0.024$. For the extreme case considered in the discussion above we found that when $\lceil d/f \rceil = 1.1$, then $E_A = 0.023$. Comparing the values of

$$\frac{d}{f} = 1.1 \rightarrow E_A = 0.023 \qquad \text{and} \qquad \frac{d}{f} = 1.0 \rightarrow E_A = 0.024$$

we see that a reduction in $[d/f]$ by an amount of 0.1 gives an increased error of only 0.001. Such a slight increase in error E_A will be found acceptable because of the advantage gained in the phase approximation to be discussed. In addition the location of an input plane at a value of $[d/f]$

can never be completely accurate and therefore for a given lens $\mathscr{F}$ number (e.g., $\mathscr{F} = 1.4$) an error in positioning the input plane from a desired value of $[d/f] = 1$ by as much as $+10$ percent will still not exceed the limitations imposed by the lens aperture ($\mathscr{F}$ number); however, the E_A will change depending on the actual value of $[d/f]$ used. The E_A will decrease as $[d/f] \to 1.1$ but E_A will increase as $[d/f]$ decreases from 1.1. For values of $[d/f]$ greater than 1.1 the $\mathscr{F} = 1.4$ lens aperture will restrict the input aperture (which we have considered to be $[r'/f] = \frac{1}{5}$.

We have shown how equations (4-115), (4-123), and (4-124) can be used to determine, or specify, the parameter limits and the accuracy of the Fourier transform representation

$$A'(x, y) = \frac{-je^{jkR(x, y)}}{\lambda(f+d)} \iint A(x_1, y_1)e^{-j2\pi(ux_1 + vy_1)} \, dx_1 \, dy_1 \quad (4\text{-}127)$$

We have shown that for maximum values

$$\left[\frac{\rho_{max}}{f}\right] = 0.14 \quad \text{and} \quad \left[\frac{r'_{max}}{f}\right] = \frac{1}{5}$$

and a desirable choice of $[d/f] = 1$, the worst possible error in amplitude values given by (4-127) is 2.4 percent. The $[\rho_{max}/f] = 0.14$ was selected to maintain the linear frequency relationship E_f to less than 1 percent. Because the neglected term in (4-66) is negative, the approximate amplitude given by equation (4-127) will be higher than the exact values by no more than 2.4 percent.

FOURIER TRANSFORM REPRESENTATION OF OPTICAL IMAGING

By limiting the area of consideration in the input plane to

$$\frac{r'_{max}}{f} < \frac{1}{5} \quad (4\text{-}125)$$

and in the back focal plane to

$$\frac{\rho_{max}}{f} < 0.14 \quad (4\text{-}89)$$

we have shown that the light amplitude distribution $A'(x, y)$ in the back focal plane of a lens is given with reasonable accuracy by

$$A'(x, y) = \frac{-je^{jkR(x, y)}}{\lambda(f+d)} F(u, v) \quad (4\text{-}69)$$

where
$$u = \frac{x}{\lambda f} \qquad (4\text{-}78)$$

$$v = \frac{y}{\lambda f} \qquad (4\text{-}79)$$

$$R(x, y) = \frac{f^2 + df + x^2 + y^2}{(f^2 + x^2 + y^2)^{1/2}}$$

$$R(x, y) = \frac{f^2 + df + \rho^2}{(f^2 + \rho^2)^{1/2}} \qquad (4\text{-}40)$$

$$F(u, v) = \iint A(x_1, y_1) e^{-j2\pi(ux_1 + vy_1)} \, dx_1 \, dy_1 \qquad (4\text{-}70)$$

As given by (4-70) $F(u, v)$ is the two-dimensional Fourier transform of the light amplitude distribution $A(x_1, y_1)$ in a plane perpendicular to the optical axis and at distance d in front of the lens.

As pointed out in the discussion of equations (4-69), (4-73), and (4-74) the phase term $e^{jkR(x, y)}$ is of concern only when a second lens is introduced to produce an image as shown in Figure 4.18. We are considering only conventional optical systems here. If holographic techniques are considered, the phase factor in (4-69) would determine the form of the interference pattern produced by $A'(x, y)$ and a reference signal.

To simplify our notation we introduced K defined as
$$K = \frac{-je^{jkR(x, y)}}{\lambda(f + d)} \qquad (4\text{-}128)$$

and wrote (4-69) in the form shown in equation (4-74) or
$$A'(x, y) = KF(u, v) \qquad (4\text{-}129)$$

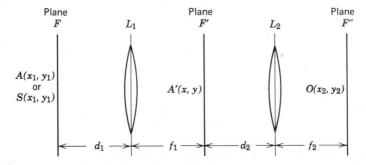

Figure 4.18   Two-lens optical imaging.

In Figure 4.18 the light amplitude distribution $A(x_1, y_1)$ in the input plane F is given as $S(x_1, y_1)$. Using equations (4-70), (4-128), and (4-129) the light amplitude distribution $A'(x, y)$ in the plane F' (back focal plane of lens L_1) is given by

$$A'(x, y) = K_1 \int \int S(x_1, y_1) e^{-j2\pi(ux_1 + vy_1)} \, dx_1 \, dy_1 \qquad (4\text{-}130)$$

where

$$u = \frac{x}{\lambda f_1} \qquad (4\text{-}131)$$

linearization

other integral gone — ie Amplitude

$$v = \frac{y}{\lambda f_1} \qquad (4\text{-}132)$$

Phase

$$K_1 = \frac{-je^{jkR_1(x,y)}}{\lambda(f_1 + d_1)} \qquad (4\text{-}133)$$

$$R_1(x, y) = \frac{f_1^2 + d_1 f_1 + \rho^2}{(f_1^2 + \rho^2)^{1/2}} \qquad (4\text{-}134)$$

Similarly $A'(x, y)$ is the input signal to the lens L_2 and the light distribution $O(x_2, y_2)$ in the output plane F'' (back focal plane L_2) can be written as

$$O(x_2, y_2) = K_2 \int \int A'(x, y) e^{-j2\pi(u'x + v'y)} \, dx \, dy \qquad (4\text{-}135)$$

where

$$u' = \frac{x_2}{\lambda f_2} \qquad v' = \frac{y_2}{\lambda f_2} \qquad (4\text{-}136)$$

$$K_2 = \frac{-je^{jkR_2(x_2,y_2)}}{\lambda(f_2 + d_2)} \qquad (4\text{-}137)$$

$$R_2(x_2, y_2) = \frac{f_2^2 + f_2 d_2 + \rho_2^2}{(f_2^2 + \rho_2^2)^{1/2}} \qquad (4\text{-}138)$$

Substituting (4-130) into (4-135) we obtain an expression for the output image $O(x_2, y_2)$ in terms of the input image $S(x_1, y_1)$

$$O(x_2, y_2) = K_2 \int \int e^{-j2\pi(u'x + v'y)} \, dx \, dy \left[K_1 \int \int S(x_1, y_1) e^{-j2\pi(ux_1 + vy_1)} \, dx_1 \, dy_1 \right]$$

$$(4\text{-}139)$$

We will assume that the function $S(x_1, y_1)$ allows the order of integration to be reversed so that we can rewrite (4-139)

$$O(x_2, y_2) = K_2 \int \int S(x_1, y_1) . dx_1 \, dy_1 \left[\int \int K_1 e^{-j2\pi(ux_1+vy_1+u'x+v'y)} \, dx \, dy \right].$$

(4-140)

The integral within the brackets is complicated by the presence of the factor K_1 which contains an exponential dependent on x and y. If we limit the values of x and y (i.e., ρ/f) so that the phase variation in K_1 can be considered negligible, the K_1 factor can be taken outside the integrals giving

$$O(x_2, y_2) = K_2 K_1 \int \int S(x_1, y_1) \, dx_1 \, dy_1 \left[\int \int e^{-j2\pi(ux_1+vy_1+u'x+v'y)} \, dx \, dy \right].$$

(4-141)

We now consider the integral within the brackets of (4-141) and substitute for u, v, u', and v' from (4-131), (4-132), and (4-136).

$$\int \int e^{-j2\pi(ux_1+vy_1+u'x+v'y)} \, dx \, dy = \int \exp\left[-j2\pi\left(\frac{x_1}{\lambda f_1} + \frac{x_2}{\lambda f_2}\right)x\right] dx$$

$$\times \int \exp\left[-j2\pi\left(\frac{y_1}{\lambda f_1} + \frac{y_2}{\lambda f_2}\right)y\right] dy \quad (4\text{-}142)$$

Up to this point we have not mentioned the limits of integration. Due to the presence of an aperture in an optical system the signals exist only over a finite range of the aperture coordinates. However, since the signals are zero outside the range (e.g., $S(x_1, y_1) = 0$ outside the aperture in the F plane), the contribution to the integral beyond the aperture limits will also be zero. Thus we can define the function from $-\infty$ to $+\infty$ (although it may only take on values within the aperture) so that the integrals can be taken from $-\infty$ to $+\infty$. These limits are in agreement with the Fourier transform integrals.

The Dirac delta function can be defined by the integral equation

$$\delta(t - b) = \int_{-\infty}^{\infty} e^{-j2\pi u(t-b)} \, du. \tag{4-143}$$

We can show this simply by referring back to equations (3-85) and (3-86) of Chapter 3; that is,

$$f(t) = \int_{-\infty}^{\infty} F(\omega) \, e^{j2\pi ft} \, df \tag{3-85}$$

$$F(\omega) = \int_{-\infty}^{\infty} f(t) \, e^{-j2\pi ft} \, dt. \tag{3-86}$$

Because $F(\omega)$ in (3-86) will not depend on the value of the variable t, we can change the variable without changing the value of $F(\omega)$; that is, when the integral is evaluated between limits, the variable t will drop out of the final result. We can now change variables in (3-86) without changing the value of $F(\omega)$ as follows:

$$F(\omega) = \int_{-\infty}^{\infty} f(\tau) \, e^{-j2\pi f\tau} \, d\tau. \tag{4-144}$$

Substituting (4-144) into (3-85) gives

$$f(t) = \int_{-\infty}^{\infty} \left[\int_{-\infty}^{\infty} f(\tau) \, e^{-j2\pi f\tau} \, d\tau \right] e^{j2\pi ft} \, df. \tag{4-145}$$

Rearranging terms

$$f(t) = \int_{-\infty}^{\infty} f(\tau) \, d\tau \left[\int_{-\infty}^{\infty} e^{-j2\pi f(\tau - t)} \, df \right] \tag{4-146}$$

Using the sifting theorem for a delta function, $f(t)$ is related to $f(\tau)$ by the integral

$$f(t) = \int_{-\infty}^{\infty} f(\tau)\delta(\tau - t) \, d\tau. \tag{4-147}$$

The description of (4-147) as a sifting property is based on the fact that the delta function is equal to zero *except* at an infinitesimally small region at $\tau = t$ and the integration gives the value of the function corresponding to this equality

$$f(t) = [f(\tau)]_{\tau=t}.$$

Comparing (4-146) and (4-147) we find that the bracketed expression of (4-146) produces the same integration result as the delta function in (4-147). We can therefore define the delta function by the equation

$$\delta(\tau - t) = \int_{-\infty}^{\infty} e^{-j2\pi f(\tau - t)} \, df. \tag{4-148}$$

Changing variables in (4-148) gives (4-143). Comparing the integral in (4-143) with each of the integrals on the right side of (4-142) we find

$$\int \int e^{-j2\pi(ux_1 + vy_1 + u'x + v'y)} \, dx \, dy = \delta\left[\frac{x_1}{\lambda f_1} + \frac{x_2}{\lambda f_2}\right] \delta\left[\frac{y_1}{\lambda f_1} + \frac{y_2}{\lambda f_2}\right]$$

$$= \lambda^2 f_1^2 \delta\left[x_1 + \frac{f_1}{f_2}x_2\right]\delta\left[y_1 + \frac{f_1}{f_2}y_2\right]. \tag{4-149}$$

In the last step of (4-149) we used the identity

$$\delta(bt) = \frac{1}{|b|}\delta(t). \tag{4-150}$$

The meaning of this identity (in fact all delta-function identities) is that when the product with any function [e.g., $f(t)$] is integrated, the result obtained by multiplying the function by $\delta(bt)$ [i.e., $\int f(t)\,\delta(bt)\,dt$] will be the same as multiplying the function by $(1/|b|)\delta(t)$ [i.e., $\int f(t)\,(1/|b|)$ $\delta(t)\,dt$]. Substituting (4-149) into (4-141) we obtain

$$O(x_2, y_2) = K_2 K_1 \lambda^2 f_1^2 \int\int S(x_1, y_1)\delta\left(x_1 + \frac{f_1}{f_2}x_2\right)\delta\left(y_1 + \frac{f_1}{f_2}y_2\right)dx_1\,dy_1. \tag{4 151}$$

Now we can make use of the sifting property of the Dirac delta function which is defined by

$$f(t) = \int_{-\infty}^{\infty} f(\tau)\,\delta(\tau - t)\,d\tau \tag{4-147}$$

or

$$f(-t) = \int_{-\infty}^{\infty} f(\tau)\,\delta(\tau + t)\,d\tau. \tag{4-152}$$

Applying the property of the delta function shown in (4-152) to (4-151) we can let $x_1 = \tau$ and $t = (f_1/f_2)x_2$ for the x_1 integration and likewise for y_1 integration, then

$$O(x_2, y_2) = K_2 K_1 \lambda^2 f_1^2 S\left(-\frac{f_1}{f_2}x_2, -\frac{f_1}{f_2}y_2\right). \tag{4-153}$$

In deriving (4-153) we assumed that K_1 was approximately constant. The factor K_2 is variable only in phase as seen by referring to (4-137). Because only intensity is seen or measured, the phase variation of K_2 can be ignored (i.e., when multiplying by the complex conjugate, the phase term will drop out) and (4-153) can be interpreted as giving the output image $O(x_2, y_2)$ in terms of a proportionality factor $(K_2 K_1 \lambda^2 f_1^2)$ multiplying the original input signal S expressed in the new coordinates $[-(f_1/f_2)x_2,$ $-(f_1/f_2)y_2]$. To clarify the significance of these new coordinates let us consider the relation between

$$S(x_1, y_1) \rightarrow S\left(-\frac{f_1}{f_2}x_2, -\frac{f_1}{f_2}y_2\right). \tag{4-154}$$

Because the two sides of relation (4-154) correspond point for point, we find that the (x_1, y_1) and (x_2, y_2) coordinates are related by

$$x_1 = -\frac{f_1}{f_2}x_2 \qquad y_1 = -\frac{f_1}{f_2}y_2. \qquad (4\text{-}155)$$

Equations (4-154) and (4-155) represent the fact that a point of the signal that was originally at the coordinates x_1 and y_1 will be imaged, respectively, to the point at

$$x_2 = -\frac{f_2}{f_1}x_1 \qquad \text{and} \qquad y_2 = -\frac{f_2}{f_1}y_1.$$

The magnification in an optical image is defined as the ratio of the imaged coordinate of a point to the original coordinate

$$m_x = \frac{x_2}{x_1} = -\frac{f_2}{f_1}$$

$$m_y = \frac{y_2}{y_1} = -\frac{f_2}{f_1}. \qquad (4\text{-}156)$$

Equation (4-156) was written separately although it is apparent that here the magnification is the same in any direction. In some cases it is possible to obtain different magnifications in different directions (e.g., cylindrical lens systems). Equations (4-153) and (4-156) show that the output image is proportional to the input image with a change in scale. The minus signs that appear in (4-153) and (4-156) represent an inversion of the image.

In many applications there is no requirement for a magnified image. In such cases we could use lenses of equal focal length $f_1 = f_2$ and obtain a magnification $m = -1$.

For the case $f_1 = f_2 = f$ equation (4-153) becomes

$$O(x_2, y_2) = K_2 K_1 \lambda^2 f^2 S(-x_2, y_2). \qquad (4\text{-}157)$$

Thus for equal focal-length lenses the output image $O(x_2, y_2)$ is proportional to an inverted replica of the input signal.

It is for this case $(f_1 = f_2 = f)$ that the optical imaging process can be described as *consecutive Fourier transforms*. This can be shown by replacing f_2 by f in (4-136) and substituting for u' and v' in (4-135) to obtain

$$O(x_2, y_2) = K_2 \int \int A'(x, y) \exp\left[-j2\pi\left(\frac{x_2 x}{\lambda f} + \frac{y_2 y}{\lambda f}\right)\right] dx\, dy \qquad (4\text{-}158)$$

by replacing f_1 by f in (4-131) and (4-132) we find

$$u = \frac{x}{\lambda f}, \quad v = \frac{y}{\lambda f}, \quad dx = \lambda f\, du, \quad dy = \lambda f\, dv. \qquad (4\text{-}159)$$

Substituting (4-159) into (4-158) we obtain the result

$$O(x_2, y_2) = K_2 \lambda^2 f^2 \int\int A'(x, y) e^{-j2\pi(ux_2 + vy_2)}\, du\, dv. \qquad (4\text{-}160)$$

Using equations (4-69), (4-129), (4-139), (4-157), and (4-160) we can express the two-step process of optical imaging as

$$A'(x, y) = K_1 F(u, v) = K_1 \int\int S(x_1, y_1) e^{-j2\pi(ux_1 + vy_1)}\, dx_1\, dy_1$$
$$\qquad (4\text{-}161)$$

$$O(x_2, y_2) = K_2 K_1 \lambda^2 f^2 S(-x_2, -y_2)$$
$$\qquad = K_2 K_1 \lambda^2 f^2 \int\int F(u, v) e^{-j2\pi(ux_2 + vy_2)}\, du\, dv \qquad (4\text{-}162)$$

where

$$u = \frac{x}{\lambda f}, \quad v = \frac{y}{\lambda f}$$

$$K_1 = \frac{-je^{jkR_1(x,y)}}{\lambda(f + d_1)} \qquad K_2 = \frac{-je^{jkR_2(x_2,y_2)}}{\lambda(f + d_2)}$$

$$R_1(x, y) = \frac{f^2 + fd_1 + \rho^2}{(f^2 + \rho^2)^{1/2}} \qquad R_2(x_2, y_2) = \frac{f^2 + fd_2 + \rho_2^2}{(f^2 + \rho^2)^{1/2}}$$

$$\rho^2 = x^2 + y^2 \qquad\qquad \rho_2^2 = x_2^2 + y_2^2.$$

The last expression in (4-162) assumes that the factor K_1 can be considered constant in phase over the range of the values u and v (i.e., x and y). This approximation is the subject we are about to consider and here we are showing the advantages of the resulting expression. To appreciate the significance of (4-161) and (4-162) let us disregard optics for a moment and consider the standard mathematical Fourier transform equation using our notation.

Fourier Transform

$$F(u, v) = \int\int S(x_1, y_1) e^{-j2\pi(ux_1 + vy_1)}\, dx_1\, dy_1 \qquad (4\text{-}163)$$

Inverse Fourier Transform

$$S(x_1, y_1) = \int\int F(u, v) e^{+j2\pi(ux_1 + vy_1)}\, du\, dv. \qquad (4\text{-}164)$$

Fourier Transform of a Fourier Transform

$$S(-x_1, -y_1) = \int \int F(u, v) e^{-j2\pi(ux_1 + vy_1)} \, du \, dv. \qquad (4\text{-}165)$$

Equation (4-163) is usually referred to as the *Fourier transform* while (4-164) is the *inverse Fourier transform*. We note that the exponent of (4-164) is positive and that of (4-165) is negative. Now because the optical transform produced by a lens always has a negative exponent [see (4-161) and (4-162)], the inverse transform defined by (4-164) never appears in optical systems; that is, in optical systems we can only take a Fourier transform but not an inverse Fourier transform. This means, for example, that we can optically take a Fourier transform of a function of x—that is,

$$T[f(x)] \rightarrow F(\omega), \qquad \text{can be done optically} \qquad (4\text{-}166)$$

—but we cannot optically perform the inverse Fourier transform—that is,

$$T^{-1}[F(\omega)] \rightarrow f(x), \qquad \text{cannot be done optically} \qquad (4\text{-}167)$$

—because only the Fourier transform can be performed optically. Then if we have a spectrum $F(\omega)$, we can optically obtain $f(-x)$ or an inverted image—that is,

$$T[F(\omega)] \rightarrow f(-x). \qquad (4\text{-}168)$$

The second lens in an optical system such as that of Figure 4.18 therefore produces a Fourier transform of a Fourier transform as represented by (4-165). We note that the inversion (or change of sign of the coordinate that appears in the image) is introduced by the second optical Fourier transform whereas if an inverse transform were taken, it would not invert the signal. Thus comparing equations (4-161) and (4-162) with (4-163) and (4-165) we note that the optical imaging process of two lenses is described by two successive Fourier transform relations. Except for determining the absolute amplitudes involved, the constants in front of the integrals of (4-161) and (4-162) do not affect the form of the variations. In most cases only the relative amplitudes are of interest and the constants are dropped.

The advantage of the Fourier transform representation described by (4-161) and (4-162) can be shown by considering the introduction of a filter in the F' plane. If we know the transmission characteristics of the filter, we can determine a function $M(u, v)$ which represents a factor that can be used to determine the fraction of the incident light amplitude passed at each coordinate corresponding to the value of u and v. The filtered output $O_f(x_2, y_2)$ is then given by (4-162) if we replace $F(u, v)$ by $M(u, v)F(u, v)$, [i.e., the product of the incident light to the filter $F(u, v)$ times the fraction of the light passed by the filter $(M(u, v))$]

$$O_f(x_2, y_2) = K_2 K_1 \lambda^2 f^2 \int \int M(u, v) F(u, v) e^{-j2\pi(ux_2+vy_2)} du \, dv. \quad (4\text{-}169)$$

Thus the specification of a filter for a particular application can be determined uniquely when the Fourier transform representation is used.

We might note at this point that the Fourier transform representation of (4-161) and (4-162) requires the use of lenses of equal focal length. The specification of a filter for the case of unequal focal lengths is exactly the same; however, the Fourier transform relation of (4-162) is modified by introducing a factor of f_1/f_2 in the exponent to account for the magnification.

This additional f_1/f_2 results in a magnified filtered image that contains the same information as the filtered image $O_f(x_2, y_2)$ given by (4-169). The only difference (neglecting the phase of K_2) is in the scale. Throughout the remaining part of this discussion we shall consider the special case of equal focal-length lenses to simplify our analysis.

ELIMINATION OF UNDESIREABLE PHASE VARIATIONS (DOUBLE FOURIER TRANSFORM)

Now that we have seen the significance of the Fourier transform representation of optical imaging we shall consider the approximation involved in the derivation of (4-162). The actual relation corresponding to (4-162) that was written for $f_1 = f_2 = f$ can be written as

$$O(x_2, y_2) = K_2 \lambda^2 f^2 \int \int K_1 F(u, v) e^{-j2\pi(ux_2+vy_2)} du \, dv. \quad (4\text{-}170)$$

The term K_1 appearing in the integral was defined as

$$K_1 = \frac{-je^{jkR_1(x,y)}}{\lambda(f+d_1)} \quad (4\text{-}171)$$

where

$$R_1(x, y) = \frac{f^2 + fd_1 + \rho^2}{(f^2 + \rho^2)^{1/2}} \quad (4\text{-}172)$$

$$\rho^2 = x^2 + y^2. \quad (4\text{-}173)$$

We have seen that (4-162) is valid only if the phase variation of K_1 can be neglected over the range of the integration variables u and v (or x and y).

We will now compare (4-170) and (4-162) to determine the validity of neglecting this phase variation in K_1.

For reasons that we shall soon see we would like to express $R_1(x, y)$ in the form of

$$R_1(x, y) = (f + d_1) + P(f - d_1) + Qf \tag{4-174}$$

$$= f(1 + P + Q) + d_1(1 - P) \tag{4-175}$$

Equation (4-172) can be rewritten as

$$R_1(x, y) = \frac{f^2 + \rho^2}{(f^2 + \rho^2)^{1/2}} + \frac{d_1 f}{(f^2 + \rho^2)^{1/2}} \tag{4-176}$$

$$R_1(x, y) = f\left(1 + \frac{\rho^2}{f^2}\right)^{1/2} + d_1\left(1 + \frac{\rho^2}{f^2}\right)^{-1/2}. \tag{4-177}$$

Equating the coefficient of f and d_1 in (4-175) and (4-177) we obtain

$$1 - P = \left(1 + \frac{\rho^2}{f^2}\right)^{-1/2}. \tag{4-178}$$

$$1 + P + Q = \left(1 + \frac{\rho^2}{f^2}\right)^{1/2}. \tag{4-179}$$

Solving for P and Q we obtain

$$P = 1 - \left(1 + \frac{\rho^2}{f^2}\right)^{-1/2} \tag{4-180}$$

$$Q = \left(1 + \frac{\rho^2}{f^2}\right)^{1/2} + \left(1 + \frac{\rho^2}{f^2}\right)^{-1/2} - 2 \tag{4-181}$$

$$Q = \left(1 + \frac{\rho^2}{f^2}\right)^{1/2} + \frac{1}{(1 + \rho^2/f^2)^{1/2}} - 2 \tag{4-182}$$

$$= \frac{(1 + \rho^2/f^2) + 1}{(1 + \rho^2/f^2)^{1/2}} - 2 = \frac{(\rho^2/f^2) + 2}{(1 + \rho^2/f^2)^{1/2}} - 2 \tag{4-183}$$

$$Q = 2\left[\frac{1 + (\rho^2/2f^2)}{(1 + \rho^2/f^2)^{1/2}} - 1\right]. \tag{4-184}$$

Substituting for P and Q equation (4-174) can be written

$$R_1(x, y) = (f + d_1) + (f - d_1)\left[1 - \frac{1}{(1 + \rho^2/f^2)^{1/2}}\right] + 2f\left[\frac{1 + \rho^2/2f^2}{(1 + \rho^2/f^2)^{1/2}} - 1\right]. \tag{4-185}$$

The K_1 term given by (4-171) can be rewritten using (4-185)

$$K_1 = \frac{-je^{jkR_1(x,y)}}{\lambda(f+d_1)} \tag{4-171}$$

$$K_1 = \left[\frac{-je^{jk(f+d_1)}}{\lambda(f+d_1)}\right] \exp\left\{jk(f-d_1)\left[1-\frac{1}{(1+\rho^2/f^2)^{1/2}}\right]\right\}$$

$$\exp\left\{jk2f\left[\frac{1+\rho^2/2f^2}{(1+\rho^2/f^2)^{1/2}}-1\right]\right\} \tag{4-186}$$

The terms grouped in the first brackets of (4-186) are constants and can therefore be taken out from under the integral sign of (4-170). The remaining exponentials in (4-186) are phase factors that depend on the variables of integration. The exponential of these remaining terms must be limited so that the phase variations can be considered negligible.

Because we are going to consider restrictions on the value of ρ/f so that the phase terms can be considered negligible, we can simplify the exponentials of (4-186) by expanding in power series of (ρ^2/f^2) and drop all but the first terms of the expansions. Expanding the exponent of the first exponential using the series expansion given in (4-76) we obtain

$$\exp\left\{jk(f-d_1)\left[1-\frac{1}{(1+e^2/f^2)^{1/2}}\right]\right\} = \exp\left\{jk(f-d_1)\right.$$

$$\left. \times\left(1-\left[1-\frac{\rho^2}{2f^2}+\frac{3}{8}\left(\frac{\rho^2}{f^2}\right)^2\cdots\right]\right)\right\} \tag{4-187}$$

$$\simeq \exp\left[j\frac{k}{2}(f-d_1)\left(\frac{\rho}{f}\right)^2\right] \tag{4-188}$$

Similarly we expand the second exponential and obtain,

$$\exp\left\{j2kf\left[\frac{1+\rho^2/2f^2}{(1+\rho^2/f^2)^{1/2}}-1\right]\right\} = \exp\left\{j2kf\left(-1+\left(1+\frac{\rho^2}{2f^2}\right)\right.\right.$$

$$\left.\left. \times\left[1-\frac{\rho^2}{2f^2}+\frac{3}{8}\left(\frac{\rho^2}{f^2}\right)^2\cdots\right]\right)\right\} \tag{4-189}$$

$$\simeq \exp\left[j\frac{kf}{4}\left(\frac{\rho}{f}\right)^4\right] \tag{4-190}$$

Substituting these approximate terms in (4-186)

$$K_1 = \left[\frac{-je^{jk(f+d_1)}}{\lambda(f+d_1)}\right] \exp\left[\frac{jk}{2}(f-d_1)\left(\frac{\rho}{f}\right)^2\right] \exp\left[j\frac{kf}{4}\left(\frac{\rho}{f}\right)^4\right] \quad (4\text{-}191)$$

For values of (ρ/f) less than our previous limit of 0.14 the approximations in each phase term in (4-191) are accurate to within 2 percent of the exact values given by (4-186). In addition it should be noted that neglecting terms in the exponentials as done here is valid only because we are considering phase changes of less than one cycle.

Because we are trying to eliminate the phase variations of K_1, equation (4-191) indicates that the optimum choice of the distance d_1 equal to f eliminates the first phase term; that is, when $d_1 = f$, equation (4-191) can be reduced to

$$K_1 = \left[\frac{-je^{j2kf}}{2\lambda f}\right] \exp\left[j\frac{kf}{4}\left(\frac{\rho}{f}\right)^4\right] \qquad \text{for } d_1 = f \qquad (4\text{-}192)$$

If we substitute (4-192) for K_1 in (4-170), we obtain

$$O(x_2, y_2) = K_2\lambda^2 f^2 \int\int K_1 F(u, v) e^{-j2\pi(ux_2 + vy_2)}\, du\, dv. \qquad (4\text{-}170)$$

$$= K_2\lambda^2 f^2 \left[\frac{-je^{j2kf}}{2\lambda f}\right] \int\int \exp\left[j\frac{kf}{4}\left(\frac{\rho}{f}\right)^4\right] F(u, v) e^{-j2\pi(ux_2 + vy_2)}\, du\, dv$$

$$(4\text{-}193)$$

$$= -j\frac{K_2\lambda f}{2} e^{j2kf} \int\int \exp\left[j\frac{kf}{4}\left(\frac{\rho}{f}\right)^4\right] F(u, v) e^{-j2\pi(ux_2 + vy_2)}\, du\, dv$$

$$(4\text{-}194)$$

When the value of (ρ/f) is limited so that the phase factor appearing under the integral of (4-194) can be neglected, the output image $O(x_2, y_2)$ is given by

$$O(x_2, y_2) = K_1' K_2 \lambda^2 f^2 \int\int F(u, v) e^{-j2\pi(ux_1 + vy_2)}\, du\, dv \qquad (4\text{-}195)$$

where

$$K_1' = \frac{-je^{j2kf}}{2\lambda f} \qquad (4\text{-}196)$$

Equation (4-162) was shown to be the desired form for the Fourier transform representation of optical imaging. Equation (4-195) is identical to (4-162) when K_1 in (4-162) can be approximated by K_1' as given by (4-196).

To derive a specification for the maximum limit on (ρ/f) that allows the variable term in K_1' (4-192) to be neglected (4-196), we shall consider the effect of the phase term for a particular $F(u, v)$; that is,

$$F(u, v) = \delta(v)\left[A_0\delta(u) + \frac{B}{2}\{\delta(u - u_0) + \delta(u + u_0)\}\right] \quad (4\text{-}197)$$

The particular function whose Fourier transform (for an infinite aperture) is represented by (4-197) is

$$A_0 + B \cos 2\pi u_0 x_1$$

and its spectrum $F(u, v)$ exist only where $v = 0$.

The locations of the frequency terms contained in equation (4-197) are diagrammed in Figure 4.19. Because by definition the delta function $\delta(v)$ is equal to zero for v unequal to zero, equation (4-197) represents the spectrum of a signal that varies only in one dimension; that is, there are no frequency components in the y direction. The signal therefore is constant with respect to the y coordinate. Rather than interpret (4-197) as the spectrum of a particular signal we can also assume that we are considering only three sample points of a more general spectrum. Because there is nothing to single out the x direction in an optical system, our analysis will apply to a set of spectral points along any radial axis in the frequency plane as indicated by the ρ axis in Figure 4.19. The terms that we will use to specify a maximum limit on (ρ/f) also apply to the general case. We will therefore simply interpret the results for the special case of (4-197) as general criteria for neglecting the undesired phase factor in (4-194).

Substituting for $F(u, v)$ as given by (4-197) into (4-194) and (4-195)

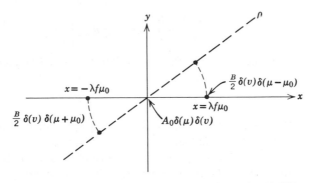

Figure 4.19 Location of frequency term of equation (4-197).

and performing the integration by applying the sifting property of the delta function $\delta(v)$ we obtain

$$\underline{O(x_2)} = K' \int \exp\left[j\frac{kf}{4}\left(\frac{x}{f}\right)^4\right]\left[A_0\delta(u) + \frac{B}{2}\{\delta(u-u_0) + \delta(u+u_0)\}\right]$$

$$\times e^{-j2\pi u x_2}\, du \qquad (4\text{-}198)$$

$$O(x_2) = K' \int \left[A_0\delta(u) + \frac{B}{2}\{\delta(u-u_0) + \delta(u+u_0)\}\right]e^{-j2\pi u x_2}\, du \qquad (4\text{-}199)$$

The new factor K' in (4-198) and (4-199) is defined as

$$K' = \frac{-jK_2\lambda f e^{j2kf}}{2} = K_1'K_2\lambda^2 f^2 \qquad (4\text{-}200)$$

In (4-198) the $\underline{O(x_2)}$ is underlined to identify the equation in which the phase term is not neglected as compared to (4-199) in which the phase term is neglected. The y_2 dependence has been dropped on the left side of equations (4-198) and (4-199) because the output image varies only with respect to the x_2 coordinate.

The first exponential in (4-198) can be rewritten using the definition

$$u \Rightarrow x/\lambda f \qquad (4\text{-}78)$$

and the integration with respect to u is performed simply by applying the sifting property of the delta function

$$\underline{O(x_2)} = K' \int \exp\left[j\frac{kf}{4}(\lambda u)^4\right]\left[A_0\delta(u) + \frac{B}{2}\{\delta(u-u_0) + \delta(u+u_0)\}\right]$$

$$\times e^{-j2\pi u x_2}\, du \qquad (4\text{-}201)$$

$$= K'\left[A_0 + \frac{B}{2}\exp\left[j\frac{kf}{4}(\lambda u_0)^4\right]\{e^{j2\pi u_0 x_2} + e^{-j2\pi u_0 x_2}\}\right] \qquad (4\text{-}202)$$

$$\underline{O(x_2)} = K'\left[A_0 + B\exp\left[j\frac{kf}{4}(\lambda u_0)^4\right]\cos 2\pi u_0 x_2\right] \qquad (4\text{-}203)$$

The result after integrating (4-199) is similar to (4-203) except for the exponential that appears in (4-203), that is,

$$O(x_2) = K'[A_0 + B\cos 2\pi u_0 x_2] \qquad (4\text{-}204)$$

Equations (4-203) and (4-204) represent the amplitude distribution of the output image produced by the frequency plane distribution $F(u, v)$ given by (4-197). Equation (4-203) represents the output $O(x_2)$ when the phase factor is considered and (4-204) represents the output $O(x_2)$ when the phase factor is neglected. Comparing (4-203) and (4-204) a criterion for neglecting the phase factor is still not very apparent since the significance of the exponential is not very clear.

If we consider the observation of the output image, we must deal with the intensity rather than the amplitude. The amplitude of the output image is given by (4-203) and (4-204) while the intensities are given by the relations

$$I(x_2) = \underline{O(x_2)}\, \underline{O^*(x_2)} = |O(x_2)|^2 \tag{4-205}$$

$$I(x_2) = O(x_2)\, O^*(x_2) = |O(x_2)|^2 \tag{4-206}$$

The starred terms in (4-205) and (4-206) represent the complex conjugate of the unstarred terms. Using equations (4-203) and (4-204) in (4-205) and (4-206), respectively, we obtain

$$I(x_2) = \underline{O(x_2)}\, \underline{O^*(x_2)} = K'\left[A_0 + B \exp\left\{j\frac{kf}{4}(\lambda u_0)^4\right\} \cos 2\pi u_0 x_2\right]$$

$$\times K'\left[A_0 + B \exp\left[-j\frac{kf}{4}(\lambda u_0)^4\right] \cos 2\pi u_0 x_2\right]$$

$$\tag{4-207}$$

$$= |K'|^2\left[A_0^2 + A_0 B \exp\left[j\frac{kf}{4}(\lambda u_0)^4\right] \cos 2\pi u_0 x_2\right.$$

$$+ A_0 B \exp\left[-j\frac{kf}{4}(\lambda u_0)^4\right] \cos 2\pi u_0 x_2$$

$$\left. + B^2 \cos^2 2\pi u_0 x_2\right] \tag{4-208}$$

$$= |K'|^2\left[A_0^2 + 2A_0 B\right.$$

$$\times \left\{\frac{\exp\left[j\frac{kf}{4}(\lambda u_0)^4\right] + \exp\left[-j\frac{kf}{4})\lambda u_0)^4\right]}{2}\right\}$$

$$\left. \times \cos 2\pi u_0 x_2 + B^2 \cos^2 2\pi u_0 x_2\right] \tag{4-209}$$

$$\underline{I(x_2)} = |K'|^2\left[A_0^2 + 2A_0 B \cos \frac{kf}{4}(\lambda u_0)^4 \cos 2\pi u_0 x_2\right.$$

$$\left. + B^2 \cos^2 2\pi u_0 x_2\right] \tag{4-210}$$

$$I(x_2) = O(x_2)O^*(x_2) = K'[A_0 + B \cos 2\pi u_0 x_2] K'[A_0 + B \cos 2\pi u_0 x_2]$$

$$(4\text{-}211)$$

$$= |K'|^2 [A_0^2 + 2A_0 B \cos 2\pi u_0 x_2 + B^2 \cos^2 2\pi u_0 x_2]$$

$$(4\text{-}212)$$

Comparing (4-210) with (4-212) we can see that the phase factor has introduced a cosine factor

$$\cos \frac{kf}{4} (\lambda u_0)^4 \qquad (4\text{-}213)$$

which attenuates the $\cos 2\pi u_0 x_2$ component in the observed image.

In optics the amplitude of the DC term must exceed the amplitude of any harmonic component in order to avoid negative values of the function (i.e., there cannot be negative light values, so functions are generally biased by a DC component A_0). The second term in (4-210) and (4-212) represents the larger of the two x_2 dependent terms in the image intensity; that is, since $B < A_0$, $B^2 < A_0 B$. It is desirable to limit the maximum value of u_0 to obtain a value of

$$\cos \frac{kf}{4} (\lambda u_0)^4 \qquad (4\text{-}213)$$

as near to one as possible so that the Fourier transform representation used in deriving (4-212) can be considered a good approximation; that is,

$$\cos \frac{kf}{4} (\lambda u_0)^4 = 1$$

then (4-210) and (4-212) are identical.

We can express the cosine term as a function of x by the definition

$$u = \frac{x}{\lambda f} \qquad (4\text{-}78)$$

and because our results will apply to the general case, we can replace x by the more general notation ρ. Thus we can express the cosine term as

$$\cos \frac{kf}{4} (\lambda u_0)^4 = \cos \frac{kf}{4} \left(\frac{\rho}{f}\right)^2 \qquad (4\text{-}213)$$

In (4-213) the frequency u_0 is given the general interpretation of a spatial frequency in the direction of a ρ axis (see Figure 4.19) and u_0 and ρ are related by

$$u_0 = \frac{\rho}{\lambda f} \qquad (4\text{-}78A)$$

that is, equation (4-213) applies to the general case of spectrum along any radial axis ρ in the back focal plane F'.

We can rewrite (4-213) by introducing a new term N

$$\cos \frac{kf}{4}\left(\frac{\rho}{f}\right)^4 = \cos N^4 \qquad (4\text{-}214)$$

where

$$N = \left(\frac{\rho}{f}\right)\left(\frac{kf}{4}\right)^{1/4}$$

We can recognize the ultimate limit on N by noting that for $N = (\pi/2)^{1/4}$ the cosine term given by (4-214) is zero (i.e., $\cos \pi/2 = 0$). For this value of N the Fourier transform result of (4-212) is completely in error with respect to the second term because the cosine term present in (4-210) is zero and the second term is eliminated; that is, equation (4-210) which does not neglect phase terms has no second term while (4-212) has a second term. Thus for $N = (\pi/2)^{1/4}$ the two-step Fourier transform representation given by (4-212) gives a term that does not exist in the actual image given by (4-210). (NOTE: this accounts for the neglecting of phase terms when very much less than $\pi/2$ which appears in most treatments of approximate solutions to diffraction problems. For values of $N < (\pi/2)^{1/4}$ the cosine of (4-214) has nonzero values as shown by the curve of Figure 4.20. Selecting a limit on N is based on specifying how accurately the second term of (4-210) and (4-212) should agree. A value of $N = 0$ is necessary to have complete agreement between (4-210) and (4-212); however, $N = 0$ corresponds to a frequency $u_0 = 0$ which corresponds to a DC term only. Thus a compromise limit must be established between the limits $N = 0$ and $N = (\pi/2)^{1/4}$.

ACCURACY OF PHASE APPROXIMATION (DOUBLE FOURIER TRANSFORM)

To determine the limitation on N it is necessary to specify the desired accuracy of the second term in (4-212) as compared to the second term in (4-210). Again the accuracy of our approximation can be given in terms of a fractional error E_ϕ defined as

$$E_\phi = \frac{1 - \cos N^4}{\cos N^4} = \frac{1}{\cos N^4} - 1 \qquad (4\text{-}215)$$

The error in the second term of (4-212) is then $+100\, E_\phi$ percent compared to the exact term in (4-210). We note that the error E_ϕ does not represent a fraction of the total image intensity. The error E_ϕ corresponds only to a particular term in the image intensity, equations (4-210) and (4-212).

We have been discussing the effect of the phase term for a particular $F(u, v)$ as given in (4-197). In this specific case the Fourier transform of the function considered was a biased cosine function, that is, $A_0 + B \cos 2\pi u_0 x_2$. In the more general case we might have a series of terms, for example,

$$A_0 + \sum_n B_n \cos 2\pi u_n x_2 \qquad (4\text{-}216)$$

which in a sense can be considered as the special case of

$$A_0 + \int B_n \cos 2\pi u_n x_2$$

To find the intensity we square the amplitude function (4-216) which for the more general case would give

$$I = |\underbrace{A_0}_{\text{first term}} + \underbrace{\sum B_n \cos 2\pi u_n x_2}_{\text{second term}}|^2 \qquad (4\text{-}217)$$

In order to square the series represented by the second term of (4-217) we shall identify the two terms making the square by different subscripts. The product of the second term in (4-217) can be represented as

$$\left(\sum_n B_n \cos 2\pi u_n x_2 \right)\left(\sum_m B_m \cos 2\pi u_m x_2 \right)$$

where m and n are integers and may or may not be equal to each other. We can therefore write the intensity I in (4-217) as

$$I = \underbrace{A_0^2}_{\substack{\text{product} \\ \text{of} \\ \text{first} \\ \text{term}}} + \underbrace{2\sum_n A_0 B_n \cos 2\pi u_n x_2}_{\text{cross product}}$$

$$\underbrace{+ \sum_n B_n^2 \cos^2 2\pi u_n x_2 + \sum\sum_{m \neq n} B_m B_n \cos 2\pi u_m x_2 \cos 2\pi u_n x_2}_{\text{product of second terms}} \qquad (4\text{-}218)$$

where we have used m and n as integers.

We have considered an amplitude function (4-216) and disregarded any phase terms. We can rewrite (4-216) to include phase terms as follows:

$$A_0 + \sum_n B_n \exp\left[j\frac{kf}{4}(\lambda u_n)^4\right] \cos 2\pi u_n x_2 \qquad (4\text{-}219)$$

The actual intensity equation therefore contains phase terms and can be written as

$$I = \left\{A_0 + \sum_n B_n \exp\left[j\frac{kf}{4}(\lambda u_n)^4\right] \cos 2\pi u_n x_2\right\}$$

$$\left\{A_0 + \sum_n B_n \exp\left[-j\frac{kf}{4}(\lambda u_n)^4\right] \cos 2\pi u_n x_2\right\} \qquad (4\text{-}220)$$

Expanding (4-219) as was done for (4-217) we obtain

$$I = A_0^2 + A_0 \sum_n B_n \exp\left[j\frac{kf}{4}(\lambda u_n)^4\right] \cos 2\pi u_n x_2$$

$$+ A_0 \sum_n B_n \exp\left[-j\frac{kf}{4}(\lambda u_n)^4\right] \cos 2\pi u_n x_2 + \sum_n B_n^2 \cos^2 2\pi u_n x_2$$

$$+ \sum_{m \neq n}\sum B_n B_m \exp\left[j\frac{kf}{4}\lambda^4 (u_n - u_m)^4\right] \cos 2\pi u_n x_2 \cos 2\pi u_m x_2$$

$$(4\text{-}221)$$

where

$$A_0 \sum_n B_n \exp\left[j\frac{kf}{4}(\lambda u_n)^4\right] \cos 2\pi u_n x_2$$

$$+ A_0 \sum_n B_n \exp\left[-j\frac{kf}{4}(\lambda u_n)^4\right] \cos 2\pi u_n x_2$$

$$= 2A_0 \sum_n B_n \cos\frac{kf}{4}(\lambda u_n)^4 \cos 2\pi u_n x_2$$

and by using Euler's formula

$$e^{jx} = \cos x + j \sin x$$

$$\sum_{m \neq n}\sum B_n B_m \exp\left[j\frac{kf}{4}\lambda^4 (u_n - u_m)^4\right] \cos 2\pi u_n x_2 \cos 2\pi u_m x^2$$

$$= \sum_{m \neq n}\sum B_n B_m \cos\frac{kf}{4}\lambda^4 (u_n - u_m)^4 \cos 2\pi u_n x_2 \cos 2\pi u_m x_2$$

$$+ j \sum_{m \neq n}\sum B_m B_m \sin\frac{kf}{4}\lambda^4 (u_n - u_m)^4 \cos 2\pi u_n x_2 \cos 2\pi u_m x_2$$

We can now write (4-221) as

$$I = A_0^2 + 2A_0 \sum_n B_n \left[\cos \frac{kf}{4}(\lambda u_n)^4 \right] \cos 2\pi u_n x_2 + \sum_n B_n 2 \cos^2 2\pi u_n x_2$$

$$+ \left[\sum_{m \neq n} \sum B_n B_m \cos \frac{kf}{4} \lambda^4 (u_n - u_m)^4 \cos 2\pi u_n x_2 \cos 2\pi u_m x_2 \right]$$

$$+ \left[j \sum_{m \neq n} \sum B_n B_m \sin \frac{kf}{4} \lambda^4 (u_n - u_m)^4 \cos 2\pi u_n x_2 \cos 2\pi u_m x_2 \right] \qquad (4\text{-}222)$$

Equation (4-222) which includes phase terms can now be compared to (4-218) which neglected the phase terms. If we examine the three bracketed terms in (4-222)

$$\cos \frac{kf}{4}(\lambda u_n)^4 \to 1 \qquad \text{by letting} \qquad \frac{kf}{4}(\lambda u_n)^4 \to 0$$

and

$$\cos \frac{kf}{4} \lambda^4 (u_n - u_m)^4 \to 1 \qquad \text{by letting} \qquad \frac{kf}{4} \lambda^4 (u_n - u_m)^4 \to 0$$

and

$$\sin \frac{kf}{4} \lambda^4 (u_n - u_m)^4 \to 0 \qquad \text{by letting} \qquad \frac{kf}{4} \lambda^4 (u_n - u_m)^4 \to 0$$

then (4-222) reduces to (4-218).

We can now see that in the general case there would be a series of terms for the image intensity. The maximum error E_ϕ would be determined by (4-215). The largest error introduced by the three terms above will correspond to that of the maximum frequency allowed as given previously for the restricted case; that is, E_ϕ represents the maximum error in the terms similar to the second term of (4-212) and would correspond to the term involving the highest frequency of interest.

APPLICATION OF ACCURACY LIMITATION

Specifying the maximum allowable error is rather arbitrary and will usually depend on the particular application considered. For an example, we can consider specifying a limit of 2 percent accuracy for our approximation; that is, $E_\phi = 0.02$. Substituting in (4-215) we obtain the condition for the maximum value of N

$$E_\phi = 0.02 \leqslant \frac{1}{\cos N^4} - 1$$

$$\cos N^4 \geqslant \frac{1}{1.02} = 0.98 \qquad (4\text{-}223)$$

From Figure 4-20 we find that the relation (4-223) requires values of N less than or equal to 0.67. Using the maximum value $N = 0.67$, we obtain the following results from (4-214).

$$N = \left(\frac{\rho}{f}\right)_{max} \left(\frac{kf}{4}\right)^{1/4} = 0.67 \tag{4-224}$$

then

$$\left(\frac{\rho}{f}\right)_{max} = 0.67 \left(\frac{4}{kf}\right)^{1/4} \tag{4-225}$$

To obtain a numerical result for comparison with our previous limit

$$\left(\frac{\rho}{f}\right)_{max} = 0.14 \tag{4-89}$$

we will again consider a wavelength $\lambda = 5461 \times 10^{-8}$ cm and a lens of focal length $f = 10$ cm. Substituting in equation (4-225) where $k = 2\pi/\lambda$ we find

$$\left(\frac{\rho}{f}\right)_{max} = 0.67 \left(\frac{4\lambda}{2\pi f}\right)^{1/4} = 0.67 \left(\frac{4 \times 5461 \times 10^{-8}}{2\pi \times 10}\right)^{1/4} \tag{4-226}$$

$$\left(\frac{\rho}{f}\right)_{max} = 0.03 \tag{4-227}$$

Equation (4-227) specifies the aperture limit in the frequency plane F' to assure an error limit of less than 2 percent due to neglecting phase variations. We note that this phase limit restricts the frequency aperture

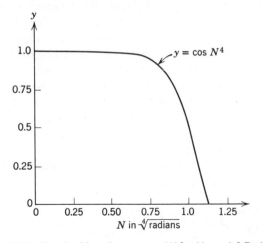

Figure 4.20 Graph of function $y = \cos N^4$ for $N = \pm 1.2$ Radians.

to approximately one fifth of the previous value of $(\rho/f)_{max} = 0.14$ which was sufficient for the linearization and amplitude approximations. The corresponding frequency limit can be found from

$$u_0 = \frac{\rho}{\lambda f} \tag{4-78A}$$

$$u_{max} = \frac{1}{\lambda}\left(\frac{\rho}{f}\right)_{max} = \frac{0.03}{5461 \times 10^{-8}} = 550 \text{ cycles/cm} \tag{4-228}$$

Thus for spatial frequencies of less than 550 cycles/cm neglecting the phase term to obtain the Fourier transform representation of (4-195) introduces an error of no more than 2 percent in terms of the form of the second term in the image intensity of (4-212). Again we can point out that for most practical cases the frequency capability of present input techniques restricts the possible frequencies to a lower value than that specified by (4-228).

We have shown how (4-214) and (4-215) are used to determine the error E_ϕ for any frequency plane aperture with a radius defined by $(\rho/f)_{max}$. Further, we have shown for a particular case ($\lambda = 5461$ Å and $f = 10$ cm) that the limit $(\rho/f)_{max} \leq 0.03$ provides an accuracy within 2 percent for the terms in which the phase variation appears. It has also been noted that this phase approximation requires a tighter restriction on the maximum frequency terms; for example, in the examples used the maximum frequency is one fifth that allowed for an accurate amplitude approximation. Of course, to obtain the desired accuracy this restriction of the frequency range will improve the accuracy of the amplitude approximation.

We can refer back to Figure 4.17 to consider the amplitude error E_A for the values $d/f = 1$ and $\rho/f = 0.03$ assuming r' max$/f = \frac{1}{5}$. The point corresponding to $d/f = 1$ and $\rho/f = 0.03$ is located below the curve corresponding to $E_A = 0.005$. Therefore the further restriction of (ρ/f) required for the phase approximation reduces the error in the amplitude approximation to a value of less than 0.5 percent. This result shows that the restriction we have considered not only provides a Fourier transform relation that is accurate in phase but also improves the accuracy of the amplitude approximation previously considered.

Within these limits the two-lens optical imaging process can be described by the following equations:

$$A'(x,y) = \left[\frac{-je^{j2kf}}{2\lambda f}\right]F(u,v) = \left[\frac{-je^{j2kf}}{2\lambda f}\right]\int\int S(x_1,y_1)e^{-j2\pi(ux_1+vy_1)}\,dx_1\,dy_1 \tag{4-229}$$

$$O(x_2, y_2) = e^{jkR_2}\left[\frac{-fe^{j2kf}}{2(f+d_2)}\right]\int\int F(u, v)e^{-j2\pi(ux_2+vy_2)}\,du\,dv. \tag{4-230}$$

Usually only the amplitude variations are of interest and the constant factors within the brackets are dropped,

$$A'(x, y) = F(u, v) = \int\int S(x_1, y_1)e^{-j2\pi(ux_1+vy_1)}\,dx_1\,dy_1 \tag{4-231}$$

$$O(x_2, y_2) = e^{jkR_2}S(-x_2, -y_2) = e^{jkR_2}\int\int F(u, v)e^{-j2\pi(ux_2+vy_2)}\,du\,dv \tag{4-232}$$

Equations (4-231) and (4-232) represent the form of the optical Fourier transform representation commonly used. These equations describe the relative amplitude and phase variations of spectrum $A'(x, y)$ and image $O(x_2, y_2)$. We note that the phase term e^{jkR_2} is retained in (4-232). This factor has no effect on the image intensity because multiplication by the complex conjugate eliminates this term. If the image $O(x_2, y_2)$ is to be processed further by another lens, however, the effect of the phase factor e^{jkR_2} must be considered. In such cases we also may not be able to neglect the variation in phase due to the factor e^{jkR_1} because it was developed based on image intensity.

OPTICAL CORRELATOR SYSTEMS

We shall now consider a three-lens optical system as shown in Figure 4.21. In this system the signal plane F is assumed to be in the front focal plane of lens L_1 and each of the other lenses (L_1 and L_2) is located so that its front focal plane coincides with the back focal plane of the preceding lens. With this configuration the amplitude distribution corresponding to the input signal to each lens is the output signal in the back focal plane of

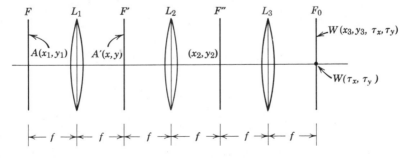

Figure 4.21 Optical correlator system.

the preceding lens (and is a focal length in front of the lens). This location of the signal planes provides the advantage of eliminating the phase terms dependent on the distance from the lens to the input plane as discussed in relation to (4-191) and (4-192). The optical system in Figure 4.21 consists of a two-lens imaging system as discussed in the preceding section followed by a third lens that produces a Fourier transform of the light amplitude distribution of the image. As pointed out at the end of the last section the processing of the image $O(x_2, y_2)$ by an additional lens involves its amplitude rather than the intensity; therefore the phase effects of each lens must be considered. We shall assume here that the aperture limitations are sufficiently restrictive so that the linear frequency and amplitude approximations developed earlier are valid. The focal lengths of the three lenses are assumed equal to simplify our analysis (in the general case unequal focal lengths would introduce magnification or demagnification).

In the optical system of Figure 4.21 lens L_1 produces a light amplitude distribution in its back focal plane F' that is proportional to the Fourier transform of the input signal $A(x_1, y_1)$ except for a multiplicative phase factor. As discussed in reference to (4-231) and (4-232) we will drop all constant factors and retain only terms that vary with respect to the coordinates in the four signal planes of interest (F, F', F'', F_0). Using only the variable exponential in (4-192) for K_1 the amplitude in the F' plane is given by (4-161) which can be written as

$$A'(x, y) = \exp\left[j\frac{kf}{4}\left(\frac{\rho}{f}\right)^4\right] F(u, v)$$

$$= \exp\left[j\frac{kf}{4}\left(\frac{\rho}{f}\right)^4\right] \int\int A(x_1, y_1) e^{-j2\pi(ux_1 + vy_1)} \, dx_1 \, dy_1 \qquad (4\text{-}233)$$

For our development of a correlator it is advantageous to introduce notation for the signal $A(x_1, y_1)$ which accounts for displacement of the signal from some reference position. Referring to Figure 4.22 we can consider

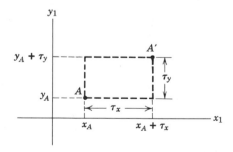

Figure 4.22   Displacement of a signal point in the input plane F.

the displacement of a signal point A to the new position A'. If A is a point of the signal $A(x_1, y_1)$, the light amplitude at A is $A(x_A, y_A)$. Since A' is the same signal point as A (it has only been moved), the light amplitude at A' must also be $A(x_A, y_A)$. The coordinates of the point A' are $x_1 = x_A + \tau_x$ and $y_1 = y_A + \tau_y$. Thus our notation for the signal must be such that if we substitute the coordinates x_1, y_1 for the point A', we obtain $A(x_A, y_A)$. The required notation is $A(x_1 - \tau_x, y_1 - \tau_y)$ as can be seen by substituting the values of x_1 and y_1 for the point A' and by substituting the values $\tau_x = \tau_y = 0$ for the point A. In either case the signal amplitude is $A(x_A, y_A)$. Using this new notation for a signal (4-233) can be rewritten as

$$A'(x, y) = \exp\left[j\frac{kf}{4}\left(\frac{\rho}{f}\right)^4\right] F(u, v)$$

$$= \exp\left[j\frac{kf}{4}\left(\frac{\rho}{f}\right)^4\right] \int\int A(x_1 - \tau_x, y_1 - \tau_y)\, e^{-j2\pi(ux_1 + vy_1)}\, dx_1\, dy_1$$

(4-234)

The displacements τ_x and τ_y are positive when the displacements are in the positive x_1 and positive y_1 directions.

In (4-234) the function $F(u, v)$ represents the Fourier transform of the displaced function $A(x_1 - \tau_x, y_1 - \tau_y)$. Thus the $F(u, v)$ in (4-234) includes all the information regarding the signal including its displacement. From Fourier transform theory the transform corresponding to a displaced signal such as in (4-234) differs from the transform $F(u, v)$ of the undisplaced signal of (4-233) by an exponential phase term, $e^{-j2\pi(\tau_x u + \tau_y v)}$. This principle need not concern us any further; it is pointed out only to emphasize that the $F(u, v)$ in (4-234) corresponds to the displaced signal $A(x_1 - \tau_x, y_1 - \tau_y)$.

The amplitude distribution given by (4-234) appears in the plane F' and represents the input signal to the lens I_2. Lens L_2 performs a Fourier transform operation on $A'(x, y)$ and the image amplitude in the plane F'' is given by

$$O(x_2, y_2) = \phi(x_2, y_2) \int\int \exp\left[j\frac{kf}{4}\left(\frac{\rho}{f}\right)^4\right] F(u, v)\, e^{-j2\pi(ux_2 + vy_2)}\, du\, dv$$

(4-235)

Equation (4-235) corresponds to (4-170) except that only the variable terms of K_1 and K_2 have been retained. The exponential appearing in the integrand corresponds to the variable term in K_1 as discussed above. The function $\phi(x_2, y_2)$ in front of the integral represents the variable part of K_2.

From the definition of K_2 given under (4-162) the variable part of K_2 is obtained from the term e^{jkR_2}

$$R_2 = \frac{f^2 + fd_2 + \rho'^2}{(f^2 + \rho'^2)^{1/2}} \quad \text{and} \quad \rho'^2 = x_2^2 + y_2^2$$

Because R_2 has the same form as R_1, R_2 can be expanded in the form of (4-185)

$$R_2(x_2, y_2) = (f + d_2) + (f - d_2)\left[1 - \frac{1}{(1 + (\rho'^2/f^2))^{1/2}}\right]$$

$$+ 2f\left[\frac{1 + (\rho'^2/2f^2)}{(1 + (\rho'^2/f^2))^{1/2}} - 1\right] \quad (4\text{-}236)$$

The first term and the -1 term in the last bracket of (4-236) are constant and can be dropped because we are interested only in the variable part. The second term vanishes because $d_2 = f$ in the system we are considering. Thus the only variable term in (4-236) is the fraction in the brackets of the third term. The function $\phi(x_2, y_2)$ is therefore given by

$$\phi(x_2, y_2) = \exp\left\{j2kf\left[\frac{1 + (\rho'^2/2f^2)}{(1 + (\rho'^2/f^2))^{1/2}}\right]\right\} \quad (4\text{-}237)$$

The variable part of K_2 given by (4-237) was derived from the complete expansion of R_2 rather than from an approximate expansion analogous to (4-191) and (4-192) because the aperture restrictions necessary for the validity of (4-191) or (4-192) would require a signal and image aperture much smaller than that normally desired in optical systems. For an image aperture defined by $(\rho'/f)_{max} = 0.14$, $f = 10$ cm, and $\lambda = 5461 \times 10^{-8}$ cm the phase term $\phi(x_2, y_2)$ can introduce phase shifts as great as 38π rad (19 cycles). It was pointed out that the phase approximation of (4-192) was accurate within 2 percent. For the image aperture considered here this phase inaccuracy can be of the order of 0.4 cycles. This magnitude of phase error may not be negligible and therefore the more complete exponential was used in defining $\phi(x_2, y_2)$ by (4-237).

Returning to the image amplitude distribution $O(x_2, y_2)$ given by (4-235) we will change the notation to take into account the possibility of image displacement corresponding to the signal displacement previously considered.

It was pointed out that the imaged amplitude $O(x_2, y_2)$ corresponds to an inverted replica of the input signal $A(x_1, y_1)$. This inverted property of the image applies to the image motion as well; that is, if the signal is displaced in the positive x_1 and y_1 direction, the image is displaced in the negative x_2 and y_2 directions. Thus if $O(x_2, y_2)$ corresponds to the inverted image of $A(x_1, y_1)$, the displaced image corresponding to $A(x_1 - \tau_x, y_1 - \tau_y)$ is obtained simply by reversing the sign of the displacement to obtain $O(x_2 + \tau_x, y_2 + \tau_y)$. Using the displacement notation for the image amplitude distribution, (4-235) can be rewritten as

$$O(x_2 + \tau_x, y_2 + \tau_y) = \phi(x_2, y_2) \iint \exp\left[j\frac{kf}{4}\left(\frac{\rho}{f}\right)^4 \right]$$

$$\times F(u, v) e^{-j2\pi(ux_2 + vy_2)} \, du \, dv \qquad (4\text{-}238)$$

The final lens L_3 in Figure 4.21 operates on the light amplitude distribution appearing in its input plane F''. For an optical correlator operation a reference signal $R(x_2, y_2)$ is inserted into the plane F'' in the form of an amplitude transmission function of a photographic transparency. In this case the light amplitude distribution operated on by lens L_3 is that which appears on the output side of the reference transparency. This light amplitude is given by the product of the incident light amplitude $O(x_2 + \tau_x, y_2 + \tau_y)$ and the reference transmission function $R(x_2, y_2)$. Thus the light amplitude distribution W in the output plane F_0 is given by the equation

$$W(x_3, y_3, \tau_x, \tau_y) = \exp\left[j\frac{kf}{4}\left(\frac{r_3}{f}\right)^4 \right] \iint R(x_2, y_2)\, O(x_2 + \tau_x, y_2 + \tau_y)$$

$$\times e^{-j2\pi(sx_2 + ty_2)} \, dx_2 \, dy_2 \qquad (4\text{-}239)$$

where

$$s = \frac{x_3}{2f}, \quad t = \frac{y_3}{2f}, \quad r_3^2 = x_3^2 + y_3^2.$$

Because we are considering a system that terminates at the F_0 plane, the intensity will be detected, measured, or recorded in the F_0 plane. The intensity in the output plane is given by the product of (4-239) and its complex conjugate. The complex conjugate product of the exponential in front of the integral results in the cancellation of the exponential. Thus we can drop the exponential in (4-239) because it will not affect the detected intensity output. Equation (4-239) therefore can be simplified to

$$W(x_3, y_3, \tau_x, \tau_y) = \iint R(x_2, y_2)\, O(x_2 + \tau_x, y_2 + \tau_y)$$

$$\times e^{-j2\pi(sx_2 + ty_2)} \, dx_2 \, dy_2 \qquad (4\text{-}240)$$

Finally if we consider only the point located at the intersection of the optical axis with the plane F_0 (back focal point of L_3), $s = t = 0$ (i.e., $x_3 = y_3 = 0$) and (4-240) reduces to

$$W(\tau_x, \tau_y) = \iint R(x_2, y_2)\, O(x_2 + \tau_x, y_2 + \tau_y)\, dx_2\, dy_2 \qquad (4\text{-}241)$$

where

$$W(\tau_x, \tau_y) = W(x_3 = 0,\, y_3 = 0,\, \tau_x, t_y)$$

Equation (4-241) corresponds to a two-dimensional correlation function which implies that the light amplitude at the back focal point ($x_3 = y_3 = 0$) of the lens L_3 is given by the cross-correlation of the reference $R(x_2, y_2)$ and the image amplitude $O(x_2, y_2)$. Thus as the input is displaced the variation of the light amplitude $W(\tau_x, \tau_y)$ corresponds to the variation of the correlation function with respect to the displacements τ_x and τ_y. We note that the correlation function defined by (4-241) involves the image amplitude $O(x_2, y_2)$ which is inverted with respect to the input signal $A(x_1, y_1)$. Therefore if $R(x_2, y_2)$ is not a symmetrical function, it must be oriented correctly with respect to the image $O(x_2, y_2)$ rather than with respect to the input signal $A(x_1, y_1)$.

We shall briefly consider the implication of the steps from (4-240) to (4-241). This step in our derivation was accomplished by stating that we would consider only the single point in the output plane F_0 which lies on the optical axis (i.e., $x_3 = y_3 = 0$). In practice it is physically impossible to isolate a single point. The best attempt we can make is to restrict our light measurement or detection to a small area about the selected point. The light amplitude at points within this area (except for the one point on the optical axis) is given by (4-240) rather than (4-241). The light amplitude distribution will not be uniform over the finite area of measurement due to the phase variation involved in the integral of (4-240). For example, if we use a pinhole aperture 10 microns in diameter to define our detection area, the phase term in (4-240) can vary as much as 4π rad (2 cycles) over the range of the image aperture (assuming $\rho'_{max} \sim 0.14f$, $f = 10$ cm, $\lambda = 5461 \times 10^{-8}$ cm). The effects of the phase term in (4-240) is to reduce the light amplitude at points off axis because the contributions to the integral are not in phase. Therefore the actual light available through a pinhole aperture located in the F_0 plane at $x_3 = y_3 = 0$ will be less than that found by assuming the light amplitude given by (4-241) appears at all points within the pinhole aperture. We will not consider this problem any further here since the analysis would depend on the type of photodetector or measurement technique used. We will assume that the variations involved are small enough so that any measurement will yield values proportional to the square of the amplitude given by (4-241).

As pointed out above the correlation function defined by (4-241) involves the image amplitude $O(x_2, y_2)$ rather than the signal amplitude $A(x_1, y_1)$. As defined by (4-238) the image amplitude contains phase terms not present in the signal. A correlation operation can be performed based on the image as given by (4-241); however, the reference signal $R(x_2, y_2)$ would have to be selected in terms of the image $O(x_2, y_2)$ including the phase terms. The correlation function obtained would correspond to a distorted signal rather than the actual signal $A(x_1, y_1)$. The presence of distortion due to the phase terms in (4-238) therefore complicates the analysis and determination of the correlation process. For example, the image amplitude $O(x_2, y_2)$ will be complex (phase variation as well as amplitude) and for complete correlation a complex reference signal is required. Such reference transparencies are difficult to produce. The phase distortions are commonly neglected and a reference signal is selected based on an ideal image (no distortion) of the input signal. We will now proceed to analyze such a system to determine the effects of the undesirable phase terms present in our equations.

Let us consider a signal that would produce an ideal image amplitude defined by

$$O(x_2 + \tau_x) = \sum_n B_n \cos 2\pi p_n(x_2 + \tau_x) \qquad (4\text{-}242)$$

Equation (4-242) defines a signal image composed of a series of cosine harmonics in one dimension. A one-dimensional signal has been chosen to simplify our analysis. Referring to (4-238) we find that each frequency term in the image has a phase term $\exp\left[j(kf/4)(\lambda p_n)^4\right]$ associated with it and the image also has a phase term $\phi(x_2, y_2)$ associated with it. Thus for the actual image we would have

$$O(x_2 + \tau_x) = \phi(x_2) \sum_n B_n \exp\left[j\frac{kf}{4}(\lambda p_n)^4\right] \cos 2\pi p_n(x_2 + \tau_x) \qquad (4\text{-}243)$$

We can consider a reference signal without phase given by

$$R(x_2) = \sum_m R_m \cos 2\pi p_m x_2 \qquad (4\text{-}244)$$

The reference signal $R(x_2)$ defined by (4-244) has been selected to have the same cosine harmonics ($p_m = p_n$ for $m = n$) as the imaged signal being considered. We note that the reference signal defined by (4-244) does not contain the phase terms present in (4-243). The product of reference and image for the ideal image of (4-242) is given by

$$R(x_2)O(x_2 + \tau_x) = \sum_{n,m} B_n R_m \cos 2\pi p_m x_2 \cos 2\pi p_n(x_2 + \tau_x) \qquad (4\text{-}245)$$

The product of reference and image for the actual image of (4-245) is given by

$$R(x_2)O(x_2+\tau_x) = \phi(x_2) \sum_{n,m} B_n R_m \exp\left[j\frac{kf}{4}(\lambda p_n)^4\right]$$

$$\times \cos 2\pi p_m x_2 \cos 2\pi p_n(x_2+x) \qquad (4\text{-}246)$$

Substituting (4-245) and (4-246) into (4-241) we obtain for the correlation function of the ideal image:

$$W(\tau_x) = \sum_{n,m} B_n R_m \int \cos 2\pi p_m x_2 \cos 2\pi p_n(x_2+\tau_x)\, dx_2 \qquad (2\text{-}247)$$

and for the correlation functions of the actual image

$$W(\tau_x) = \sum_{n,m} B_n R_m \exp\left[j\frac{kf}{4}(2p_n)^4\right]\int \phi(x_2)\cos 2\pi p_m x_2$$

$$\times \cos 2\pi p_n(x_2+\tau x)\, dx_2 \qquad (4\text{-}248)$$

Comparing (4-247) and (4-248) we find that the term $\exp[j(kf/4)(\lambda p_n)^4]$ affects the phase of each term in the double summation. From our previous discussion of frequency limitations with respect to this phase term we can show that applying the limitation developed for imaged intensity limits the phase variation of this term to approximately 7 degrees. In summing terms that are not in phase the result will be less than summing the same terms in amplitude only. Thus the presence of the phase term $\exp[j(kf/4)(\lambda p_n)^4]$ has the effect of reducing the value of $W(\tau_x)$ in (4-248) as compared to (4-247). Because the maximum phase will be about 7 degrees, however, the difference due to this term will be small. The phase term $\phi(x_2)$ appears in the integral of each term in the sum and has the same effect on the integral (can be considered as summation) as the phase term discussed above had on the summation. As discussed above, however, the phase variations of $\phi(x_2)$ ranges over 19 cycles and the effect on the value of integral will be correspondingly greater. The actual magnitude of the reduction in $W(\tau_x)$ caused by these phase terms is difficult to evaluate in general because the reduction will depend on the form of the signals involved. From our discussion here, however, it is apparent that the actual correlation function observed will be smaller in amplitude than that predicted using an ideal image. This result is obvious if we consider that the presence of the phase terms in the actual image produce a mismatch between the signal and reference and therefore the correlation will be reduced. The effects of these phase terms can be reduced by further restricting the frequency range p_{max} (or ρ_{max}) and signal and image aperture size that would limit the variation of the phase terms. We will not proceed with an analysis of the required limitations because the analysis will depend to a large extent on the type of signals involved and the correlation results desired. Here we have developed the equations necessary for such an evaluation and hopefully have pointed out the significance of the various effects which appear in an optical system.

PHASE CORRECTIONS

We have discussed the effects of undesirable phase terms in optical systems and have demonstrated that these effects can be minimized by restricting the size of signal apertures and the spectral range of the signals. An alternative approach can be pursued by inserting phase corrections into the optical system. Such phase corrections can be implemented by inserting thin sheets or plates of transparent materials whose thickness or index of refraction has variations that introduce phase terms opposite in sign to those introduced by the system.

The basic equation representing the Fourier transform operation of a lens was given by (4-68) as

$$A'(x, y) = \frac{-je^{jkR(x, y)}}{\lambda(f+d)} \iint A(x_1, y_1) e^{-j2\pi(ux_1 + vy_1)} \, dx_1 \, dy_1 \qquad (4\text{-}68)$$

Rewriting this equation retaining only the variable part of the terms outside the integral we obtain

$$A'(x, y) = \phi(x, y) \iint A(x_1, y_1) e^{-j2\pi(ux_1 + vy_1)} \, dx_1 \, dy_1 \qquad (4\text{-}249)$$

where

$$\phi(x, y) = \exp\left\{ j2kf\left[\frac{1 + (\rho^2/2f^2)}{(1 + (\rho^2/f^2))^{1/2}} \right] \right\}$$

assuming $d = f$, as derived in (4-237). As pointed out in all our discussions the phase term $\phi(x, y)$ destroys the simple Fourier transform representation of lens focusing properties since the integral part of (4-249) corresponds to a Fourier transform by itself. Let us consider inserting a phase correction plate into the back focal plane of a lens with transmission properties given by

$$P(x, y) = A_0 e^{jc} \phi^*(x, y) \qquad (4\text{-}250)$$

$$\phi^*(x, y) = \exp\left\{ -j2kf\left[\frac{1 + (\rho^2/2f^2)}{(1 + (\rho^2/f^2))^{1/2}} \right] \right\}$$

where A_0 is a constant and C is a constant.

The light amplitude distribution appearing at the output side of the plate will be

$$A'(x, y)P(x, y) = \iint A(x_1, y_1)e^{-j2\pi(ux_1 + vy_1)}\, dx_1\, dy_1 \qquad (4\text{-}251)$$

where the constant term $A_0 e^{jC}$ has been dropped and $\phi(x, y)\, \phi^*(x, y) = 1$. Thus by inserting a phase plate with transmission properties given by (4-250) in the back focal plane of each lens in an optical system the phase terms are eliminated. From the definition given by (4-250) we find that the phase correction depends on the focal length f of the lens and the wavelength λ $(k = 2\pi/\lambda)$ of the light. The phase correction is not dependent on the signal used and therefore a phase plate can be made for the lens and wavelength to be used in the system. Of course, the correction of phase by this method requires an accurate technique for producing the phase plate and positioning the plate in the optical system. In any case the elimination of undesirable phase terms is possible at least in theory. Any inaccuracies in production or location of the phase plate may be acceptable as long as the phase terms are appreciably less than before the plate was introduced. Assuming the phase plate is an accurate representation of the transmission function of (4-250), the Fourier transform relation of (4-251) will be valid. With the relationship given by (4-251) the operation of spectrum analyzer, imaging and optical correlation systems can be described by the ideal cases used in the respective discussions and no undesirable phase terms appear in the equations.

PHASE TERMS WHEN $d \neq f$

In our consideration of phase terms we considered the special case of an input plane coincident with the front focal plane of a lens $(d = f)$. This special case was chosen to eliminate the phase effects of a term proportional to $(f - d)$. From (4-236) we can write a complete expression for the variable part of the exponential term e^{ikR} as

$$\phi(x, y) = \exp\left[jk\frac{(d-f)}{(1 + (\rho^2/f^2))^{1/2}}\right]\exp\left\{j2kf\left[\frac{1 + (\rho^2/2f^2)}{(1 + (\rho^2/f^2))^{1/2}}\right]\right\} \qquad (4\text{-}252)$$

Equation (4-252) reduces to the form of (4-237) when $d = f$. If we can restrict our consideration to a rather limited range in the back focal plane of a lens, we have shown that (4-252) can be given to a good approximation in the form of (4-191)

$$\phi(x, y) = \exp\left[\frac{jk}{2}(f-d)\left(\frac{\rho}{f}\right)^2\right]\exp\left[j\frac{kf}{4}\left(\frac{\rho}{f}\right)^4\right] \qquad (4\text{-}253)$$

We can consider (4-253) as a representation of the phase in a back focal plane containing a frequency spectrum while (4-252) is a more accurate representation that applies in a back focal plane containing an image of the input signal. This application of (4-252) and (4-253) is based on the relatively larger apertures commonly used in the signal and image planes.

In the systems that we have considered here the complete phase variations as given by (4-252) appear only in the correlator system. This can be seen by noting the presence of $\phi(x_2, y_2)$ in (4-238). Because this phase factor is expressed in terms of the coordinates x_2 and y_2 of the image phase, we cannot use a very restrictive aperture limitation without severely affecting our signal handling capability. Therefore the approximation of (4-253) will not be valid and $\phi(x_2, y_2)$ in (4-238) will have the form of (4-252) with d_2 and ρ_2 replacing d and r, respectively,

$$\phi(x_2, y_2) = \exp\left[jk\frac{(d_2-f)}{(1+(\rho_2^2/f^2))^{1/2}}\right]\exp\left\{j2kf\left[\frac{1+(\rho_2^2/2f^2)}{(1+(\rho_2^2/f^2))^{1/2}}\right]\right\} \quad (4\text{-}254)$$

where $\rho_2^2 = x_2^2 + y_2^2$ and d_2 is the distance from the spectrum plane F' to the lens L_2 (see Figure 4.21). In the sample correlation function of (4-248) we can see that the additional phase term dependent on (d_2-f) will increase the effect of $\phi(x_2)$ on the integrals. In practice a system would be specified on the basis of locating lens L_2 so that $d_2 = f$. The exact positioning of the lenses in an optical system, however, is obviously a practical impossibility. Thus the additional phase term containing (d_2-f) represents the phase distortion introduced by inaccuracies in the implementation of the system. Because the quantity (d_2-f) represents an inaccuracy, its value will usually be undetermined. Therefore the first term in (4-254) represents an undetermined phase error in the optical correlator system. If a guess or estimate of the tolerances in the system can be made, this error term can be used to determine the maximum distortion of the correlation function by analysis similar to that implied by (4-248).

Because the variation of the phase term containing (d_2-f) in (4-254) is not known specifically, the elimination of this term by a phase correction plate is not possible. Thus in a system containing phase correction plates only the second term of (4-254) can be eliminated. In such systems $\phi(x_2, y_2)$ is completely given by the position error term

$$\phi(x_2, y_2) = \exp\left[jk\frac{(d_2-f)}{(1+(\rho_2^2/f^2))^{1/2}}\right] \quad (4\text{-}255)$$

The distortion of the correlation function in a phase-corrected system is therefore completely dependent upon the positioning errors. Again referring to the sample of (4-248) $\phi(x_2)$ would be given in the form of (4-255). The phase term in front of the integral of (4-248) would also be replaced by an error term from an expression such as (4-253) as will be discussed below.

The phase term given by (4-253) represents the variable part of (4-191). The exponential dependent on $(f-d_1)$ represents an error term due to lens positioning. To account for this error the complete phase approximation of (4-253) must be used in place of the K_1 exponential of (4-192). Thus the error phase term will appear throughout our previous analysis wherever we have used the K_1 term.

We considered the effect of the K_1 phase term on the image intensity and on the correlation function in earlier sections. In our correlator discussion the variable phase term of K_1 appears in the integral used to define the image amplitude distribution in (4-238). To account for errors in placement of lens L_1 (see Figure 4.21) the exponential $\exp[j(kf/4)(\rho/f)^4]$ in (4-238) must be replaced by a phase term of the form of (4-253) that can be written

$$\phi(x, y) = \exp\left[\frac{jk}{2}(f-d_1)\left(\frac{\rho}{f}\right)^2\right]\exp\left[j\frac{kf}{4}\left(\frac{\rho}{f}\right)^4\right] \qquad (4\text{-}256)$$

Referring to our sample correlation of (4-248) the phase term given by (4-256) will replace the $\exp[j(kf/4)(\rho/f)^4]$ term in front of the integral. The error phase term has the effect of adding on additional variation to the phase of the terms in the summation. For a phase corrected system the $\exp[j(kf/4)(\rho/f)^4]$ term is eliminated and the undesirable phase difference of terms in the summation will be dependent only on the accuracy of the system implementation.

In our discussion of imaging systems we defined a factor N by (4-214) and developed a method for determining the accuracy of the image intensity based on this parameter. To extend this method to include the case for $d_1 \neq f$ we merely redefine N by the equation

$$N^4 = \frac{k}{2}\left(\frac{\rho}{f}\right)^2\left[(f-d_1)+\frac{f}{2}\left(\frac{\rho}{f}\right)^2\right] \qquad (4\text{-}257)$$

which is obtained from the exponents of the terms in (4-256). For d_1 less than f the limits on N defined in our previous discussion will apply to (4-257) for the maximum value of ρ. It is noted that since $(f-d_1)$ is a positive quantity when d_1 is less than f, the required limit on ρ_{max}/f will be less than that determined for the case $d_1 = f$. When d_1 is greater than f, $(f-d_1)$ is a negative quantity that would imply that the value of ρ_{max}/f can be greater than that for the case $d_1 = f$. This is true except for cases in which d_1 is sufficiently greater than f so that for some value of ρ less than ρ_{max}/f the value of (4-257) is greater in absolute value than for ρ_{max}/f; that is, since $(f-d_1)$ is negative, the right side of (4-257) is zero at $\rho = 0$, becomes negative as ρ increases until it reaches a maximum negative value, and then increases to positive values. Depending on the value of d_1 and the limit ρ_{max}/f it is possible that the phase at the maximum negative value is greater than that at the aperture limit ρ_{max}/f. In such cases the maximum negative value must be considered rather than the end value at ρ_{max}/f and the aperture may have to be restricted to values below this maximum. In any case the brackets on the right side of (4-257) must be considered as an absolute value symbol when the quantity within is negative so that N will have real values. In other words we are concerned with the magnitude of the phase variations and not the sign.

For systems with phase correction only the first term of (4-256) will remain and (4-257) will be simplified to

$$N^4 = \frac{k}{4}\left(\frac{\rho}{f}\right)^2 (f-d_1) \qquad (4\text{-}258)$$

Equation (4-258) can then be used with the image intensity criteria developed earlier to consider the effects of positioning error for lens L_1 in phase-corrected systems.

In most of the literature the phase-corrected form of (4-256)

$$\phi(x, y) = \exp\left[j\frac{k}{2}(f-d_1)\left(\frac{\rho}{f}\right)^2\right] \qquad (4\text{-}259)$$

is used even though phase-correction techniques may not be employed. This application of (4-259) requires that the frequency limitation be sufficient so that the $\exp\left[j(kf/4)(\rho/f)^4\right]$ term can be neglected. This application also implies that the term $(f-d_1)$ is much greater than the maximum value of $(f/2)(\rho/f)^2$. If this condition does not hold the neglected term will contribute a phase comparable to that of (4-259) which would then be in error. Conversely if $(f-d_1)$ is not greater than $(f/2)(\rho/f)^2$ and the $\exp\left[j(kf/4)(\rho/f)^4\right]$ term is considered negligible, then the term given by (4-259) is also negligible because it is comparable to the neglected term.

In this section we have outlined the procedure for taking into account the additional phase term arising from inaccuracies in the positioning of lenses. It was pointed out that because these terms result from inaccuracies, they are generally not completely specified. The worse case, however, can be specified by estimating the maximum error in the position of a lens. From this extreme estimate the necessary aperture limitation or the evaluation of errors in the desired optical outputs can be determined for a worse case analysis. Unfortunately, because of the undetermined nature of these terms, phase correction cannot be used to eliminate their effects.

5

Optical Spectrum Analysis

OPTICAL DATA PROCESSORS SIGNAL INPUT

The medium most often used to provide an input signal to optical data processors is photographic film. Photographic film can provide images that accurately describe the input signal in optical terms, permitting optical processing. An image on a film is seen because of the film's ability to block the passage of light. The image can be measured by either its transmittance or its opacity. Transmittance is defined by the ratio I_t/I_0 where I_t is the intensity of the light transmitted and I_0 is the intensity of the incident light. The transmittance function therefore gives the fraction of the incident light transmitted through each portion of the film. Opacity is the reciprocal of the transmittance (that is, I_0/I_t) and therefore the opacity can also be determined for each point on the film. The opacity of each point on the film plotted as a function of position is called the *opacity function*.

Let us consider one point on a film that is made up of a layer of developed silver particles in a transparent support. When incident light strikes this point, it will reduce the intensity of the transmitted light by a value $1/p$ of the incident light intensity. If we consider a second identical layer of silver particles on top of the first layer, this layer will also block the same fraction of the incident light upon it as the first layer. The light transmitted by two layers therefore would be

$$\frac{1}{p} \times \frac{1}{p} = \left(\frac{1}{p}\right)^2$$

of the incident light to the film. For n identical layers the intensity of the light transmitted will be $(1/p)^n$ of the incident light. This concept will be used later to show that the resultant transmission function of two or more superimposed transmission functions is the *product* of the super-imposed transmission functions.

SPECTRUM ANALYZER

The spectrum of a uniformly illuminated transmission function located in the front focal plane of a lens is produced in the back focal plane of the lens. This spectrum is formed by the Fourier transform characteristics of diffraction. A review of the basic principles involved in focusing the diffraction pattern formed by slits will be helpful in understanding the diffraction patterns formed by transmission functions. Figure 5.1 shows how diffracted wavefronts produced by uniformly spaced slits are focused in the back focal plane of a lens. Initially we will assume slits to be in the front focal plane because they are easy to visualize. The zero-order wave-fronts are focused to a point at the back focal plane of the lens on the optical axis. This zero-order image point is usually referred to as the DC component of the diffraction pattern (or spectrum). The two first-order wavefronts (that is, the first order up and the first order down) will be focused to two points located symmetrically about the zero-order image point. The distance between the zero-order and the first-order image points will be proportional to the spatial frequency $1/S$ of the slits, where S is the distance between slits. It is important to note that the images formed in the back focal plane are *points* whose locations are dependent on the spatial frequency of the slits. The diffraction pattern from slits can be related to a Fourier transform (spectrum) as was shown in Chapter 4.

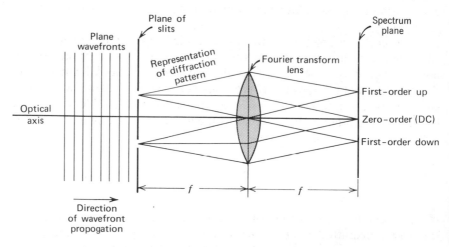

Figure 5.1 Diffraction pattern focused in a focal plane.

A sinusoidal and a square-wave transmission function are shown in Figure 5.2. In the case of a sinusoidal transmission function the slits are not as clearly defined as with a square wave but the similarity is evident. It is assumed that the transmission functions in Figure 5.2 are formed on film; that is, the amplitude of the light through the film will vary propor-

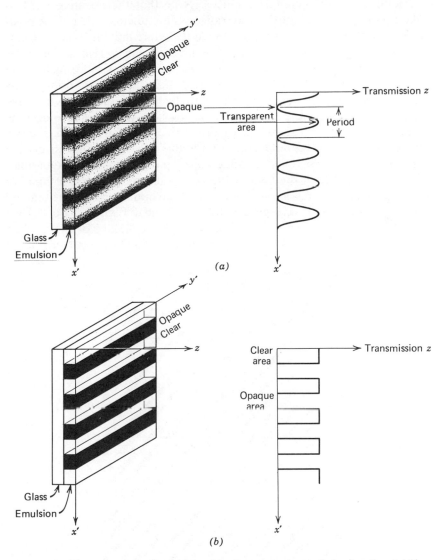

Figure 5.2 Comparison of a sinusoidal and a square wave transmission function. (*a*) Sinusoidal transmission function, (*b*) square wave transmission function.

tionally to the transmission function. Figure 5.3(a) shows a reproduction of a square-wave type of transmission function. When the opaque areas are equal to the spacing of the transparent areas, the transparency is called a *Ronchi ruling;* that is, a Ronchi ruling has a transmission function (along a line perpendicular to the lines of the ruling) that can be represented by a periodic wave of rectangular pulses. Note that the negative used to make Figure 5.3(a) has the desired transmission function and not the positive print of it shown in the figure. Since the white and black portions do not transmit light through the paper (the print is seen by reflected light), it is just a representation of the negative and not the desired transmission function. In addition the fact that Figure 5.3(a) is a positive print means that the black areas in the print were actually clear areas on the negative and the white areas on the print were black (opaque) areas on the negative.

The zero-order term shown in Figure 5.1 corresponds to a bias and is referred to as the DC component. Later in this chapter we shall discuss in more detail why a bias must be added to a sine-wave transmission function. The amplitude of the light at the zero-order point is proportional to this bias. Figure 5.3(b) is a positive image of the optical spectrum obtained from the Ronchi ruling of Figure 5.3(a). Referring to Figures 5.1 or 5.5 we can see that a spectrum is obtained in the back focal plane of a lens when a transmission function is uniformly illuminated by plane wavefronts in

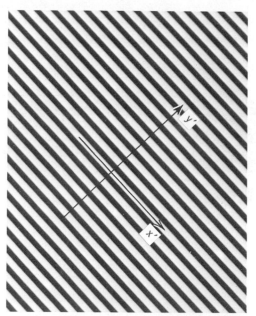

Figure 5.3   (*a*) A Ronchi ruling.

Figure 5.3 (*b*) Optical spectrum of Ronchi ruling in (*a*).

the front focal plane of the lens. The center spot (which is the brightest spot) is the DC component. An Airy pattern (a diffraction pattern of concentric rings of successively lesser intensity) is formed around each spectral point (see Appendix 7). Because of the film's sensitivity to intensity levels, the spectral points corresponding to high intensities are recorded larger than spectral points of lower intensity. In addition an effect known as halation can cause the spectral points to appear larger than they actually are. When the light that forms the image passes through the film into air beyond, some of the light is reflected back into the emulsion. Since light diffuses, it will spread out beyond the boundaries of the image. Causing the image points to appear larger than they actually are.

We can see from Figure 5.4 that the first-harmonic spectral points (one on each side of the DC spectral point) has an intensity almost as high as the DC terms. The intensity of the other odd harmonics drop off rapidly as evidenced by a smaller recorded Airy pattern around them. It is also evident that there are no even harmonics present in this spectrum; that is, there is no harmonics whose distance from the DC is an even multiple

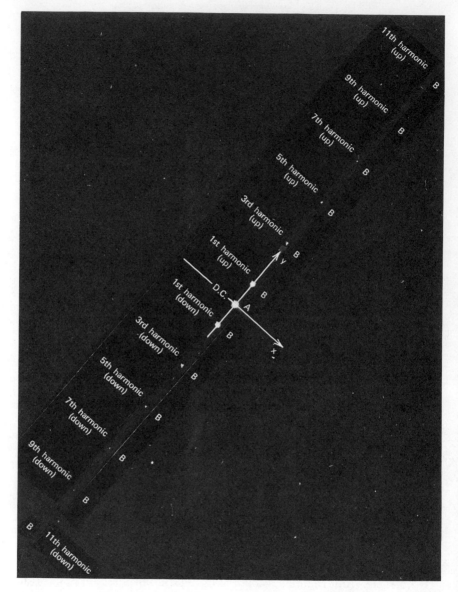

Figure 5.4 Optical spectrum of Ronchi ruling in Fig. 5.3

of the distance between the DC and the first harmonic. The spectrum formed by the diffraction pattern of the transmission function is related to the Fourier series of the transmission function.

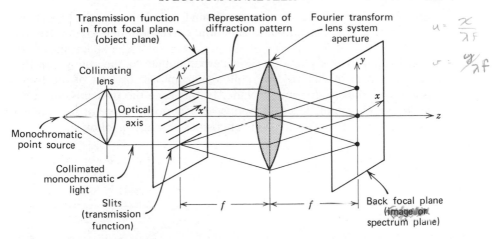

$$u = \frac{x}{\lambda f}$$

$$\sigma = \frac{y}{\lambda f}$$

Figure 5.5 Optical spectrum analyzer.

The first-order terms of the diffraction pattern (see Figure 5.5) correspond to the fundamental frequency of the Fourier series that can be used to represent the transmission function. The wavelength of this fundamental (lowest) frequency for a square wave or Ronchi ruling function is equal to the (maximum) distance between opaque areas of the transmission function. The higher-order terms of the diffraction pattern correspond to the higher-frequency components in the Fourier expansion of the transmission function. Referring to Figure 5.2(a) we can see that the sinusoidal transmission function (true single-frequency sine wave) has a maximum distance between opaque regions equal to the period of the sine wave. Such a transmission function will produce only three spectral points; that is, points located at positions corresponding to the DC, first-order up and down terms. The DC term (corresponding to the input film bias) appears as a point on the lens optical axis in the back focal plane. Two points located symmetrically about this DC point correspond to the frequency of the sine wave. The distance of these points from the DC point will increase if the frequency of the transmission function is increased (spacing between opaque areas reduced).

Figure 5.5 shows an optical spectrum analyzer in which the transmission function opaque lines are parallel to the x' axis (perpendicular to the y' axis) and in the object plane ($x'y'$ plane). The object plane is the front focal plane of the Fourier transform lens. The spectrum of the transmission function (in the object plane) is formed in the back focal plane of the lens. The spectral points of the spectrum formed are on the y axis, which is parallel to the y' axis.

In Figure 5.4 we have seen the spectrum formed by an optical spectrum analyzer when a transmission function corresponding to the Ronchi ruling in Figure 5.3(a) is put in the object plane. The spectrum of the Ronchi ruling forms along an axis perpendicular to the rulings, in this case the y axis of Figure 5.5. This spectrum will be that of the transmission function in the object plane. In this case the transmission function in the object plane corresponds to a periodic wave of rectangular pulses that can be written in terms of cosines or sines as follows:

$$f(y') = \frac{1}{2}\left(1 + \frac{4}{\pi}\cos\omega y' - \frac{4}{3\pi}\cos 3\omega y' + \frac{4}{5\pi}\cos 5\omega y' - \cdots\right) \quad (5\text{-}1)$$

$$\underset{\substack{\uparrow \\ \text{DC}}}{} \quad \underset{\substack{\uparrow \\ \text{first} \\ \text{harmonic}}}{} \quad \underset{\substack{\uparrow \\ \text{third} \\ \text{harmonic}}}{} \quad \underset{\substack{\uparrow \\ \text{fifth} \\ \text{harmonic}}}{}$$

$$f(y') = \frac{1}{2}\left(1 + \frac{4}{\pi}\sin\omega y' + \frac{4}{3\pi}\sin 3\omega y' + \frac{4}{5\pi}\sin 5\omega y' + \cdots\right) \quad (5\text{-}2)$$

The only difference between the rectangular pulse trains described by (5-1) and (5-2) is the location of the origin. As shown in Figure 5.6 equations (5-1) and (5-2) represent identical pulse trains displaced one quarter of a cycle with respect to each other. The curves in Figure 5.6 show the first harmonics plus the bias of one half. The result of adding the bias and all the harmonics is represented by the rectangular pulses in Figure 5.6.

When we consider a transmission function, we usually consider that it should have a range that varies between zero and one; that is, zero corresponds to an opaque area where no light is transmitted and one corresponds to a clear area where the maximum value of the light amplitude is transmitted. Values of the transmission function between zero and one correspond to varying amounts of attenuation of the light amplitude. We can see therefore that when (5-1) and (5-2) are multiplied out and when all terms are added together the result will be a transmission function of values of either zero or one (rectangular pulse train) as shown in Figure 5.6. If we consider only (5-1) which can be written as

$$f(y') = \frac{1}{2} + \frac{2}{\pi}\cos\omega y' - \frac{2}{3\pi}\cos 3\omega y' + \frac{2}{5\pi}\cos 5\omega y' - \cdots \quad (5\text{-}3)$$

the amplitude of each of the components in the Fourier series is given by the coefficients of the respective terms. Equation (5-3) describes the amplitude transmission function of the Ronchi ruling as well as its amplitude spectral content. It is apparent from equations (5-1), (5-2), and (5-3) that there should be *no* even harmonics in the Fourier series for a periodic wave of rectangular pulses. It can also be seen from (5-1) and (5-2) that regardless of whether the function is considered a cosine

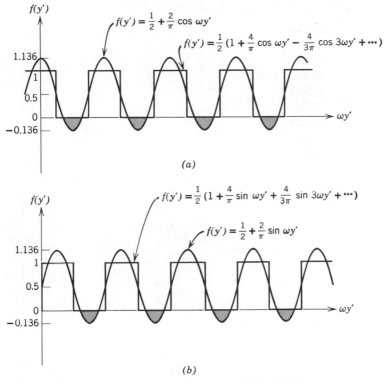

$$f(y') = \frac{1}{2} + \frac{2}{\pi} \cos \omega y'$$

$$f(y') = \frac{1}{2}(1 + \frac{4}{\pi} \cos \omega y' - \frac{4}{3\pi} \cos 3\omega y' + \cdots)$$

(a)

$$f(y') = \frac{1}{2}(1 + \frac{4}{\pi} \sin \omega y' + \frac{4}{3\pi} \sin 3\omega y' + \cdots)$$

$$f(y') = \frac{1}{2} + \frac{2}{\pi} \sin \omega y'$$

(b)

Figure 5.6   Graphs of equations (5-1) and (5-2).

or a sine function, the amplitude of any specific harmonic will be the same.

Figure 5.7(a) shows just the object and back focal planes of the optical spectrum analyzer of Figure 5.5. In the object plane we see a Ronchi ruling as an input function. The object plane can therefore be called the *input* or *function plane*. The output plane in Figure 5.7 is labeled the *spectrum plane*. Distances along the y axis represent frequency; that is, the location of a spectral point on the y axis corresponds to a particular frequency component being present in the original function. From Figure 5.7(a) we can see that the Ronchi ruling forms horizontal lines. Incident plane wavefronts (collimated light) on these horizontal lines will be diffracted in a vertical direction. This means that horizontal lines form a vertical diffraction pattern (spectrum) and vertical lines form a horizontal diffraction pattern. The amplitude of each harmonic is plotted in the z-axis direction (the direction of the z axis is taken as positive amplitude). We have previously determined that it is desirable to have the input transmission function vary between zero and one; therefore using (5-3)

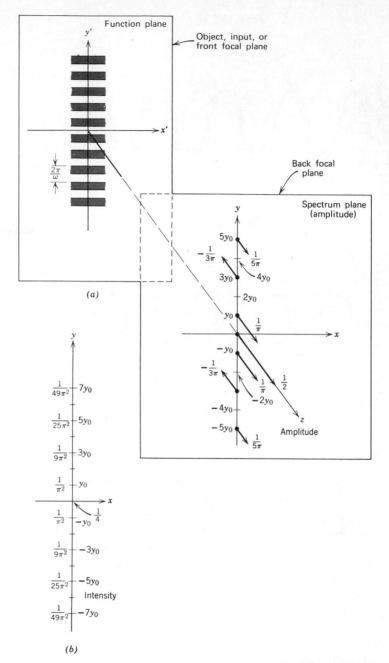

Figure 5.7 Ronchi ruling and its amplitude and intensity spectrums. (*a*) Ronchi ruling in function plane and its amplitude spectrum in the spectrum plane. (*b*) Intensity spectrum of Ronchi ruling.

we obtain the amplitudes of the various harmonics as shown in Figure 5.7(a). The first harmonic up and the first harmonic down each have an amplitude of $1/\pi$. We can see this by examining (5-1) and noting that

$$\cos y' = \frac{e^{jy'} + e^{-jy'}}{2}$$

then

$$f(y') = \frac{1}{2}\left(1 + \frac{4}{\pi}\cos\omega y' - \cdots\right) = \frac{1}{2}\left[1 + \frac{4}{\pi}\left\{\frac{e^{j\omega y'}}{2} + \frac{e^{-j\omega y'}}{2}\right\} - \cdots\right]$$

$$= \frac{1}{2} + \frac{1}{\pi}e^{j\omega y'} + \frac{1}{\pi}e^{-j\omega y'} - \cdots \tag{5-4}$$

From (5-4) we can see that the first harmonic of (5-1) is composed of two exponential terms. Each of these exponential terms accounts for a different spectral point. From (5-4) we see that if the amplitude of the DC bias is one half, then the amplitude of the first harmonic up is $1/\pi$ and the amplitude of the first harmonic down is $1/\pi$. Using a similar procedure the values of the other harmonics were found and plotted in Figure 5.7. Because the intensity is the square of the amplitude, we can find the actual light intensity that would expose a film (if it were placed in the spectral plane) by squaring the amplitude of each term in (5-4). In the previous chapter it was shown that the location of spectral points is related to the frequency by

$$v = \frac{\omega_y}{2\pi} = \frac{y}{\lambda f} \tag{4-79}$$

$$y_0 = \frac{\lambda f}{2\pi}\omega_y \; = \; \frac{\lambda f}{T} \leftarrow \text{period of Ronchi grating}$$

and therefore the location of the spectral points can be plotted as shown in Figure 5.7(a) and the intensity of these spectral points is shown in Figure 5.7(b).

Figure 5.3 shows a high-frequency Ronchi ruling and its spectrum. Careful examination of the Ronchi ruling shows it to have many imperfections. The imperfections, however, are only a small part of the overall pattern. When we carefully examine the spectrum of this Ronchi ruling, we find that there are even harmonics present. Although these even harmonics are obviously of much smaller magnitude than the DC and odd harmonics, they are nevertheless present. It is the imperfections in the Ronchi ruling (such as poor edges on the opaque lines, dust, scratches, and transparent areas of slightly varying density) that have produced even harmonic spectral points. The same comments that were made above for Figure 5.3 apply to Figures 5.8 and 5.9. Comparing Figure 5.8 to that of

Figure 5.8   High-frequency Ronchi ruling.

Figure 5.3 we see that Figure 5.3 is a ruling of a lower frequency; that is, there are less lines per inch in Figure 5.3 than in Figure 5.8. The spectrum of Figure 5.8 (shown in Figure 5.9) has spectral points more widely separated than that of Figure 5.3(b). The distance between the DC spectral point (the center, and brightest spectral point) and the first harmonic (spectral points on each side of the DC point) is inversely proportional to the Ronchi ruling spacing. We can therefore determine the spacing of the Ronchi ruling by measurement of the spacing in the spectrum.

It is obvious from examination of Figures 5.8 and 5.9 that the spectrum of the transmission function is quite exact. The spectrum, in fact, is so exact that at times it can be somewhat misleading; for example, consider the spectrum shown in Figure 5.10 which was made with a reasonably good Ronchi ruling in the image plane. We know that the spectrum formed should have little or no even harmonics. Actually if we did not know that there should be no even harmonics in the spectrum of a good Ronchi ruling, we might have assumed that the even harmonics in Figure 5.10 were part of the actual spectrum we were looking for. We know from the mathematics—for example, (5-4)—that this is not the spectrum we are looking for. There must be some other phenomena occurring that has introduced the even harmonics. Let us now consider what caused the even harmonics in the spectrum of Figure 5.10.

Figure 5.9   Optical spectrum of Ronchi ruling of Fig. 5.8.

Photographic film has the characteristic of shrinking and expanding during and after the development process. A negative will tend to be thicker in transparent areas than in opaque areas (depending on the development process). This means that although the light is correctly attenuated by the film, its relative phases will be changed. For example, we know that a plane wave must be incident on the object plane to produce a Fourier transform; that is, a wave of constant phase is incident on the object plane. Therefore when we take the Fourier transform, it is of the phase and amplitude of the light leaving the object plane. In the case of a Ronchi ruling we desire the phase of the light leaving the object plane to be constant but the amplitude of the light to vary in accordance with the desired function; that is, the amplitude of the light leaving the object plane should vary in accordance with (5-4). The light amplitude should therefore go from a high value to a zero value as indicated in (5-4) and as shown in Figure 5.2(b). In order to avoid the difficulties of phase changes due to film shrinkage an oil of matching index of refraction is usually used. A drop of oil (usually of the type used with oil emersion lenses with microscopes but even high-grade machine oil will work) is

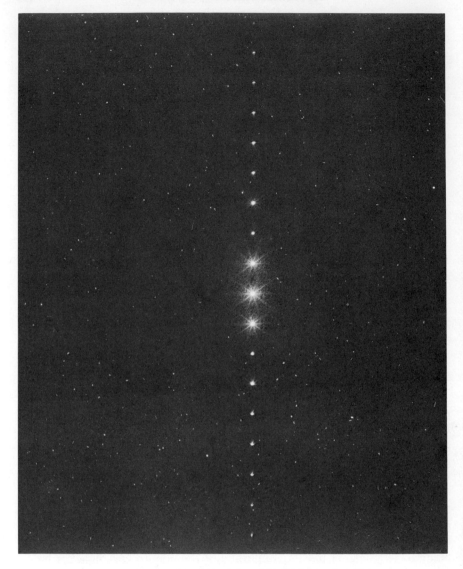

Figure 5.10   Spectrum formed by a Ronchi ruling of varying thickness.

placed on each side of the film and then the film is sandwiched between two optical flats. This method, in effect, produces a transmission function that varies in amplitude in accordance with the function recorded on the film but will have no phase shifts introduced by the varying thicknesses of the film (since in effect the thickness is now constant). It is therefore clear that for accurate results the transmission function in the object plane must be of excellent quality or the results will suffer. This can be considered from another point of view; that is, whatever the transmission function in the object plane (phase and amplitude) the spectrum formed will be of this function.

PERPENDICULARLY CROSSED SPECTRUMS

We have seen how a spectrum can be obtained of a Ronchi ruling by using an optical spectrum analyzer similar to that shown in Figure 5.5. Let us now consider the spectrum obtained by placing two similar Ronchi rulings in a plane but perpendicular to each other. Figure 5.13(a) shows a reproduction of two such perpendicularly crossed Ronchi rulings. From Figure 5.13(a) it can be seen that there are two series of evenly spaced slits formed at right angles to each other. Each transmission function still retains its own identity. When the two transmission functions are put together, the (total) combined transmission function is the *product* of the two individual transmission functions. Figure 5.11 shows two transparencies of Ronchi rulings placed at right angles to each other. The transmission function of each is shown in Figure 5.2(b). If we consider a point P_1 on transparency number 1, we can consider its transmission equal to one since it is in the transparent area. The corresponding point on transparency number 2 is P_1', which is also in a transparent area and

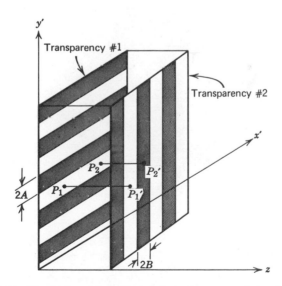

Figure 5.11 Illustration showing the product of two transmission functions.

can also be considered to have a transmission of one. We can see, there-fore, that light coming from the left of transparency number 1 will pass through point P_1 on transparency number 1 to point P_1' on transparency number 2. Point P_1' also having a transmission of one means the light leaving point P_1' will leave without being effected; that is, the product of the transmission of P_1 and P_1' is $1 \times 1 = 1$. The point P_2 on transparency number 1 also has a transmission of one; however, the corresponding point P_1' on transparency number 2 can be considered to be zero because it is an opaque area. The product of the transmission of P_2 and P_2' is $1 \times 0 = 0$, which again shows that the resultant transmission function is the product of the two individual transmission functions. If a point on transparency number 1 should transmit $\frac{1}{2}$ the incident light to it and the corresponding point on transparency number 2 transmits $\frac{3}{4}$ of the incident light to it, the light transmitted by the combination will be $\frac{3}{8}$ of the incident light; that is, the product $\frac{1}{2} \times \frac{3}{4} = \frac{3}{8}$.

In order to determine the spectrum of two crossed Ronchi rulings such as shown in Figure 5.13(a) let us consider the square-wave representation of the two Ronchi rulings. Figure 5.12 shows the two Ronchi rulings

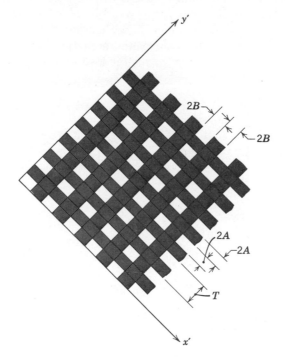

Figure 5.12   Two Ronchi rulings at right angles to each other.

Figure 5.13 Two perpendicularly crossed Ronchi rulings.

plotted on an x' and y' axis. A rectangular wave can be represented as a transmission function by the expression

$$f(t) = \frac{1}{2}\left(1 + \frac{4}{\pi}\cos \omega t - \frac{4}{3\pi}\cos 3\omega t + \frac{4}{5\pi}\cos 5\omega t - \frac{4}{7\pi}\cos 7\omega t + \cdots\right)(5.5)$$

We can rewrite this to represent the transmission function of each of the two Ronchi rulings of Figure 5.11. The product of each of the two individual transmission functions will be the transmission function of the two combined. The equation for the transmission function of the Ronchi ruling with slits parallel to the x' axis was given by (5-1)

$$f(y') = \frac{1}{2}\left(1 + \frac{4}{\pi}\cos\omega y' - \frac{4}{3\pi}\cos 3\omega y' + \frac{4}{5\pi}\cos 5\omega y' - \cdots\right) \quad (5\text{-}1)$$

where $\omega = 2\pi f$. From Figure 5.12 the period of the transmission function can be seen to be $4A$; therefore

$$T = 4A = \frac{1}{f}$$

or

$$f = \frac{1}{4A}$$

then $\omega = 2\pi/4A$ and

$$f(y') = \frac{1}{2}\left(1 + \frac{4}{\pi}\cos\frac{2\pi}{4A}y' - \frac{4}{3\pi}\cos\frac{2\pi}{4A}3y' + \frac{4}{5\pi}\cos\frac{2\pi}{4A}5y' - \cdots\right)$$

$$= \frac{1}{2}\left[1 + \sum_{n=0}^{\infty}\frac{4(-1)^n}{(2n+1)\pi}\cos\frac{(2n+1)\pi y'}{2A}\right] \quad (5\text{-}6)$$

Likewise the transmission function for the Ronchi ruling with slits parallel to the y' axis can be written

$$f(x') = \frac{1}{2}\left[1 + \sum_{p=0}^{\infty}\frac{4(-1)^p}{(2p+1)\pi}\cos\frac{(2p+1)\pi x'}{2B}\right] \quad (5\text{-}7)$$

The functions $f(y')$ and $f(x')$ are the respective transmission functions of the two Ronchi rulings. The product of these two transmission functions is the transmission function of the two combined Ronchi rulings; that is,

$$f(t) = f(x')f(y') \quad (5\text{-}8)$$

The spectrum of $f(t)$ is $F(\omega_{x'}, \omega_{y'})$ which can be written

$$F(\omega_{x'}, \omega_{y'}) = \int_{-\infty}^{\infty}\int_{-\infty}^{\infty} f(t)\exp\left[-j(\omega_{x'}x' + \omega_{y'}y')\right]dx'\,dy' \quad (5\text{-}9)$$

$$F(\omega_{x'}, \omega_{y'}) = \int_{-\infty}^{\infty}\int_{-\infty}^{\infty} f(x')f(y')\exp\left[-j(\omega_{x'}x' + \omega_{y'}y')\right]dx'\,dy'$$

$$(5\text{-}10)$$

$$= \int_{-\infty}^{\infty}\int_{-\infty}^{\infty}\left\{\frac{1}{2}\left[1 + \sum_{n=0}^{\infty}\frac{4(-1)^n}{(2n+1)\pi}\cos\frac{(2n+1)\pi y'}{2A}\right]\right\}$$

$$\times \left\{ \frac{1}{2} \left[1 + \sum_{p=0}^{\infty} \frac{4(-1)^p}{(2p+1)\pi} \cos \frac{(2p+1)\pi x'}{2B} \right] \right\}$$

$$\times \left\{ \exp\left[-j(\omega_{x'}x' + \omega_{y'}y') \right] \right\} dx' \, dy'$$

$$= \frac{1}{4} \int_{-\infty}^{\infty} \int_{-\infty}^{\infty} \left\{ 1 + \sum_{n=0}^{\infty} \frac{4(-1)^n}{(2n+1)\pi} \cos \frac{(2n+1)\pi y'}{2A} \right\}$$

$$\times \left\{ 1 + \sum_{p=0}^{\infty} \frac{4(-1)^p}{(2p+1)\pi} \cos \frac{(2p+1)\pi x'}{2B} \right\}$$

$$\times \exp\left[-j(\omega_{x'}x' + \omega_{y'}y') \right] dx' \, dy' \qquad (5\text{-}11)$$

To solve this for the resultant spectrum we will make use of the fact that the Fourier transform of a sum is the sum of the Fourier transforms. Let us call the summation in each expression

$$\eta_1 = \sum_{n=0}^{\infty} \frac{4(-1)^n}{(2n+1)\pi} \cos \frac{(2n+1)\pi y'}{2A}$$

$$\eta_2 = \sum_{p=0}^{\infty} \frac{4(-1)^p}{(2p+1)\pi} \cos \frac{(2p+1)\pi x'}{2B}$$

We can now write (5-11) as

$$F(\omega_{x'}, \omega_{y'}) = \tfrac{1}{4} \int_{-\infty}^{\infty} \int_{-\infty}^{\infty} \left\{ (1+\eta_1)(1+\eta_2) \right\}$$

$$\times \exp\left[-j(\omega_{x'}x' + \omega_{y'}y') \right] dx' \, dy' \qquad (5\text{-}12)$$

Multiplying the two brackets gives four terms as follows

$$F(\omega_{x'}, \omega_{y'}) = \tfrac{1}{4} \int_{-\infty}^{\infty} \int_{-\infty}^{\infty} (1+\eta_1+\eta_2+\eta_1\eta_2)$$

$$\times \exp\left[-j(\omega_{x'}x' + \omega_{y'}y') \right] dx' \, dy' \qquad (5\text{-}13)$$

The Fourier transform is the sum of each of the Fourier transforms, so we can write (5-13) as

$$F(\omega_{x'}, \omega_{y'}) = \tfrac{1}{4} \int_{-\infty}^{\infty} \int_{-\infty}^{\infty} \exp\left[-j(\omega_{x'}x' + \omega_{y'}y') \right] dx' \, dy' \qquad \textbf{Term A}$$

$$+ \tfrac{1}{4} \int_{-\infty}^{\infty} \int_{-\infty}^{\infty} \eta_1 \exp\left[-j(\omega_{x'}x' + \omega_{y'}y') \right] dx' \, dy' \qquad \textbf{Term B}$$

$$+\tfrac{1}{4}\int_{-\infty}^{\infty}\int_{-\infty}^{\infty}\eta_2\exp\left[-j(\omega_{x'}x'+\omega_{y'}y')\right]dx'\,dy' \qquad \textbf{Term C}$$

$$+\tfrac{1}{4}\int_{-\infty}^{\infty}\int_{-\infty}^{\infty}\eta_1\eta_2\exp\left[-j(\omega_{x'}x'+\omega_{y'}y')\right]dx'\,dy' \qquad \textbf{Term D}$$

$$(5\text{-}14)$$

where we have used the notation $\omega_{x'}$, and $\omega_{y'}$ for the axes in the Fourier transform and we can relate these to and x and y axes by

$$\omega_{x'}=\frac{2\pi x}{\lambda f}$$

$$\omega_{y'}=\frac{2\pi x}{\lambda f}$$

Each term in (5-14) will give rise to components in the resultant spectrum. We can solve for each of the components in the spectrum in either of two ways: (1) use the table of Fourier transforms, (Appendix 13); (2) mathematically solve for each of the components. We shall use both methods to demonstrate the two techniques.

COMPONENTS IN THE SPECTRUM

Term A

Term A from (5-14) is given as

$$\textbf{Term A}=\tfrac{1}{4}\int_{-\infty}^{\infty}\int_{-\infty}^{\infty}\exp\left[-j(\omega_{x'}x'+\omega_{y'}y')\right]dx'\,dy'$$

$$=\tfrac{1}{4}\left\{\int_{-\infty}^{\infty}\exp\left[-j\omega_{x'}x'\right]dx'\int_{-\infty}^{\infty}\exp\left[-j\omega_{y'}y'\right]dy'\right\}$$

$$(5\text{-}15)$$

A term such as $\int_{-\infty}^{\infty}\exp\left[-j\omega_0 t\right]dt$ is, in effect, the Fourier transform of a constant. The integral form of an exponential from $-\infty$ to $+\infty$ is a δ function; that is,

$$e^{-jAx}\,dx=2\pi\delta(A)$$

The **Term A** can therefore be evaluated as follows

$$\textbf{Term A}=\tfrac{1}{4}\left\{\int_{-\infty}^{\infty}\exp\left[-j\omega_{x'}x'\right]dx'\int_{-\infty}^{\infty}\exp\left[-j\omega_{y'}y'\right]dy'\right\}$$

$$=\tfrac{1}{4}\left\{\left[2\pi\delta(\omega_{x'})\right]\left[2\pi\delta(\omega_{y'})\right]\right\}$$

$$=\tfrac{1}{4}\left[4\pi^2\delta(\omega_{x'})\delta(\omega_{y'})\right]$$

$$=\pi^2\left[\delta(\omega_{x'})\delta(\omega_{y'})\right] \qquad (5\text{-}16)$$

Since $\delta(\omega_{y'}) \neq 0$ only when $\omega_{y'} = 0$, and $\delta(\omega_{x'}) \neq 0$ only when $\omega_{x'} = 0$, term $A \neq 0$ only when $\omega_{y'} = 0$ and $\omega_{x'} = 0$. The spectrum obtained here is the amplitude density spectrum and although the amplitude $\delta(\omega_{y'})$ and $\delta(\omega_{x'})$ are not finite at $\omega_{y'} = 0$ and $\omega_{x'} = 0$, the relative amplitude of this term will be taken as the coefficient π^2. This can be interpreted as the contribution of this term when the amplitude density spectrum is integrated over the infinitesimal frequency interval centered about $\omega_{x'} = 0$ and $\omega_{y'} = 0$.

The same result is naturally obtained from Appendix 13 which shows that when $f(t) = 1$, $F(\omega) = 2\pi\delta(\omega)$. Thus we can obtain immediately

$$F(\omega_{x'})F(\omega_{y'}) = \pi^2\delta(\omega_{x'})\delta(\omega_{y'}) \tag{5-16}$$

The intensity at this spectral point will be proportional to the amplitude squared (π^4).

Term B

$$\textbf{Term B} = \frac{1}{4} \int_{-\infty}^{\infty} \int_{-\infty}^{\infty} \eta_1 \exp\left[-j(\omega_{x'}x' + \omega_{y'}y')\right] dx' \, dy'$$

$$= \frac{1}{4} \int_{-\infty}^{\infty} \int_{-\infty}^{\infty} \sum_{n=0}^{\infty} \frac{4(-1)^n}{(2n+1)\pi} \cos \frac{(2n+1)\pi y'}{2A}$$

$$\times \exp\left[-j(\omega_{x'}x' + \omega_{y'}y')\right] dx' \, dy' \tag{5-17}$$

Collection terms in x' and y' gives

$$\textbf{Term B} = \frac{1}{4}\left\{ \int_{-\infty}^{\infty} \exp\left[-j\omega_{x'}x'\right] dx' \int_{-\infty}^{\infty} \sum_{n=0}^{\infty} \frac{4(-1)^n}{(2n+1)\pi} \cos\left[\frac{(2n+1)\pi y'}{2A}\right] \right.$$

$$\left. \times \exp\left[-j\omega_{y'}y'\right] dy' \right\}$$

$$= \frac{1}{4}\left\{ \int_{-\infty}^{\infty} \exp\left[-j\omega_{x'}x'\right] dx' \sum_{n=0}^{\infty} \frac{4(-1)^n}{(2n+1)\pi} \int_{-\infty}^{\infty} \cos\left[\frac{(2n+1)\pi y'}{2A}\right] \right.$$

$$\left. \times \exp\left[-j\omega_{y'}y'\right] dy' \right\}$$

Since

$$\cos\theta = \frac{e^{j\theta} + e^{-j\theta}}{2}$$

then **Term B**

$$= \frac{1}{4}\left[\int_{-\infty}^{\infty} \exp\left[-j\omega_{x'}x'\right] dx' \sum_{n=0}^{\infty} \frac{4(-1)^n}{(2n+1)\pi}\right.$$

$$\times \int_{-\infty}^{\infty} \left\{\frac{\exp\left\{j\left[\frac{(2n+1)\pi y'}{2A}\right]\right\} + \exp\left\{-j\left[\frac{(2n+1)\pi y'}{2A}\right]\right\}}{2}\right\}$$

$$\left.\times \exp\left[-j\omega_{y'}y'\right] dy'\right]$$

$$= \frac{1}{4}\left[\int_{-\infty}^{\infty} \exp\left[-j\omega_{x'}x'\right] dx' \sum_{n=0}^{\infty} \frac{4(-1)^n}{(2n+1)\pi}\right.$$

$$\left.\times \int_{-\infty}^{\infty} \left\{\frac{\exp\left\{j\left[\frac{(2n+1)\pi y'}{2A} - \omega_{y'}y'\right]\right\} + \exp\left\{-j\left[\frac{(2n+1)\pi y'}{2A} + \omega_{y'}y'\right]\right\}}{2}\right\}dy'\right]$$

$$= \frac{1}{4}\left[\int_{-\infty}^{\infty} \exp\left[-j\omega_{x'}x'\right] dx' \sum_{n=0}^{\infty} \frac{4(-1)^n}{(2n+1)2\pi}\right.$$

$$\left.\times \int_{-\infty}^{\infty}\left\{\exp\left\{j\left[\frac{(2n+1)\pi y'}{2A} - \omega_{y'}y'\right]\right\}dy' + \exp\left\{-j\left[\frac{(2n+1)\pi y'}{2A} + \omega_{y'}y'\right]\right\}\right\}dy'\right]$$

$$= \frac{1}{4}\left[\int_{-\infty}^{\infty} \exp\left[-j\omega_{x'}x'\right] dx' \sum_{n=0}^{\infty} \frac{4(-1)^n}{(2n+1)2\pi}\right.$$

$$\left.\times\left\{\int_{-\infty}^{\infty}\exp\left\{jy'\left[\frac{(2n+1)\pi}{2A} - \omega_{y'}\right]\right\}dy' + \int_{-\infty}^{\infty}\exp\left\{-jy'\left[\frac{(2n+1)\pi}{2A} + \omega_{y'}\right]\right\}dy'\right\}\right]$$

The integral of a complex exponential from $-\infty$ to $+\infty$ will give a δ function as follows

Term B $= \frac{1}{4}\left[2\pi\delta(\omega_{x'}) \sum_{n=0}^{\infty} \frac{4(-1)^n}{2\pi(2n+1)}\left\{2\pi\delta\left(\omega_{y'} - \frac{(2n+1)\pi}{2A}\right)\right.\right.$

$$\left.\left.+ 2\pi\delta\left(\omega_{y'} + \frac{(2n+1)\pi}{2A}\right)\right\}\right]$$

$$= \frac{1}{4}\left[\sum_{n=0}^{\infty} \frac{4(-1)^n}{2\pi(2n+1)}[2\pi\delta(\omega_{x'})]\left\{2\pi\delta\left(\omega_{y'} - \frac{(2n+1)\pi}{2A}\right)\right.\right.$$

$$\left.\left.+ 2\pi\delta\left(\omega_{y'} + \frac{(2n+1)\pi}{2A}\right)\right\}\right]$$

$$= \sum_{n=0}^{\infty} \frac{2(-1)^n \pi}{(2n+1)} [\delta(\omega_{x'})] \left\{ \delta\left(\omega_{y'} - \frac{(2n+1)\pi}{2A}\right) + \delta\left(\omega_{y'} + \frac{(2n+1)\pi}{2A}\right) \right\}.$$

$$(5\text{-}18)$$

This can be interpreted to mean that any particular **Term B** spectral point has an amplitude given by the coefficient of (5-18) or

$$A = \frac{2(-1)^n \pi}{(2n+1)}.$$

The intensity (amplitude squared) is given by

$$I = A^2 = \frac{4\pi^2}{(2n+1)^2}$$

since $[(-1)^n]^2 = 1$. The location of any particular spectral point can be found by setting the δ function variables in (5-18) equal to zero; that is, set

$$\omega_{x'} = 0$$

$$\omega_{y'} - \frac{(2n+1)\pi}{2A} = 0$$

and

$$\omega_{y'} + \frac{(2n+1)\pi}{2A} = 0.$$

This gives spectral points at

$$\omega_{x'} = 0 \qquad\qquad (5\text{-}19)$$

$$\omega_{y'} = \pm \frac{(2n+1)\pi}{2A}. \qquad\qquad (5\text{-}20)$$

The spectral points formed by **Term B** can only exist on the line $\omega_{x'} = 0$ and at locations where

$$\omega_{y'} = \pm \frac{(2n+1)\pi}{2A}.$$

Term C

$$\textbf{Term C} = \frac{1}{4}\left[\int_{-\infty}^{\infty} \int_{-\infty}^{\infty} \eta_2 \exp\left\{-j(\omega_{x'}x' + \omega_{y'}y')\right\} dx'\, dy' \right]$$

$$= \frac{1}{4}\left[\int_{-\infty}^{\infty} \int_{-\infty}^{\infty} \left[\sum_{p=0}^{\infty} \frac{4(-1)^p}{(2p+1)\pi} \cos \frac{(2p+1)\pi x'}{2B} \right] \right.$$

$$\left. \times \exp\left\{-j(\omega_{x'}x' + \omega_{y'}y')\right\} dx'\, dy' \right] \qquad (5\text{-}21)$$

The equation (5-21) for **Term C** is identical in form to that for **Term B** as given by (5-17) except the constant B replaces A, p replaces n, and x' and y' are interchanged. Thus we can write the results for **Term C** immediately from (5-18) by replacing A by B, and n by p, and interchanging x' and y'

$$\textbf{Term C} = \left[\sum_{p=0}^{\infty} \frac{2(-1)^p \pi}{(2p+1)} [\delta(\omega_{y'})] \right] \left\{ \delta\left(\omega_{x'} - \frac{(2p+1)\pi}{2B}\right) \right.$$
$$\left. + \delta\left(\omega_{x'} + \frac{(2p+1)\pi}{2B}\right) \right\}. \qquad (5\text{-}22)$$

This can be interpreted to mean that any particular **Term C** spectral point has an amplitude given by the coefficient of (5-22) or

$$A = \frac{2(-1)^p \pi}{(2p+1)}.$$

The intensity (amplitude squared) is given by

$$I = A^2 = \frac{4\pi^2}{(2p+1)^2}$$

since $[(-1)^p]^2 = 1$. The location of any particular spectral point can be found by setting the δ functions in (5-22) equal to zero; that is,

$$\frac{(2p+1)\pi}{2B} - \omega_{x'} = 0$$

$$\frac{-(2p+1)\pi}{2B} - \omega_{x'} = 0$$

$$\omega_{y'} = 0.$$

This gives spectral points at

$$\omega_{x'} = \pm \frac{(2p+1)\pi}{2B} \qquad (5\text{-}23)$$

$$\omega_{y'} = 0. \qquad (5\text{-}24)$$

Term C spectral points can exist only on the line $\omega_{y'} = 0$ and at locations where

$$\omega_{x'} = \pm \frac{(2p+1)\pi}{2B}.$$

Term D

$$\textbf{Term D} = \frac{1}{4} \int_{-\infty}^{\infty} \int_{-\infty}^{\infty} \eta_1 \eta_2 \exp\left[-j(\omega_{x'}x' + \omega_{y'}y')\right] dx'\, dy'$$

$$= \frac{1}{4} \int_{-\infty}^{\infty} \int_{-\infty}^{\infty} \left[\sum_{n=0}^{\infty} \frac{4(-1)^n}{(2n+1)\pi} \cos\frac{(2n+1)\pi y'}{2A} \right]\left[\sum_{p=0}^{\infty} \frac{4(-1)^p}{(2p+1)\pi} \right.$$

$$\times \left. \cos\frac{(2p+1)\pi x'}{2B}\right][\exp\{-j(\omega_{x'}x' + \omega_{y'}y')\}]\, dx'\, dy' \qquad (5\text{-}25)$$

$$= \frac{1}{4} \sum_{n=0}^{\infty} \sum_{p=0}^{\infty} \frac{16(-1)^n(-1)^p}{(2n+1)(2p+1)\pi^2} \int_{-\infty}^{\infty} \int_{-\infty}^{\infty} \left[\cos\frac{(2n+1)\pi y'}{2A}\right]$$

$$\times \left[\cos\frac{(2p+1)\pi x'}{2B}\right][\exp\{-j(\omega_{x'}x' + \omega_{y'}y')\}]\, dx'\, dy'$$

$$= \frac{1}{4} \sum_{n=0}^{\infty} \sum_{p=0}^{\infty} \frac{16(-1)^n(-1)^p}{(2n+1)(2p+1)\pi^2}$$

$$\times \int_{-\infty}^{\infty} \int_{-\infty}^{\infty} \left[\frac{\exp\left\{j\dfrac{(2n+1)}{2A}\pi y'\right\} + \exp\left\{-j\dfrac{(2n+1)}{2A}\pi y'\right\}}{2} \right]$$

$$\times \left[\frac{\exp\left\{j\dfrac{(2p+1)}{2B}\pi x'\right\} + \exp\left\{-j\dfrac{(2p+1)}{2B}\pi x'\right\}}{2} \right]$$

$$\times [\exp\{-j(\omega_{x'}x' + \omega_{y'}y')\}]\, dx'\, dy'$$

$$= \sum_{n=0}^{\infty} \sum_{p=0}^{\infty} \frac{(-1)^n(-1)^p}{(2n+1)(2p+1)\pi^2} \int_{-\infty}^{\infty} \int_{-\infty}^{\infty} \left[\exp\left\{j\frac{(2n+1)}{2A}\pi y'\right\}\right.$$

$$\left. + \exp\left\{-j\frac{(2n+1)}{2A}\pi y'\right\}\right]$$

$$\times \left[\exp\left\{j\frac{(2p+1)}{2B}\pi x'\right\} + \exp\left\{-j\frac{(2p+1)}{2B}\pi x'\right\}\right]$$

$$\qquad\qquad\qquad\qquad\qquad\qquad\qquad (5\text{-}26)$$

$$\times [\exp\{-j(\omega_{x'}x' + \omega_{y'}y')\}]\, dx'\, dy'$$

which is in the form

$$(a+b)(c+d)(f) = acf + bcf + adf + bdf$$

where

$$a = \exp\left[j\frac{(2n+1)}{2A}\pi y'\right]$$

$$b = \exp\left[-j\frac{(2n+1)}{2A}\pi y'\right]$$

$$c = \exp\left[j\frac{(2p+1)}{2B}\pi x'\right]$$

$$d = \exp\left[-j\frac{(2p+1)}{2B}\pi x'\right]$$

$$f = \exp\left[-j(\omega_{x'}x' + \omega_{y'}y')\right].$$

Then (5-26) becomes

$$\textbf{Term D} = \sum_{n=0}^{\infty}\sum_{p=0}^{\infty}\frac{(-1)^n(-1)^p}{(2n+1)(2p+1)\pi^2}\int_{-\infty}^{\infty}\int_{-\infty}^{\infty}\left\{\exp\left[j\frac{(2n+1)}{2A}\pi y'\right.\right.$$

$$\left.+j\frac{(2p+1)}{2B}\pi x' - j(\omega_{x'}x' + \omega_{y'}y')\right]$$

$$+\exp\left[-j\frac{(2n+1)}{2A}\pi y' + j\frac{(2p+1)}{2B}\pi x' - j(\omega_{x'}x' + \omega_{y'}y')\right]$$

$$+\exp\left[j\frac{(2n+1)}{2A}\pi y' - j\frac{(2p+1)}{2B}\pi x' - j(\omega_{x'}x' + \omega_{y'}y')\right]$$

$$\left.+\exp\left[-j\frac{(2n+1)}{2A}\pi y' - j\frac{(2p+1)}{2B}\pi x' - j(\omega_{x'}x' + \omega_{y'}y')\right]\right\}dx'\,dy'$$

$$(5\text{-}27)$$

$$= \sum_{n=0}^{\infty}\sum_{p=0}^{\infty}\frac{(-1)^n(-1)^p}{(2n+1)(2p+1)\pi^2}$$

$$\times\left[\int_{-\infty}^{\infty}\exp\left\{jy'\left[\frac{(2n+1)\pi}{2A} - \omega_{y'}\right]\right\}dy'\right.$$

$$\times\int_{-\infty}^{\infty}\exp\left\{jx'\left[\frac{(2p+1)\pi}{2B} - \omega_{x'}\right]\right\}dx'$$

$$+\int_{-\infty}^{\infty}\exp\left\{-jy'\left[\frac{(2n+1)\pi}{2A} + \omega_{y'}\right]\right\}dy'$$

$$\times\int_{-\infty}^{\infty}\exp\left\{jx'\left[\frac{(2p+1)\pi}{2B} - \omega_{x'}\right]\right\}dx'$$

$$+ \int_{-\infty}^{\infty} \exp\left\{ jy'\left[\frac{(2n+1)\pi}{2A} - \omega_{y'} \right] \right\} dy'$$

$$\times \int_{-\infty}^{\infty} \exp\left\{ -jx'\left[\frac{(2p+1)\pi}{2B} - \omega_{x'} \right] \right\} dx'$$

$$+ \int_{-\infty}^{\infty} \exp\left\{ -jy'\left[\frac{(2n+1)\pi}{2A} + \omega_{y'} \right] \right\} dy'$$

$$\times \int_{-\infty}^{\infty} \exp\left\{ -jx'\left[\frac{(2p+1)\pi}{2B} + \omega_{x'} \right] \right\} dx' \Bigg] \tag{5-28}$$

The complex exponential integrated from $-\infty$ to $+\infty$ can be written as a delta function. **Term D** can be written in terms of delta functions as

$$\textbf{Term D} = \sum_{n=0}^{\infty} \sum_{p=0}^{\infty} \frac{(-1)^n (-1)^p}{(2n+1)(2p+1)\pi^2}$$

$$\times \Bigg[\left\{ 2\pi\delta\left(\frac{(2n+1)\pi}{2A} - \omega_{y'} \right) \right\} \left\{ 2\pi\delta\left(\frac{(2p+1)\pi}{2B} - \omega_{x'} \right) \right\}$$

$$+ \left\{ 2\pi\delta\left(-\frac{(2n+1)\pi}{2A} - \omega_{y'} \right) \right\} \left\{ 2\pi\delta\left(\frac{(2p+1)\pi}{2B} - \omega_{x'} \right) \right\}$$

$$+ \left\{ 2\pi\delta\left(\frac{(2n+1)\pi}{2A} - \omega_{y'} \right) \right\} \left\{ 2\pi\delta\left(-\frac{(2p+1)\pi}{2B} - \omega_{x'} \right) \right\}$$

$$+ \left\{ 2\pi\delta\left(-\frac{(2n+1)\pi}{2A} - \omega_{y'} \right) \right\} \left\{ 2\pi\delta\left(-\frac{(2p+1)\pi}{2B} - \omega_{x'} \right) \right\} \Bigg] \tag{5-29}$$

$$= \sum_{n=0}^{\infty} \sum_{p=0}^{\infty} \frac{(-1)^n (-1)^p 4\pi^2}{(2n+1)(2p+1)\pi^2}$$

$$\times \Bigg[\left\{ \delta\left(\frac{(2n+1)\pi}{2A} - \omega_{y'} \right) \delta\left(\frac{(2p+1)}{2B} - \omega_{x'} \right) \right\}$$

$$+ \left\{ \delta\left(-\frac{(2n+1)\pi}{2A} - \omega_{y'} \right) \delta\left(\frac{(2p+1)\pi}{2B} - \omega_{x'} \right) \right\}$$

$$+ \left\{ \delta\left(\frac{(2n+1)\pi}{2A} - \omega_{y'} \right) \delta\left(-\frac{(2p+1)\pi}{2B} - \omega_{x'} \right) \right\}$$

$$+ \left\{ \delta\left(-\frac{(2n+1)\pi}{2A} - \omega_{y'} \right) \delta\left(-\frac{(2p+1)\pi}{2B} - \omega_{x'} \right) \right\} \Bigg] \tag{5-30}$$

Equation (5-30) shows that **Term D** spectral points are formed at locations given by

$$\omega_{x'} = \pm \frac{(2p+1)\pi}{2B} \tag{5-31}$$

$$\omega_{y'} = \pm \frac{(2n+1)\pi}{2A} \tag{5-32}$$

with an intensity given by

$$A^2 = \frac{16}{(2n+1)^2(2p+1)^2}.$$

The spectrum of two perpendicular crossed Ronchi rulings has been found to contain four types of components, namely:

Term A, which is the DC component.

Term B which is a spectrum of the horizontal lines located on the

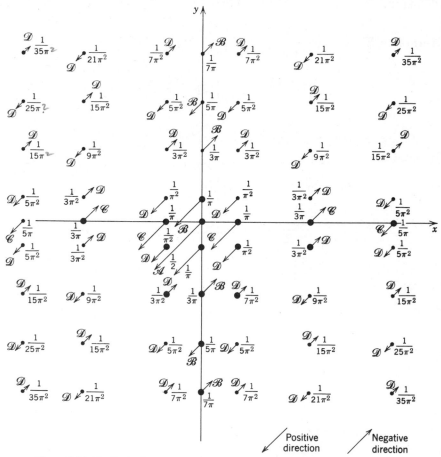

Figure 5.14 (a) Optical spectrum of two perpendicularly crossed Ronchi rulings.

$1/25(49\pi^4)$ •	$1/9(49\pi^4)$ •	$1/49\pi^4$ •	• $1/49\pi^2$ • $1/49\pi^4$	• $1/9(49\pi^4)$	• $1/25(49\pi^4)$	
$\{1/25\pi^2\}^2$ •	$1/9(25\pi^4)$ •	$1/25\pi^4$ •	• $1/25\pi^2$ • $1/25\pi^4$	• $1/9(25\pi^4)$	• $\{1/25\pi^2\}^2$	
$1/9(25\pi^4)$ •	$\{1/9\pi^2\}^2$ •	$1/9\pi^4$ •	• $1/9\pi^2$ • $1/9\pi^4$	• $\{1/9\pi^2\}^2$	• $1/9(25\pi^4)$	
$1/25\pi^4$ •	$1/9\pi^4$ •	$1/\pi^4$ •	• $1/\pi^2$ • $1/\pi^4$	• $1/9\pi^4$	• $1/25\pi^4$	
$1/25\pi^2$	$1/9\pi^2$	$1/\pi^2$	$1/4$ • $1/\pi^2$	$1/9\pi^2$	$1/25\pi^2$	
$1/25\pi^4$ •	$1/9\pi^4$ •	$1/\pi^4$ •	• $1/\pi^2$ • $1/\pi^4$	• $1/9\pi^4$	• $1/25\pi^4$	
$1/9(25\pi^4)$ •	$\{1/9\pi^2\}^2$ •	$1/9\pi^4$ •	• $1/9\pi^2$ • $1/9\pi^4$	• $\{1/9\pi^2\}^2$	• $1/9(25\pi^4)$	
$\{1/25\pi^2\}^2$ •	$1/9(25\pi^4)$ •	$1/25\pi^4$ •	• $1/25\pi^2$ • $1/25\pi^4$	• $1/9(25\pi^4)$	• $\{1/25\pi^2\}^2$	
$1/25(49\pi^4)$ •	$1/9(49\pi^4)$ •	$1/49\pi^4$ •	• $1/49\pi^2$ • $1/49\pi^4$	• $1/9(49\pi^4)$	• $1/25(49\pi^4)$	

Figure 5.14 (b) Intensity of points shown in Figure 5.14(a).

$\omega_{y'}$ axis. The horizontal lines of Figure 5.5 correspond to the **Term B** of the spectrum of two perpendicularly crossed Ronchi rulings. The amplitude of the spectral points of the horizontal lines of Figure 5.5 and that of **Term B** are proportional. This can be seen by comparing the values of the spectral components plotted in Figure 5.7 with the corresponding **Term B** spectral points plotted in Figure 5.14(a) after normalizing (dividing by $1/2\pi^2$); that is, we can normalize Figure 5.14(a) by dividing all the spectral-point amplitudes by $1/2\pi^2$, which will make the DC term in both figures have an amplitude of $\frac{1}{2}$. After normalizing we can see that the relative amplitudes of the spectral points of Figure 5.7 and the **Term B** spectral points of Figure 5.14(a) are proportional to each other and will be found at identical locations in the spectrum plane. Figure 5.14(b) shows the intensity of the respective points shown in Figure 5.14(a).

Term C which is a spectrum of the vertical lines located on the $\omega_{x'}$ axis.

Term D which are spectral points located off axis.

The following chart summarizes the results of the spectrum formed by **two** perpendicularly crossed Ronchi rulings.

Term	Location of Spectral Points	Amplitude	Normalized Amplitude	Intensity	Normalized Intensity
A	$\omega_{x'} = 0$ $\omega_{y'} = 0$	π^2	$\frac{1}{2}$	π^4	$\frac{1}{4}$
B	$\omega_{x'} = 0$ $\omega_{y'} = \pm\dfrac{(2n+1)\pi}{2A}$	$\dfrac{2(-1)^n\pi}{(2n+1)}$	$\dfrac{(-1)^n}{\pi(2n+1)}$	$\dfrac{4\pi^2}{(2n+1)^2}$	$\dfrac{1}{\pi^2(2n+1)^2}$
C	$\omega_{x'} = \pm\dfrac{(2p+1)\pi}{2B}$ $\omega_{y'} = 0$	$\dfrac{2(-1)^p\pi}{(2p+1)}$	$\dfrac{(-1)^p}{\pi(2p+1)}$	$\dfrac{4\pi^2}{(2p+1)^2}$	$\dfrac{1}{\pi^2(2p+1)^2}$
D	$\omega_{x'} = \pm\dfrac{(2p+1)\pi}{2B}$ $\omega_{y'} = \pm\dfrac{(2n+1)\pi}{2A}$	$\dfrac{4(-1)^n(-1)^p}{(2n+1)(2p+1)}$	$\dfrac{2(-1)^n(-1)^p}{\pi^2(2n+1)(2p+1)}$	$\dfrac{16}{(2n+1)^2(2p+1)^2}$	$\dfrac{4}{\pi^4(2n+1)^2(2p+1)^2}$

Figure 5.15(a) shows a photograph of the spectrum of two crossed Ronchi rulings, and Figure 5.15(b) shows the points labeled corresponding to the above table.

SPECTRUM OF THREE CROSSED RONCHI RULINGS

We have just considered the case of the spectrum formed from two perpendicularly crossed Ronchi rulings. We shall now add a third Ronchi ruling to the two perpendicularly crossed Ronchi rulings and

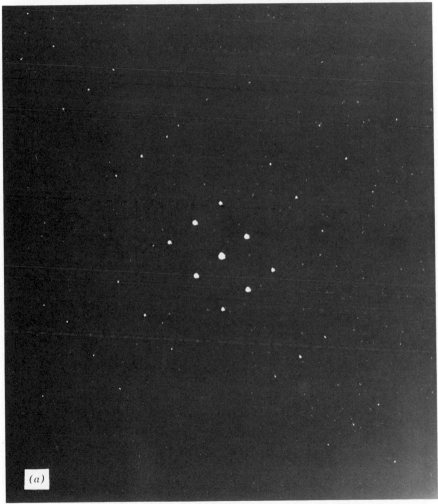

(a)

Figure 5.15 (a)   Spectrum of the two perpendicularly crossed Ronchi rulings of Figure 5.13.

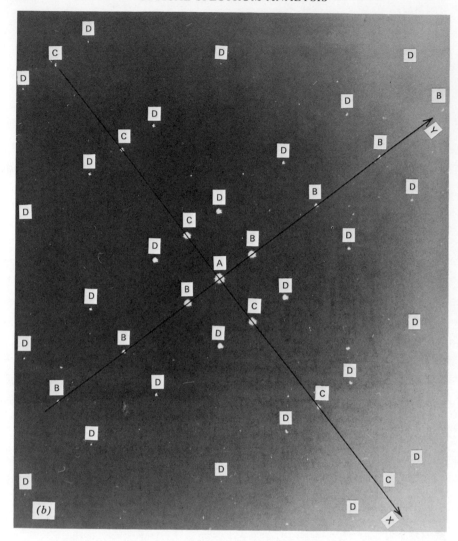

Figure 5.15 (b)   Identification of points corresponding to table on page 266.

determine the spectrum formed by them. In the previous example the transmission function of the Ronchi ruling that had its lines perpendicular to the y' axis was given by

$$f(y') = \frac{1}{2}\left[1 + \sum_{n=0}^{\infty} \frac{4(-1)^n}{(2n+1)\pi} \cos \frac{(2n+1)\pi y'}{2A}\right] \qquad (5\text{-}6)$$

The transmission function of the Ronchi ruling that had its lines perpendicular to the x' axis was given by

$$f(x') = \frac{1}{2}\left[1 + \sum_{p=0}^{\infty} \frac{4(-1)^p}{(2p+1)\pi} \cos\frac{(2p+1)\pi x'}{2B}\right] \qquad (5\text{-}7)$$

We shall now consider a third Ronchi ruling superimposed on the other two at an angle to both the x' and y' axes. Figure 5.16 shows the orientation of the three Ronchi rulings with respect to each other. The product of the three amplitude transmission functions is the combined amplitude transmission function of the three Ronchi rulings.

Let us examine the third Ronchi ruling in more detail. Figure 5.17 shows this Ronchi ruling when drawn on the x' and y' axes. The transmission function of this ruling can be written in terms of the x' and y' axes by referring to Figure 5.17(b). Line 1 of the Ronchi ruling is shown drawn through the origin for simplicity in the derivation. The equation of a line is given by

$$y' = mx' + b$$

where m is the slope and b is the y' intercept. The equation of line 1 therefore can be written as

$$y' = x' \tan\varphi$$

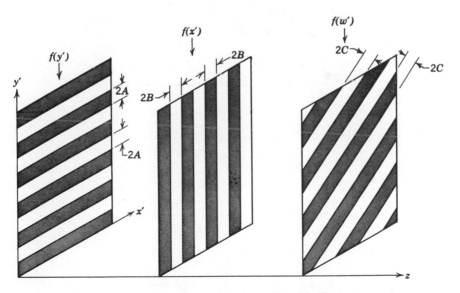

Figure 5.16 Orientation of three Ronchi rulings.

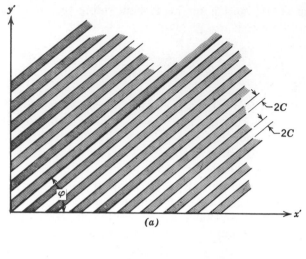

(a)

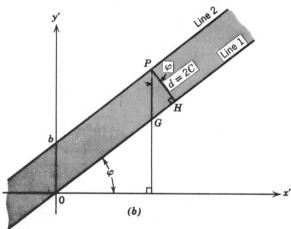

(b)

Figure 5.17 Ronchi ruling at right angle φ with the x axis.

since the y' intercept is zero (the origin). Let us now consider any point P with respect to its distance from line 1. Let us call the distance from point P to line 1, d. Through point P let us draw line 2 parallel to line 1. This line intercepts the y' axis at point b. In triangle GHP the line $\overline{PG}$ is equal to $\overline{Ob}$ (where the value of b equals the y' intercept). From triangle GHP we can write

$$\cos \varphi = \frac{d}{\overline{PG}} \qquad \text{or} \qquad \overline{PG} = \frac{d}{\cos \varphi}$$

where $\overline{PG}$ is the y' intercept. Because line 2 is parallel to line 1, the equation of line 2 can be written

$$y' = mx' + b = x' \tan \varphi + b = x' \tan \varphi + \overline{PG}$$

$$= x' \frac{\sin \varphi}{\cos \varphi} + \frac{d}{\cos \varphi}$$

$$y' \cos \varphi = x' \sin \varphi + d$$

$$d = y' \cos \varphi - x' \sin \varphi \qquad (5\text{-}33)$$

which is the location of any point P with respect to the lines of the ruling that are referred to the x' and y' axes. When $y' \cos \varphi - x' \sin \varphi$ increases by $2C$, the point will have moved half the spatial period. The spatial frequency is therefore a function of both x' and y'.

The transmission function $f(w')$ for the third Ronchi ruling can be written as

$$f(w') = \frac{1}{2}\left[1 + \sum_{q=0}^{\infty} \frac{4(-1)^q}{(2q+1)\pi} \cos\left\{\frac{(2q+1)\pi}{2C}(y' \cos \varphi - x' \sin \varphi)\right\}\right] \quad (5\text{-}34)$$

Equation (5-34) can be obtained directly by referring to (5-6) or (5-7), substituting the value of d in (5-33) for the variable, and making the appropriate changes for the constants.

The functions $f(x)$, $f(y')$, and $f(w)$ are the respective amplitude transmission functions of the three Ronchi rulings shown in Figure 5.16. The product of these three amplitude transmission functions is the amplitude transmission function of the three combined Ronchi rulings; that is,

$$f(t) = f(x')f(y')f(w) \qquad (5\text{-}35)$$

where $f(t)$ is the combined transmission function. The spectrum of $f(t)$ is $F(\omega_{x'},\omega_{y'})$ which can be written

$$F(\omega_{x'}, \omega_{y'}) = \int_{-\infty}^{\infty} \int_{-\infty}^{\infty} f(t) \exp\left[-j(\omega_{x'}x' + \omega_{y'}y')\right] dx' \, dy' \qquad (5\text{-}36)$$

$$= \int_{-\infty}^{\infty} \int_{-\infty}^{\infty} f(x')f(y')f(w) \exp\left[-j(\omega_{x'}x' + \omega_{y'}y')\right] dx' \, dy'$$

$$(5\text{-}37)$$

$$= \frac{1}{8} \int_{-\infty}^{\infty} \int_{-\infty}^{\infty} \left[1 + \sum_{n=0}^{\infty} \frac{4(-1)^n}{(2n+1)\pi} \cos \frac{(2n+1)\pi y'}{2A} \right]$$

$$\times \left[1 + \sum_{p=0}^{\infty} \frac{4(-1)^p}{(2p+1)\pi} \cos \frac{(2p+1)\pi x'}{2B} \right]$$

$$\times \left[1 + \sum_{q=0}^{\infty} \frac{4(-1)^q}{(2q+1)\pi} \cos \left(\frac{(2q+1)\pi}{2C} (y' \cos \varphi - x' \sin \varphi) \right) \right]$$

$$\times \exp \left[-j(\omega_{x'} x' + \omega_{y'} y') \right] dx' \, dy' \tag{5-38}$$

Just as in the previous example of two perpendicularly crossed Ronchi rulings we shall let

$$\eta_1 = \sum_{n=0}^{\infty} \frac{4(-1)^n}{(2n+1)\pi} \cos \frac{(2n+1)\pi y'}{2A} \tag{5-39}$$

$$\eta_2 = \sum_{p=0}^{\infty} \frac{4(-1)^p}{(2p+1)\pi} \cos \frac{(2p+1)\pi x'}{2B} \tag{5-40}$$

Now in (5-38) we will let

$$\eta_3 = \sum^{\infty} \frac{4(-1)^q}{(2q+1)\pi} \cos \left\{ \frac{(2q+1)\pi}{2C} (y' \cos \varphi - x' \sin \varphi) \right\} \tag{5-41}$$

We can now write (5-38) as

$$F(\omega_{x'}, \omega_{y'}) = \frac{1}{8} \int_{-\infty}^{\infty} \int_{-\infty}^{\infty} (1+\eta_1)(1+\eta_2)(1+\eta_3)$$

$$\times \exp \left[-j(\omega_{x'} x' + \omega_{y'} y') \right] dx' \, dy' \tag{5-42}$$

Multiplying the three bracketed terms in (5-42) the following eight terms are obtained:

$$F(\omega_{x'}, \omega_{y'}) = \frac{1}{8} \int_{-\infty}^{\infty} \int_{-\infty}^{\infty} (1+\eta_1+\eta_2+\eta_3+\eta_1\eta_2+\eta_2\eta_3+\eta_1\eta_3+\eta_1\eta_2\eta_3)$$

$$\times \exp \left[-j(\omega_{x'} x' + \omega_{y'} y') \right] dx' \, dy' \tag{5-43}$$

The Fourier transform is the sum of each of the Fourier transforms, so we can write (5-43) as

$$F(\omega_{x'}, \omega_{y'}) = \frac{1}{8} \int_{-\infty}^{\infty} \int_{-\infty}^{\infty} \exp\left[-j(\omega_{x'}x' + \omega_{y'}y')\right] dx' \, dy' \qquad \text{A}$$

$$+ \frac{1}{8} \int_{-\infty}^{\infty} \int_{-\infty}^{\infty} \eta_1 \exp\left[-j(\omega_{x'}x' + \omega_{y'}y')\right] dx' \, dy' \qquad \text{B}$$

$$+ \frac{1}{8} \int_{-\infty}^{\infty} \int_{-\infty}^{\infty} \eta_2 \exp\left[-j(\omega_{x'}x' + \omega_{y'}y')\right] dx' \, dy' \qquad \text{C}$$

$$+ \frac{1}{8} \int_{-\infty}^{\infty} \int_{-\infty}^{\infty} \eta_1\eta_2 \exp\left[-j(\omega_{x'}x' + \omega_{y'}y')\right] dx' \, dy' \qquad \text{D}$$

$$+ \frac{1}{8} \int_{-\infty}^{\infty} \int_{-\infty}^{\infty} \eta_3 \exp\left[-j(\omega_{x'}x' + \omega_{y'}y')\right] dx' \, dy' \qquad \text{E}$$

$$+ \frac{1}{8} \int_{-\infty}^{\infty} \int_{-\infty}^{\infty} \eta_2\eta_3 \exp\left[-j(\omega_{x'}x' + \omega_{y'}y')\right] dx' \, dy' \qquad \text{F}$$

$$+ \frac{1}{8} \int_{-\infty}^{\infty} \int_{-\infty}^{\infty} \eta_1\eta_3 \exp\left[-j(\omega_{x'}x' + \omega_{y'}y')\right] dx' \, dy' \qquad \text{G}$$

$$+ \frac{1}{8} \int_{-\infty}^{\infty} \int_{-\infty}^{\infty} \eta_1\eta_2\eta_3 \exp\left[-j(\omega_{x'}x' + \omega_{y'}y')\right] dx' \, dy' \qquad \text{H}$$

$$(5\text{-}44)$$

Each term in (5-44) will give rise to components in the resulting spectrum. We can see that the first four terms **A, B, C,** and **D** are identical (except for a factor of $\frac{1}{2}$) to the corresponding terms in the previous example of two perpendicularly crossed Ronchi rulings (which, in effect, means that these terms will result in spectrums whose spectral points will have the same relative intensities). This is an important result since it shows that the addition of a transmission function to an existing transmission function does not change the spectrum of the original transmission function; it just adds additional points to the resulting spectrum and changes the amplitude of the original spectral points by a constant. Since we have already determined the components **A, B, C,** and **D** for the spectrum of two perpendicularly crossed Ronchi rulings, we need only take those values and divide them by the constant factor of $\frac{1}{2}$ to find the corresponding amplitudes of the corresponding spectral points when a third Ronchi ruling is added. These results are summarized in the chart at the end of this section. We shall now determine the components **E, F, G,** and **H** of (5-44).

Term E

Term E $= \dfrac{1}{8} \int_{-\infty}^{\infty} \int_{-\infty}^{\infty} \eta_3 \exp\left[-j(\omega_{x'}x' + \omega_{y'}y')\right] dx' \, dy' \qquad (5\text{-}45)$

$$= \frac{1}{8} \int_{-\infty}^{\infty} \int_{-\infty}^{\infty} \sum_{q=0}^{\infty} \frac{4(-1)^q}{(2q+1)\pi} \cos \left\{ \frac{(2q+1)\pi}{2C} (y' \cos \varphi - x' \sin \varphi) \right\}$$

$$\times \exp \left[-j(\omega_{x'} + \omega_{y'}) \right] dx' \, dy' \qquad (5\text{-}46)$$

Using the relationship

$$\cos \theta = \frac{e^{j\theta} + e^{-j\theta}}{2}$$

equation (5-46) can be written

Term E $= \dfrac{1}{8} \displaystyle\sum_{q=0}^{\infty} \dfrac{4(-1)^q}{2(2q+1)\pi}$

$$\times \left[\int_{-\infty}^{\infty} \int_{-\infty}^{\infty} \exp \left\{ j \left[\frac{(2q+1)\pi}{2C} (y' \cos \varphi - x' \sin \varphi) \right] \right\} \right.$$

$$\times \exp \left[-j(\omega_{x'} x' + \omega_{y'} y') \right] dx' \, dy'$$

$$+ \int_{-\infty}^{\infty} \int_{-\infty}^{\infty} \exp \left\{ -j \left[\frac{(2q+1)\pi}{2C} (y' \cos \varphi - x' \sin \varphi) \right] \right\}$$

$$\left. \times \exp \left[-j(\omega_{x'} x' + \omega_{y'} y') \right] dx' \, dy' \right] \qquad (5\text{-}47)$$

$$= \frac{1}{8} \sum_{q=0}^{\infty} \frac{4(-1)^q}{2\pi(2q+1)} \left[\int_{-\infty}^{\infty} \exp \left\{ jy' \left[\frac{(2q+1)\pi}{2C} (\cos \varphi) - \omega_{y'} \right] \right\} dy' \right.$$

$$\times \int_{-\infty}^{\infty} \exp \left\{ -jx' \left[\frac{(2q+1)\pi}{2C} (\sin \varphi) + \omega_{x'} \right] \right\} dx'$$

$$+ \int_{-\infty}^{\infty} \exp \left\{ -jy' \left[\frac{(2q+1)\pi}{2C} (\cos \varphi) + \omega_{y'} \right] \right\} dy'$$

$$\left. \times \int_{-\infty}^{\infty} \exp \left\{ jx' \left[\frac{(2q+1)\pi}{2C} (\sin \varphi) - \omega_{x'} \right] \right\} dx' \right] \qquad (5\text{-}48)$$

$$= \frac{1}{8} \sum_{q=0}^{\infty} \frac{4(-1)^q}{2\pi(2q+1)} \left[\left\{ 2\pi\delta \left(\frac{(2q+1)\pi}{2C} (\cos \varphi) - \omega_{y'} \right) \right\} \right.$$

$$\times \left\{ 2\pi\delta \left(\frac{(2q+1)\pi}{2C} (\sin \varphi) + \omega_{x'} \right) \right\}$$

$$+ \left\{ 2\pi\delta\left(\frac{(2q+1)\pi}{2C}(\cos\varphi) + \omega_{y'} \right) \right\}$$

$$\times \left\{ 2\pi\delta\left(\frac{(2q+1)\pi}{2C}(\sin\varphi) - \omega_{x'} \right) \right\} \Bigg] \tag{5-49}$$

$$= \sum_{q=0}^{\infty} \frac{(-1)^q\pi}{(2q+1)} \Bigg[\left\{ \delta\left(\frac{(2q+1)\pi}{2C}(\cos\varphi) - \omega_{y'} \right) \right\}$$

$$\times \left\{ \delta\left(\frac{(2q+1)\pi}{2C}(\sin\varphi) + \omega_{x'} \right) \right\}$$

$$+ \left\{ \delta\left(\frac{(2q+1)\pi}{2C}(\cos\varphi) + \omega_{y'} \right) \right\} \left\{ \delta\left(\frac{(2q+1)\pi}{2C}(\sin\varphi) - \omega_{x'} \right) \right\} \Bigg]$$

$$\tag{5-50}$$

$$\omega_{x'} = \pm \frac{(2q+1)\pi}{2C} \sin\varphi$$

$$\omega_{y'} = \pm \frac{(2q+1)\pi}{2C} \cos\varphi$$

but when $\omega_{x'}$ is positive, $\omega_{y'}$ is negative, which can be seen from (5-50). This can be written

$$\omega_{x'} = \epsilon \frac{(2q+1)\pi}{2C} \sin\varphi$$

$$\omega_{y'} = -\epsilon \frac{(2q+1)\pi}{2C} \cos\varphi$$

where $\epsilon = \pm 1$.

The amplitude of each spectral component is given by the coefficient of (5-50) or,

$$A = \frac{(-1)^q\pi}{(2q+1)} \tag{5-51}$$

and the intensity of each spectral point is given by the square of (5-51)

$$I = \left[\frac{\pi}{(2q+1)} \right]^2 = \frac{\pi^2}{(2q+1)^2} \tag{5-52}$$

Note that the square of $(-1)^q$ in (5-51) becomes $+1$ when squared.

Term F

$$\text{Term F} = \frac{1}{8} \int_{-\infty}^{\infty} \int_{-\infty}^{\infty} \eta_2 \eta_3 \exp\left[-j(\omega_{x'}x' + \omega_{y'}y')\right] dx' \, dy' \tag{5-53}$$

$$= \frac{1}{8} \int_{-\infty}^{\infty} \int_{-\infty}^{\infty} \left[\sum_{p=0}^{\infty} \frac{4(-1)^p}{(2p+1)\pi} \cos\frac{(2p+1)\pi x'}{2B}\right]$$

$$\times \left[\sum_{q=0}^{\infty} \frac{4(-1)^q}{(2q+1)\pi} \cos\frac{(2q+1)\pi}{2C}(y'\cos\varphi - x'\sin\varphi)\right]$$

$$\times \exp\left[-j[\omega_{x'}x' + \omega_{y'}y')\right] dx' \, dy' \tag{5-54}$$

$$= \sum_{p=0}^{\infty} \sum_{q=0}^{\infty} \frac{2(-1)^p(-1)^q}{\pi^2(2p+1)(2q+1)} \int_{-\infty}^{\infty} \int_{-\infty}^{\infty} \left[\cos\frac{(2p+1)\pi x'}{2B}\right]$$

$$\times \left[\cos\frac{(2q+1)\pi}{2C}(y'\cos\varphi - x'\sin\varphi)\right]$$

$$\times \exp\left[-j(\omega_{x'}x' + \omega_{y'}y')\right] dx' \, dy' \tag{5-55}$$

$$= \sum_{p=0}^{\infty} \sum_{q=0}^{\infty} \frac{2(-1)^p(-1)^q}{\pi^2(2p+1)(2q+1)}$$

$$\times \int_{-\infty}^{\infty} \int_{-\infty}^{\infty} \left[\frac{\exp\left\{jx'\left[\frac{(2p+1)\pi}{2B}\right]\right\} + \exp\left\{-jx'\left[\frac{(2p+1)\pi}{2B}\right]\right\}}{2}\right]$$

$$\times \left[\left\{\frac{\exp\left\{j\left[\frac{(2q+1)\pi}{2C}\right](y'\cos\varphi - x'\sin\varphi)\right\}}{2}\right\}\right.$$

$$\left. + \frac{\exp\left\{-j\left[\frac{(2q+1)\pi}{2C}\right](y'\cos\varphi - x'\sin\varphi)\right\}}{2}\right]$$

$$\times \left[\exp\left\{-j(\omega_{x'}x' + \omega_{y'}y')\right\}\right] dx' \, dy' \tag{5-56}$$

$$= \sum_{p=0}^{\infty} \sum_{q=0}^{\infty} \frac{(-1)^p(-1)^q}{2\pi^2(2p+1)(2q+1)} \int_{-\infty}^{\infty} \int_{-\infty}^{\infty} \left[\exp\left\{jx'\left[\frac{(2p+1)\pi}{2B}\right]\right.\right.$$

$$+j\left[\frac{(2q+1)\pi}{2C}\right](y'\cos\varphi-x'\sin\varphi)\bigg\}+\exp\bigg\{-jx'\left[\frac{(2p+1)\pi}{2B}\right]$$

$$+j\left[\frac{(2q+1)\pi}{2C}\right](y'\cos\varphi-x'\sin\varphi)\bigg\}+\exp\bigg\{jx'\left[\frac{(2p+1)\pi}{2B}\right]$$

$$-j\left[\frac{(2q+1)\pi}{2C}\right](y'\cos\varphi-x'\sin\varphi)\bigg\}+\exp\bigg\{-jx'\left[\frac{(2p+1)\pi}{2B}\right]$$

$$-j\left[\frac{(2q+1)\pi}{2C}\right](y'\cos\varphi-x'\sin\varphi)\bigg\}\bigg]$$

$$\times\left[\exp\left\{-j(\omega_{x'}x'+\omega_{y'}y')\right\}\right]dx'\,dy' \tag{5-57}$$

$$=\sum_{p=0}^{\infty}\sum_{q=0}^{\infty}\frac{(-1)^{p}(-1)^{q}}{2\pi^{2}(2p+1)(2q+1)}\int_{-\infty}^{\infty}\int_{-\infty}^{\infty}\left[\exp\left\{jx'\left[\frac{(2p+1)\pi}{2B}\right]\right.\right.$$

$$+j\left[\frac{(2q+1)\pi}{2C}\right](y'\cos\varphi-x'\sin\varphi)-j(\omega_{x'}x'+\omega_{y'}y')\bigg\}$$

$$+\exp\left\{-jx'\left[\frac{(2p+1)\pi}{2B}\right]+j\left[\frac{(2q+1)\pi}{2C}\right]\right.$$

$$\times(y'\cos\varphi-x'\sin\varphi)-j(\omega_{x'}x'+\omega_{y'}y')\bigg\}$$

$$+\exp\left\{jx'\left[\frac{(2p+1)\pi}{2B}\right]-j\left[\frac{(2q+1)\pi}{2C}\right]\right.$$

$$\times(y'\cos\varphi-x'\sin\varphi)-j(\omega_{x'}x'+\omega_{y'}y')\bigg\}$$

$$+\exp\left\{-jx'\left[\frac{(2p+1)\pi}{2B}\right]-j\left[\frac{(2q+1)\pi}{2C}\right]\right.$$

$$\times(y'\cos\varphi-x'\sin\varphi)-j(\omega_{x'}x'+\omega_{y'}y')\bigg\}\bigg]dx'\,dy' \tag{5-58}$$

$$=\sum_{p=0}^{\infty}\sum_{q=0}^{\infty}\frac{(-1)^{p}(-1)^{q}}{2\pi^{2}(2p+1)(2q+1)}$$

$$\times\int_{-\infty}^{\infty}\int_{-\infty}^{\infty}\left\{\left[\exp\left\{jx'\left[\frac{(2p+1)\pi}{2B}\right]-j\left[\frac{(2q+1)\pi}{2C}\right](x'\sin\varphi)-j(\omega_{x'}x')\right\}\right]\right.$$

$$\times\left[\exp\left\{j\left[\frac{(2q+1)\pi}{2C}\right](y'\cos\varphi)-j(\omega_{y'}y')\right\}\right]$$

$$+ \left[\exp \left\{ -jx' \left[\frac{(2p+1)\pi}{2B} \right] - j \left[\frac{(2q+1)\pi}{2C} \right] (x' \sin \varphi) - j(\omega_{x'}x') \right\} \right]$$

$$\times \left[\exp \left\{ j \left[\frac{(2q+1)\pi}{2C} \right] (y' \cos \varphi) - j(\omega_{y'}y') \right\} \right]$$

$$+ \left[\exp \left\{ jx' \left[\frac{(2p+1)\pi}{2B} \right] + j \left[\frac{(2q+1)\pi}{2C} \right] (x' \sin \varphi) - j(\omega_{x'}x') \right\} \right]$$

$$\times \left[\exp \left\{ -j \left[\frac{(2q+1)\pi}{2C} \right] (y' \cos \varphi) - j(\omega_{y'}y') \right\} \right]$$

$$+ \left[\exp \left\{ -jx' \left[\frac{(2p+1)\pi}{2B} \right] \right. \right.$$

$$+ j \left[\frac{(2q+1)\pi}{2C} \right] (x' \sin \varphi) - j(\omega_{x'}x') \right\} \bigg]$$

$$\times \left[\exp \left\{ -j \left[\frac{(2q+1)\pi}{2C} \right] (y' \cos \varphi) - j(\omega_{y'}y') \right\} \right] \right\} dx' \, dy' \qquad (5\text{-}59)$$

$$= \sum_{p=0}^{\infty} \sum_{q=0}^{\infty} \frac{(-1)^p (-1)^q}{2\pi^2 (2p+1)(2q+1)}$$

$$\times \left[\int_{-\infty}^{\infty} \exp \left\{ jx' \left[\frac{(2p+1)\pi}{2B} - \frac{(2q+1)\pi}{2C} \right] (\sin \varphi) - \omega_{x'} \right] \right\} dx'$$

$$\times \int_{-\infty}^{\infty} \exp \left\{ jy' \left[\frac{(2q+1)\pi}{2C} (\cos \varphi) - \omega_{y'} \right] \right\} dy'$$

$$+ \int_{-\infty}^{\infty} \exp \left\{ -jx' \left[\frac{(2p+1)\pi}{2B} + \frac{(2q+1)\pi}{2C} (\sin \varphi) + \omega_{x'} \right] \right\} dx'$$

$$\times \int_{-\infty}^{\infty} \exp \left\{ jy' \left[\frac{(2q+1)\pi}{2C} (\cos \varphi) - \omega_{y'} \right] \right\} dy'$$

$$+ \int_{-\infty}^{\infty} \exp \left\{ jx' \left[\frac{(2p+1)\pi}{2B} + \frac{(2q+1)\pi}{2C} (\sin \varphi) - \omega_{x'} \right] \right\} dx'$$

$$\times \int_{-\infty}^{\infty} \exp \left\{ -jy' \left[\frac{(2q+1)\pi}{2C} (\cos \varphi) + \omega_{y'} \right] \right\} dy'$$

$$+ \int_{-\infty}^{\infty} \exp\left\{-jx'\left[\frac{(2p+1)\pi}{2B} - \frac{(2q+1)\pi}{2C}(\sin\varphi) + \omega_{x'}\right]\right\} dx'$$

$$\times \int_{-\infty}^{\infty} \exp\left\{-jy'\left[\frac{(2q+1)\pi}{2C}(\cos\varphi) + \omega_{y'}\right]\right\} dy'\Bigg] \tag{5-60}$$

$$= \sum_{p=0}^{\infty} \sum_{q=0}^{\infty} \frac{(-1)^p(-1)^q}{2\pi^2(2p+1)(2q+1)}$$

$$\times \Bigg[\left\{2\pi\delta\left(\frac{(2q+1)\pi}{2B} - \frac{(2q+1)\pi}{2C}(\sin\varphi) - \omega_{x'}\right)\right\}$$

$$\times \left\{2\pi\delta\left(\frac{(2q+1)\pi}{2C}(\cos\varphi) - \omega_{y'}\right)\right\}$$

$$+ \left\{2\pi\delta\left(-\frac{(2p+1)\pi}{2B} - \frac{(2q+1)\pi}{2C}(\sin\varphi) - \omega_{x'}\right)\right\}$$

$$\times \left\{2\pi\delta\left(\frac{(2q+1)\pi}{2C}(\cos\varphi) - \omega_{y'}\right)\right\}$$

$$+ \left\{2\pi\delta\left(\frac{(2p+1)\pi}{2B} + \frac{(2q+1)\pi}{2C}(\sin\varphi) - \omega_{x'}\right)\right\}$$

$$\times \left\{2\pi\delta\left(-\frac{(2q+1)\pi}{2C}(\cos\varphi) - \omega_{y'}\right)\right\}$$

$$+ \left\{2\pi\delta\left(-\frac{(2p+1)\pi}{2B} + \frac{(2q+1)\pi}{2C}(\sin\varphi) - \omega_{x'}\right)\right\}$$

$$\times \left\{2\pi\delta\left(-\frac{(2q+1)\pi}{2C}(\cos\varphi) - \omega_{y'}\right)\right\}\Bigg] \tag{5-61}$$

$$= \sum_{p=0}^{\infty} \sum_{q=0}^{\infty} \frac{(-1)^p(-1)^q 4\pi^2}{2\pi^2(2p+1)(2q+1)}$$

$$\times \Bigg[\left\{\delta\left(\frac{(2p+1)\pi}{2B} - \frac{(2q+1)\pi}{2C}(\sin\varphi) - \omega_{x'}\right)\right\}$$

$$\times \left\{\delta\left(\frac{(2q+1)\pi}{2C}(\cos\varphi) - \omega_{y'}\right)\right\}$$

$$+\left\{\delta\left(-\frac{(2p+1)\pi}{2B}-\frac{(2q+1)}{2C}(\sin\varphi)-\omega_{x'}\right)\right\}$$

$$\times\left\{\delta\left(\frac{(2q+1)\pi}{2C}(\cos\varphi)-\omega_{y'}\right)\right\}$$

$$+\left\{\delta\left(\frac{(2p+1)\pi}{2B}+\frac{(2q+1)\pi}{2C}(\sin\varphi)-\omega_{x'}\right)\right\}$$

$$\times\left\{\delta\left(-\frac{(2q+1)\pi}{2C}(\cos\varphi)-\omega_{y'}\right)\right\}$$

$$+\left\{\delta\left(-\frac{(2p+1)\pi}{2B}+\frac{(2q+1)\pi}{2C}(\sin\varphi)-\omega_{x'}\right)\right\}$$

$$\left.\times\left\{\delta\left(-\frac{(2q+1)\pi}{2C}(\cos\varphi)-\omega_{y'}\right)\right\}\right] \tag{5-62}$$

From (5-62) we can see that

$$\omega_{x'}=\pm\frac{(2p+1)\pi}{2B}\pm\frac{(2q+1)\pi}{2C}(\sin\varphi)$$

$$\omega_{y'}=\pm\frac{(2q+1)\pi}{2C}\cos\varphi$$

In additional we can see from each of the terms in (5-62) that when $\omega_{y'}$ is positive (the first two terms in (5-62)); that is, set

$$\frac{(2q+1)\pi}{2C}\cos\varphi-\omega_{y'}=0$$

then

$$\omega_{y'}=\frac{(2q+1)\pi}{2C}\cos\varphi$$

and the sign of the term

$$\frac{(2q+1)\pi}{2C}\sin\varphi$$

in the $\omega_{x'}$ expression is negative. When $\omega_{y'}$ is negative (last two terms in (5-62); that is, set

$$-\frac{(2q+1)\pi}{2C}\cos\varphi-\omega_{y'}=0$$

then

$$\omega_{y'} = -\frac{(2q+1)\pi}{2C}\cos\varphi$$

and the sign of the term $[(2q+1)\pi/2C]\sin\varphi$ in the $\omega_{x'}$ equation is positive. We can therefore write the spectral components from (5-62) in the form

$$\omega_{x'} = \pm\frac{(2p+1)\pi}{2B} - \epsilon\frac{(2q+1)\pi}{2C}\sin\varphi$$

$$\omega_{y'} = \epsilon\frac{(2q+1)\pi}{2C}\cos\varphi$$

where $\epsilon = \pm 1$. The amplitude of each spectral point is given by the coefficient of the δ function of (5-62) or

$$A = \frac{2(-1)^p(-1)^q}{(2p+1)(2q+1)} \tag{5-63}$$

The intensity of each spectral point is given by the square of (5-63).

$$I = \frac{4}{(2p+1)^2(2q+1)^2} \tag{5-64}$$

Note the $(-1)^p$ and $(-1)^q$ in (5-63) each becomes $+1$ when squared.

Term G $= \frac{1}{8}\int_{-\infty}^{\infty}\int_{-\infty}^{\infty}\eta_1\eta_3\exp\left[-j(\omega_{x'}x'+\omega_{y'}y')\right]dx'\,dy'$ $\tag{5-65}$

$$= \frac{1}{8}\int_{-\infty}^{\infty}\int_{-\infty}^{\infty}\left[\sum_{n=0}^{\infty}\frac{4(-1)^n}{(2n+1)\pi}\cos\frac{(2n+1)\pi y'}{2A}\right]$$

$$\times\left[\sum_{q=0}^{\infty}\frac{4(-1)^q}{(2q+1)\pi}\cos\frac{(2q+1)u}{2C}(y'\cos\varphi - x'\sin\varphi)\right]$$

$$\times\left[\exp\{-j(\omega_{x'}x'+\omega_{y'}y')\}\right]dx'\,dy' \tag{5-66}$$

$$= \frac{1}{8}\sum_{n=0}^{\infty}\sum_{q=0}^{\infty}\frac{16(-1)^n(-1)^q}{\pi^2(2n+1)(2q+1)}\int_{-\infty}^{\infty}\int_{-\infty}^{\infty}\left[\cos\frac{(2n+1)\pi y'}{2A}\right]$$

$$\times\left[\cos\frac{(2q+1)\pi}{2C}(y'\cos\varphi - x'\sin\varphi)\right]$$

$$\times\left[\exp\{-j(\omega_{x'}x'+\omega_{y'}y')\}\right]dx'\,dy' \tag{5-67}$$

$$= \frac{1}{8} \sum_{n=0}^{\infty} \sum_{q=0}^{\infty} \frac{16(-1)^n(-1)^q}{\pi^2(2n+1)(2q+1)}$$

$$\times \int_{-\infty}^{\infty} \int_{-\infty}^{\infty} \left[\frac{\exp\left\{j\left(\frac{(2n+1)\pi y'}{2A}\right)\right\} + \exp\left\{-j\left(\frac{(2n+1)\pi y'}{2A}\right)\right\}}{2} \right]$$

$$\times \left[\left\{ \frac{\exp\left\{j\left(\frac{(2q+1)\pi}{2C}(y'\cos\varphi - x'\sin\varphi)\right)\right\}}{2} \right\} \right.$$

$$\left. + \left\{ \frac{\exp\left\{-j\left(\frac{(2q+1)\pi}{2C}(y'\cos\varphi - x'\sin\varphi)\right)\right\}}{2} \right\} \right]$$

$$\times \left[\exp\left\{-j(\omega_{x'}x' + \omega_{y'}y')\right\} \right] dx'\, dy' \qquad (5\text{-}68)$$

Since

$$(a+b)(c+d)g = acg + bcg + adg + bdg \qquad (5\text{-}69)$$

then in (5-68) we can let

$$a = \exp\left\{j\left[\frac{(2n+1)\pi y'}{2A}\right]\right\}$$

$$b = \exp\left\{-j\left[\frac{(2n+1)\pi y'}{2A}\right]\right\}$$

$$c = \exp\left\{j\left[\frac{(2q+1)\pi}{2C}\right](y'\cos\varphi - x'\sin\varphi)\right\}$$

$$d = \exp\left\{-j\left[\frac{(2q+1)\pi}{2C}\right](y'\cos\varphi - x'\sin\varphi)\right\}$$

$$g = \exp\left\{-j(\omega_{x'}x' + \omega_{y'}y')\right\}$$

we can now rewrite (5-68) using the form of (5-69) as

Term G $= \dfrac{1}{8} \sum_{n=0}^{\infty} \sum_{q=0}^{\infty} \dfrac{4(-1)^n(-1)^q}{\pi^2(2n+1)(2q+1)}$

$$\times \left[\int_{-\infty}^{\infty} \int_{-\infty}^{\infty} \left[\exp\left\{j\left(\frac{(2n+1)\pi y'}{2A}\right) \right. \right. \right.$$

$$+j\left(\frac{(2q+1)\pi}{2C}\right)(y'\cos\varphi-x'\sin\varphi)$$

$$-j(\omega_{x'}x'+\omega_{y'}y')\bigg\}\bigg]+\bigg[\exp\bigg\{-j\left(\frac{(2n+1)\pi y'}{2A}\right)$$

$$+j\left(\frac{(2q+1)\pi}{2C}\right)(y'\cos\varphi-x'\sin\varphi)$$

$$-j(\omega_{x'}x'+\omega_{y'}y')\bigg\}\bigg]+\bigg[\exp\bigg\{j\left(\frac{(2n+1)\pi y'}{2A}\right)$$

$$-j\left(\frac{(2q+1)\pi}{2C}\right)(y'\cos\varphi-x'\sin\varphi)$$

$$-j(\omega_{x'}x'+\omega_{y'}y')\bigg\}\bigg]+\bigg[\exp\bigg\{-j\left(\frac{(2n+1)\pi y'}{2A}\right)$$

$$-j\left(\frac{(2q+1)\pi}{2C}\right)(y'\cos\varphi-x'\sin\varphi)$$

$$-j(\omega_{x'}x'-\omega_{y'}y')\bigg\}\bigg]\bigg]dx'\,dy' \tag{5-70}$$

$$=\frac{1}{8}\sum_{n=0}^{\infty}\sum_{q=0}^{\infty}\frac{4(-1)^n(-1)^q}{\pi^2(2n+1)(2q+1)}$$

$$\times\bigg[\int_{-\infty}^{\infty}\exp\bigg\{jy'\bigg[\left(\frac{(2n+1)\pi}{2A}\right)+\left(\frac{(2q+1)\pi}{2C}\right)(\cos\varphi)-\omega_{y'}\bigg]\bigg\}dy'$$

$$\times\int_{-\infty}^{\infty}\exp\bigg\{-jx'\bigg[\left(\frac{(2q+1)\pi}{2C}\right)(\sin\varphi)+\omega_{x'}\bigg]\bigg\}dx'$$

$$+\int_{-\infty}^{\infty}\exp\bigg\{-jy'\bigg[\left(\frac{(2n+1)\pi}{2A}\right)-\left(\frac{(2q+1)\pi}{2C}\right)(\cos\varphi)+\omega_{y'}\bigg]\bigg\}dy'$$

$$\times\int_{-\infty}^{\infty}\exp\bigg\{-jx'\bigg[\left(\frac{(2q+1)\pi}{2C}\right)(\sin\varphi)+\omega_{x'}\bigg]\bigg\}dx'$$

$$+\int_{-\infty}^{\infty}\exp\bigg\{jy'\bigg[\left(\frac{(2n+1)\pi}{2A}\right)-\left(\frac{(2q+1)\pi}{2C}\right)(\cos\varphi)-\omega_{y'}\bigg]\bigg\}dy'$$

$$\times \int_{-\infty}^{\infty} \exp\left\{jx'\left[\left(\frac{(2q+1)\pi}{2C}\right)(\sin\varphi) - \omega_{x'}\right]\right\} dx'$$

$$+ \int_{-\infty}^{\infty} \exp\left\{-jy'\left[\left(\frac{(2n+1)\pi}{2A}\right) + \left(\frac{(2q+1)\pi}{2C}\right)(\cos\varphi) + \omega_{y'}\right]\right\} dy'$$

$$\times \left. \int_{-\infty}^{\infty} \exp\left\{jx'\left[\left(\frac{(2q+1)\pi}{2C}\right)(\sin\varphi) - \omega_{x'}\right]\right\} dx' \right] \tag{5-71}$$

$$= \frac{1}{8} \sum_{n=0}^{\infty} \sum_{q=0}^{\infty} \frac{4(-1)^n(-1)^q}{\pi^2(2n+1)(2q+1)}$$

$$\times \left[\left\{2\pi\delta\left(\left(\frac{(2n+1)\pi}{2A}\right) + \left(\frac{(2q+1)\pi}{2C}\right)(\cos\varphi) - \omega_{y'}\right)\right\} \right.$$

$$\times \left\{2\pi\delta\left(-\left(\frac{(2q+1)\pi}{2C}\right)(\sin\varphi) - \omega_{x'}\right)\right\}$$

$$+ \left\{2\pi\delta\left(-\left(\frac{(2n+1)\pi}{2A}\right) + \left(\frac{(2q+1)\pi}{2C}\right)(\cos\varphi) - \omega_{y'}\right)\right\}$$

$$\times \left\{2\pi\delta\left(-\left(\frac{(2q+1)\pi}{2C}\right)(\sin\varphi) - \omega_{x'}\right)\right\}$$

$$+ \left\{2\pi\delta\left(+\left(\frac{(2n+1)\pi}{2A}\right) - \left(\frac{(2q+1)\pi}{2C}\right)(\cos\varphi) - \omega_{y'}\right)\right\}$$

$$\times \left\{2\pi\delta\left(\left(\frac{(2q+1)\pi}{2C}\right)(\sin\varphi) - \omega_{x'}\right)\right\}$$

$$+ \left\{2\pi\delta\left(-\left(\frac{(2n+1)\pi}{2A}\right) - \left(\frac{(2q+1)\pi}{2C}\right)(\cos\varphi) - \omega_{y'}\right)\right\}$$

$$\times \left. \left\{2\pi\delta\left(\left(\frac{(2q+1)\pi}{2C}\right)(\sin\varphi) - \omega_{x'}\right)\right\} \right] \tag{5-72}$$

$$= \sum_{n=0}^{\infty} \sum_{q=0}^{\infty} \frac{2(-1)^n(-1)^q}{(2n+1)(2q+1)}$$

$$\times \left[\delta\left(\left(\frac{(2n+1)\pi}{2A}\right) + \left(\frac{(2q+1)\pi}{2C}\right)(\cos\varphi) - \omega_{y'}\right) \right.$$

$$\times \delta\left(-\left(\frac{(2q+1)\pi}{2C}\right)(\sin\varphi) - \omega_{x'}\right)$$

$$+ \delta\left(-\left(\frac{(2n+1)\pi}{2A}\right)+\left(\frac{(2q+1)\pi}{2C}\right)(\cos\varphi)-\omega_{y'}\right)$$

$$\times \delta\left(-\left(\frac{(2q+1)\pi}{2C}\right)(\sin\varphi)-\omega_{x'}\right)$$

$$+ \delta\left(\left(\frac{(2n+1)\pi}{2A}\right)-\left(\frac{(2q+1)\pi}{2C}\right)(\cos\varphi)-\omega_{y'}\right)$$

$$\times \delta\left(\left(\frac{(2q+1)\pi}{2C}\right)(\sin\varphi)-\omega_{x'}\right)$$

$$+ \delta\left(-\left(\frac{(2n+1)\pi}{2A}\right)-\left(\frac{(2q+1)\pi}{2C}\right)(\cos\varphi)-\omega_{y'}\right)$$

$$\left. \times \delta\left(\left(\frac{(2q+1)\pi}{2C}\right)(\sin\varphi)-\omega_{x'}\right)\right] \tag{5-73}$$

From this we can see that

$$\omega_{x'} = \pm\frac{(2q+1)\pi}{2C}\sin\varphi$$

$$\omega_{y'} = \pm\frac{(2n+1)\pi}{2A}\pm\frac{(2q+1)\pi}{2C}\cos\varphi$$

In addition we can see that when $\omega_{x'}$ is negative (first two terms in (5-73)) the $[(2q+1)\pi/2C]\cos\varphi$ term is positive. The $(2n+1)/2A$ term can be either positive or negative. When $\omega_{x'}$ is positive (last two terms in (5-73) the $[(2q+1)\pi/2C]\cos\varphi$ is negative and the $(2n+1)/2A$ term can be either positive or negative. From this we can write the spectral terms as

$$\omega_{x'} = \epsilon\frac{(2q+1)\pi}{2C}\sin\varphi$$

$$\omega_{y'} = \pm\frac{(2n+1)\pi}{2A}-\epsilon\frac{(2q+1)}{2C}\cos\varphi$$

where $\epsilon = \pm 1$. Each spectral point has an amplitude that is given by the coefficient of (5-73)

$$A = \frac{2(-1)^n(-1)^q}{(2n+1)(2q+1)} \tag{5-74}$$

The intensity of each spectral point is given by the square of (5-74)

$$I = \frac{4}{(2n+1)^2(2q+1)^2} \tag{5-75}$$

Term H

$$\textbf{Term H} = \frac{1}{8} \int_{-\infty}^{\infty} \int_{-\infty}^{\infty} \eta_1 \eta_2 \eta_3 \exp\left[-j(\omega_{x'} x' + \omega_{y'} y')\right] dx'\, dy' \tag{5-76}$$

$$= \frac{1}{8} \int_{-\infty}^{\infty} \int_{-\infty}^{\infty} \left[\sum_{n=0}^{\infty} \frac{4(-1)^n}{(2n+1)\pi} \cos\frac{(2n+1)\pi y'}{2A}\right]$$

$$\times \left[\sum_{p=0}^{\infty} \frac{4(-1)^p}{(2p+1)\pi} \cos\frac{(2p+1)\pi x'}{2B}\right]$$

$$\times \left[\sum_{q=0}^{\infty} \frac{4(-1)^q}{(2q+1)\pi} \cos\frac{(2q+1)\pi}{2C} (y'\cos\varphi - x'\sin\varphi)\right]$$

$$\times \left[\exp\left\{-j(\omega_{x'} x' + \omega_{y'} y')\right\}\right] dx'\, dy' \tag{5-77}$$

$$= \frac{1}{8} \sum_{n=0}^{\infty} \sum_{p=0}^{\infty} \sum_{q=0}^{\infty} \frac{64(-1)^n(-1)^p(-1)^q}{(2n+1)(2p+1)(2q+1)\pi^3}$$

$$\times \int_{-\infty}^{\infty} \int_{-\infty}^{\infty} \left[\cos\frac{(2n+1)\pi y'}{2A}\right]\left[\cos\frac{(2p+1)\pi x'}{2B}\right]$$

$$\times \left[\cos\frac{(2q+1)\pi}{2C} (y'\cos\varphi - x'\sin\varphi)\right]$$

$$\times \left[\exp\left\{-j(\omega_{x'} x' + \omega_{y'} y')\right\}\right] dx'\, dy' \tag{5-78}$$

$$= \frac{1}{8} \sum_{n=0}^{\infty} \sum_{p=0}^{\infty} \sum_{q=0}^{\infty} \frac{64(-1)^n(-1)^p(-1)^q}{8\pi^3(2n+1)(2p+1)(2q+1)}$$

$$\times \int_{-\infty}^{\infty} \int_{-\infty}^{\infty} \left[\exp\left\{j\left(\frac{(2n+1)\pi y'}{2A}\right)\right\} + \exp\left\{-j\left(\frac{(2n+1)\pi y'}{2A}\right)\right\}\right]$$

$$\times \left[\exp\left\{j\left(\frac{(2p+1)\pi x'}{2B}\right)\right\} + \exp\left\{-j\left(\frac{(2p+1)\pi x'}{2B}\right)\right\}\right]$$

$$\times \left[\exp\left\{j\left(\frac{(2q+1)\pi}{2C}\right)(y'\cos\varphi - x'\sin\varphi)\right\}\right.$$

$$+ \left.\exp\left\{-j\left(\frac{(2q+1)\pi}{2C}\right)(y'\cos\varphi - x'\sin\varphi)\right\}\right]$$

$$\times \left[\exp\left\{-j(\omega_{x'} x' + \omega_{y'} y')\right\}\right] dx'\, dy' \tag{5-79}$$

We can multiply this out by first noting that the product

$$(a+b)(c+d)(f+g)h = acfh + bcfh + adfh + bdfh$$
$$+ acgh + bcgh + adgh + bdgh \qquad (5\text{-}80)$$

and making the following substitutions:

$$a = \exp\left\{j\left[\frac{(2n+1)\pi y'}{2A}\right]\right\}$$

$$b = \exp\left\{-j\left[\frac{(2n+1)\pi y'}{2A}\right]\right\}$$

$$c = \exp\left\{j\left[\frac{(2p+1)\pi x'}{2B}\right]\right\}$$

$$d = \exp\left\{-j\left[\frac{(2p+1)\pi x'}{2B}\right]\right\}$$

$$f = \exp\left\{j\left[\frac{(2q+1)}{2C}\right](y' \cos\varphi - x' \sin\varphi)\right\}$$

$$g = \exp\left\{-j\left[\frac{(2q+1)}{2C}\right](y' \cos\varphi - x' \sin\varphi)\right\}$$

$$h = \exp\left\{-j(\omega_{x'}x' + \omega_{y'}y')\right\}$$

We can now rewrite (5-79) using the form of (5-80) as

$$\textbf{Term H} = \sum_{n=0}^{\infty}\sum_{p=0}^{\infty}\sum_{q=0}^{\infty} \frac{(-1)^n(-1)^p(-1)^q}{\pi^3(2n+1)(2p+1)(2q+1)}$$

$$\times \int_{-\infty}^{\infty}\int_{-\infty}^{\infty}\left[\left[\exp\left\{j\frac{(2n+1)\pi y'}{2A} + j\frac{(2p+1)\pi x'}{2B}\right.\right.\right.$$

$$\left.\left.+ j\frac{(2q+1)\pi}{2C}(y' \cos\varphi - x' \sin\varphi) - j(\omega_{x'}x' + \omega_{y'}y')\right\}\right]$$

$$+ \left[\exp\left\{-j\frac{(2n+1)\pi y'}{2A} + j\frac{(2p+1)\pi x'}{2B}\right.\right.$$

$$\left.\left.+ j\frac{(2q+1)\pi}{2C}(y' \cos\varphi - x' \sin\varphi) - j(\omega_{x'}x' + \omega_{y'}y')\right\}\right]$$

$$+\left[\exp\left\{j\frac{(2n+1)\pi y'}{2A}-j\frac{(2p+1)\pi x'}{2B}\right.\right.$$

$$\left.\left.+j\frac{(2q+1)\pi}{2C}(y'\cos\varphi-x'\sin\varphi)-j(\omega_{x'}x'+\omega_{y'}y')\right\}\right]$$

$$+\left[\exp\left\{-j\frac{(2n+1)\pi y'}{2A}-j\frac{(2p+1)\pi x'}{2B}\right.\right.$$

$$\left.\left.+j\frac{(2q+1)\pi}{2C}(y'\cos\varphi-x'\sin\varphi)-j(\omega_{x'}x'+\omega_{y'}y')\right\}\right]$$

$$+\left[\exp\left\{j\frac{(2n+1)\pi y'}{2A}+j\frac{(2p+1)\pi x'}{2B}\right.\right.$$

$$\left.\left.-j\frac{(2q+1)\pi}{2C}(y'\cos\varphi-x'\sin\varphi)-j(\omega_{x'}x'+\omega_{y'}y')\right\}\right]$$

$$+\left[\exp\left\{-j\frac{(2n+1)\pi y'}{2A}+j\frac{(2p+1)\pi x'}{2B}\right.\right.$$

$$\left.\left.-j\frac{(2q+1)\pi}{2C}(y'\cos\varphi-x'\sin\varphi)-j(\omega_{x'}x'+\omega_{y'}y')\right\}\right]$$

$$+\left[\exp\left\{j\frac{(2n+1)\pi y'}{2A}-j\frac{(2p+1)\pi x'}{2B}\right.\right.$$

$$\left.\left.-j\frac{(2q+1)\pi}{2C}(y'\cos\varphi-x'\sin\varphi)-j(\omega_{x'}x'+\omega_{y'}y')\right\}\right]$$

$$+\left[\exp\left\{-j\frac{(2n+1)\pi y'}{2A}-j\frac{(2p+1)\pi x'}{2B}\right.\right.$$

$$\left.\left.-j\frac{(2q+1)\pi}{2C}(y'\cos\varphi-x'\sin\varphi)-j(\omega_{x'}x'+\omega_{y'}y')\right\}\right]\right]dx'\text{dy}' \quad (5\text{-}81)$$

$$\textbf{Term H} = \sum_{n=0}^{\infty}\sum_{p=0}^{\infty}\sum_{q=0}^{\infty}\frac{(-1)^n(-1)^p(-1)^q}{\pi^3(2n+1)(2p+1)(2q+1)}$$

$$\times\left[\int_{-\infty}^{\infty}\exp\left\{jy'\left[\left(\frac{(2n+1)\pi}{2A}\right)+\left[\frac{(2q+1)\pi}{2C}\right](\cos\varphi)-\omega_{y'}\right]\right\}dy'\right.$$

$$\times \int_{-\infty}^{\infty} \exp\left\{jx'\left[\left(\frac{(2p+1)\pi}{2B}\right) - \left[\frac{(2q+1)\pi}{2C}\right](\sin\varphi) - \omega_{x'}\right]\right\} dx'$$

$$+ \int_{-\infty}^{\infty} \exp\left\{-jy'\left[\left(\frac{(2n+1)\pi}{2A}\right) - \left[\frac{(2q+1)\pi}{2C}\right](\cos\varphi) + \omega_{y'}\right]\right\} dy'$$

$$\times \int_{-\infty}^{\infty} \exp\left\{jx'\left[\left(\frac{(2p+1)\pi}{2B}\right) - \left[\frac{(2q+1)\pi}{2C}\right](\sin\varphi) - \omega_{x'}\right]\right\} dx'$$

$$+ \int_{-\infty}^{\infty} \exp\left\{jy'\left[\left(\frac{(2n+1)\pi}{2A}\right) + \left[\frac{(2q+1)\pi}{2C}\right](\cos\varphi) - \omega_{y'}\right]\right\} dy'$$

$$\times \int_{-\infty}^{\infty} \exp\left\{-jx'\left[\left(\frac{(2p+1)\pi}{2B}\right) + \left[\frac{(2q+1)\pi}{2C}\right](\sin\varphi) + \omega_{x'}\right]\right\} dx'$$

$$+ \int_{-\infty}^{\infty} \exp\left\{-jy'\left[\left(\frac{(2n+1)\pi}{2A}\right) - \left[\frac{(2q+1)\pi}{2C}\right](\cos\varphi) + \omega_{y'}\right]\right\} dy'$$

$$\times \int_{-\infty}^{\infty} \exp\left\{-jx'\left[\left(\frac{(2p+1)\pi}{2B}\right) + \left[\frac{(2q+1)\pi}{2C}\right](\sin\varphi) + \omega_{x'}\right]\right\} dx'$$

$$+ \int_{-\infty}^{\infty} \exp\left\{jy'\left[\left(\frac{(2n+1)\pi}{2A}\right) - \left[\frac{(2q+1)\pi}{2C}\right](\cos\varphi) - \omega_{y'}\right]\right\} dy'$$

$$\times \int_{-\infty}^{\infty} \exp\left\{jx'\left[\left(\frac{(2p+1)\pi}{2B}\right) + \left[\frac{(2q+1)\pi}{2C}\right](\sin\varphi) - \omega_{x'}\right]\right\} dx'$$

$$+ \int_{-\infty}^{\infty} \exp\left\{-jy'\left[\left(\frac{(2n+1)\pi}{2A}\right) + \left[\frac{(2q+1)\pi}{2C}\right](\cos\varphi) + \omega_{y'}\right]\right\} dy'$$

$$\times \int_{-\infty}^{\infty} \exp\left\{jx'\left[\left(\frac{(2p+1)\pi}{2B}\right) + \left[\frac{(2q+1)\pi}{2C}\right](\sin\varphi) - \omega_{x'}\right] dx'\right.$$

$$+ \int_{-\infty}^{\infty} \exp\left\{jy'\left[\left(\frac{(2n+1)\pi}{2A}\right) - \left[\frac{(2q+1)\pi}{2C}\right](\cos\varphi) - \omega_{y'}\right]\right\} dy'$$

$$\times \int_{-\infty}^{\infty} \exp\left\{-jx'\left[\left(\frac{(2p+1)\pi}{2B}\right)-\left[\frac{(2q+1)\pi}{2C}\right](\sin\varphi)+\omega_{x'}\right]\right\}dx'$$

$$+\int_{-\infty}^{\infty} \exp\left\{-jy'\left[\left(\frac{(2n+1)\pi}{2A}\right)+\left[\frac{(2q+1)\pi}{2C}\right](\cos\varphi)+\omega_{y'}\right]\right\}dy'$$

$$\times \int_{-\infty}^{\infty} \exp\left\{-jx'\left[\left(\frac{(2p+1)\pi}{2B}\right)-\left[\frac{(2q+1)\pi}{2C}\right](\sin\varphi)+\omega_{x'}\right]\right\}dx'\right] \quad (5\text{-}82)$$

Term H $= \displaystyle\sum_{n=0}^{\infty}\sum_{p=0}^{\infty}\sum_{q=0}^{\infty} \frac{(-1)^{n}(-1)^{p}(-1)^{q}\,(4\pi^{2})}{\pi^{3}(2n-1)(2p+1)(2q+1)}$

$$\times\left[\delta\left(\frac{(2n+1)\pi}{2A}+\frac{(2q+1)\pi}{2C}\cos\varphi-\omega_{y'}\right)\right.$$

$$\times\delta\left(\frac{(2p+1)\pi}{2B}-\frac{(2q+1)\pi}{2C}\sin\varphi-\omega_{x'}\right)$$

$$+\delta\left(-\frac{(2n+1)\pi}{2A}+\frac{(2q+1)\pi}{2C}\cos\varphi-\omega_{y'}\right)$$

$$\times\delta\left(\frac{(2p+1)\pi}{2B}-\frac{(2q+1)\pi}{2C}\sin\varphi-\omega_{x'}\right)$$

$$+\delta\left(\frac{(2n+1)\pi}{2A}+\frac{(2q+1)\pi}{2C}\cos\varphi-\omega_{y'}\right)$$

$$\times\delta\left(-\frac{(2p+1)\pi}{2B}-\frac{(2q+1)\pi}{2C}\sin\varphi-\omega_{x'}\right)$$

$$+\delta\left(-\frac{(2n+1)\pi}{2A}+\frac{(2q+1)\pi}{2C}\cos\varphi-\omega_{y'}\right)$$

$$\times\delta\left(-\frac{(2p+1)\pi}{2B}-\frac{(2q+1)\pi}{2C}\sin\varphi-\omega_{x'}\right)$$

$$+\delta\left(\frac{(2n+1)\pi}{2A}-\frac{(2q+1)\pi}{2C}\cos\varphi-\omega_{y'}\right)$$

$$\times\delta\left(\frac{(2p+1)\pi}{2B}+\frac{(2q+1)\pi}{2C}\sin\varphi-\omega_{x'}\right)$$

$$+ \delta\left(-\frac{(2n+1)\pi}{2A} - \frac{(2q+1)\pi}{2C} \cos\varphi - \omega_{y'}\right)$$

$$\times \delta\left(\frac{(2p+1)\pi}{2B} + \frac{(2q+1)\pi}{2C} \sin\varphi - \omega_{x'}\right)$$

$$+ \delta\left(\frac{(2n+1)\pi}{2A} - \frac{(2q+1)\pi}{2C} \cos\varphi - \omega_{y'}\right)$$

$$\times \delta\left(-\frac{(2p+1)\pi}{2B} + \frac{(2q+1)\pi}{2C} \sin\varphi - \omega_{x'}\right)$$

$$+ \delta\left(-\frac{(2n+1)\pi}{2A} - \frac{(2q+1)\pi}{2C} \cos\varphi - \omega_{y'}\right)$$

$$\times \delta\left(-\frac{(2p+1)\pi}{2B} + \frac{(2q+1)\pi}{2C} \sin\varphi - \omega_{x'}\right)\Bigg] \tag{5-83}$$

From (5-83) it can be deduced that when the term

$$\frac{(2q+1)\pi}{2C} \sin\varphi$$

is positive, the term $[(2q+1)\pi/2C]\cos\varphi$ is negative and vice versa. The spectral components formed by **Term H** will therefore be found at

$$\omega_{x'} = \epsilon\frac{(2p+1)\pi}{2B} + \epsilon'\frac{(2q+1)\pi}{2C} \sin\varphi$$

$$\omega_{y'} = \epsilon''\frac{(2n+1)\pi}{2A} - \epsilon'\frac{(2q+1)\pi}{2C} \cos\varphi$$

where

$$\epsilon = \pm1, \epsilon' = \pm1 \quad \text{and} \quad \epsilon'' = \pm1.$$

The amplitude of the spectral points is given by the coefficient of (5-83) or

$$A = \frac{4(-1)^n(-1)^p(-1)^q}{\pi(2n+1)(2p+1)(2q+1)} \tag{5-84}$$

and the intensity (amplitude squared) is given by

$$I = \frac{16}{\pi^2(2n+1)^2(2p+1)^2(2q+1)^2} \tag{5-85}$$

Referring to Figure 5.18 we can see that the **Term B** spectrum is oriented coincident with the y axis and the **Term C** spectrum is coincident with the x axis. The terms corresponding to the **Terms D** are located at points on lines through the **Terms B** parallel to the x axis and intersecting lines through the **Terms C** parallel to the y axis. The spectrum corresponding to the **Terms E** are along a line perpendicular to the orientation of the third Ronchi ruling. Points corresponding to **Term F** will be on lines through the spectral points corresponding to **Term E** points on lines parallel to the x axis. The spacing of the **Term F** points will be similar to

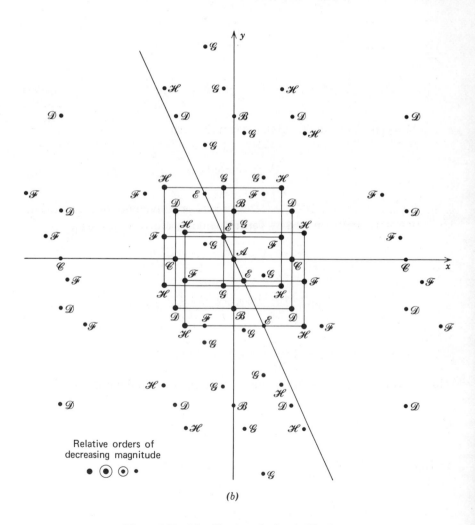

Relative orders of decreasing magnitude

(b)

Figure 5.18 Identification of points in Fig. 5.17.

that of **Term C** points except each will be centered on a **Term E** point whereas the **Term C** points are centered about the **Term A** (DC) point. The spectrum corresponding to the **Term G** points is along lines through the spectral points corresponding to **Term E** on lines parallel to the y axis. The spacing of the **Term G** points on each of these lines will be similar to that of the **Term B** points except each will be centered on a **Term E** point, whereas the **Term B** points are centered about the **Term A** (DC) point. The point corresponding to **Term H** will be on lines parallel to the x axis through **Term G** points intersecting lines parallel to the y axis through **Term F** points.

We can conclude that the Ronchi ruling producing the spectrum corresponding to **Terms B** will also diffract light from the other spectral points in a corresponding manner, thus producing points D and G. Likewise the spectrum C diffracts light corresponding to it from points B and E producing points D and F. Corresponding diffraction of light from points F and G corresponding to spectra B and C produces points H.

Figure 5.19 shows two Ronchi rulings of different frequencies perpen-

Figure 5.19 Two perpendicularly crossed Ronchi rulings of different frequencies crossed with a third Ronchi ruling of different frequency.

dicular to each other with a third frequency Ronchi ruling inclined approximately 30 degrees to the x' axis. The spectrum of Figure 5.19 is shown in Figures 5.20 and 5.21. Figure 5.21 shows the identification of

Figure 5.20 Optical spectrum of Figure 5.19.

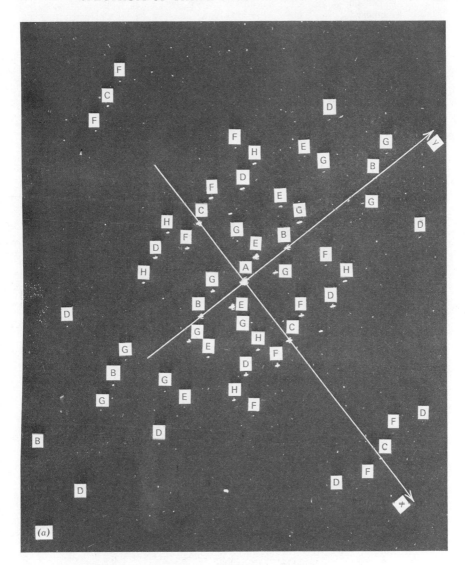

Figure 5.21   Enlargement of Figure 5.20 and identification of points.

points in the spectrum shown in Figure 5.20. Figure 5.22 shows two Ronchi rulings of different frequencies, one horizontal and the other inclined approximately 45 degrees to the y' axis. Figure 5.23 shows the optical spectrum of Figure 5.22.

Summary

Term	Location of Spectral Points	Amplitude
A	$\omega_{x'} = 0$ $\omega_{y'} = 0$	$\dfrac{\pi^2}{2}$
B	$\omega_{x'} = 0$ $\omega_{y'} = \pm \dfrac{(2n+1)\pi}{2A}$	$\dfrac{(-1)^n \pi}{(2n+1)}$
C	$\omega_{x'} = \pm \dfrac{(2p+1)\pi}{2B}$ $\omega_{y'} = 0$	$\dfrac{(-1)^p \pi}{(2p+1)}$
D	$\omega_{x'} = \pm \dfrac{(2p+1)\pi}{2B}$ $\omega_{y'} = \pm \dfrac{(2n+1)\pi}{2A}$	$\dfrac{2(-1)^n(-1)^p}{(2n+1)(2p+1)}$
E	$\omega_{x'} = \epsilon \dfrac{(2q+1)\pi}{2C} \sin\varphi$ $\omega_{y'} = -\epsilon \dfrac{(2q+1)\pi}{2C} \cos\varphi$ $\epsilon = \pm 1$	$\dfrac{(-1)^q \pi}{(2q+1)}$
F	$\omega_{x'} = \pm \dfrac{(2p+1)\pi}{2B} - \epsilon\dfrac{(2q+1)}{2C}\sin\varphi$ $\omega_{y'} = \epsilon \dfrac{(2q+1)\pi}{2C}\cos\varphi$ $\epsilon = \pm 1$	$\dfrac{2(-1)^p(-1)^q}{(2p+1)(2q+1)}$
G	$\omega_{x'} = \epsilon \dfrac{(2q+1)\pi}{2C}\sin\varphi$ $\omega_{y'} = \pm \dfrac{(2n+1)\pi}{2A} - \epsilon\dfrac{(2q+1)\pi}{2C}\cos\varphi$ $\epsilon = \pm 1$	$\dfrac{2(-1)^n(-1)^q}{(2n+1)(2q+1)}$
H	$\omega_{y'} = \epsilon''\dfrac{(2n+1)\pi}{2A} - \epsilon'\dfrac{(2q+1)\pi}{2C}\cos\varphi$ $\omega_{x'} = \epsilon \dfrac{(2p+1)\pi}{2B} + \epsilon'\dfrac{2q+1)\pi}{2C}\sin\varphi$ $\epsilon = \pm 1,\ \epsilon' = \pm 1,\ \text{and}\ \epsilon'' = \pm 1$	$\dfrac{4(-1)^n(-1)^p(-1)^q}{\pi(2n+1)(2p+1)(2q+1)}$

Normalized Amplitude	Intensity	Normalized Intensity
$\dfrac{1}{2}$	$\dfrac{\pi^4}{4}$	$\dfrac{1}{4}$
$\dfrac{(-1)^n}{\pi(2n+1)}$	$\dfrac{\pi^2}{(2n+1)^2}$	$\dfrac{1}{\pi^2(2n+1)^2}$
$\dfrac{(-1)^p}{\pi(2p+1)}$	$\dfrac{\pi^2}{(2p+1)^2}$	$\dfrac{1}{\pi^2(2p+1)^2}$
$\dfrac{2(-1)^n(-1)^p}{\pi^2(2n+1)(2p+1)}$	$\dfrac{4}{(2n+1)^2(2p+1)^2}$	$\dfrac{4}{\pi^4(2n+1)^2(2p+1)^2}$
$\dfrac{(-1)^q}{\pi(2q+1)}$	$\dfrac{\pi^2}{(2q+1)^2}$	$\dfrac{1}{\pi^2(2q+1)^2}$
$\dfrac{2(-1)^p(-1)^q}{\pi^2(2p+1)(2q+1)}$	$\dfrac{4}{(2p+1)^2(2q+1)^2}$	$\dfrac{4}{\pi^4(2p+1)^2(2q+1)^2}$
$\dfrac{2(-1)^n(-1)^q}{\pi^2(2n+1)(2q+1)}$	$\dfrac{4}{(2n+1)^2(2q+1)^2}$	$\dfrac{4}{\pi^4(2n+1)^2(2q+1)^2}$
$\dfrac{4(-1)^n(-1)^p(-1)^q}{\pi^3(2n+1)(2p+1)(2q+1)}$	$\dfrac{16}{\pi^2(2n+1)^2(2p+1)^2(2q+1)^2}$	$\dfrac{16}{\pi^6(2n+1)^2(2p+1)^2(2q+1)^2}$

Figure 5.22   Two different Ronchi rulings inclined at approximately 45 degrees to each other.

ADDING TWO TRANSMISSION FUNCTIONS

We have just considered the effects of multiplying together various transmission functions. For the examples considered we have seen that the product of two (or more) transmission functions which are Ronchi

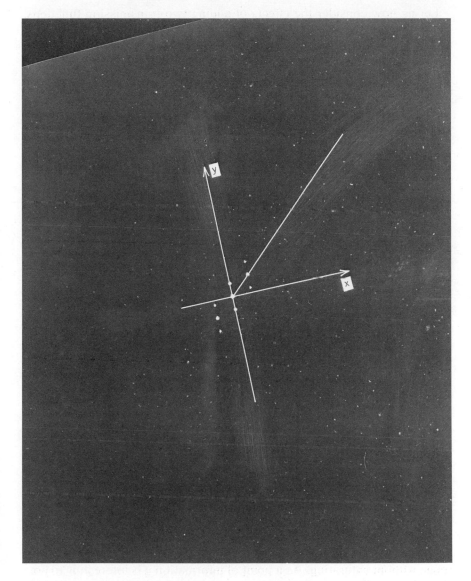

Figure 5.23   Spectrum of the two Ronchi rulings in Fig. 5.22.

rulings will result in spectra that have cross-product terms; that is, terms will appear that are off axis and are related to the product of the on-axis spectral points. We have previously designated the cross product terms as **Terms D, E, F, G,** and **H.**

Let us now consider the effects of adding two transmission functions. For simplicity we shall consider the sum of two transmission functions that are Ronchi rulings and oriented 90 degrees to each other. The Fourier transform of the sum of two functions is equal to the sum of the Fourier transforms of the two functions. We can write this mathematically as

$$F\left[f(x') + f(y')\right] = \int_{-\infty}^{\infty} \int_{-\infty}^{\infty} \left[f(x') + f(y')\right] \exp\left[-j(\omega_{x'}x' + \omega_{y'}y')\right] dx' \, dy'$$

Multiplying out

$$F\left[f(x') + f(y')\right] = \int_{-\infty}^{\infty} \int_{-\infty}^{\infty} f(x') \exp\left[-j(\omega_{x'}x' + \omega_{y'}y')\right] dx' \, dy'$$
$$+ f(y') \exp\left[-j(\omega_{x'}x' + \omega_{y'}y')\right] dx' \, dy'$$

$$= \int_{-\infty}^{\infty} f(x') \exp\left[-j(\omega_{x'}x' + \omega_{y'}y')\right] dx' \, dy'$$
$$+ \int_{-\infty}^{\infty} f(y') \exp\left[-j(\omega_{x'}x' + \omega_{y'}y')\right] dx' \, dy'$$

$$= 2\pi\delta(\omega_{y'}) \int_{-\infty}^{\infty} f(x') \exp\left[-j\omega_{x'}x'\right] dx$$
$$\downarrow$$

This δ function
restricts this term
to the $\omega_{x'}$ axis
(i.e., $\omega_{y'} = 0$)

$$+ 2\pi\delta(\omega_{x'}) \int_{-\infty}^{\infty} f(y') \exp\left[-j\omega_{y'}y'\right] dy'$$
$$\downarrow$$

This δ function
restricts this term
to the $\omega_{y'}$ axis
(i.e., $\omega_{x'} = 0$)

We have thus shown that the Fourier transform of the sum of two perpendicular Ronchi rulings (transmission functions) will result in two on-axis spectra perpendicular to each other without any cross-product terms.

Figure 5.24 shows a photograph of two perpendicular Ronchi rulings. The transmission function of each of these Ronchi rulings is added to the other; that is, the transmission function is produced by double exposing where each exposure is controlled so that the film records in its linear range without saturating. As a point of information it is interesting to note that if we allowed the film to saturate on each exposure, then double exposing the film could be made to be the product of the two exposing transmission functions.

Figure 5.25 shows the spectrum formed by the sum of the two transmission functions of Figure 5.24 (note the absence of cross-product terms and both spectrums are on axis).

Figure 5.24   Sum of two Ronchi Ruling.

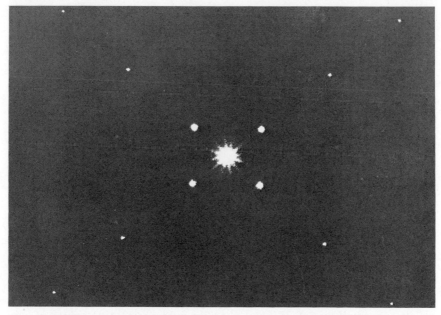

Figure 5.25   Spectrum of Figure 5.24 showing absence of cross-product terms.

6

Characteristics of Photographic Film

INTENSITY-AMPLITUDE COMPLICATION

It has been shown that optical data processing is based on the diffraction principle that produces a Fourier transform of the transmission function. This Fourier transform and transmission function are based on *light amplitude*. Practically the complex amplitude of light cannot be measured. The characteristics of physical photodetectors (e.g., film, human eye, photocell) are such that they can only sense the *intensity* of light (proportional to amplitude squared) and they cannot sense the amplitude of light directly. Measured values of light are therefore expressed in terms of intensity rather than amplitude. For optical data processing it is necessary to consider how a desired signal can be expressed as an amplitude of light. Since only intensity of light can be measured, the square root of the measured intensity (which is proportional to amplitude) should be proportional to the desired signal, that is,

$$I \propto A^2$$
$$I = KA^2$$

and if $K = 1$,

$$I = A^2.$$

The following discussion will clarify this process and we shall assume $K = 1$.

As an example let us assume that a desired signal is to be the sine wave given by

$$y = A \sin x \tag{6-1}$$

as shown in Figure 6.1(a). It is desired that the *amplitude* of the light is to vary in accordance with this equation. The sine wave has negative values on alternate half cycles. It is not possible to have negative values of light

302

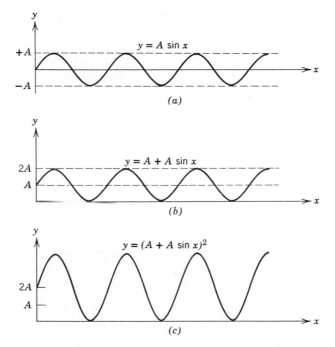

Figure 6.1 Sine-wave amplitude, biased sine wave, and intensity.

amplitude. Either there is no light (zero amplitude) or there is some light (positive amplitude). It should be noted here that the transmission function is *never* negative. In order to represent this signal (sine wave) as a variation in light amplitude, it is necessary to introduce a bias to the amplitude as shown in Figure 6.1(b). It is seen that adding a bias of $+A$ to the original wave eliminates negative values. Should a photocell be used to monitor a light signal of the form shown in Figure 6.1(b), it will not produce a sine wave as an output. The photocell responding to *light intensity* will produce an output that will be of the form in Figure 6.1(c) which is the square of Figure 6.1(b). In other words if a photocell is used to monitor a light signal of *amplitude*

$$y = A + A \sin x \tag{6-2}$$

its output will correspond to the intensity of this light and will be

$$y = [A + A \sin x]^2 \tag{6-3}$$

In order to find the amplitude of the input light signal, it is necessary to take the square root of the photocell output.

To summarize, it has been shown that optical processing deals with the Fourier transform of light amplitudes. A desired sine-wave amplitude variation must be modified by the addition of a bias to form an amplitude variation that does not require a negative light amplitude. It was further pointed out that it is necessary to take the square root of a photocell (or any other photosensor) output to determine the amplitude of the light input since only *light intensity can be sensed.*

Once again let us consider the light intensity given by

$$y = [A + A \sin x]^2 \tag{6-3}$$

A voltage waveform

$$y = [A + A \sin x]^2 \tag{6-3}$$

applied to the vertical deflection input circuit of an oscilloscope would be seen on a cathode-ray tube face as shown in Figure 6.2(a). If a photocell were to monitor such a presentation, there would be no variation in light intensity corresponding to the waveform being monitored. The light intensity output from a cathode-ray tube is approximately linear with respect to the input voltage on the accelerating grid. Thus when a voltage waveform

$$y = [A + A \sin x]^2 \tag{6-3}$$

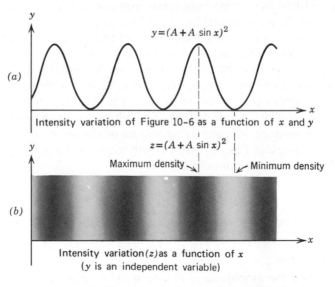

Figure 6.2 Oscilloscope waveform and intensity displays.

is applied to the accelerator grid (z-axis modulation) instead of vertical deflection, the light intensity on the face of the cathode-ray tube will be proportional to $[A + A \sin x]^2$. Figure 6.2(b) shows the oscilloscope presentation when the applied voltage $y = (A + A \sin x)^2$ appears as a light-intensity variation. Comparing the waveform of Figure 6.2(a) with the intensity variation of Figure 6.2(b), it is seen that the areas in which the applied signal is maximum correspond to areas of maximum intensity in the display.

The characteristics of transmission functions must also be considered in terms of intensity since photosensor characteristics are expressed in terms of intensity. Using Figure 6.2 and the discussion above, the relation between the transmission function and intensity characteristics is easily shown. The intensity variation obtained by z-axis modulation of an oscilloscope can also be obtained by inserting an appropriate variable-density film in the path of a constant-intensity light beam. The transmission characteristic of film expressed in terms of intensities is called the *transmittance*. If the intensity of the incident light is I_0 and the intensity of the light after passing through the film is I_T, the transmittance is given by

$$T = \frac{I_T}{I_0} \tag{6-4}$$

Figure 6.3 shows a film of transmittance $T(x)$; that is, the transmittance varies as a function of x along the film. Let us assume that the intensity I_T of the transmitted light varies in accordance with

$$I_T = (A + A \sin x)^2 \tag{6-5}$$

that is, we can assume we have moved a photocell (which can only sense intensity) to various positions along the x axis of the film and noted its

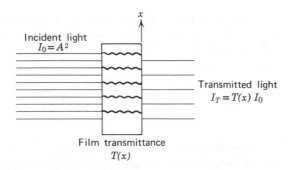

Figure 6.3   Transmittance of film.

output to be $(A + A \sin x)^2$. If we further assume that the incident light has an intensity of

$$I_0 = 4A^2$$

by substituting in (6-4) we can obtain

$$T = \frac{I_T}{I_0} \tag{6-4}$$

$$T(x) = \frac{(A + A \sin x)^2}{4A^2} = \frac{A^2(1 + \sin x)^2}{4A^2}$$

$$T(x) = (\tfrac{1}{2} + \tfrac{1}{2} \sin x)^2 \tag{6-6}$$

In this example we have shown that the transmittance is proportional to the intensity of the transmitted light (I_T) and inversely proportional to the intensity of the incident light (I_0). In optical data processing, however, we are usually interested in the transmission function and not the transmittance function. The difference being that the transmission function is proportional to the amplitude of the transmitted light and inversely proportioned to the incident amplitude; that is, for the example given above the transmission function is given by

$$f(x) = \frac{A_T}{A_0} = \frac{A + A \sin x}{2A} = \tfrac{1}{2}(1 + \sin x) \tag{6-7}$$

where $f(x)$ is the transmission function of film, A_0 is the amplitude of light incident on film, and A_T is the amplitude of light transmitted through film.

Since in optical data processing the optical Fourier transforms are made of the amplitude of the light distributions (transmission functions), we usually will deal only with them. The relationship between transmission function, transmittance, and intensity is given by

$$f(x) = [T(x)]^{1/2} = \left[\frac{I_T}{I_0}\right]^{1/2} = \frac{A_T}{A_0} \tag{6-8}$$

Although this relationship was derived for a specific case, the same result is obtained in the general case; that is, we must retain the relationship that the intensity is equal to the amplitude squared.

PHOTOGRAPHIC FILM BASICS

At the present time photographic film is one of the most convenient and versatile mediums available with which it is possible to modulate light for optical data processing. Photographic film offers the possibility of great information storage as well as reasonable possibilities for real time capabilities. The use of photographic film in optical data processing requires a rather thorough knowledge of the characteristics and limitations of photographic film. For this reason in this section we will discuss some of the detailed characteristics of photographic films.

The term *photographic film* usually refers to a light-sensitive emulsion on a base support. If a plastic base is used to support the emulsion, the end product is called the *photographic film*. When a glass plate is used to support the emulsion, the end product is called a *photographic plate*. The light-sensitive emulsion is basically a mixture of silver halide crystals in gelatin. The sensitivity of the silver halide crystals can be affected by the addition of various sensitizing agents. The size of the silver halide crystals that are photosensitive varies from less than one micron to several microns. Usually the larger the silver halide crystal, the greater its sensitivity to light; that is, a larger crystal generally requires less light to expose (make it developable) it. An emulsion that contains relatively large crystals is generally called a *fast* emulsion. A *fast* emulsion, however, will lack the capability of registering fine details. The limits of resolution of the emulsion will be determined by the size of the crystals that become exposed. The smaller the crystals, the greater the detail that can be captured in the film. The smaller the crystals, however, the slower the speed of the film. For a given emulsion there is a definite exposure latitude that is determined by the size of the crystals composing the emulsion. Below a certain level of exposure the smaller crystals will not be exposed. This effect influences the contrast of the film. The photographic emulsion must be properly selected to permit the greatest resolution to be captured on the film as well as the highest contrast ratios. Film must be selected, therefore, to render a final product that registers the resolution required as well as the tonal range. In general both of these requirements cannot be met and some compromise between them is usually made.

Any light that will produce a photographic effect on silver halide must be absorbed by the sensitive material. When light energy (photons) is absorbed by silver halide crystals, there is usually no visible change in the appearance of the emulsion. The exposed emulsion, however, contains an invisible *latent image* that can be converted into a visible silver image by the action of the developer. The developer chemically changes the exposed silver halide crystals to metallic silver, thus making the exposed portions of the emulsion opaque.

The development process is possible because certain chemicals react with the exposed silver halide in preference to that which is unexposed. If the developing process is continued for a sufficient length of time, all the silver halide crystals in a photographic emulsion will be reduced to metallic silver. This tends to produce a very slight uniform silver coating over the entire image. This thin coating is generally referred to as *fog*. Photographic fog is generally considered undesirable as it adversely affects both the resolution and the contrast capabilities of the film. When the amount of light absorbed by the silver halide crystals is insufficient to expose them, the developer (initially) has no effect on them. The purpose of a fixing bath is to dissolve out any silver halide crystals remaining in the emulsion.

PHOTOGRAPHIC TECHNIQUES FOR OPTICAL PROCESSING

A photographic *negative* is produced by exposing, developing, and fixing the photographic film. A photographic *positive* can be made from the negative by direct contact printing. Contact printing is accomplished by placing the developed negative over and in direct contact with an unexposed film and exposing the unexposed film by projecting light through the negative. The ability of developed photographic images to block the passage of light can be used as a measure of the photographic quality of the reproduced image. The ratio of the intensity, I_T, of the light transmitted through the film to the intensity, I_0, of the incident light is a measure of the photographic film's ability to transmit light. This ratio, I_T/I_0, has previously been defined as the transmittance, T. The ratio of incident light to transmitted light is defined as opacity, I_0/I_T. The common logarithm of the opacity is the optical density (or density). The equation defining the density is

$$\text{density} = \log_{10} \frac{I_0}{I_T} = -\log_{10} T \qquad (6\text{-}9)$$

The following nomograph shows the relation between density, transmission, and opacity.

Now let us assume that it is possible to produce photographic film such that only a single layer of metallic silver salts is developed. When this developed film is placed in front of a light source, it will reduce the intensity of the incident light, I_0, by the factor, T_1 (that is, $I_T = T_1 I_0$). If a second identical film were placed over the first, the intensity of the transmitted light would be

$$I_T = (T_1)^2 I_0 \qquad (6\text{-}10)$$

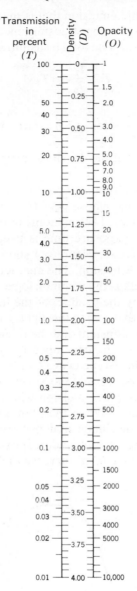

which indicates that the transmittance of superimposed photographic films is given by the product of the individual transmittances. When n identical layers are used, the intensity of the transmitted light will be

$$I_T = (T_1)^n I_0 \tag{6-11}$$

We assume a photographic emulsion that is made up of n different layers of metallic silver, each having transmittance T_1 and corresponding density D_1 as given by (6-9). The total transmittance is then

$$T = \frac{I_T}{I_0} = (T_1)^n \tag{6-12}$$

And the total density is

$$D = -\log_{10}T = -\log_{10}(T_1)^n = n(-\log_{10}T_1) = nD_1 \tag{6-13}$$

It is evident from (6-13) that the density is proportional to the number of layers n in the emulsion.

A photographic emulsion that is exposed to a constant intensity of light will have a developed image whose density will increase with increasing exposure time (within limits). The relationship between the density of the developed image and the exposure time is usually indicated by a graph called the *characteristic curve*. Figure 6.4 shows a typical characteristic curve for a photographic emulsion. The characteristics curve is obtained by plotting the density versus the common logarithm of the exposure. The exposure is determined by the product of the intensity of the light I_0 and the time t that the film is exposed to this light. From this we can see that if the exposure time is kept constant and the intensity of the exposing light is increased, the density of the developed image will be increased. The characteristics curve is the S-type curve commonly found in engineering. Referring to Figure 6.5 we see that the lower limit of the curve indicates the fog level. The photographic emulsion will not increase in density for exposure values less than a. When the photographic film is exposed to values between a and b, the change in density will be nonlinear. This area is usually considered the area of underexposure. Exposure values between b and c will result in a *linear* increase in density with respect to the

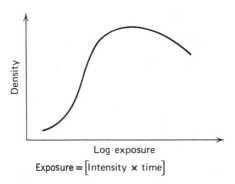

Exposure = [Intensity × time]

Figure 6.4   Typical characteristic curve for a photographic emulsion.

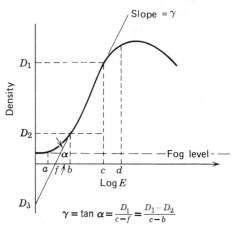

Figure 6.5   Characteristic curve of a photographic emulsion.

logarithm (base 10) of exposure. Exposure in the area c and d results in very slight changes in density with exposure (i.e., slight increase in density for increased exposure). This area of the curve is generally considered the area of overexposure. The area of the curve corresponding to an exposure greater than d shows a decrease in density with increasing exposure time. This region of the curve is usually associated with what is called *solarization* of the film.

On the straight line portion of the curve (i.e., corresponding to exposures between b and c) the change in density is proportional to the change in the logarithm of exposure. From Figure 6.5 we can see that if the straight line portion of the curve is extended, it intersects the log E axis at point f. This point f is often used as a measure of the film speed. The slope of this line is given by

$$\gamma = \tan \alpha = \frac{D_1}{c-f} = \frac{D_1 - D_2}{c-b} \tag{6-14}$$

where γ is the slope of the straight line portion of the curve.

(NOTE: The angle relation $\tan \alpha = \gamma$ holds only when the two axes (density versus log E) are plotted to the same scale. When the two axes are not to the same scale, $\tan \alpha$ will not equal the slope γ. However, the slope γ will always equal $\Delta D / \Delta \log_{10} E$.)

The characteristic curve of photographic emulsions depends not only on the nature of the emulsion, but also on the method by which it was developed. Different emulsions will generally have different characteristic curves and therefore will have different γ. Each of these emulsions in turn can have a different γ depending on the development process. The type of developing solutions used, the time of processing, the temperature of the developer, all in turn reflect on the value of γ. Figure 6.6 shows a family of characteristic curves for a single type of film for which only the development time has been changed. It can be seen from the plots in Figure 6.6 that in general the γ will increase as the development time increases.

LINEARIZING PHOTOGRAPHIC FILM

From the characteristic curves of photographic films we can see that the photographic process is extremely nonlinear. The very fact that the density is plotted against the *logarithm* of the exposure is an indication of this nonlinearity. There are methods by which the photographic process can be linearized. The following discussion will illustrate one of these methods.

We shall consider the case where *amplitude* variations of light corresponding to a given signal E_s are desired. Initially a photographic film is exposed to a varying light *intensity* corresponding to the signal E_s as shown in Figure 6.7(a). This can be accomplished by moving a constant-intensity light source across the film at varying speeds so that the product at any point of the intensity and time gives the required value of E_s. Assuming the exposure of the photographic film is accomplished in the linear region of the characteristic curve shown in Figure 6.5, the density will be given by the equation

$$D_1 = \gamma_1(c - f) = \gamma_1(\log_{10}E_c - \log_{10}E_f) \tag{6-15}$$

where E_c is the product of intensity and time (exposure) corresponding to point c; and where E_f is the product of intensity and time corresponding to point f. For the case we are considering this equation can be written as

$$D_1 = \gamma_1[(\log_{10}t_1 I_s) - (f)] \tag{6-16}$$

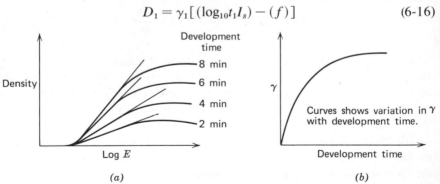

(a)   *(b)*

Figure 6.6 Affect of development time on gamma. (*a*) Family of characteristic curves showing change with development time. (*b*) Plot of γ versus development time.

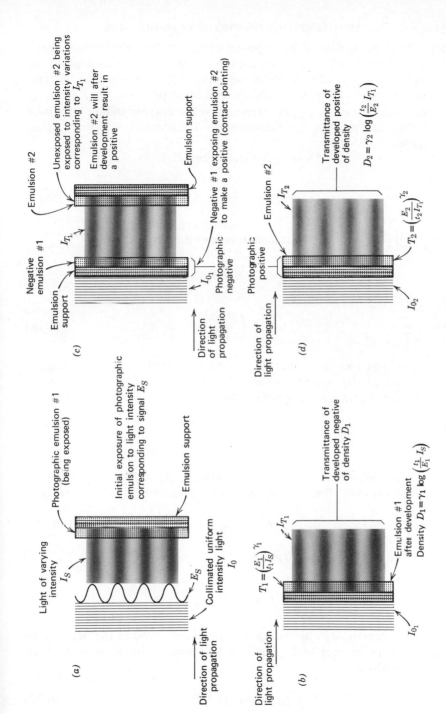

Figure 6.7 The linearized photographic process.

313

where D_1 is the density of the exposed film; I_s is the light intensity variation corresponding to E_s; t_1 is the exposure time; γ_1 is the slope of the straight line portion of the characteristics curve as determined by the type of emulsion, type of exposure, and development process; and $f = \log_{10} E_f$ as noted above. Let E_1 be the exposure value corresponding to point f (i.e., $f = \log_{10} E_1$ or $E_f = E_1$). Substituting for f in (6-16) and rearranging terms gives

$$D_1 = \gamma_1 [\log_{10} t_1 I_s - \log_{10} E_1]$$

or

$$D_1 = \gamma_1 \log_{10} \left(\frac{t_1}{E_1} I_s \right) \tag{6-17}$$

From (6-9) $D = -\log_{10} T$ the transmittance of this film will be

$$T_1 = 10^{-D_1} = 10^{-\gamma_1 \{ \log_{10} [(t_1/E_1) I_s] \}} = 10^{\log_{10} \left[\left(\frac{E_1}{t_1} \right) \left(\frac{1}{I_s} \right) \right]^{\gamma_1}}$$

or

$$T_1 = \left(\frac{E_1}{t_1 I_s} \right)^{\gamma_1} \tag{6-18}$$

It is quite obvious from (6-18) that the transmittance is not linearly proportional to the amplitude variation of the signal E_s (or I_s). After development the transmittance of this film can also be determined by

$$T_1 = \frac{I_{T_1}}{I_{0_1}} \tag{6-19}$$

where I_{0_1} is the intensity of uniform incident light, and I_{T_1} is the intensity of transmitted light. Rearranging (6-19) the transmitted light intensity is given by

$$I_{T_1} = T_1 I_{0_1}. \tag{6-20}$$

Substituting (6-18) into (6-20) gives

$$I_{T_1} = I_{0_1} T_1 = I_{0_1} \left\{ \frac{E_1}{t_1 I_s} \right\}^{\gamma_1} \tag{6-21}$$

Thus the intensity, I_{T_1}, of the light transmitted through the negative is nonlinear with respect to I_s. For simplification we let $K_1 = (E_1/t_1)^{\gamma_1}$ in (6-21) to produce

$$I_{T_1} = K_1 I_{0_1} [I_s]^{-\gamma_1} \tag{6-22}$$

We now assume the negative is used to expose a second film (positive) by contact printing as shown in Figure 6.7(c). As shown the light exposing the second film must pass through the initial negative. If the incident light of constant intensity, I_{0_1}, illuminates the negative (Figure 6.7(b), the intensity transmitted is I_1. The intensity of the light exposing the second film will be I_{T_1} as defined above. The density of this second film (after development) will be (using the form of (6-17) and referring to Figure 6.7(c)

$$D_2 = \gamma_2 \log_{10} \left(\frac{t_2}{E_2} I_{T_1} \right)$$

(6-23)

Continuing as above (6-18) the transmittance of this second film is

$$T_2 = \left(\frac{E_2}{t_2 I_{T_1}} \right)^{\gamma_2}$$

(6-18)

Letting $K_2 = (E_2/t_2)^{\gamma_2}$ the transmittance can be given as

$$T_2 = K_2 (I_{T_1})^{-\gamma_2}$$

(6-24)

When this positive is developed and illuminated by uniform light intensity, I_{0_2}, the transmitted light intensity, I_{T_2}, will be given by (see Figure 6.7(d)

$$I_{T_2} = T_2 I_{0_2} = K_2 I_{0_2} (I_{T_1})^{-\gamma_2}$$

(6-25)

Substituting for I_{T_1} as given by (6-22) gives

$$I_{T_2} = K_2 I_{0_2} (K_1 I_{0_1})^{-\gamma_2} I_s^{\gamma_1 \gamma_2}$$

(6-26)

Defining a new constant $K = K_2 I_{0_2} (K_1 I_{0_1})^{-\gamma_2}$, equation (6-26) becomes

$$I_{T_2} = K I_s^{\gamma_1 \gamma_2}$$

(6-27)

Equation (6-27) shows that the intensity of light transmitted through the positive is porportional to the original signal light intensity raised to the $\gamma_1 \gamma_2$ power. When the product of γ_1 and γ_2 is one (i.e., $\gamma_1 \gamma_2 = 1$), the intensity of the light transmitted through the positive (illuminated by a uniform light intensity) is proportional to the original light intensity.

The amplitude, A, of the light transmitted through the positive is, by definition, the square root of the intensity given by (6-27) or

$$A = \sqrt{I_{T_2}} = K_1^{1/2} I_s^{(\gamma_1 \gamma_2)/2}$$

(6-28)

when $\gamma_1\gamma_2 = 1$, the amplitude, A, is proportional to $I_s^{1/2}$. This means that although the intensity variations in the transmitted light vary as the original signal intensity, the amplitude variations do not.

In an earlier discussion it was pointed out that if the original signal intensity is the square of the desired signal, the amplitude will be proportional to the desired signal. It is apparent from (6-28) however, that when $\gamma_1\gamma_2 = 2$, the *amplitude* of the light transmitted through the positive is proportional to the original signal intensity (desired signal). It is apparent from this discussion that a desired input signal can be represented by the intensity used to expose a film and if the gamma product is set equal to two (i.e., $\gamma_1\gamma_2 = 2$), the resulting *amplitude* transmission function will correspond to the desired signal.

Figures 6.8, 6.9, and 6.10 are graphical representations of the mathematics discussed above. It should be noted that the $\gamma_1\gamma_2$ product is different for each figure, so a graphical comparison of the effects of this term is possible. Only Figure 6.8, which is the case where $\gamma_1\gamma_2 = 2$, will be explained in detail because the same explanation applies for all values of $\gamma_1\gamma_2$. Diagram A of Figure 6.8 shows a sine wave that is assumed to be an input signal for an optical processor. As shown earlier an input signal must be represented by a (amplitude) transmission function for optical processing purposes. Using a photographic plate to implement the transmission function Figure 6.8 shows the steps required to represent the input signal of Diagram A as a transmission function. As in previous examples only one dimension will be considered (x coordinate) but the basic process applies to the two-dimensional case (x and y coordinates). Diagram A of Figure 6.8 shows the input signal as a function of x represented by an intensity ($I_s = 51 + 50 \sin x$). This intensity signal is used to expose a photographic plate for 10 sec ($t_1 = 10$). The exposure of the plate is given by the product ($t_1 I_s$). Therefore the exposure (E_{s_1}) is given by

$$E_{s_1} = t_1 I_s = 510 + 500 \sin x$$

The waveform for this exposure (E_{s_1}) is also given by Diagram A except the scale is multiplied by a factor of 10.

Diagram C represents the linear part of the characteristic curve for the film being exposed. It is assumed that the plate is developed so that $\gamma_1 = 2$ and $f = 1$ (see Figure 6.5) as shown. The characteristic curve is a plot of density versus the logarithm (base 10) of exposure. Therefore in order to project values of exposure E_{s_1} (Diagram A) onto the characteristic curve of Diagram C, it is necessary to plot $\log_{10} E_{s_1}$ as a function of x as shown in Diagram B. Using Diagrams B and C it is possible to determine graphically the density as a function x shown in Diagram D. For

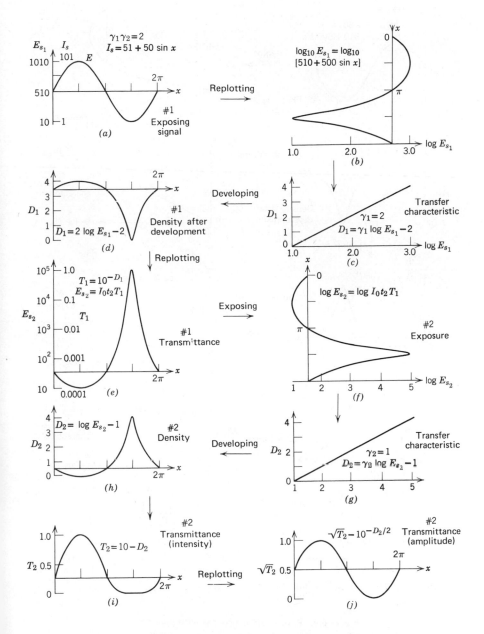

Figure 6.8 Photographic process with $\gamma_1\gamma_2 = 2$.

317

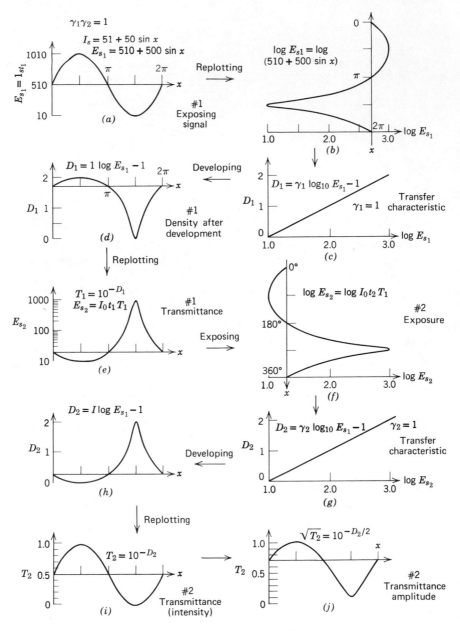

Figure 6.9 Photographic process with $\gamma_1\gamma_2 = 1$.

318

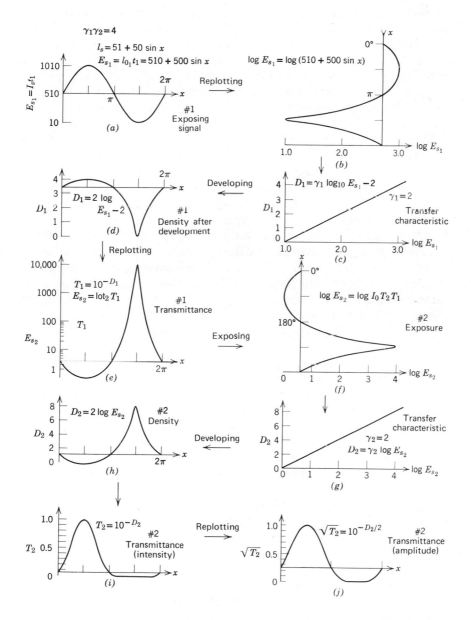

Figure 6.10 Photographic process with $\gamma_1\gamma_2 = 4$.

319

example, we can choose a value of x in Diagram B and from this point project a straight line vertically downward to intersect the straight line of Diagram C. From this point of intersection projecting a point horizontally to the chosen x coordinate in Diagram D gives the value of the density for that value of x. Repeating this procedure for all values of x will give the complete density curve of Diagram D. Diagram D shows that the variation in density is proportional to the common logarithm of the original light intensity, I_s.

Diagram E is a plot of the transmittance, T_1, corresponding to the density shown in Diagram D ($T_1 = 10^{-D_1}$). The negative with transmittance, T_1 shown in Diagram E can be used to expose a new plate by contact printing. If the exposing light used has uniform intensity I_0 (in this case 10^4), the exposure of the second plate will be given by relation $E_{s_2} = I_0 t_2 T_1$. This exposure E_{s_2} is plotted in Diagram E for $t_2 = 10$ sec. The $\log_{10} E_{s_2}$ is plotted in Diagram F. We shall assume that the second film is developed to a $\gamma_2 = 1$ as shown by the characteristic curve of Diagram G. If we project values for $\log_{10} E_{s_2}$ onto the characteristic curve, the plot of the density shown in Diagram H can be obtained. This procedure is the same as that used above for obtaining Diagram D; that is, the density D_2 shown in Diagram H was produced in the second film by an exposure E_{s_2} developed to a $\gamma_2 = 1$. The transmittance corresponding to the density of Diagram H is shown in Diagram I.

It has previously been shown that the transmission function $f(x)$ is equal to the square root of the transmittance. In Figure 6.8 Diagram J shows the transmission function obtained by taking the square root of the transmittance given in Diagram I. From a comparison of the transmission function given by Diagram J with the signal expressed as an intensity in Diagram A it is apparent that the transmission function corresponds to the desired input signal. It is important to note that the correspondence between Diagram J and A in Figure 6.8 is due to the gamma product being equal to two ($\gamma_1 \gamma_2 = 2$).

Figure 6.8 shows that when $\gamma_1 \gamma_2 = 2$, the square root of the final transmittance of the film will correspond to the original exposing signal intensity and therefore the final transmission function $f(x)$ will be equal to the original signal.

Figure 6.9 shows that when $\gamma_1 \gamma_2 = 1$, the final transmittance of the film will correspond to the original exposing signal intensity and therefore the final transmission function $f(x)$ will be equal to the square root of the original signal.

Figure 6.10 shows that when $\gamma_1 \gamma_2 = 4$, neither the final transmittance nor the transmission function of the film will correspond to the original exposing intensity.

Comparing Diagram J of Figures 6.8, 6.9, and 6.10 demonstrates the requirement for a product $\gamma_1\gamma_2 = 2$. Although the same desired input signal was represented by the exposing signal intensity in each case, only when $\gamma_1\gamma_2 = 2$ is the transmission function of the second film a representation of the desired signal.

In the above discussion we have assumed that the photographic film thickness is constant. If the film thickness is not constant, the optical path through the film is not constant and phase shifts will be introduced into the signal. In the normal photographic process unexposed silver halide is removed from the emulsion. It was previously stated that portions of photographic film will shrink on drying. This shrinkage will cause phase shifts to be introduced into the signal (which distort the input signal). To eliminate these undesirable phase shifts the signal film is usually placed between optical flats — that is, uniformly thick and flat sheets of glass. The surface of the film is coated with an oil (such as used for oil immersion lenses on microscopes) to fill the voids in the film with a material matching the index of refraction of the film as shown in Figure 6.11. This in effect produces a uniformly thick film (constant phase).

In some cases it is possible to use only phase changes to introduce an input signal to an optical processor. Transparent mediums whose thickness changes in accordance to the desired signal can be used as signal inputs. Such signal inputs can be used with lower light levels since the light amplitude is not being alternated by the film (only its phase). It is possible to have the input signal vary in both amplitude and phase but such signals are difficult to produce, measure, and control accurately. In

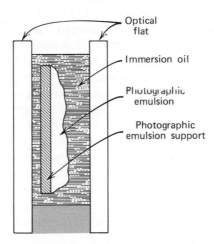

Figure 6.11   Use of immersion oil to correct photographic film phase shifts.

general the simplest and most convenient input signals (easily reproduced, measured, and controlled) are those varying only in amplitude (and therefore usually requiring the use of optical flats and immersion oil).

OTHER RECORDING TECHNIQUES

Photochromic and Electrochromic Materials

Photochromic materials are materials that darken (and stay darkened for relatively long periods of time) after exposure to certain wavelengths (colors of light) and lighten as exposed to other wavelengths of light. These materials can display resolutions and contrasts comparable to the best photographic films but they do not require the processing liquids and drying times (with usual emulsion shrinkage) of the photographic films. Electrochromic materials are similar to photochromic materials but are capable of being written on by an electron beam.

Photochromic materials are presently being investigated for use in optical data processors. Certain of these materials have extremely high resolution and darken as exposed to certain wavelengths (colors) of light and lighten when exposed to other wavelengths (colors) of light. This permits writing a signal with very fine detail on the material with one color of light and reading the signal (without appreciably affecting it) with another color of light. When desired, the signal can be erased using the appropriate "erase" wavelength of light for the material. In addition the electrochromic materials can be written on with an electron beam, which makes it possible to incorporate these qualities in cathode-ray tubes; that is, a signal can be written on the face of a CRT with an electron beam or by exposure from an external light source, erased by exposure from an external light source of proper wavelength, and read by (transmitted light) energizing appropriate colored phosphors (behind the electrochromic material) on the faceplate or by (reflected light) illuminating the faceplate from an external light source.

It is therefore possible to "write" television-type images on the electrochromic faceplate using an electron beam. The electron beam can then be used to read the image (by energizing the appropriate colored phosphors). It is therefore possible to erase, read, and write either the whole or portions of an image. For viewing television types of images on an electrochromic faceplate high room (external) illumination is desirable for viewing the image by reflected light. When viewed by transmitted light (from phosphors) the electrochromic faceplate appears similar to a normal television-type CRT; that is, low room illumination is desirable for the image to be seen.

By scanning the stored electrochromic image with the electron beam a photocell can be used to pick up the image point by point. The photocell output can be used to transmit the image to either a computer memory or another remote electrochromic display tube for storage.

We can see that although photographic film is presently widely used in optical processors, other materials such as photochromics and electrochromics may eventually replace photographic films.

Thermoplastics

Certain systems use a thermoplastic (not photographic film) recording technique to eliminate the chemical processing. The technique requires an electron gun to deposit electrons where information is to be recorded. An infrared heater liquifies the tape where the electrons are being deposited. The thermoplastic is then deformed by electrostatic forces (proportional to the deposited electrons). The distorted and deflected thermoplastic forms grooves similar to a phonograph record which remain after the plastic has cooled. A Schlieren or other type of coherent optical system can then be used to convert the information (which is now in the form of grooves) to light intensities representing processed data. The resolution that can be obtained using this phase type of input when projected on a screen is about 2500 picture elements per inch, 16 gray levels, and a range of about 100 to 1 in brightness.

7

Optical Filtering and Correlation

OPTICAL FILTERING

Figure 7.1 shows how the basic principles set forth in a spectrum analyzer can be used to develop an optical filter by the addition of a single lens that images the spectrum into an image plane. We have seen how lens 1 acts as the spectrum analyzer producing in its back focal plane (transform plane) the spectrum of the amplitude transmission function in its front focal plane (object plane). If the points of light in the transform plane are used as an object in the front focal plane of lens 2, the back focal plane (image plane) of lens 2 will contain the spectrum of the light in the transform plane. Lens 2 thus produces a transform of a transform that is in effect the inverse transform giving the original transmission function. To be more specific any object in the object plane of lens 1 will be imaged by the two lens into the image plane of the system.

We have seen how sinusoidal transmission functions produce points of light in the transform plane. For example, a 100-cps sine-wave transmission function will produce a DC point of light and two equally spaced points of light (corresponding to 100 cps) centered about the DC term. If the light from the pair of equally spaced points is blocked in the transform plane (that is, this light is not allowed to reach lens 2), the light in the image plane will not have variations corresponding to the frequency blocked. If we have a transmission function representing the sum of a 100-cps and a 150-cps sine wave, two pairs of light points are produced in the spectrum. One pair corresponds to the 100-cps sine wave and the other to the 150-cps sine wave. If we block the light from the pair of dots corresponding to 100 cps in the transform plane, the light in the image plane will have only variations corresponding to the 150-cps sine wave. Since no light corresponding to the 100-cps wave is allowed to pass the transform plane, variations corresponding to this wave will not appear in the image plane. We have thus effectively filtered 100 cps from an input singal.

324

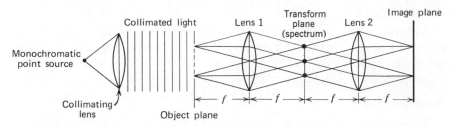

Figure 7.1   Optical filter.

Extension of this basic idea can be used to make optical filters pass or reject any desired range of spatial frequencies. Not only can a range of spatial frequencies be passed or eliminated but specific inclinations can also be filtered. A few typical examples of optical filters are shown in Figure 7.2 and discussed below.

High-Pass Filter

Low frequencies can be removed by placing an obstruction in the transform plane that will block the area around the DC component (low frequencies) while a transparent area permits all light corresponding to frequencies greater than a specified value to be passed to lens 2.

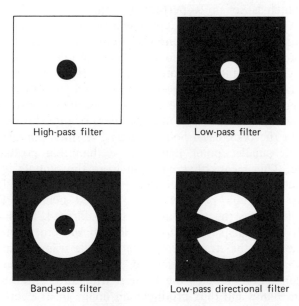

Figure 7.2   Optical filters.

Low-Pass Filter

A low-pass filter can be formed by putting a photographic iris in the transform plane. This iris will permit all light corresponding to frequencies below a specific value to be passed to lens 2 while blocking the light corresponding to frequencies above this specific value.

Band-Pass Filter

A band-pass filter can be constructed to block the light corresponding to the low frequencies and permit light corresponding to a specific range of frequencies to be passed. Any frequencies above the desired pass band can also be eliminated by blocking the light corresponding to these higher frequencies. It can be seen that we can effectively produce a pass band filter by combining a high-pass filter with a low-pass filter.

Directional Filter

Additional flexibility in optical filtering can be achieved by selecting specifically oriented transmission functions for filtering. By blocking the light corresponding to certain specific orientations we can in effect select these orientations for filtering. High- and low-pass filters can be combined with directional filters to allow only light from transmission functions of specific frequency and orientation to pass to lens 2.

EXAMPLE OF OPTICAL FILTERING. Figure 7.3 is a photograph of an optical correlator which is set up to perform optical filtering. The light source in Figure 7.3 is a mercury arc. Lens 1 is a focal length from the arc and therefore produces what is essentially a collimated light beam. A heat shield (to protect the lenses and filters) is mounted in the collimated light beam. This heat filter is coated to pass most of the visible light but reflects heat. A second interference filter, which transmits only the 5461 Å green line from the mercury arc source, is also mounted in the collimated light beam. The collimated light beam is brought to a focus by lens 2. At the back focal point of lens 2 a small pinhole is inserted to produce a smaller point light source than that produced by the mercury arc. This means that the light from the pinhole is essentially monochromatic and spatially coherent. Lens 3 is a focal length from the pinhole and thus produces a collimated beam of light which uniformly illuminates the transparency to be filtered. Lens 4 is a focal length from the transparency to be filtered and therefore produces the spectrum of the transparency amplitude transmission function in its back focal plane. The desired spatial filter is inserted in the spectrum plane. Lens 5 takes the Fourier transform of the light leaving the spectrum plane and thus produces an image in its back focal plane (i.e., an image of the transparency in the front focal plane of lens 4). This

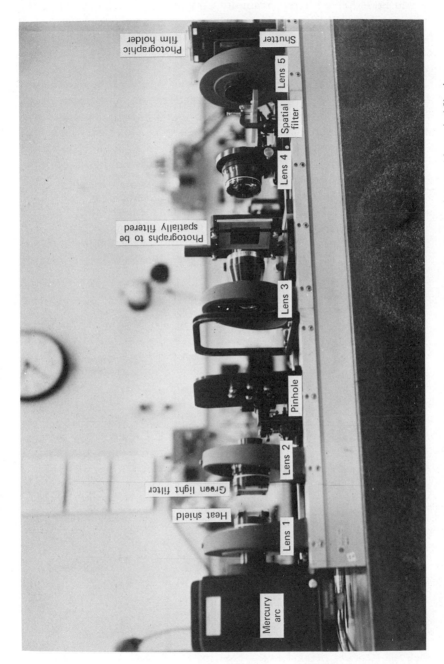

Figure 7.3 Portion of an optical correlator (manufactured by Conductron Corporation) used for optical filtering.

image can be either a filtered or unfiltered image of the transparency in the front focal plane of lens 4. In the back focal plane of lens 5 there is a photographic film holder that permits the image to be photographed. Between lens 5 and the photographic film holder there is a shutter to control the photographic film exposure time.

Figure 7.4 is a picture taken of a color television image, printed using incoherent light. Figure 7.5 is the image formed by the transparency used in Figure 7.4 when imaged through lenses 4 and 5 in the optical filter of Figure 7.3 (using spatially coherent light). The television raster lines are evident in both Figures 7.4 and 7.5. In Figure 7.5 small circle-type blemishes can be seen that are not in Figure 7.4. These small blemishes are caused by diffraction patterns formed by small dust particles on the lenses, or in the air, while others may be caused by small bubbles or imperfections in the lenses. These small circle-type blemishes are characteristic with the use of coherent light and are difficult to eliminate completely. A dust-free atmosphere and careful cleaning of the lenses minimize these effects. Use of a larger pinhole (which makes the light less coherent

Figure 7.4 Picture of television image to be processed. (Made with incoherent light.)

Figure 7.5 Picture of a television image (unfiltered).

spatially) will minimize these imperfections but will also reduce the capability of performing a coherent type of optical filtering. Other techniques for minimizing these blemishes are to move the light source slightly (keeping it in the front focal plane) during exposure. Because the imaging properties of lenses 1 and 2 in Figure 7.1 do not depend on the use of coherent light, moving the point source slightly in the front focal plane of the collimating lens will not tend to blur the image. The fact that the blemishes are caused by diffraction of coherent light by dust particles means that moving the light source slightly during exposure will tend to blur out the diffraction patterns formed by the coherent properties of the light (since the light will tend to be incoherent). Another technique is to image only one spectral point from the transform plane instead of the entire spectrum. This will tend to reduce the blemishes caused by dust on the transparency in the object plane or on lens 1.

Figure 7.6 is the image formed of the transparency of Figure 7.4 using incoherent light. This image was intentionally blurred (by defocusing) in an attempt to remove the raster lines. This picture represents the best

Figure 7.6   Blurred image of Fig. 7.4.

compromise between a sharp image and removal of the raster lines. Obviously removal of the raster lines by dofocusing is not a satisfactory technique. Figure 7.7 is an image formed of the transparency of Figure 7.4 when the spatial filter is extremely small. In this case the spatial filter was a 25-micron pinhole that eliminated essentially all the high spatial frequencies; that is, the spatial filter was a very low-pass filter. Since the Fourier transform of a DC (zero frequency) corresponds to a uniform illumination, we can see that some frequencies other than the DC are being passed but not of sufficiently high frequency to obtain a reasonably good quality image.

Figure 7.8(a) and 7.8(b) show the optical spectrum of Figure 7.4. Figure 7.8(a) shows the spectrum developed linearly; that is, except for saturation of the DC term (in the center), the amplitudes of the harmonics are properly proportioned to each other. The low amplitudes of the higher frequencies will therefore not be visible. Figure 7.8(b) has been over-developed so that the high-frequency components are made visible. Because of the limited range of photographic paper (photographic film has a greater range than photographic paper) when the low-amplitude high-

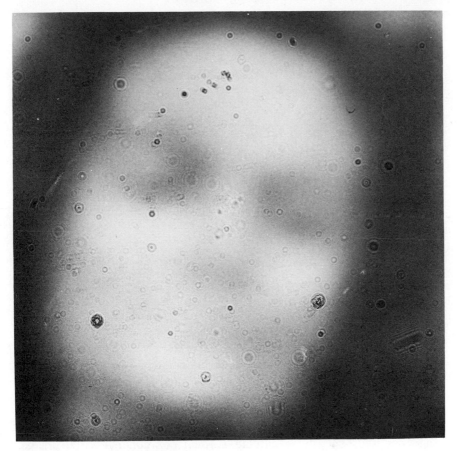

Figure 7.7 Twenty-five micron pinhole spatial filter.

frequency components are made visible, the high-amplitude low-frequency components will be saturated. The high-intensity vertical line and dots in the center correspond to the raster line frequencies. The light 45° line going up to the left through the DC point is caused by the color dot pattern (color mask) spatial frequency. This dot pattern can be seen by looking carefully at the raster lines in Figure 7.4. Figure 7.9 shows how the major portion of the vertical diffraction pattern (from the horizontal raster lines) can be blocked (some of the higher order vertical diffraction patterns are still being passed). Figure 7.10 shows that most of the horizontal raster lines have been eliminated (there is still a slight trace of raster line in the hair on the left) but the dot pattern spatial frequency has been accentuated. This accentuation is evidenced by new lines running

Figure 7.8(a)   Optical spectrum of Fig. 7.4 (Linear development but DC saturated.)

vertically up to the right (which is most apparent near the chin on the right). Figure 7.11 shows how a horizontal slit filter (which eliminates all vertical components in the spectrum) will eliminate the raster lines in the image but will tend to accentuate vertical components in the image. Figure 7.12 shows the image formed when a relatively large diameter (but thick, that is, a long hole) pinhole is used. In this case the imperfections (burrs, dust, etc.) in the pinhole have produced a great many blemishes in the image. Figure 7.13 shows the results of using the same diameter pinhole as was used in Figure 7.12 but in this case the material used was very thin. The result is a filtered picture with relatively few blemishes. The aperture size selected for Figures 7.12 and 7.13 is such that it passes the DC component with a sufficient amount of the higher frequencies (in all directions equally) without passing the fundamental raster line frequency. The quality of the reconstructed image is therefore determined by many parameters but for best overall filtered picture quality the techniques used in Figure 7.13 will give the best results; however, it was made without attempting to use the blemish removal technique of moving the light source, which would remove most of the

Figure 7.8(b)   Optical spectrum of Fig. 7.4 (overdeveloped).

Figure 7.9   Prong filter.

Figure 7.10　Prong filter of Fig. 7.9.

undesirable diffraction patterns. Appendix 15 shows another example of optical filtering.

OPTICAL SIGNAL CORRELATOR

The mathematical operation of correlation is the comparison of two signals to determine the degree of correspondence between them. For two periodic functions of the same frequency, $f_1(t)$ and $f_2(t)$, the correlation function is expressed by the equation

$$\phi_{12}(\tau) = \frac{1}{T} \int_{-T/2}^{T/2} f_1(t) f_2(t+\tau)\, dt \qquad (7\text{-}1)$$

When $f_1(t)$ and $f_2(t)$ are not the same function, the correlation function $\phi_{12}(\tau)$ is called a *cross-correlation*. The second subscript always corresponds to the function that is displaced. The order of the subscripts is important since $\phi_{12}(\tau)$ is not necessarily the same as $\phi_{21}(\tau)$.

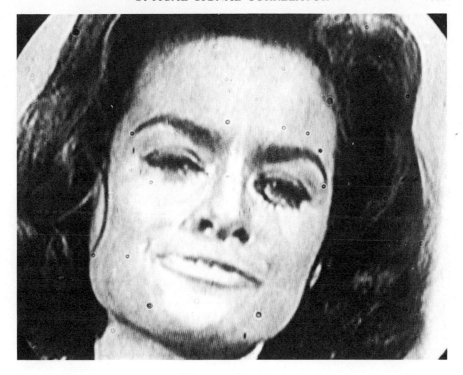

Figure 7.11 Slit filter.

For the special case when the two periodic functions are the same the correlation function can be expressed by

$$\phi_{11}(\tau) = \frac{1}{T} \int_{-T/2}^{T/2} f_1(t) f_1(t+\tau) \, dt \qquad (7\text{-}2)$$

In this case the correlation function $\phi_{11}(\tau)$ is called an _autocorrelation_.

The subscripts $_{11}$ indicate that an autocorrelation is obtained by comparing a function with itself.

The above equations define the correlation function for periodic functions. In practice one never encounters the truly periodic function that exists over the range of its variable from $-\infty$ to $+\infty$. If a function exists over a finite range of its variable (say from $-T/2$ to $T/2$) and is zero outside this range, it can be considered to have a period that approaches ∞. Such functions are called _aperiodic_ and the cross-correlation function corresponding to such aperiodic functions is defined as

$$\phi_{12}(\tau) = \int_{-\infty}^{\infty} f_1(t) f_2(t+\tau) \, dt \qquad (7\text{-}3)$$

Figure 7.12   0.014 in. diameter thick pinhole filter.

and the autocorrelation function corresponding to an aperiodic function is defined as

$$\phi_{11}(\tau) = \int_{-\infty}^{\infty} f_1(t)f_1(t+\tau)\ dt \tag{7-4}$$

The correlation expressed mathematically above can be performed optically. A photograph of an optical signal correlator is shown in Figure 7.3. Schematically an optical correlator is shown in Figure 7.14. An optical correlator is basically an extension of the principles discussed in optical filtering. It is quite apparent that lenses 1, 2, and 3 in Figure 7.14 form the basic optical filter in Figure 7.1. Lens 1 collimates the light from a point source (i.e., lens 1 changes the spherical wavefronts of a point source to plane wavefronts). To produce collimated light the point source must be at the front focal point of lens 1. The plane wavefronts incident upon the object plane uniformly illuminate it and a diffraction pattern is produced by the object (in the object plane). When the object plane is in the front focal plane of lens 2, a Fourier transform of the object amplitude transmission function is produced in the back focal plane of lens 2

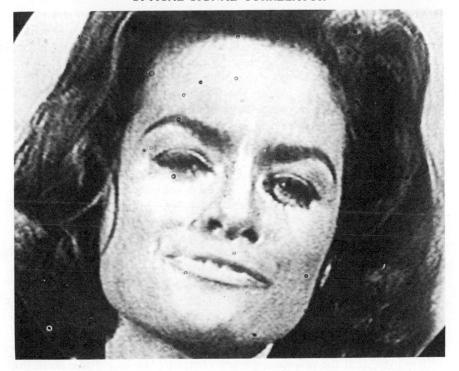

Figure 7.13   0.014 in. diameter thin pinhole filter.

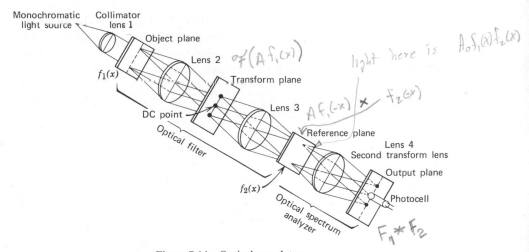

Figure 7.14   Optical correlator.

(transform plane). Lens 3 is positioned so that the transform plane is in its front focal plane. The Fourier transform of the Fourier transform in the transform plane is produced in the back focal plane (reference plane) of lens 3. By performing an inverse Fourier transform, the original object transmission function has been reproduced; lenses 2 and 3 therefore form a projection system that images the transmission function of the object plane onto the reference plane. To this point the operation of the optical correlator is identical to that of an optical filter. As will be shown later this optical filtering characteristic, which is inherent in an optical correlator, can be used to advantage.

Rather than photographing the image in the reference plane as would be done in an optical filter, the image is used as the incident light on a second transmission function inserted at the reference plane. The amplitude of the light transmitted by the reference plane will be the product of the incident light amplitude and the transmission function in the reference plane. At this point a mathematical derivation will be useful to demonstrate the properties of the light amplitude transmitted by the reference plane. (For simplicity only one dimensional signal will be considered.) We let the constant amplitude of the plane waves incident on the object plane (Figure 7.14) be given by A_0. The light transmitted by the object plane is then the product $A_0 f_1(x)$, where $f_1(x)$ is the transmission function in the object plane. It is this transmitted light at the object plane that is imaged at the reference plane. Therefore the amplitude, A, of the incident light at the reference plane is given by

$$A = A_0 f_1(x) \qquad (7\text{-}5)$$

Now, as in the case of the object plane, the light transmitted by the reference plane will be the product of the incident light amplitude $A_0 f_1(x)$ and the transmission function $f_2(x)$ in the reference plane. The expression for the transmitted light from the reference plane then becomes $[A_0 f_1(x)]$ $[f_2(x)]$. From this last expression it is apparent that the light present beyond the reference plane is the same as that which would be present if the original light (A_0) were incident on a transmission function given by the product of $f_1(x) f_2(x)$. Now returning to the correlator shown in Figure 7.14, lens 4 is located a focal length from the reference plane and therefore produces a Fourier transform of the light transmitted by the reference plane. As was explained in the section of Fourier transforms the amplitude in the output plane (back focal plane of lens 4) is then given by the Fourier transform of the product $f_1(x) f_2(x)$ or

$$\text{fourier transform } \{A_0 f_1(x) f_2(x)\} = A_0 \int_{x_1}^{x_2} f_1(x) f_2(x) e^{-j\omega x} \, dx. \qquad (7\text{-}6)$$

In (7-3) the displaced function was $f_2(t)$. Changing the variable from a time coordinate t to a spatial coordinate x permits us to write (7-3) as

$$\phi_{12}(\tau) = \int_{-\infty}^{\infty} f_1(x) f_2(x+\tau)\, dx \tag{7-7}$$

where $f_2(x)$ is the function being displaced. This means that in Figure 7.14 the reference would be the function displaced. This is usually undesirable so we can rewrite (7-7) as

$$\phi_{21}(\tau) = \int_{-\infty}^{\infty} f_2(x) f_1(x+\tau)\, dx \tag{7-8}$$

making $f_1(x)$ the displaced function. To displace a function we, in effect, slide one function over the other. In Figure 7.15 we have two identical functions (in this case two identical photographs). In order to displace one from the other we move them along a base line; that is, we cannot rotate, change scale, or move vertically one function with respect to the other. The only permissible movement (τ) is along the base line as shown in Figure 7.15. The correlation operation shown in Figure 7.15 is auto-correlation because the reference and signal functions are identical functions; that is $f_1(x) = f_2(x)$. As one picture slides over its duplicate the amount of light transmitted will vary. The maximum amount of light will be transmitted when the images of the two negatives coincide. This point corresponds to maximum correlation. A similar situation occurs in the optical correlator. The amplitude in the output plane is then given by

$$\text{fourier transform } \{A_0 f_2(x) f_1(x+\tau) = A_0 \int_{x_1}^{x_2} f_2(x) f_1(x+\tau) e^{-j\omega x}\, dx.$$
$$\tag{7-9}$$

The equation is similar in form to the correlation function

$$\phi_{21}(\tau) = \int_{x_1}^{x_2} f_2(x) f_1(x+\tau)\, dx. \tag{7-10}$$

The constants in front of the integrals can be ignored because they only represent a scale factor with respect to intensity. The exponential in the Fourier transform (7-9) can be set equal to one by setting ω equal to zero. Dropping the constant factor and equating ω to zero gives

$$\text{fourier transform } \{f_2(x) f_1(x+\tau)\}_{\omega=0} = \phi_{21}(\tau) = \int_{x_1}^{x_2} f_2(x) f_1(x+\tau)\, dx.$$
$$\tag{7-11}$$

Because setting $\omega = 0$ is equivalent to eliminating all output terms except the zero-order term, the above relation amounts to the statement that the

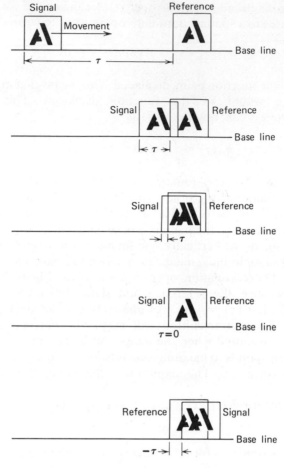

Figure 7.15   Illustration of correlation.

zero-order spectral term in the output plane of the optical correlator *is the correlation function* of the transmission functions in the object and reference planes.

For any particular displacement (τ) the zero-order output term gives the value of the correlation function corresponding to the specific value of τ. If the signal film in the object plane of the correlator is continuously displaced, the zero-order term varies as the correlation function. The zero-order term can be monitored by a photocell as shown in Figure 7.14. Because the amplitude of the zero-order term is the correlation function, the intensity sensitive photocell output will be the square of the correlation function. It is also possible to record the correlation function on film by replacing the photocell with a recording film. By moving the signal film synchronously it is possible to record a continuous picture of the correlation function; that is, for any position of the signal film with respect to the reference film (τ) there is a corresponding position of the film in the output plane and therefore the correlation function on the film is plotted against an abscissa of displacement (τ).

The operation of an optical correlator (Figure 7.14) can be summarized as the cascade operation of an optical filter and an optical spectrum analyzer with a second transmission function inserted in the plane common to the two sections. In other words an optical correlator can be made by taking the output of an optical filter and using it as the input to an optical spectrum analyzer with a transmission function inserted in the optical filter output plane (spectrum analyzer input plane). The light amplitude at the zero-order spectral point in the correlator output plane (spectrum analyzer output) is a measure of the correlation function between the transmission functions present in the object and reference planes.

In practice it is difficult to make transmission functions exactly represent desired signals. As shown in Figure 7.16 a desired sinusoidal signal variation $f_1'(x)$ can not be exactly represented as a transmission function because it would require negative values for the transmission function. In order to avoid requiring negative values for the transmission function a bias must be added to the signal. The transmission function therefore is the signal function $f_1'(x)$ plus a bias B_1. The optical correlation therefore

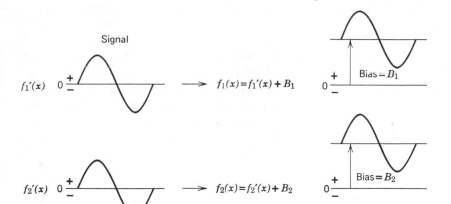

Figure 7.16 Correlation function — signal versus transmission function.

is not the correlation of the signals but the correlation of the signals represented by the transmission functions. Figure 7.16 shows the differences between the desired signals (with negative values) and the representation of these signals as transmission functions.

From Figure 7.16 we can see that the correlation function of the two signals $f_1'(x)$ and $f_2'(x)$ is given by (7-10)

$$\phi_{21}'(\tau) = \int_{x_1}^{x_2} f_2'(x) f_1'(x+\tau)\, dx \qquad (7\text{-}10\text{A})$$

The correlation of the two transmission functions representing these signals can be expressed as

$$\phi_{21}(\tau) = \int_{x_1}^{x_2} f_2(x) f_1(x+\tau)\, dx \qquad (7\text{-}10)$$

The primes are used in (7-10A) to distinguish between the actual signals that are to be correlated and their representation as transmission functions. We can now substitute in (7-10) to determine what correlation function is found when the signals are introduced with a bias (transmission functions).

$$\phi_{21}(\tau) = \int_{x_1}^{x_2} f_2(x) f_1(x+\tau)\, dx \qquad (7\text{-}10)$$

$$= \int_{x_1}^{x_2} [f_2'(x) + B_2][f_1'(x+\tau) + B_1]\, dx \qquad (7\text{-}12)$$

$$= \int_{x_1}^{x_2} f_2'(x) f_1'(x+\tau)\, dx + B_1 \int_{x_1}^{x_2} [f_2'(x) + B_2]\, dx$$

$$+ B_2 \int_{x_1}^{x_2} f_1'(x+\tau)\, dx \qquad (7\text{-}13)$$

$$\phi_{21}(\tau) = \phi_{21}'(\tau) + B_1 \int_{x_1}^{x_2} f_2'(x)\, dx + B_2 \int_{x_1}^{x_2} f_1'(x+\tau)\, dx + B_1 B_2 (x_2 - x_1)$$

$$(7\text{-}14)$$

From (7-14) we can see that the correlation of the transmission functions includes bias terms in addition to the desired correlation function $\phi_{21}(\tau)$. For a constant (τ) the bias terms in (7-14) in effect only determine the DC level of the output light amplitude; when the term $B_1 \int_{x_1}^{x_2} f_2'(x)\, dx$ is integrated and the limits $(x_1 \rightarrow x_2)$ inserted, the final value of this term will be a constant. In order to have the correlation of the transmission functions (signals plus bias) equal to the correlation function of the signals we would like

$$\phi_{21}(\tau) = \phi_{21}'(\tau). \qquad (7\text{-}15)$$

We can accomplish this by blocking the light corresponding to the three undesirable terms in (7-14). Examination of the three undesirable terms indicates that a DC stop in the transform plane of the optical filter section will eliminate all but the desired correlation function; that is, each of these three undesirable terms is a constant.

In order to make the point more clear let us examine (7-14), assuming the optical correlator configuration as shown in Figure 7.14; that is, $f_1(x)$ is in the object plane. With $f_1(x)$ in the object plane a DC stop in the transform plane will block all the light corresponding to the bias term B_1 [see Figure (7.16)]. The DC stop therefore eliminates the two terms in (7-14) containing B_1 since $B_1 = 0$. The remaining undesirable term in (7-14) is $B_2 \int_{x_1}^{x_2} f_1'(x+\tau)\, dx$ which is a constant. This term corresponds to the constant value of the signal $f_1'(x)$ and is therefore also blocked (made equal to zero) by the DC stop. By using a DC stop in the transform plane, the optical correlator output will be equal to the correlation function of the two signals inserted as transmission functions (signals plus bias) regardless of which signal is placed in the object plane and which is placed in the reference plane.

The light source shown in Figure 7.14 is a monochromatic point source of light. We can therefore assume that when collimating lens 1 is a focal distance from the point source, a collimated light beam will (uniformly) illuminate the object plane. This light is therefore temporally and spatially coherent (because it is monochromatic light coming from a point source). In Figure 7.17 a mercury arc is shown as a light source. Although the light from a mercury arc is not monochromatic, we can filter this light to produce essentially monochromatic light. The mercury arc is placed at the focal point of a spherical reflector so as to increase the light illuminating

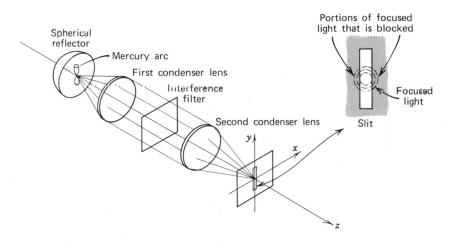

Figure 7.17   Coherence of a slit.

the slit. The first condenser lens is a focal length from the arc (point source) and thus produces collimated light. A filter is introduced into this beam of light so that only essentially monochromatic light illuminates the second condenser lens. The second condenser lens images the monochromatic light down to a point (at its focal length) in a slit. Because the focal point of the lens is not perfect (it has dimensions and depth), the slit will block some of the light (as shown in Figure 7.17) so that in the x direction the light appears to be from a point source (and therefore has a high degree of spatial coherence). In the y direction the spatial coherence is very poor. The slit therefore provides a monochromatic source of light that is spatially coherent in the x direction. A channelized optical correlator uses a slit source rather than a point source because spatial coherence is only required in one (not two) directions.

A big advantage of optical correlators over the conventional electronic correlator is the ability to handle many input signals simultaneously. A channelized optical correlator is shown in Figure 7.18. Comparison of the optical correlators of Figures 7.14 and 7.18 indicates that the basic difference (other than the type of light source) is the addition of a cylindrical lens for channelized operation.

The details of the channelized section of the optical correlator are shown in Figure 7.18. We have previously mentioned that a two-dimensional optical process can be treated as two separate one-dimensional processes. The channelized section can therefore be represented by the two processes shown as the top and side views in Figure 7.19. The top view shows that the cylindrical lens has no focusing effect (no curvature)

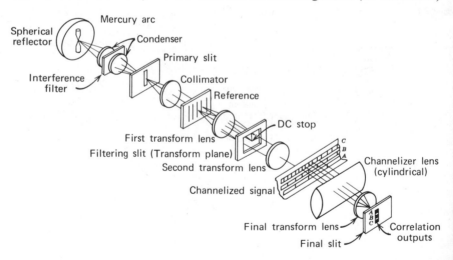

Figure 7.18 Channelized optical correlator.

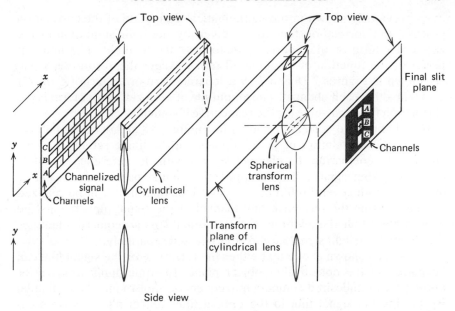

Figure 7.19 Channelized section of optical correlator.

for the x coordinate. The top view, therefore, has the configuration of a spectrum analyzer with the spherical lens producing a Fourier transform in the final slit plane because the cylindrical lens is, in effect, not being used. This Fourier transform will correspond to information signal variations along the x direction (e.g., vertical slits in channelized signal plane). Examination of the side view shows that both the cylindrical and spherical lenses have a focusing action for the y coordinates. The side view has the configuration of an optical filter system that images the information in the y direction (e.g., horizontal channels) into the final slit plane. The back focal plane of the cylindrical lens (transform plane in Figure 7.19) corresponds to the filter plane of the optical filter configuration discussed in a previous section. In effect, the cylindrical lens produces a Fourier transform in the y direction and the spherical lens produces a Fourier transform of the transform (i.e., as inverse transform) which results in an image of the original signal in the y direction.

It has been shown that the Fourier transform of the information (e.g., vertical lines) in each channel is produced by the spherical transform lens and that each channel is imaged in the final slit plane by the cylindrical lens. The imaging of the horizontal channels assures that the Fourier transform of each channel is obtained without intermixing information between channels.

As previously shown in the mathematical derivation of the correlation process, the correlation function is given by the light amplitude of the zero-order term of the Fourier transform at the final slit. The final slit performs the function of blocking all terms except the zero-order term. As shown in Figures 7.18 and 7.19 a separate zero-order term (A, B, C) is obtained for each channel. The amplitude of each zero-order term is the correlation function value for the respective channel.

The channelized correlator shown in Figure 7.18 and discussed above involves the correlation of many signals with a single reference. This amounts to comparing the channel signals with the reference signal to detect matching signals. In some applications it may be desirable to match one signal with several reference signals. This operation can be realized by channelizing the reference film instead of the signal film. It must be kept in mind that the addition of a cylindrical lens is required wherever the channelized light signals are to be processed separately.

It has been shown above that either the reference or the signal film can be placed in the optical filter object plane. In some applications prior knowledge of undesired component frequencies (noise) may be available. By placing the signal film in the optical filter object plane, noise components can be eliminated (along with the DC term) by using an appropriate optical filter. This application of optical filtering in the correlator system provides a means for improving the signal-to-noise ratio of the signal inputs.

In either of the cases considered above the amplitude of the zero-order terms in the final slit represents the correlation function. These correlation functions can be monitored by separate photocells or recorded on photographic film. The photographic record would consist of intensity modulated bands where each band corresponds to a channel. Whatever means is used to measure or record the correlation function, it must be kept in mind that the amplitude of the light in the zero-order term corresponds to the correlation function. Since the detection must be accomplished by an intensity sensitive device, the square root of the detected value must be taken if the value (amplitude) of the correlation function is desired.

MULTIPLE-IMAGE RECORDING TECHNIQUE

The principles described earlier in this chapter can possibly be used to record multiple superimposed images on a photographic film and still permit each image to be viewed separately without crosstalk (i.e., different images interfering with each other).

Let us consider the modified camera shown in Figure 7.20. The modification of the camera consists of a Ronchi ruling placed between the lens

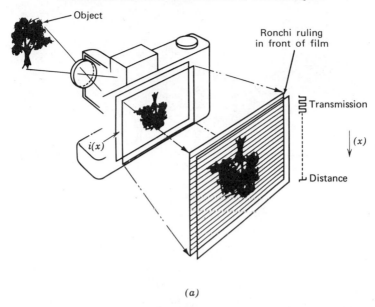

Object

Ronchi ruling
in front of film

Transmission

(x)

Distance

$i(x)$

(a)

Figure 7.20.

and the film. The Ronchi ruling is placed directly over the film so that the photographic film is exposed as the product of the Ronchi ruling and the image. The spatial frequency of the Ronchi ruling is selected so that it is higher than the highest spatial frequency required to be seen in the image. The developed photographic film, which is the product of the image $i(x)$ and the Ronchi ruling $r(x)$, can now be placed in a coherent optical processor as shown in Figure 7.21.

Figure 7.21 shows the Fourier transform of the product $i(x)r(x)$ being taken by lens 1. The Fourier transform $I(x) * R(x)$ is shown in the Fourier transform plane. Lens 2 takes the Fourier transform of the Fourier transform $I(x) * R(x)$ which will produce an inverted image of $i(x)r(x)$. In the spectrum plane the spectral points of the Ronchi ruling are obtained as described in detail in Chapter 5. Convolved about each spectral point is the spectrum $I(x)$ of the image $i(x)$.

If the DC is blocked in the spectral plane, the image formed by lens 2 will be the $i(x)$ without the Ronchi ruling superimposed. The reason for this can be seen by considering the biased square wave train shown in Figure 7.22(a). Because the light transmitted by a Ronchi ruling will vary from zero to a maximum as shown in Figure 7.22(a), it, in effect, corresponds to a biased square wave. Naturally if the bias is removed, the signal will vary between a negative and a positive maximum value as shown in

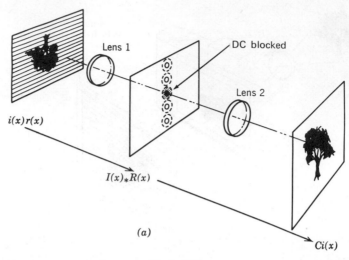

$i(x)r(x)$

$I(x)_*R(x)$

(a)

$Ci(x)$

Figure 7.21.

Figure 7.22(b). Because the Ronchi ruling corresponds to the biased pulse train, we can see that removing the bias will correspond to imaging a pulse train that has negative and positive values. Because a photographic film only responds to intensity, we must square the values of the waveform of Figure 7.22(b) to determine the image produced. We can see from Figure 7.22(c) that the square of the pulse train of Figure 7.22(b) is a constant C. We can therefore see that the Fourier transform of the DC blocked spectrum of a Ronchi ruling will be a constant. The Fourier transform of the combined spectrum $I(x) * R(x)$ with the DC blocked will produce an image $i(x)$ times a constant (the constant is formed by the square of the Fourier transform $r(x)$ with the DC blocked). Therefore by blocking the DC we can reproduce a replica of the original image without the superimposed Ronchi ruling.

We can now make other exposures as shown in Figure 7.23. Each image is made with the Ronchi ruling at a different orientation. Each exposure on the photographic film is added to the previous exposure, where each exposure is the product of a new image with the Ronchi

0 ⊓⊔⊓⊔ Block DC gives ⟹ 0 ⊓⊔⊓⊔ squaring ⟶ $\frac{C}{2}$ ▭

(a)   (b)   (c)

Figure 7.22.

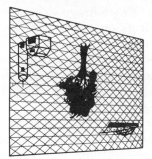

Multiple exposures are sums, but each exposure is to be the product of the Ronchi ruling (at a new orientation) with an image

Multiple exposure
$$E_1 + E_2 + \cdots E_n$$
where
$$E_1 = i_1(x) r_1(x)$$
$$\underline{E_2 = i_2(x) r_2(x)}$$
$$\overline{E_n = i_n(x) r_n(x)}$$

Figure 7.23.

ruling at a new orientation. A series of multiple exposures can be placed on the photographic film as long as the film does not saturate.

We can see from Figure 7.24 that the spectrum of the series of multiple exposures will be spectral points along lines perpendicular to their respective Ronchi rulings (all passing through the DC point). Around

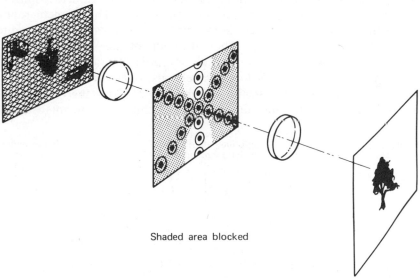

Shaded area blocked

Figure 7.24.

each spectral point the spectrum of each image will be formed; that is, the spectrum of a Ronchi ruling forms on a line (through the DC point) perpendicular to lines of the Ronchi ruling and around each spectral point (in a particular Ronchi ruling spectrum) is the spectrum of the respective image. Any particular image can be viewed by taking the Fourier transform of a particular spectrum (without the DC since the DC is convolved with the spectrums of all the images).

Different images can be taken on one photographic plate as long as the exposures are made with the Ronchi ruling oriented at angles sufficiently great so that the convolved image spectrums (around each Ronchi ruling spectral point) do not overlap and the photographic emulsion does not saturate. In addition by having three Ronchi rulings in the primary colors oriented $120°$ to each other permits a single exposure to be made simultaneously on black and white film, but each of the primary colored Ronchi rulings will expose the black and white film proportional to its color. On reconstruction the three superimposed multiple images on black and white film are illuminated with the primary colors and the three corresponding spectrums formed are therefore illuminated in the three primary colors. A filter is then used on each spectrum permitting only its corresponding color to be passed through the filter. The reconstructed image is therefore three superimposed images, in perfect registration with each other, each in a primary color.

An alternative method for reconstruction is illustrated in Figure 7.25. In this case a Ronchi ruling is oriented parallel to the Ronchi ruling in the image that it is desired to have reconstructed. This Ronchi ruling is illuminated with collimated light and lens 1 forms the spectrum in Fourier transform plane 1. The DC point of the spectrum formed is blocked so that lens 2 constructs first-order up and down wavefronts, second-order up and down wavefronts, third-order up and down wavefronts, etc., but because the DC point is blocked, there will be no zero-order wavefront. The multiple exposed film is illuminated by all

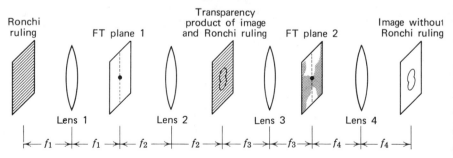

Figure 7.25 Alternate reconstruction method.

constructed wavefronts but not the zero-order wavefront. The location of the spectrum formed by lens 3 is therefore different than if the multiple exposed film were illuminated with a zero-order wavefront (as is usually done in a spectrum analyzer or correlator). The spectrum formed, for example, by the first-order down wavefront (illuminating the multiple exposed image) will be a spectrum shifted down one order with respect to the spectrum produced by a zero-order illumination wavefront; that is, its DC point appears at the normal first-order down point and its first-order up point appears at the normal DC point. We can see therefore that at the usual DC point we will have superimposed image spectrums in registration, formed by each of the wavefronts illuminating the multiple exposed image. Taking the Fourier transform of the DC point by lens 4 reconstructs the desired image without the superimposed Ronchi ruling or crosstalk. The zero order in the Fourier transform plane of lens 1 was blocked to prevent crosstalk when multiple images are recorded. Techniques similar to those described previously can be used to reconstruct a colored image.

Figures 7.26 through 7.35 illustrate the multiple imaging technique. Figure 7.26 shows a print of a negative which had four superimposed

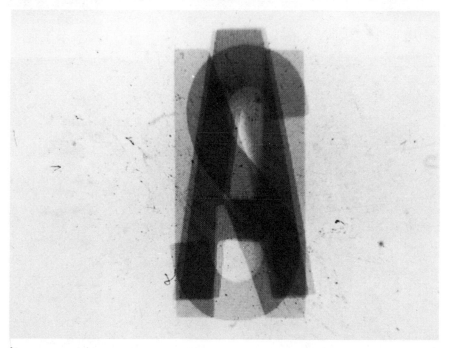

Figure 7.26 Four superimposed images printed using incoherent light.

Figure 7.27 The four superimposed images of Fig. 7.26 printed using coherent light.

Figure 7.28 Defocused spectrum of image in Fig. 7.27.

Figure 7.29   Image 1 of Fig. 7.27.

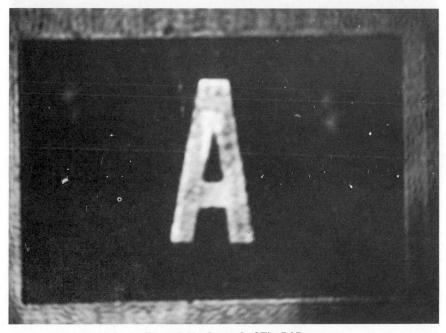

Figure 7.30   Image 2 of Fig. 7.27.

Figure 7.31 Image 3 of Fig. 7.27.

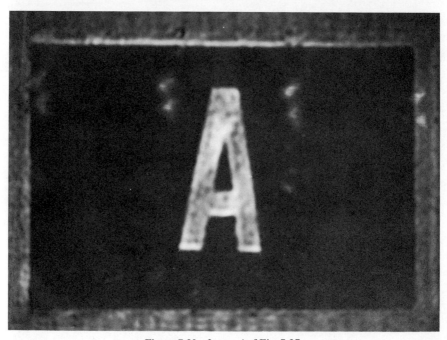

Figure 7.32 Image 4 of Fig. 7.27.

Figure 7.33 A double exposure of two continuous tone pictures.

images on it (four separate overlapping exposures). This photograph was printed using incoherent light. Figure 7.27 is a print of the same negative used for Figure 7.26 but printed using coherent light without any attempts to remove the undesirable diffraction patterns (caused by dust, lens bubbles, etc.). Figure 7.28 shows the defocused spectrum of the super-imposed image of Figure 7.27 (that is this image was not taken in the

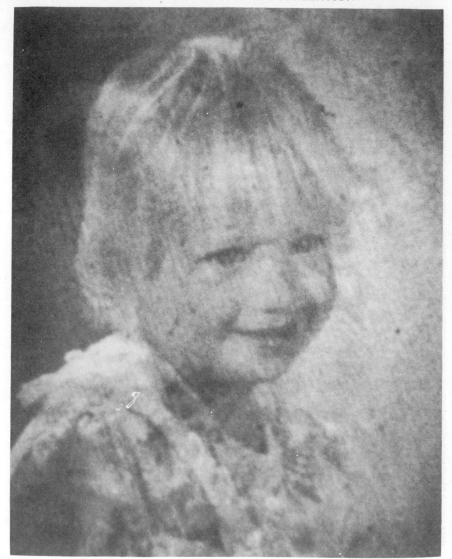

Figure 7.34   One image separated out from negative of Fig. 7.33.

transform plane). Each of the letters in this defocused spectrum would appear as a point of light when in focus. Figures 7.29 through 7.32 are the separated images of each of the four exposures. The letters NASA (two separate A exposures were made) in each of the separated images do show some crosstalk (evidence of the other exposures). The reason for this is primarily because of the fact that the original four exposures were

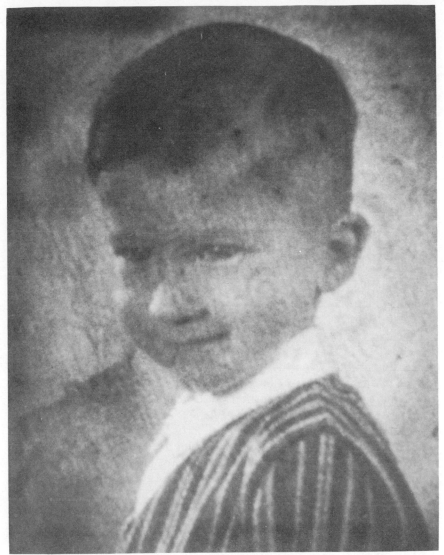

Figure 7.35 One image separated out from negative of Fig. 7.33.

not balanced and there was some saturating of the film. These images
were basically of a binary nature; that is, either clear or opaque. Figure
7.33 is a double exposure of two continuous tone images printed using
coherent light. Figures 7.34 and 7.35 show the two images separated.
Again no effort was made to eliminate the undesirable diffraction patterns.

8

Analysis of Optical Data Processing Systems

DEVELOPMENT OF BLOCK-DIAGRAM APPROACH

Now that we have completed a basic review of optics and optical data processing techniques we will attempt a general analysis of optical transform systems. In this analysis a building-block approach is used to facilitate the mathematical development of some of the principles involved in optical systems.

The reversibility characteristics of Fourier transforms are shown by the flow diagram in Figure 8.1. The double path represented by oppositely directed arrows between the functions $f(x)$ and $F(\omega)$ indicates that each is given by the Fourier transform of the other. In other words the Fourier transform of $f(x)$ is $F(\omega)$ and the Fourier transform (inverse transform) of $F(\omega)$ is $f(x)$. A single arrow from the amplitude spectrum $F(\omega)$ to the power spectrum $P(\omega)$ indicates that this process is not reversible. When going from the complex amplitude spectrum to the power spectrum, the phase is lost. Since no phase information is available in the power spectrum, it is impossible to determine the phase of the amplitude spectrum from the power spectrum. This irreversibility holds in any system that uses a square law detector to monitor the power of a signal with complex amplitude (e.g., in optics, detectors such as photographic films, photocells, the eye, etc.) because the phase is lost in the detection process.

Figure 8.1 can be redrawn using the optical terms discussed in earlier chapters. An optical flow diagram is shown in Figure 8.2. This flow diagram has the same reversibility characteristics as the Fourier transform diagram shown in Figure 8.1. The Fourier transform operation is performed by the focusing action of a lens on the diffraction pattern as explained earlier. The photodetector senses only intensity and therefore acts on the complex amplitude spectrum $[F(\omega)$ to give the power spectrum $P(\omega)]$. The operation of the photodetector on the complex

358

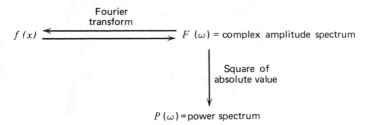

$f(x)$ $\xleftarrow{\text{Fourier transform}}$ $F(\omega)$ = complex amplitude spectrum

Square of
absolute value

$P(\omega)$ = power spectrum

Figure 8.1   Fourier transform flow diagram.

amplitude spectrum is irreversible because it loses the phase information contained in the complex amplitude spectrum.

From the previous discussions of diffraction it can be seen that the operations represented by the oppositely directed arrows in Figures 8.1 and 8.2 are performed when light passes from the front focal plane to the back focal plane of a lens. This immediately suggests that a block-diagram representation of an optical system can be developed as shown in Figure 8.3. A typical optical transform arrangement is shown in Figure 8.3A. Figure 8.3B shows a block diagram corresponding to this optical system. In this diagram regular arrows ($\rightarrow$) represent the flow of information (as light amplitude) into or out of a function block and diamond head arrows ($\downarrow$) represent information (e.g., films) inserted into, or extracted from, a focal plane. Examination of the system illustrated in Figure 8.3A will show that the signal film plane is in the front focal plane of the transform lens. The signal film is also shown in the back focal plane of the collimator lens. It is not necessary to have the signal film at a focal length from the collimator lens (i.e., the requirement for the Fourier transform is that the signal film be uniformly illuminated, which will occur at any distance from the collimator lens because the light is collimated by it). In Figure 8.3B it is not clear to which lens the front focal plane label

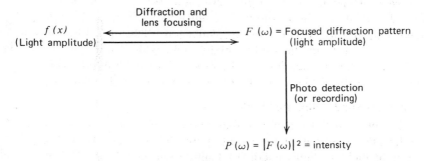

$f(x)$
(Light amplitude) $\xleftarrow{\text{Diffraction and lens focusing}}$ $F(\omega)$ = Focused diffraction pattern
(light amplitude)

Photo detection
(or recording)

$P(\omega) = |F(\omega)|^2$ = intensity

Figure 8.2   Optical flow diagram.

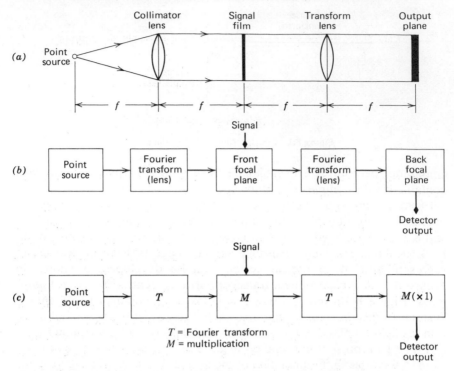

Figure 8.3 Development of an optical block diagram. (*a*) Typical optical transform arrangement. (*b*) Block diagram of (*a*). (*c*) Functional block diagram.

refers. To eliminate this confusion and at the same time improve the functional significance of the block-diagram approach, it is only necessary to recall that the light amplitude transmitted by a signal film is the *product* of the incident light amplitude and the signal transmission function. It is immediately apparent that the focal plane notation used in Figure 8.3B can be replaced by a "multiply" notation. We can now represent our typical optical system as shown in Figure 8.3C. Note that if there is no signal film in a focal plane, the transmission function is equal to one as indicated by (*x*1) in the output plane represented in Figure 8.3C. To facilitate further applications of the block-diagram symbols developed in Figure 8.3, Table 8.1 lists all symbols required in our following discussions.

Point Source and Collimated Light

It has previously been shown that a point source at the front focal point of a lens produces collimated light beyond the lens. This principle is used to obtain incident light of constant amplitude and phase in the optical systems discussed here. We will now investigate this principle referring

TABLE 8.1 OPTICAL BLOCK DIAGRAM SYMBOLS

Symbol	Description
	Signal Input. Usually inserted as a film transmission function. Can only be inserted into a multiplier block.
	Signal Output. Photographically recorded or photodetected. Note that output is Intensity = Amplitude^2 and must be taken from a multiplier block.
	Flow Path. Arrow indicates direction of information flow (direction of light propagation).
( ) Source	*Light Source.* Label specifies type of source (e.g., point source). Note that there are no inputs.
T	*Fourier Transform.* Output light amplitude is the Fourier Transform of the input light amplitude.
M	*Multiplier.* Output light amplitude is the product of the input light amplitude and the inserted transmission function. Note that if no film is present, the transmission function is equal to one.

to the block-diagram representation shown in Figure 8.4, which is used to obtain light of constant amplitude and phase. We can see in the table of Fourier transforms (Appendix 13) that a delta function at $x = 0$ transforms as shown in Figure 8.5. We can relate the Fourier transform of a delta function shown in Figure 8.5 to that of the Fourier transform of a point source. If we consider the point source to be a delta function, its Fourier transform is light of constant amplitude and phase.

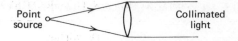

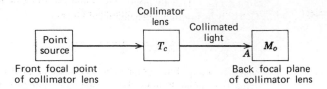

Figure 8.4 Point source and collimator lens.

We will now consider in more detail an impulse function (delta function) that can be used to describe mathematically an ideal point light source. An impulse is defined as the limit approached when the width of a square pulse is made to approach zero while the area of the pulse is held constant. If we consider a square pulse of width a and amplitude A/a, the area of the pulse will be A (constant). This type of square pulse is shown in Figure 8.6. In order to maintain constant area the amplitude A/a increases as the width a decreases. As the width a approaches zero the amplitude becomes infinite. Mathematically the amplitude $S(x)$ of an impulse may be expressed as a function of x by the expression

$$S(x) = \lim_{a \to 0} \frac{A}{a} W(x) \tag{8-1}$$

where A is a constant; a is the width of pulse; $W(x)$ is one where $|x|$ is less than $a/2$ and zero where $|x|$ is greater than $a/2$. The term $W(x)$

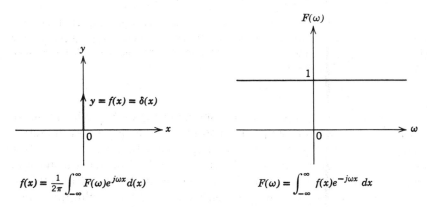

Figure 8.5 Fourier transform of a delta function.

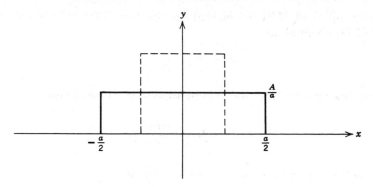

Figure 8.6 Square wave approaches impulse as $a \to 0$.

defines the width of the pulse since the amplitude $S(x)$ given by (8-1) will be zero when $W(x)$ is zero.

Because an ideal point source has zero width, we can use the impulse function given by (8-1) to describe the light amplitude of an ideal point source. The Fourier transform of the light amplitude can then be written

$$S(\omega) = \int_{-\infty}^{\infty} S(x)e^{-j\omega x}\,dx$$

$$S(\omega) = \int_{-\infty}^{\infty} \lim_{a \to 0} \frac{A}{a} W(x)e^{-j\omega x}\,dx \qquad (8\text{-}2)$$

Because the integration is with respect to x, we can take the limit and amplitude terms outside the integral sign

$$S(\omega) = \lim_{a \to 0} \frac{A}{a} \int_{-\infty}^{\infty} W(x)e^{-j\omega x}\,dx \qquad (8\text{-}3)$$

Because $W(x)$ is equal to zero for $|x| > a/2$, we need consider the integral only between the limits $-a/2$ and $a/2$, where $W(x)$ is equal to one

$$S(\omega) = \lim_{a \to 0} \frac{A}{a} \int_{-a/2}^{a/2} e^{-j\omega x}\,dx \qquad (8\text{-}4)$$

Performing the integration and evaluating at the limits we obtain

$$S(\omega) = \lim_{a \to 0} \frac{A}{a}\left[-\frac{1}{j\omega}\Big|e^{-j\omega x}\Big|_{-a/2}^{a/2}\right]$$

$$S(\omega) = \lim_{a \to 0} \frac{A}{a}\left[\frac{-e^{-j(\omega a/2)} + e^{j(\omega a/2)}}{j\omega}\right] \qquad (8\text{-}5)$$

We can simplify (8-5) by noting that the sine function can be expressed in terms of exponentials as

$$\sin \theta = \frac{e^{j\theta} - e^{-j\theta}}{2j}.$$

Substituting $\sin (a\omega/2)$ for the exponentials in (8-5) we obtain

$$S(\omega) = \lim_{a \to 0} A \frac{\sin a\omega/2}{a\omega/2} \qquad (8\text{-}6)$$

If we apply the limit as a goes to zero, we obtain

$$S(\omega) = A \qquad (8\text{-}7)$$

Since

$$\frac{\sin a\omega/2}{a\omega/2} \to 1$$

as a approaches zero. This can be seen by recalling that $\sin \theta \cong \theta$ for small angles, that is,

$$\frac{\sin a\omega/2}{a\omega/2} = \frac{a\omega/2}{a\omega/2} = 1$$

We have thus shown that an ideal point source described as an impulse

$$S(x) = \lim_{a \to 0} \frac{A}{a} W(x) \qquad (8\text{-}8)$$

will produce light of constant amplitude A and phase when passed through a collimator T_c as shown in Figure 8.4.

It was shown that in the case of an ideal point source described by (8-7) as the source width a decreases, the amplitude A/a increases. It should be obvious that in practice the amplitude of a source does not increase as the width of the source is decreased. The practical case for amplitude is more closely approximated by the expression $AW(x)$, where it is assumed that the amplitude A is constant in magnitude and phase within the source width. This expression is more realistic than (8-7) because it implies that the amplitude of light in the source remains constant as the width of the source is decreased. The Fourier transform of this light amplitude $AW(x)$ results in the light amplitude

$$S(\omega) = A \int_{-a/2}^{a/2} W(x)e^{-j\omega x} \, dx = \frac{aA}{2} \frac{\sin a\omega/2}{a\omega/2} \qquad (8\text{-}9)$$

For small a

$$\frac{\sin a\omega/2}{a\omega/2} \cong \frac{a\omega/2}{a\omega/2} = 1$$

and the amplitude A can be written as

$$S(\omega) = \frac{aA}{2} \tag{8-10}$$

Now as a is made small the amplitude $S(\omega)$ in the back focal plane decreases due to the factor a that appears in (8-10).

It has been shown that constant light amplitude across the back focal plane of an ideal source is described by (8-7) above. A practical source produces approximately constant light amplitude in the back focal plane of a collimator lens and is described by (8-10). This approximation was obtained by (a) neglecting amplitude and phase variations across the width of the source and (b) neglecting the variations due to the coordinate ω in the back focal plane [by assuming the approximation $\sin(a\omega/2)/(a\omega/2) = 1$]. These approximations improve as the source width is decreased. In our following analysis we will assume that the collimator lens produces constant light amplitude and phase across the back focal plane of the collimator lens.

OPTICAL SPECTRUM ANALYZER

The simplest configuration for an optical data processor is that of a spectrum analyzer as shown in Figure 8.3 and repeated in Figure 8.7 for convenience. The output of the collimator, T_c, is assumed to be a constant A as discussed in the last section. We shall designate the signal input to the multiplier M_0 as $f(x)$. The output of M_0 will then be given by the product $Af(x)$. This product $Af(x)$ is operated on by a lens producing the Fourier transform (T_1) of the input $Af(x)$. The Fourier transform, T_1, of $Af(x)$ is

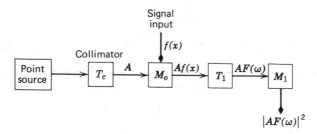

Figure 8.7 Optical spectrum analyzer.

$AF(\omega)$. The output of T_1 is shown to be $AF(\omega)$ and is then the input to multiplier M_1. Since no other signal is inserted into M_1, the output of M_1 is also $AF(\omega)$—that is, multiplication by one. The detected output will then be $|AF(\omega)|^2$ (see Figure 8.2) which is the power spectrum for the signal $f(x)$.

We shall now consider some of the details of the transform operation by developing the mathematics for a sample input signal. Let us consider a film with a transmission function of the form

$$f(x) = \tfrac{1}{2}[1 + \cos \omega_0(x + x_0)] \tag{8-11}$$

where ω_0 is the spatial angular frequency, x is the spatial coordinate, and x_0 is the displacement with respect to $x = 0$.

The graph of the transmission function $f(x)$ given by (8-11) is shown in Figure 8.8. A cosine function can be expressed in terms of exponentials as

$$\cos \theta = \frac{e^{j\theta} + e^{-j\theta}}{2}$$

Expressing the cosine term in exponential form the transmission function of (8-11) can be rewritten

$$f(x) = \tfrac{1}{2}[1 + \tfrac{1}{2} e^{j\omega_0(x+x_0)} + \tfrac{1}{2} e^{-j\omega_0(x+x_0)}] \tag{8-12}$$

Figure 8.9 shows an infinitely long cosine function reproduced from the table of Fourier transforms. The Fourier transform of such a single-frequency, infinitely long cosine transmission function produces two spectral points as shown in Figure 8.9. Because the cosine function we are considering (in Figure 8.8) has a DC component, we can expect it to have three spectral points (a) DC component and the two shown in Figure 8.9 at $\pm \omega_0$).

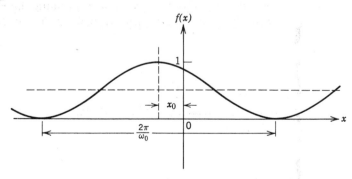

Figure 8.8 Graph of $f(x) = \tfrac{1}{2}[1 + \cos \omega_0(x + x_0)]$.

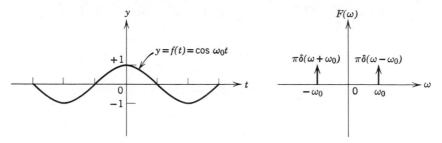

Figure 8.9 Fourier transform of a cosine function $y = \cos \omega_0 t$.

   Intuitively we might expect each of the three terms in (8-12) to correspond to one of the spectral points. We can prove that this is so by taking the Fourier transform of $Af(x)$. We will maintain the order of the three terms in (8-12) so that we may examine our results for correspondence between each term and a spectral point. The Fourier transform operation on $Af(x)$ is represented by T_1 and is given by the expression

$$AF(\omega) = \int_{-\infty}^{\infty} Af(x) \, e^{-j\omega x} \, dx$$

Letting A equal one and substituting the expression for $f(x)$ given by (8-12) we obtain

$$AF(\omega) = \int_{-\infty}^{\infty} \tfrac{1}{2}\left(1 + \tfrac{1}{2} e^{j\omega_0(x+x_0)} + \tfrac{1}{2} e^{-j\omega_0(x+x_0)}\right) e^{-j\omega x} \, dx \qquad (8\text{-}13)$$

It was previously explained that the limits of integration can be replaced by the aperture limits because $f(x)$ is zero beyond the aperture dimensions. Therefore letting $2a$ equal the length of the aperture and multiplying out the terms in (8-13) we get

$$AF(\omega) = \int_{-a}^{a} \tfrac{1}{2} e^{-j\omega x} \, dx + \int_{-a}^{a} \tfrac{1}{4} e^{j\omega_0(x+x_0) - j\omega x} \, dx + \int_{-a}^{a} \tfrac{1}{4} e^{-j\omega_0(x+x_0)} \, j\omega x \, dx$$

$$(8\text{-}14)$$

We can now factor the functions appearing under the integral and place constant factors outside the integral

$$AF(\omega) = \frac{1}{2} \int_{-a}^{a} e^{-j\omega x} \, dx + \frac{e^{j\omega_0 x_0}}{4} \int_{-a}^{a} e^{jx(\omega_0 - \omega)} \, dx + \frac{e^{-j\omega_0 x_0}}{4} \int_{-a}^{a} e^{-jx(\omega_0 + \omega)} \, dx$$

$$(8\text{-}15)$$

Performing the integration and evaluating at the limit we obtain

$$AF(\omega) = \frac{-e^{-ja\omega} + e^{ja\omega}}{2j\omega} + \frac{e^{j\omega_0 x_0}}{4}\left[\frac{e^{ja(\omega_0 - \omega)} - e^{-ja(\omega_0 - \omega)}}{j(\omega_0 - \omega)}\right]$$

$$+ \frac{e^{-j\omega_0 x_0}}{4}\left[\frac{-e^{-ja(\omega_0 + \omega)} + e^{ja(\omega_0 + \omega)}}{j(\omega_0 + \omega)}\right] \qquad (8\text{-}16)$$

We can simplify (8-16) by noting that the sine function can be written in terms of exponentials as

$$\sin \theta = \frac{e^{j\theta} - e^{-j\theta}}{2j}$$

Using the exponential definition for sine we can rewrite (8-16) as

$$AF(\omega) = \frac{\sin a\omega}{\omega} + \frac{e^{j\omega_0 x_0}}{2} \frac{\sin a(\omega_0 - \omega)}{(\omega_0 - \omega)} + \frac{e^{-j\omega_0 x_0}}{2} \frac{\sin a(\omega_0 + \omega)}{(\omega_0 + \omega)} \tag{8-17}$$

We can change each term in (8-17) into the familiar form $\sin \theta/\theta$ by multiplying $AF(\omega)$ by a/a

$$AF(\omega) = a \left[\frac{\sin a\omega}{a\omega} \right] + \frac{ae^{j\omega_0 x_0}}{2} \left[\frac{\sin a(\omega_0 - \omega)}{a(\omega_0 - \omega)} \right] + \frac{ae^{-j\omega_0 x_0}}{2} \left[\frac{\sin a(\omega_0 + \omega)}{a(\omega_0 + \omega)} \right] \tag{8-18}$$

The curve for $\sin ax/ax$ versus x is shown in Figure 8.10. It is seen that the function has its greatest peak for x equal to zero. The amplitudes of the higher-order peaks are seen to drop off rapidly. It can therefore be assumed that an approximation for the function $\sin ax/ax$ can be made by considering only the maximum peak centered at $x = 0$ with a total width of $2\pi/a$. Noting that we maintained the order of terms in our development of (8-18) we can match corresponding terms between (8-18) and (8-12),

$$AF(\omega) = a \left[\frac{\sin a\omega}{a\omega} \right] + \frac{ae^{j\omega_0 x_0}}{2} \left[\frac{\sin a(\omega_0 - \omega)}{a(\omega_0 - \omega)} \right] + \frac{ae^{-j\omega_0 x_0}}{2} \left[\frac{\sin a(\omega_0 + \omega)}{a(\omega_0 + \omega)} \right] \tag{8-18}$$

$$f(x) = \tfrac{1}{2} \left[1 + \tfrac{1}{2} e^{j\omega_0(x+x_0)} + \tfrac{1}{2} e^{-j\omega_0(x+x_0)} \right] \tag{8-12}$$

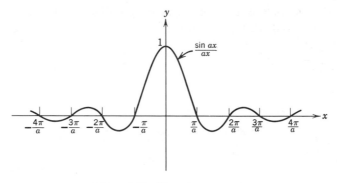

Figure 8.10   Sin ax/ax versus x.

to see that each term of (8-12) does correspond to a particular spectral point given in (8-18). If we note that the constant term $(+1)$ in (8-12) can be expressed as $e^{\pm j0}$, then we can say that each exponential term, $e^{j\omega_n x}$, in the expansion of $f(x)$ produces a spectral point at $\omega = \omega_n$. It should be apparent that this statement applies to the general case since no restrictions were placed on ω_0 in the above example. Any signal consisting of component frequencies can be expanded similarly to (8-12). Each component frequency, ω_n, will result in exponentials of the form $e^{j\omega_n(x + x_n)}$ and $e^{-j\omega_n(x + x_n)}$ as in (8-12). These exponential terms produce spectral points at $\omega = \omega_n$ and $\omega = -\omega_n$ as proved above. Thus each frequency component ω_n produces two spectral points, one at $\omega = \omega_n$ and another at $\omega = -\omega_n$.

The resolution of spectral points is dependent on the aperture dimension $2a$. It is shown in Figure 8.10 that the width of a spectral point can be considered to be $2\pi/a$—that is, the main peak. This width specification means that two adjacent spectral points will begin to overlap when the difference between the corresponding frequencies becomes less than $2\pi/a$. In conclusion it should be noted that the detected output of a spectrum analyzer will be the square of the amplitude curves illustrated in Figure 8.10 (since intensity equals the amplitude squared).

Figure 8.11 is a diagram of the transform operation. If we use only the

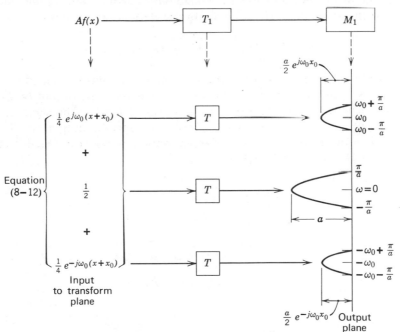

Figure 8.11 Diagram of transform operation.

center main peaks we can see how each spectral point is produced. If we consider more than just the main peaks of the respective $(\sin x/x)$ curves, we will obtain other points in the spectral plane corresponding to the higher-order peaks. In addition we could have originally used a finite interval cosine function (more representative of a cosine function in an aperture) rather than the infinitely long cosine function and arrived at similar results. Both the finite interval and infinite interval cosine functions are shown in the table of Fourier transforms (Appendix 13).

Optical Filtering

An optical filter system is diagramed in Figure 8.12. This optical filter configuration is an extension of the spectrum analyzer shown in Figure 8.7. The addition of a transform lens, T_2, and a multiplier, M_2, converts the spectrum analyzer to an optical filter system. We have seen in the description of the spectrum analyzer (Figure 8.7) that the output of T_1 is $AF(\omega)$. In the optical filter configuration the output of T_1 is not detected by M_1 as in the spectrum analyzer but is multiplied by a function $G(\omega)$. The function $G(\omega)$ is a filter function that is inserted into M_1 as shown in Figure 8.12. The output of M_1 is the product of its two inputs—that is, $AF(\omega) \times G(\omega)$. The output product of M_1 is $AF(\omega)G(\omega)$ and is applied to the input of the transform T_2. To simplify our notation we will adopt the shorthand notation $AF_f(\omega)$ to designate $AF(\omega)G(\omega)$; that is,

$$AF_f(\omega) = AF(\omega)G(\omega)$$

where the subscript f indicates that the signal $F(\omega)$ has been modified or filtered. Continuing this notation we can designate the output of T_2 as $AF_f(x)$ where $f_f(x)$ is the Fourier transform of $F_f(\omega)$. The significance of the subscript f becomes clear if we consider the output amplitude $AF_f(x)$ as the filtered form of the input amplitude $AF(x)$. The detected output of M_2 will be $|Af_f(x)|^2$.

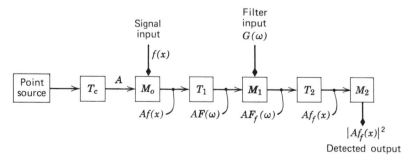

Figure 8.12 Optical filter system.

The basic operation of the optical filter system shown in Figure 8.12 can be explained by examining the light amplitude $Af_f(x)$ present at multiplier M_2. In order to simplify the discussion let us first let the amplitude of the light A equal one. Let us also assume that there is no filter signal present, that is, $G(\omega) = 1$. With these assumptions, the output of M_1 will be $F_f(\omega) = F(\omega)$ and the output of T_2 will be $f_f(x) = f(x)$. Therefore when there is no filter signal present, the light amplitude variations at the input of M_2 will be identical to the signal input $f(x)$ and is imaged onto the input of the multiplier M_2. Referring to Figure 8.3 we can see that the input to M_2 is in effect the output plane. In this case the detected output of M_2 would be $|f(x)|^2$. It is apparent that if a filter signal $G(\omega)$ is inserted into M_1, the output of M_1 is effectively a modified (or filtered) spectrum of $f(x)$. The usual filter input signal $G(\omega)$ is inserted as a film with a transmission function that has values of one or zero depending on the coordinate ω. In effect the filter input signal $G(\omega)$ blocks (multiplies by zero) any frequency component of $f(x)$ that corresponds to the values of ω for which $G(\omega)$ is equal to zero. Thus the light amplitude $f_f(x)$ at M_2 represents the filtered form of the input signal $f(x)$.

To examine some of the details of optical filter operation we can again consider the input signal $f(x)$ given as

$$f(x) = \tfrac{1}{2}[1 + \cos \omega_0(x + x_0)] = \tfrac{1}{2}[1 + \tfrac{1}{2}e^{j\omega_0(x + x_0)} + \tfrac{1}{2}e^{-j\omega_0(x + x_0)}] \qquad (8\text{-}19)$$

The light amplitude at the output of M_0 (for any constant A) will be

$$Af(x) = \frac{A}{2}[1 + \cos \omega_0(x + x_0)] = \frac{A}{2}[1 + \tfrac{1}{2}e^{j\omega_0(x + x_0)} + \tfrac{1}{2}e^{-j\omega_0(x + x_0)}] \qquad (8\text{-}20)$$

The transformed input to M_1 (as shown for (8-18) of the optical spectrum analyzer section) will be

$$AF(\omega) = aA\left[\frac{\sin a\omega}{a\omega} + \frac{e^{j\omega_0 x_0}}{2}\frac{\sin a(\omega_0 - \omega)}{a(\omega_0 - \omega)} + \frac{e^{-j\omega_0 x_0}}{2}\frac{\sin a(\omega_0 + \omega)}{a(\omega_0 + \omega)}\right] \qquad (8\text{-}21)$$

If we use the approximation indicated by the amplitude curves shown in Figure 8.11 (neglect all peaks except the maximum), we need consider only the values of the filter function $G(\omega)$ in the neighborhood of each of the spectral points. To abbreviate our notation we will designate the value of $G(\omega)$ in a region $2\pi/a$ wide around the center point of the region — that is, $\pm\pi/a$ from the center point. When $G(\omega)$ is equal to zero at $\omega = b$, this implies that $G(\omega) = 0$ for ω in the region between $b - (\pi/a)$ and $b + (\pi/a)$. If $G(\omega)$ is equal to zero at $\omega = b$, a spectral point of $AF(\omega)$ that has its peak centered at $\omega = b$ will be blocked (multiplied by zero). Because there is a direct term for term correspondence between $AF(\omega)$

and $Af(x)$, blocking a point of $AF(\omega)$ centered at $\omega = b$ eliminates the corresponding term (e^{jbx}) in $Af(x)$. As an example we can consider the case where $G(\omega) = 0$ at $\omega = 0$. We know that because $\omega = 0$, this means that the function $G(\omega)$ is blocking the DC term. In (8-21) only the term $(\sin a\omega)/a\omega$ is centered at $\omega = 0$ and therefore this term in (8-21) is multiplied by zero. Because the $(\sin a\omega)/a\omega$ term of $AF(\omega)$, equation (8-21) is blocked (multiplied by zero), the corresponding term of (8-20) is eliminated; that is, the $A/2$ term of (8-20) is eliminated. Equation (8-21) with the $(\sin a\omega)/a\omega$ term eliminated becomes the filtered transform $AF_f(\omega)$ and can be written

$$AF_f(\omega) = \frac{aA}{2}\left[e^{j\omega_0 x_0}\left(\frac{\sin a(\omega_0 - \omega)}{a(\omega_0 - \omega)}\right) + e^{-j\omega_0 x_0}\left(\frac{\sin a(\omega_0 + \omega)}{a(\omega_0 + \omega)}\right)\right] \qquad (4\text{-}22)$$

The light amplitude $Af_f(x)$ at the input to M_2 becomes

$$Af_f(x) = \frac{A}{2}\cos \omega_0(x + x_0) \qquad (8\text{-}23)$$

since the constant term $A/2$ was eliminated by blocking the corresponding term $(\sin a\omega)/a\omega$ of $AF(\omega)$. The significance of the filter operation can be seen by comparing (8-21) and (8-22) and by comparing (8-20) and (8-23); that is

$$AF(\omega) = aA\left[\frac{\sin a\omega}{a\omega} + \frac{e^{j\omega_0 x_0}}{2}\frac{\sin a(\omega_0 - \omega)}{a(\omega_0 - \omega)} + \frac{e^{-j\omega_0 x_0}}{2}\frac{\sin a(\omega_0 + \omega)}{a(\omega_0 + \omega)}\right] \qquad (8\text{-}21)$$

$$AF_f(\omega) = \frac{aA}{2}\left[e^{j\omega_0 x_0}\frac{\sin a(\omega_0 - \omega)}{a(\omega_0 - \omega)} + e^{-j\omega_0 x_0}\frac{\sin a(\omega_0 + \omega)}{a(\omega_0 + \omega)}\right] \qquad (8\text{-}22)$$

$$Af(x) = \frac{A}{2}\left[1 + \cos \omega_0(x + x_0)\right] \qquad (8\text{-}20)$$

$$Af_f(x) = \frac{A}{2}\left[\cos \omega_0(x + x_0)\right] \qquad (8\text{-}23)$$

Some of the possible filtering operations that can be performed on the function

$$f(x) = \tfrac{1}{2}(1 + \cos \omega_0(x + x_0))$$

which we have been considering are given in Table 8.2. It should be noted that the results given in Table 8.2 are approximate because only the maximum peaks have been considered in the development. Other filtering operations such as filter functions of $G(\omega)$ with values other than just one or zero or the effects of blocking only a fraction of the region covered by a spectral point were not considered when obtaining the results shown in Table 8.2.

TABLE 8.2 EXAMPLE OF FILTERING

$$Af(x) = A[1 + \cos \omega_0(x + x_0)] = A[1 + \tfrac{1}{2}e^{j\omega_0(x + x_0)} + \tfrac{1}{2}e^{-j\omega_0(x + x_0)}]$$

| Filter Function $G(w)$ | | | Spectral Terms Blocked | | | Output at M_2 | |
At $\omega = -\omega_0$	At $\omega = 0$	At $\omega = \omega_0$	Lower Sideband	DC	Upper Sideband	Amplitude	Intensity (Detected)
1	1	1				$\dfrac{A}{2}[1 + \cos \omega_0(x + x_0)]$	$\dfrac{A^2}{4}[1 + \cos \omega_0(x + x_0)]^2$
0	0	0	X	X	X	0	0
0	1	0	X		X	$\dfrac{A}{2}$	$\dfrac{A^2}{4}$
1	0	0		X	X	$\dfrac{A}{4}e^{-j\omega_0(x + x_0)}$	$\dfrac{A^2}{16}$
0	0	1	X	X		$\dfrac{A}{4}e^{j\omega_0(x + x_0)}$	$\dfrac{A^2}{16}$
1	0	1		X		$\dfrac{A}{2}\cos \omega_0(x + x_0)$	$\dfrac{A^2}{4}\cos^2 \omega_0(x + x_0)$
1	1	0			X	$\dfrac{A}{2}[1 + \tfrac{1}{2}e^{-j\omega_0(x + x_0)}]$	$\dfrac{A^2}{4}[\tfrac{5}{4} + \cos \omega_0(x + x_0)]$
0	1	1	X			$\dfrac{A}{2}[1 + \tfrac{1}{2}e^{+j\omega_0(x + x_0)}]$	$\dfrac{A^2}{4}[\tfrac{5}{4} + \cos \omega_0(x + x_0)]$

In order to illustrate how Table 8.2 was produced let us consider the case where $G(\omega)$ is equal to zero only at $\omega = -\omega_0$ (which is the condition shown on the last line of Table 8.2). Since the $\tfrac{1}{2}e^{-j\omega_0(x + x_0)}$ term is blocked by the filtering operation of $G(\omega)$, the output amplitude as given by (8-2) becomes

$$Af(x) = \frac{A}{2}[1 + \cos \omega_0(x + x_0)] = \frac{A}{2}[1 + \tfrac{1}{2}e^{j\omega_0(x + x_0)} + \tfrac{1}{2}e^{-j\omega_0(x + x_0)}] \qquad (8\text{-}20)$$

$$Af_f(x) = \frac{A}{2}[1 + \tfrac{1}{2}e^{j\omega_0(x + x_0)}] \qquad (8\text{-}24)$$

We have already determined that the output intensity is given by the square of the absolute value (magnitude) of the amplitude. The output amplitude (input to M_2 in Figure 8.12 is shown as $Af_f(x)$. It is important to note that the output intensity (detected output of Figure 8.12) is given by the square of the absolute value (magnitude) of $Af_f(x)$ and *not* by the square of $Af_f(x)$.

The square of $Af_f(x)$ is

$$[Af_f(x)]^2 = \left(\frac{A}{2}\right)^2 [1 + \tfrac{1}{2}e^{j\omega_0(x+x_0)}]^2$$

$$[Af_f(x)]^2 = \frac{A^2}{4}[1 + e^{j\omega_0(x+x_0)} + \tfrac{1}{4}e^{j\,2\omega_0(x+x_0)}] \tag{8-25}$$

We know the intensity has a real value. By (8-25) we can see that $[Af_f(x)]^2$ has both magnitude and phase. We can therefore determine that the value of intensity is not given by (8-25). We can rewrite (8-24) making use of the fact that

$$e^{jkx} = \cos kx + j \sin kx$$

Equation (8-24) can be written

$$Af_f(x) = \frac{A}{2}[1 + \tfrac{1}{2}e^{j\omega_0(x+x_0)}] \tag{8-24}$$

$$Af_f(x) = \frac{A}{2}[1 + \tfrac{1}{2}\cos \omega_0(x+x_0) + j\tfrac{1}{2}\sin \omega_0(x+x_0)] \tag{8-26}$$

Equation (8-26) is in the form,

$$A(x) = R + jX \tag{8-27}$$

where $A(x)$ is the amplitude dependent on $x[Af_f(x)$ in (8-26)], R is the real component . . . $[A/2 + A/4 \cos \omega_0(x+x_0)$ in (8-26)], and X is the imaginary component . . . $[A/4 \sin \omega_0(x+x_0)$ in (8-26)]. The magnitude of $A(x)$ in (8-27) is given by

$$|A(x)| = \sqrt{R^2 + X^2} \tag{8-28}$$

where $|A(x)|$ means *magnitude of $A(x)$*. Now if we define intensity as the square of the magnitude of the complex light amplitude, we may write

$$I = |A(x)|^2 = R^2 + X^2 \tag{8-29}$$

Because R and X are real, the intensity given by (8-29) is real and therefore agrees with physical measurements. The right side of (8-29) can be factored giving

$$R^2 + X^2 = (R + jX)(R - jX) \tag{8-30}$$

The first term $(R + jX)$ is recognized as our amplitude $A(x)$ as given by (8-27). The second term differs only in the sign of the imaginary term and is defined as the complex conjugate of $A(x)$. The complex conjugate is usually indicated by a star; that is, $A^*(x)$ is the complex conjugate of $A(x)$. Equation (8-29) may now be written

$$I = |A(x)|^2 = A(x)A^*(x) \tag{8-31}$$

We can now determine the intensity of the complex amplitude expressed by (8-24). The intensity of the complex light amplitude given in (8-24) is

$$I = |Af_f(x)|^2 = [Af_f(x)][Af_f(x)]^* \tag{8-32}$$

When a function is written as a complex exponential, the complex conjugate is formed by changing the sign of the imaginary exponent. Using the expression for $Af_f(x)$ given in (8-24), the equation (8-32) can be written

$$I = |Af_f(x)|^2 = \left\{\frac{A}{2}[1 + \tfrac{1}{2}e^{j\omega_0(x+x_0)}]\right\}\left\{\frac{A}{2}[1 + \tfrac{1}{2}e^{-j\omega_0(x+x_0)}]\right\} \tag{8-33}$$

Carrying out the multiplication of (8-33) we obtain

$$I = |Af_f(x)|^2 = \frac{A^2}{4}[1 + \tfrac{1}{2}e^{+j\omega_0(x+x_0)} + \tfrac{1}{2}e^{-j\omega_0(x+x_0)} + \tfrac{1}{4}] \tag{8-34}$$

Because

$$\cos\theta = \frac{e^{j\theta} + e^{-j\theta}}{2}$$

then

$$\cos\omega_0(x+x_0) = \frac{e^{j\omega_0(x+x_0)} + e^{-j\omega_0(x+x_0)}}{2}$$

and (8-34) becomes,

$$I = |Af_f(x)|^2 = \frac{A^2}{4}[\tfrac{5}{4} + \cos\omega_0(x+x_0)] \tag{8-35}$$

Equation (8-35) is the expression for the detected output (intensity) for an input signal to an optical filter given by (8-19) with a filter function such that its value is zero at $\omega = -\omega_0$. We can also see that the use of the complex conjugate to determine the square of the absolute value results in an intensity expression that is real (contains no imaginary part and therefore no phase information).

The usual application of filtering is to eliminate undesired frequencies. It is apparent from examination of the output amplitude column of Table 8.2 that all traces of a given frequency are eliminated only when *both* the upper and lower spectral points (sidebands) are blocked, that is, in order to eliminate a frequency ω_0, the filter function $G(\omega)$ must be equal to zero at both $\omega = \omega_0$ and $\omega = -\omega_0$.

OPTICAL CORRELATOR

An optical correlator is shown in Figure 8.13. This optical correlator configuration is obtained by adding a transform lens, T_3, and multiplier, M_3, to the optical filter arrangement of Figure 8.12. The output of T_2 is $Af_{1f}(x)$ which corresponds to the input $Af_f(x)$ to M_2 shown in Figure 8.12. Instead of detecting an output from M_2 as in an optical filter, a reference signal $f_2(x)$ is inserted into M_2 and the output product $Af_{1f}(x)f_2(x)$ is applied to the input of the transform T_3. The transform T_3 produces the spectrum of the product, $Af_{1f}(x)f_2(x)$, in the plane represented by M_3.

In the previous discussion of the optical correlator it was pointed out that the correlation function $f_{1f}(x)f_2(x)$ appears at the output of the correlator in a position corresponding to $\omega = 0$. This means that in the output plane, M_3, of the optical correlator the portion of the spectrum, $Af_{1f}(x)f_2(x)$, which appears at the DC term is also the correlation function. The selection of the portion of the spectrum at $\omega = 0$ is performed by a low-pass optical filter. The low-pass optical filter is shown in Figure 8.13 as an input $L(\omega)$ to M_3. This low-pass filter provides an input signal $L(\omega)$ to M_3 which has a value equal to one in a very small region about $\omega = 0$ and a value equal to zero elsewhere. The light amplitude at M_3 is given by the product of $L(\omega)$ and the transform of $Af_{1f}(x)f_2(x)$. This product is the correlation function ϕ_{21} when $f_1(x)$ is the displaced signal. The detected output at M_3 will be $|\phi_{21}|^2$—that is, the absolute magnitude of the correlation function squared.

Let us examine some of the details of optical correlation by assuming an input signal of

$$f_1(x) = \tfrac{1}{4}[2 + \cos \omega_1(x + x_1) + \cos \omega_2(x + x_2)] \tag{8-36}$$

where ω_1 is not equal to ω_2 (that is, $f_1(x)$ has two frequency components).

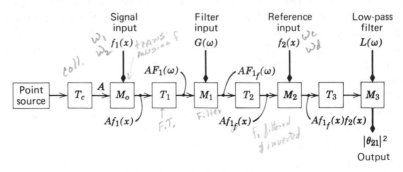

Figure 8.13   Optical correlator.

Recalling that a cosine function can be expressed in terms of exponentials, we can write (8-36) as

$$f_1(x) = \tfrac{1}{4}[2 + \tfrac{1}{2} e^{j\omega_1(x+x_1)} + \tfrac{1}{2} e^{-j\omega_1(x+x_1)} + \tfrac{1}{2} e^{j\omega_2(x+x_2)} + \tfrac{1}{2} e^{-j\omega_2(x+x_2)}] \qquad (8\text{-}37)$$

In processing data usually only the variable terms are of interest and therefore eliminating the DC term can actually be desirable. If only the variable terms of $f_1(x)$ are of interest, a DC optical stop can be inserted into M_1 of Figure 8.13. A DC stop corresponds to a filter function $G(\omega)$ that is equal to zero at $\omega = 0$. As shown in the previous section on optical filters this type of filter function eliminates the constant term of the input signal. Multiplying (8-37) by the light amplitude A and eliminating the constant term (since we are blocking it by the filter input to M_1) permit us to write (8-37) as

$$Af_{1f}(x) = \frac{A}{4} [\tfrac{1}{2} e^{j\omega_1(x+x_1)} + \tfrac{1}{2} e^{-j\omega_1(x+x_1)} + \tfrac{1}{2} e^{j\omega_2(x+x_2)} + \tfrac{1}{2} e^{-j\omega_2(x+x_2)}] \qquad (8\text{-}38)$$

From Figure 8.13 we can see that $Af_{1f}(x)$ is the signal input into multiplier M_2 (or the output of T_2). A reference signal $f_2(x)$ is also applied to M_2. The product of $Af_{1f}(x)$ times $f_2(x)$ is the output of M_2; that is, the output of M_2 is $Af_{1f}(x)f_2(x)$, where $f_2(x)$ is a reference signal. If we assume the reference signal $f_2(x)$ to be

$$f_2(x) = \tfrac{1}{4}[2 + \cos \omega_c x + \cos \omega_d x]$$
$$= \tfrac{1}{4}[2 + \tfrac{1}{2} e^{j\omega_c x} + \tfrac{1}{2} e^{-j\omega_c x} + \tfrac{1}{2} e^{j\omega_d x} + \tfrac{1}{2} e^{-j\omega_d x}] \qquad (8\text{-}39)$$

then the product $Af_{1f}(x)f_2(x)$ is given by the product of (8-38) times (8-39). Then

$$Af_{1f}(x)f_2(x) = \left\{\frac{A}{4} [\tfrac{1}{2} e^{j\omega_1(x+x_1)} + \tfrac{1}{2} e^{-j\omega_1(x+x_1)} + \tfrac{1}{2} e^{j\omega_2(x+x_2)} + \tfrac{1}{2} e^{-j\omega_2(x+x_2)}] \right\}$$

$$\times \{\tfrac{1}{4}[2 + \tfrac{1}{2} e^{j\omega_c x} + \tfrac{1}{2} e^{-j\omega_c x} + \tfrac{1}{2} e^{j\omega_d x} + \tfrac{1}{2} e^{-j\omega_d x}]\}$$

$$= \left\{\frac{A}{8} [e^{j\omega_1(x+x_1)} + e^{-j\omega_1(x+x_1)} + e^{j\omega_2(x+x_2)} + e^{-j\omega_2(x+x_2)}] \right\}$$

$$\times \{\tfrac{1}{8}[4 + e^{j\omega_c x} + e^{-j\omega_c x} + e^{j\omega_d x} + e^{-j\omega_d x}]\}$$

$$Af_{1f}(x)f_2(x) = \frac{A}{64} e^{j\omega_1(x+x_1)} [4 + e^{j\omega_c x} + e^{-j\omega_c x} + e^{j\omega_d x} + e^{-j\omega_d x}]$$

$$+ \frac{A}{64} e^{-j\omega_1(x+x_1)} [4 + e^{j\omega_c x} + e^{-j\omega_c x} + e^{j\omega_d x} + e^{-j\omega_d x}]$$

$$+ \frac{A}{64} e^{j\omega_2(x+x_2)} [4 + e^{j\omega_c x} + e^{-j\omega_c x} + e^{j\omega_d x} + -j\omega_d x]$$

$$+ \frac{A}{64} e^{-j\omega_2(x+x_2)} [4 + e^{j\omega_c x} + e^{-j\omega_c x} + e^{j\omega_d x} + e^{-j\omega_d x}] \qquad (8\text{-}40)$$

We can let A equal 16 and factor $e^{j\omega_1 x}$ and $e^{j\omega_2 x}$ as follows:

$$Af_{1f}(x)f_2(x) = \frac{e^{j\omega_1 x_1}}{4}[4e^{j\omega_1 x} + e^{j(\omega_1+\omega_c)x} + e^{j(\omega_1-\omega_c)x} + e^{j(\omega_1+\omega_d)x} + e^{j(\omega_1-\omega_d)x}]$$

$$+\frac{e^{-j\omega_1 x_1}}{4}[4e^{-j\omega_1 x} + e^{-j(\omega_1-\omega_c)x} + e^{-j(\omega_1+\omega_c)x} + e^{-j(\omega_1-\omega_d)x} + e^{-j(\omega_1+\omega_d)x}]$$

$$+\frac{e^{j\omega_2 x_2}}{4}[4e^{j\omega_2 x} + e^{j(\omega_2+\omega_c)x} + e^{j(\omega_2-\omega_c)x} + e^{j(\omega_2+\omega_d)x} + e^{j(\omega_2-\omega_d)x}]$$

$$+\frac{e^{-j\omega_2 x_2}}{4}[4e^{-j\omega_2 x} + e^{-j(\omega_2-\omega_c)x} + e^{-j(\omega_2+\omega_c)x} + e^{-j(\omega_2-\omega_d)x} + e^{-j(\omega_2+\omega_d)x}]$$

$$(8\text{-}41)$$

As shown in Figure 8.13 $Af_{1f}(x)f_2(x)$ as given in (8-41) is transformed by T_3 and then multiplied by $L(\omega)$ in M_3. Because (8-41) is rather cumbersome to carry through completely, we will make use of the filter function $L(\omega)$ to simplify this equation. It was stated above that $L(\omega)$ was zero for all values of ω other than zero; that is, the operation of filter $L(\omega)$ is to block effectively all spectral points not centered at $\omega = 0$. It was pointed out that blocking a spectral point effectively eliminates the corresponding exponential in the original signal. Because (8-41) is our signal, we can eliminate any exponential that does not produce a spectral point at $\omega = 0$. From our discussion of spectrum analyzers we found that each exponential term e^{jkx} produces a spectral distribution

$$\left[\frac{\sin a(k-\omega)}{a(k-\omega)}\right]$$

centered at $k = \omega$. We can therefore eliminate all exponentials for which the term corresponding to k is not zero. For example, in the first bracket of (8-41) the terms corresponding to k are

$$\omega_1, \; (\omega_1+\omega_c), \; (\omega_1-\omega_c), \; (\omega_1+\omega_d), \; (\omega_1-\omega_d)$$

The first, second, and fourth terms are definitely not equal to zero because we can assume that

$$\omega_1 \neq 0, \quad \omega_c \neq 0, \quad \text{and} \quad \omega_d \neq 0$$

We can therefore eliminate the first, second, and fourth terms. We can also assume that $\omega_c \neq \omega_d$ and therefore ω_1 cannot be equal to both ω_c and ω_d. Let us assume that $\omega_1 \neq \omega_d$. This implies that $(\omega_1-\omega_d) \neq 0$ and therefore it can also be eliminated. The only exponential remaining is $e^{jx(\omega_1 - \omega_c)}$ which may or may not be eliminated depending on whether ω_1 is equal to ω_c. Applying the same elimination technique to the second bracket of (8-41) we can eliminate all terms except $e^{-jx(\omega_1 - \omega_c)}$ by the same arguments. Because ω_2 cannot be equal to both ω_c and ω_d (ω_c and ω_d are unequal), we can assume ω_2 is not equal to ω_c. Proceeding with the same argument as for the first and second brackets the third and fourth brackets can be reduced to $e^{j(\omega_2 - \omega_d)x}$ and $e^{-j(\omega_2 - \omega_d)x}$, respectively. Equation (8-41) can now be written

$$Af_{1f}(x)f_2(x) = \frac{e^{j\omega_1 x_1}}{4}\left[e^{j(\omega_1 - \omega_c)x}\right] + \frac{e^{-j\omega_1 x_1}}{4}\left[e^{-j(\omega_1 - \omega_c)x}\right]$$

$$+ \frac{e^{j\omega_2 x_2}}{4}\left[e^{j(\omega_2 - \omega_d)x}\right] + \frac{e^{-j\omega_2 x_2}}{4}\left[e^{-j(\omega_2 - \omega_d)x}\right]$$

$$\text{if } \omega_1 = \omega_c$$
$$\& \ \omega_2 = \omega_d$$

(8-42)

The location of the spectral points produced by the bracketed terms in (8-42) depends on the values of ω_1 and ω_2. If ω_1 is not equal to ω_c, the first two terms produce spectral points at $\omega = \omega_1 - \omega_c$ and $\omega = -(\omega_1 - \omega_c)$, respectively. Because these points are not at $\omega = 0$, they are eliminated by the action of $L(\omega)$ and will contribute nothing to the correlator output. Likewise if ω_2 is not equal to ω_d, the last two terms are eliminated by $L(\omega)$ and contribute nothing to the correlator output. Thus if ω_1 is not equal to ω_c and ω_2 is not equal to ω_d, all terms in (8-42) are eliminated and the output of the correlator will be zero (that is, $\phi_{21} = 0$).

If ω_1 is equal to ω_c, the first two brackets in (8-42) produce spectral points at $\omega = 0$ since

$$\omega_1 - \omega_c = -(\omega_1 - \omega_c) = 0$$

Likewise if ω_2 is equal to ω_d, the last two brackets in (8-42) produce spectral points at $\omega = 0$ since

$$\omega_2 - \omega_d = -(\omega_2 - \omega_d) = 0$$

Thus for the case when $\omega_1 = \omega_c$ and $\omega_2 = \omega_d$, equation (8-42) can be written

$$Af_{1f}(x)f_2(x) = \frac{e^{j\omega_c x_1}}{4} + \frac{e^{-j\omega_c x_1}}{4} + \frac{e^{j\omega_d x_2}}{4} + \frac{e^{-j\omega_d x_2}}{4}$$

(8-43)

Because the coordinate x does not appear in any of the terms on the right side of (8-43), the transform T, produced at the output of T_3 will be the transform of a constant (that is, constant over the length of the aperture).

From the table of Fourier transforms (Appendix 13) we can see that the Fourier transform of a constant C over a limited aperture is $(2aC \sin a\omega)/(a\omega)$ where $2a$ is the length of the aperture. The transform T of $Af_{1f}(x) \times f_2(x)$ can be written directly from (8-43) as

$$T = \left[\frac{e^{j\omega_c x_1}}{2} + \frac{e^{-j\omega_c x_1}}{2} + \frac{e^{j\omega_d x_2}}{2} + \frac{e^{-j\omega_d x_2}}{2} \right] \frac{a \sin a\omega}{a\omega} \tag{8-44}$$

where the terms inside the bracket can be considered constants. Equation (8-44) is the expression for the transform performed by the third-transform lens T_3 of Figure 8.13 when only the $\omega = 0$ terms are considered; that is, equation (8-44) is the product of the transform applied to the input M_3 and the filter $L(\omega)$. The filter $L(\omega)$ has a value of one in a very small region about $\omega = 0$, which therefore eliminates all higher-order terms from the transform input to M_3 of Figure 8.13 and gives

$$TL(\omega) = \left[\frac{e^{j\omega_c x_1}}{2} + \frac{e^{-j\omega_c x_1}}{2} + \frac{e^{j\omega_d x_2}}{2} + \frac{e^{-j\omega_d x_2}}{2} \right] \frac{a \sin a\omega}{a\omega}$$

Since

$$\cos \theta = \frac{e^{j\theta}}{2} + \frac{e^{-j\theta}}{2}$$

$$TL(\omega) = [\cos \omega_c x_1 + \cos \omega_d x_2] \frac{a \sin a\omega}{a\omega} \tag{8-45}$$

as the product at M_3.

Because $L(\omega)$ is one only in a very small region about $\omega = 0$, the ω appearing in (8-45) is restricted to very small values and

$$\frac{\sin a\omega}{a\omega} \approx 1$$

[since $\sin a\omega \approx a\omega$ for small $a\omega$]. Applying this approximation to (8-45) and noting that the product TL (ω) is the correlation function ϕ_{21} we can write

$$\phi_{21} = TL(\omega) = a[\cos \omega_c x_1 + \cos \omega_d x_2] \tag{8-46}$$

The detected output $|\phi_{21}|^2$ for this case will be

$$|\phi_{21}|^2 = a^2[\cos \omega_c x_1 + \cos \omega_d x_2]^2 \tag{8-47}$$

From the result obtained above we can develop a general rule for determining correlator outputs. It was shown that a frequency component of the signal function that *does not match* a frequency component of the reference function *contributes nothing* to the correlation function (that is, contribution to ϕ_{21} is zero). In addition it was shown that a frequency component of the signal that *does match* a frequency component of the reference *contributes a cosine term* (signal components originally expressed as cosines) to the correlation function. In other words the correlation contribution of a signal component ω_n is *zero* if the frequency is *not* present in the reference and the contribution is cos $\omega_n x_n$ if the frequency ω_n *is* present in the reference. Table 8.3 lists correlation functions for various combinations of signal frequencies ω_1, ω_2 and reference frequencies ω_c, ω_d. Direct application of the rules just described produce the results given in Table 8.3. A dash in the signal frequency column indicates that the particular signal frequency is not equal to a frequency in the reference frequency. The dash was used in such cases since any frequency not in the reference would produce the same (zero) contribution. Examination of the correlation function listed in Table 8.3 shows that when the reference contains more than one frequency the correlation function can be used to determine which particular frequency, or combination of frequencies is present in the signal function. As a final note, it should be pointed out that in our treatment we have assumed all frequency components to be of the same amplitude. In practice this is not the case and an amplitude factor will appear in front of each term in all the expressions considered here.

Single Sideband Correlation

Interesting correlation results are obtained when we consider blocking (with an appropriate $G(\omega)$ in the multiplier M_1) one of the two spectral points corresponding to a signal frequency (in addition to the DC point).

TABLE 8.3 SAMPLE CORRELATION FUNCTIONS

Reference Frequencies	Signal Frequencies		Correlation Function	
	ω_1	ω_2	ϕ_{21} Amplitude	$\lvert\phi_{21}\rvert^2$ Intensity-(Detected)
ω_c & ω_d	—	—	0	0
ω_c & ω_d	ω_c	—	$a \cos \omega_c x_1$	$a^2 \cos^2 \omega_c x_1$
ω_c & ω_d	—	ω_d	$a \cos \omega_d x_2$	$a^2 \cos^2 \omega_d x_2$
ω_c & ω_d	ω_c	ω_d	$a(\cos \omega_c x_1 + \cos \omega_d x_2)$	$a^2(\cos \omega_c x_1 + \cos \omega_d x_2)^2$
ω_c	—	—	0	0
ω_c	ω_c	—	$a \cos \omega_c x_1$	$a^2 \cos_2 \omega_c x_1$
ω_d	—	—	0	0
ω_d	—	ω_d	$a \cos \omega_d x_2$	$a^2 \cos_2 \omega_d x_2$

The dash — indicates the signal frequencies are not equal to the reference frequencies

If we call the up-and-down spectral points the upper and lower sidebands, the case we are about to consider can be called *single-sideband correlation.*

We will only consider the development of the case for upper-sideband correlation. The lower-sideband case would follow the same development except for a change in the sign of the exponents. Single-sideband correlation is produced by inserting a special filter function $G(\omega)$ into M_1 (shown in Figure 8.13). For upper-sideband correlation $G(\omega)$ would have a value of one for all values of ω greater than zero and a value of zero for ω equal to or less than zero. In effect this filter function $G(\omega)$ blocks the DC spectral component plus all spectral terms on one side of the DC point. We will use the same signal function as in our previous development for (double-sideband) correlation; that is, signal function $f_1(x)$ is given as

$$f_1(x) = \tfrac{1}{4}[2 + \cos \omega_1(x + x_1) + \cos \omega_2(x + x_2)] \qquad (8\text{-}36)$$

In the previous discussion we used a $G(\omega)$ that was zero only at $\omega = 0$ (DC stop) and obtained a filtered signal at the output of T_2 (see Figure 8.13) that was given by

$$Af_{1f}(x) = \frac{A}{4}[\tfrac{1}{2}e^{j\omega_1(x + x_1)} + \tfrac{1}{2}e^{-j\omega_1(x + x_1)} + \tfrac{1}{2}e^{j\omega_2(x + x_2)} + \tfrac{1}{2}e^{-j\omega_2(x + x_2)}] \qquad (8\text{-}38)$$

In our present example $G(\omega)$, is zero at $\omega = 0$ and also at all negative values of ω. The second and fourth terms in the bracket of (8-38) produce spectral points at $\omega = -\omega_1$ and $\omega = -\omega_2$. These points are blocked by the filter $G(\omega)$ and therefore the second and fourth terms in (8-38) can be eliminated and $Af_{1f}(x)$ for single-sideband operation can be written

$$Af_{1f}(x) = \frac{A}{4}[\tfrac{1}{2}e^{j\omega_1(x + x_1)} + \tfrac{1}{2}e^{j\omega_2(x + x_2)}] \qquad (8\text{-}48)$$

We will use the reference signal $f_2(x)$ which was given in equation (8-39)

$$f_2(x) = \tfrac{1}{4}[2 + \tfrac{1}{2}e^{j\omega_c x} + \tfrac{1}{2}e^{-j\omega_c x} + \tfrac{1}{2}e^{j\omega_d x} + \tfrac{1}{2}e^{-j\omega_d x}] \qquad (8\text{-}39)$$

The product at the output of M_2 for single-sideband correlation is given by the product of (8-39) and (8-48) as

$$Af_{1f}(x)f_2(x) = \tfrac{1}{4}[2 + \tfrac{1}{2}e^{j\omega_c x} + \tfrac{1}{2}e^{-j\omega_c x} + \tfrac{1}{2}e^{j\omega_d x} + \tfrac{1}{2}e^{-j\omega_d x}]$$

$$\times \frac{A}{4}[\tfrac{1}{2}e^{j\omega_1(x + x_1)} + \tfrac{1}{2}e^{+j\omega_2(x + x_2)}]$$

$$= \frac{A}{64}[4 + e^{j\omega_c x} + e^{-j\omega_c x} + e^{j\omega_d x} + e^{-j\omega_d x}]$$

$$\times [e^{j\omega_1(x+x_1)} + e^{j\omega_2(x+x_2)}]$$

$$= \frac{A}{64} e^{j\omega_1(x+x_1)} [4 + e^{j\omega_c x} + e^{-j\omega_c x} + e^{j\omega_d x} + e^{-j\omega_d x}]$$

$$+ \frac{A}{64} e^{j\omega_2(x+x_2)} [4 + e^{j\omega_c x} + e^{-j\omega_c x} + e^{j\omega_d x} + e^{-j\omega_d x}]$$

$$(8\text{-}49)$$

If we compare (8-49) to (8-40),

$$Af_{1f}(x) f_2(x) = \frac{A}{64} e^{j\omega_1(x+x_1)} [4 + e^{j\omega_c x} + e^{-j\omega_c x} + e^{j\omega_d x} + e^{-j\omega_d x}] \quad \checkmark$$

$$+ \frac{A}{64} e^{-j\omega_1(x+x_1)} [4 + e^{j\omega_c x} + e^{-j\omega_c x} + e^{j\omega_d x} + e^{-j\omega_d x}]$$

$$+ \frac{A}{64} e^{j\omega_2(x+x_2)} [4 + e^{j\omega_c x} + e^{-j\omega_c x} + e^{j\omega_d x} + e^{-j\omega_d x}] \quad \checkmark$$

$$+ \frac{A}{64} e^{-j\omega_2(x+x_2)} [4 + e^{j\omega_c x} + e^{-j\omega_c x} + e^{j\omega_d x} + e^{-j\omega_d x}] \quad (8\text{-}40)$$

we can note that blocking a spectral term eliminates all terms dependent on the exponential corresponding to the particular spectral term. For example, the second and fourth brackets in (8-40) are multiplied by $e^{-j\omega_1(x+x_1)}$ and $e^{-j\omega_2(x_2+x_2)}$, respectively. Because these exponentials are eliminated by $G(\omega)$ (corresponding spectral terms are blocked), the second and fourth brackets of (8-40) are not present in (8-49). In order to simplify our development from this point we will make use of this elimination principle directly on the complete expression derived for the normal (double-sideband) correlation. The equation for the correlation function ϕ_{21} was given by (8-46) as

$$\phi_{21} = a [\cos \omega_c x_1 + \cos \omega_d x_2] \quad (8\text{-}46)$$

which can be written as

$$\phi_{21} = a [\tfrac{1}{2} e^{j\omega_c x_1} + \tfrac{1}{2} e^{-j\omega_c x_1} + \tfrac{1}{2} e^{j\omega_d x_2} + \tfrac{1}{2} e^{-j\omega_d x_2}] \quad (8\text{-}50)$$

when $\omega_1 = \omega_c$ and $\omega_2 = \omega_d$. In the development of (8-46) each of the terms in (8-50) corresponded to a spectral term of the input signal, $f_1(x)$. As pointed out in the comparison between equations (8-40) and (8-49),

$G(\omega)$ equal to zero for negative values of ω eliminates all terms dependent on negative exponential terms in $f_1(x)$. Thus the second and fourth
terms in the bracket of (8-50) can be eliminated and ϕ_{21} is given as

$$\phi_{21} = \frac{a}{2}[e^{j\omega_c x_1} + e^{j\omega_d x_2}] \tag{8-51}$$

when $\omega_1 = \omega_c$ and $\omega_2 = \omega_d$;

$$\phi_{21} = 0$$

when $\omega_1 \neq \omega_c$ and $\omega_2 \neq \omega_d$.

The two cases given in (8-51) are represented by the light amplitude at
the correlator output (output of M_3). The first case in (8-51) follows from
elimination of terms in (8-50) due to the action of the filter $G(\omega)$. The
second case corresponds directly to the results obtained for normal
(double-sideband) correlation. The zero correlation results apply to both
single-sideband and double-sideband correlation because the development considered the elimination of individual exponential terms.

The basic difference between single-sideband correlation and the full
double-sideband correlation of the last section can be seen by comparing
equations (8-50) and (8-51). In the single-sideband correlation given by
(8-51) we see that each frequency that appears in both the signal and the
reference contributes a single exponential to the correlation function. In
the double-sideband correlation given by (8-50) a frequency that appears
in both the signal and the reference contributes two exponentials (which
can be combined into a cosine term) to the correlation function.

For a more complete comparison the (double-sideband) correlation
functions that were listed in Table 8.3 are given for single-sideband
correlation in Table 8.4 (upper sideband) and Table 8.5 (lower sideband).
A comparison of Tables 8.4 and 8.5 shows that the single-sideband
correlation function has the same form for upper- or lower-sideband
operation. Opposite signs appear in the exponents of the correlation
functions ϕ_{21} listed in the amplitude column; a plus sign appears for
upper-sideband operation and a minus sign appears for lower-sideband
operation. Because this sign difference is eliminated in determining the
intensity (square of the absolute value of the amplitude), the detected output is exactly the same for either upper- or lower-sideband correlation.

The differences between double-sideband correlation (Table 8.3) and
single-sideband correlation (Tables 8.4 and 8.5) can be seen by direct
comparison. The most significant difference appears in the detected output for the case of one signal frequency matching one of two reference
frequencies. In single-sideband correlation a detected value of $a^2/4$ is

TABLE 8.4. SINGLE (UPPER)-SIDEBAND CORRELATION

Reference Frequencies	Signal Frequencies ω_1	ω_2	Correlation Functions ϕ_{21} Amplitude	$|\phi_{21}|^2$ Intensity (Detected)
$\omega_c \,\&\, \omega_d$	—	—	0	0
$\omega_c \,\&\, \omega_d$	ω_c	—	$\frac{a}{2}e^{j\omega_c x_1}$	$\frac{a^2}{4}$
$\omega_c \,\&\, \omega_d$	—	ω_d	$\frac{a}{2}e^{j\omega_d x_2}$	$\frac{a^2}{4}$
$\omega_c \,\&\, \omega_d$	ω_c	ω_d	$\frac{a}{2}(e^{j\omega_c x_1}+e^{j\omega_d x_2})$	$\frac{a^2}{2}[1+\cos(\omega_c x_1 - \omega_d x_2)]$
ω_c	—	—	0	0
ω_c	ω_c	—	$\frac{a}{2}e^{j\omega_c x_1}$	$\frac{a^2}{4}$
ω_d	—	—	0	0
ω_d	—	ω_d	$\frac{a}{2}e^{j\omega_d x_2}$	$\frac{a^2}{4}$

TABLE 8.5. SINGLE (LOWER)-SIDEBAND CORRELATION

Reference Frequencies	Signal Frequencies ω_1	ω_2	Correlation Functions ϕ_{21} Amplitude	$|\phi_{21}|^2$ Intensity (Detected)
$\omega_c \,\&\, \omega_d$	—	—	0	0
$\omega_c \,\&\, \omega_d$	ω_c	—	$\frac{a}{2}e^{-j\omega_c x_1}$	$\frac{a^2}{4}$
$\omega_c \,\&\, \omega_d$	—	ω_d	$\frac{a}{2}e^{j\omega_d x_2}$	$\frac{a^2}{4}$
$\omega_c \,\&\, \omega_d$	ω_c	ω_d	$\frac{a}{2}(e^{-j\omega_c x_1}+e^{-j\omega_d x_2})$	$\frac{a^2}{2}[1+\cos(\omega_c x_1 - \omega_d x_2)]$
ω_c	—	—	0	0
ω_c	ω_c	—	$\frac{a}{2}e^{-j\omega_c x_1}$	$\frac{a^2}{4}$
ω_d	—	—	0	0
ω_d	—	ω_d	$\frac{a}{2}e^{-j\omega_d x_2}$	$\frac{a^2}{4}$

obtained for a match of either frequency; That is, regardless of which frequency matches the correlator output will be the constant $a^2/4$. In double-sideband correlation the detected value depends on the matching frequency; that is, if ω_c is the matching frequency, $a^2 \cos^2 \omega_c x_1$ will be the output; and if ω_d is the matching frequency, $a^2 \cos^2 \omega_d x_2$ will be the output. Because the spacing between peaks of a squared cosine function depends on the ω of the function, it is possible to distinguish frequencies by the peak spacing of the $a^2 \cos^2 \omega x$ outputs. Thus *single-sideband correlation* indicates that *a match* exists (by constant intensity); *double-sideband correlation* not only indicates that a match exists but it also indicates *which match* exists (by the spacing between peaks). When several (more than one) signal frequencies match reference frequencies, *both correlation* methods produce usable results. The results depend on the matching frequencies but the form of the detected correlation function $(|\phi_{21}|^2)$ is different for each method. This can be seen by comparing the results given in Tables 8.3 and 8.4 for ω_c and ω_d present in both reference and signal; that is, when ω_c and ω_d are both present in the input signal, the single-sideband correlation gives a function that is a cosine function of the difference between $\omega_c x_1$ and $\omega_d x_2$ while in double-sideband correlation the result is the sums of two cosine functions of $\omega_c x_1$ and $\omega_d x_2$.

9

Zone Plates

A zone plate is usually referred to as a plate that is capable of producing image points (focuses) when illuminated by spherical or plane wavefronts of monochromatic light. It will be shown in a later chapter that there is a similarity between a hologram reconstruction (and formation) of a real and virtual image and those reconstructed by zone plates. This similarity permits easy visualization of the holographic phenomena. A detailed look at zone plates and their characteristics and capabilities is therefore of interest.

FRESNEL DIFFRACTION

Fresnel diffraction of a plane wave can be explained by considering Figure 9.1. The plane T contains points that will cause the light to diffract around them. Suppose it is desired to determine the amplitude of the light

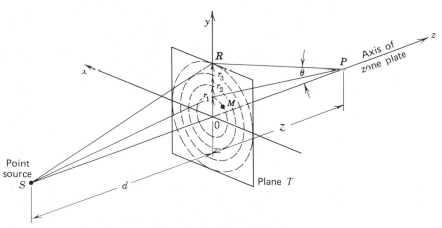

Figure 9.1 The zone plate.

at point P caused by the plane wave striking the diffracting surface (plane T) which is a distance z from point P. We can imagine a plane wavefront striking the plane surface (T) in phase at each point on the plane and consider two path lengths (such as $0P$ and r_1P) from the plane T to point P. Depending on the difference in path lengths between $0P$ and r_1P the light can arrive at point P in phase, out of phase, or anything in between as shown in Figure 9.3. Now let us suppose we desire to determine all those points in plane T that will diffract light in such a way as to add constructively (be essentially in phase) at point P. We can do this by imagining the plane wave striking the plane T that is divided into a series of zones bounded by circles of radius $r_1, r_2, r_3, \ldots$, where r_1 is the smallest circle. The circles are such that the distances

$$r_1P = z + \frac{\lambda}{2}$$

$$r_2P = z + \frac{2\lambda}{2}$$

$$r_3P = z + \frac{3\lambda}{2}$$

$$\vdots$$

$$r_nP = z + \frac{n\lambda}{2}$$

By the Pythagorean theorem the radius of the nth zone is

$$r_n^2 + z^2 = \left(z + \frac{n\lambda}{2}\right)^2 \qquad (9\text{-}1)$$

where n is an integer

Multiplying out we get

$$r_n^2 + z^2 = z^2 + n\lambda z + \frac{n^2\lambda^2}{4} \qquad (9\text{-}2)$$

$$r_n^2 = n\lambda z + \frac{n^2\lambda^2}{4} \qquad (9\text{-}3)$$

If we consider (as is the usual case) that $n\lambda$ is small compared with z, we can write (9-3) as

$$r_n^2 = n\lambda z \qquad (9\text{-}4)$$

$$r_n = \sqrt{n\lambda z} \qquad (9\text{-}5)$$

Equation (9-5) can be used to construct a zone plate. It states that for a given focal length z the radius of the zones (for a given wavelength λ) is proportional to the square root of natural numbers. Figure 9.2 shows photographs of drawing of typical Fresnel zone plates. Both the negative and positive images are shown. Both the negative and positive images act as zone plates since they are drawn so that their radii are in the ratio of the square roots of natural number. This makes the area of all zones in the circular zone plate Figures 9.2(c) and (d) the same. The linear zone plate is drawn so that the lines separating zones are a distance from the center which is proportional to the square root of natural numbers. The linear zone plate Figures 9.2(a) and (b) produces a line as a focus (similar to a cylindrical lens) whereas the circular zone plate produces a point focus. (We shall see later in this chapter that zone plates similar to those shown in Figure 9.2 actually produce more than one focal point.)

DERIVATION OF THE LIGHT DISTRIBUTION ON THE AXIS

Referring to Figure 9.1 it can be seen that the axis of the zone plate is the perpendicular to the plane T of the zone plate through point 0. We shall only consider the case of a point source S at infinity on the axis. We can then consider that the light falling on the plate consists of plane waves, of wavelength λ (or wavenumber $k = 2\pi/\lambda$) and amplitude A_0.

The variations of transparency in the plate causes diffraction of the plane wave. We shall derive the complex amplitude of this diffracted light at a point P on the axis for a zone plate.

The complex amplitude is determined from the diffraction formula that has the following form for a point on the axis

$$U(P) = \int_{zp} \frac{A_0}{j\lambda} t(r) \frac{\exp(-jk\sqrt{z^2 + r^2})}{\sqrt{z^2 + r^2}} 2\pi r \, dr \tag{9-6}$$

If we let R equal the maximum radius of the zone plate, then

$$U(P) = \int_0^R \frac{A_0}{j\lambda} t(r) \frac{\exp(-jk\sqrt{z^2 + r^2})}{\sqrt{z^2 + r^2}} 2\pi r \, dr \tag{9-7}$$

$$U(P) = \frac{2\pi A_0}{j\lambda} \int_0^R \frac{t(r)r}{\sqrt{z^2 + r^2}} \exp(-jk\sqrt{z^2 + r^2}) \, dr \tag{9-8}$$

Usually in diffraction theory there is a factor $(1 + \cos\theta)/2$ in the expression for the complex amplitude, $U(P)$, that is called the *obliquity factor*. This factor accounts for the obliquity of the plane waves with respect to the axis after diffraction. We have considered that the source S and the point P are far from the zone plate and the distances are large relative to the size of the zone plate (i.e., $d \gg R$ and $z \gg R$). Under these conditions the angle θ (i.e., the angle the rays make with the axis) must be close to zero and therefore $\cos\theta = 0° = 1$ making the value of $(1 + \cos\theta)/2$ nearly one, so that the obliquity factor term can be ignored.

Figure 9.2 Photographs of Fresnel zone plates. (*a*) Linear zone plate (clear center), (*c*) circular zone plate (clear center).

Figure 9.2 (b) Linear zone plate (black center), (d) circular zone plate (black center).

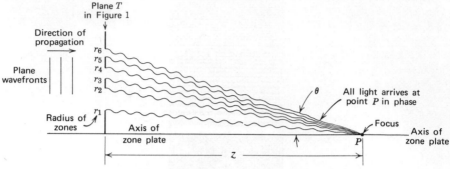

Figure 9.3 Operation of a zone plate.

For (9-8) to hold we must make the assumption that plane wavefronts reach the plane of the zone plate with a constant phase e^{-jk}. We can observe that if the point source S is not at infinity, spherical waves will strike the zone plate. Equation (9-8) will hold only when these spherical wavefronts can be considered as approximations of plane wavefronts; that is, when d is large, the differences of phase on the zone plate will be small and approximately the constant phase produced by a plane wavefront. The maximum differences are between the instantaneous phase at 0 and the corresponding phase at any point R at the edge of the zone plate (where R is the maximum radius of this zone plate). The optical path-length difference can be calculated as follows:

$$\overline{SR} - \overline{S0} = \sqrt{d^2 + R^2} - d$$

$$= \sqrt{d^2 + R^2} - d\frac{\sqrt{d^2 + R^2} + d}{\sqrt{d^2 + R^2} + d}$$

$$= \frac{d^2 + R^2 - d^2}{\sqrt{d^2 + R^2} + d} = \frac{R^2}{\sqrt{d^2 + R^2} + d}$$

but $d \gg R$ and $d^2 \gg R^2$ then

$$\overline{SR} - \overline{S0} = \frac{R^2}{\sqrt{d^2} + d} = \frac{R^2}{2d} \tag{9-9}$$

In order for the phase differences along the zone plate to be small we would like the differences in path length between the line $\overline{SR}$ and the line $\overline{S0}$ to be very much less than one half a wavelength; that is, we would like $R^2/2d \ll \lambda/2$ or $d \gg R^2/\lambda$. For example, a zone plate with a maximum radius $R = 1$ cm and light of wavelength $\lambda = 0.6$, microns $= 6 \times 10^{-5}$ cm,

$$d \gg \frac{\text{cm}^2}{6 \times 10^{-5}\,\text{cm}} = 16 \times 10^3\,\text{cm}\frac{1\text{M}}{10^2\,\text{cm}} = 160\,\text{meters}$$

This means that the point source S must be at least 160 meters from the zone plate if the assumption is to be made that it is being struck by plane wavefronts.

We can now assume that we can make plane wavefronts incident on the zone plate when the point source S is sufficiently far from the zone plate and is on the zone-plate axis. Let us now determine the effect of moving the point source S to a position S' off the axis of the zone plate as shown in Figure 9.4. We can see from Figure 9.4(a) that plane wavefronts can be incident on the zone plate but if they are striking it at an angle α, the condition of equal phase over the zone plate may no longer exist. Under these conditions the maximum phase difference will occur between two

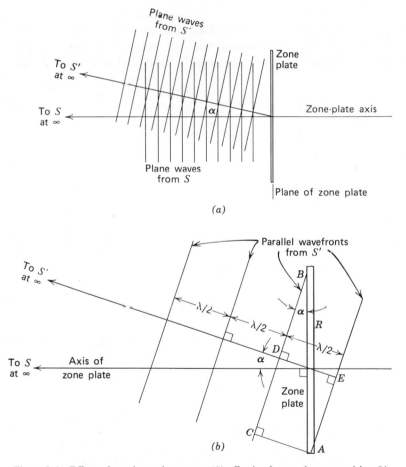

Figure 9.4   Effect of moving point source (S) off axis of zone plate to position S^1.

points at opposite ends of the diameter. In Figure 9.4(b) we see parallel wavefronts perpendicular to line $S'E$ making an angle α with the axis of the zone plate. In triangle ABC the line AC represents the additional optical path length that a wavefront must travel after it first strikes the zone plate at B before it leaves the zone plate at C. The wavefront is traveling parallel to line $S'E$. When the wavefront reaches point D, it will just intersect the zone plate at B and as it travels along line $S'E$, this wavefront will intersect different portions of the zone plate. When the wavefront reaches point E on line $S'E$, it will also be just leaving the zone plate at point A. From Figure 9.4(b) we can see

$$\sin \alpha = \frac{\overline{AC}}{2R} \tag{9-10}$$

In order for the phase difference along the zone plate to be a minimum we would like the angle α to be such that the distance AC (which represents the optical path-length difference) should be a minimum; that is, since

$$\sin \alpha = \frac{AC}{2R} \tag{9-10}$$

$$AC = 2R \sin \alpha$$

and we desire $AC \ll \lambda/2$; therefore

$$\frac{\lambda}{2} \gg AC = 2R \sin \alpha$$

$$\frac{\lambda}{2} \gg 2R \sin \alpha$$

$\sin \alpha \ll \lambda/4R$ and if α is small $\sin \alpha = \alpha$ and

$$\alpha \ll \frac{\lambda}{4R} \tag{9-11}$$

Let us consider the example we used previously where $\lambda = 0.6$ microns and $R = 1$ cm. Then

$$\alpha \ll \frac{\lambda}{4R} = \frac{0.6 \times 10^{-4} \, \text{cm}}{4 \, \text{cm}} \tag{9-11}$$

$$\alpha \ll 0.15 \times 10^{-4} \, \text{rad}$$

$$\alpha \ll \frac{0.15 \times 10^{-4} \, \text{rad}}{1} \times \frac{360°}{2\pi \, \text{rad}} \times \frac{3600 \, \text{sec}}{1°} \approx 3 \, \text{arc sec}$$

This means that the source must be less than 3 sec of arc off axis and at least 160 meters from the zone plate if the assumption is to be made that plane waves strike the zone plate.

In order to get a better understanding of (9-8),

$$U(P) = \frac{2\pi A_0}{j\lambda} \int_0^R \frac{t(r)}{\sqrt{z^2+r^2}} \exp\left(-jk\sqrt{z^2+r^2}\right) dr \qquad (9\text{-}8)$$

we shall transform it for the case where $z \gg R$ by making some simplifying approximations. In order to simplify (9-8) we must first consider the binomial expansion for

$$\frac{1}{\sqrt{z^2+r^2}} = (z^2+r^2)^{-1/2} = z^{-1}(1+r^2/z^2)^{-1/2}$$

by first noting that

$$(1+x)^{-n} = 1 - nx + \frac{n(n+1)x^2}{2!} - \frac{n(n+1)(n+2)x^3}{3!} + \cdots \qquad (9\text{-}12)$$

where $(x^2 < 1)$. Therefore expanding

$$z^{-1}\left(1+\frac{r^2}{z^2}\right)^{-1/2} = z^{-1}\left(1 - \frac{r^2}{2z^2} + \frac{3r^4}{8z^4} - \frac{5r^6}{16z^6} + \cdots\right)$$

$$= \left(\frac{1}{z} - \frac{r^2}{2z^3} + \frac{3r^4}{8z^5} - \frac{5r^6}{16z^7} + \cdots\right) \qquad (9\text{-}13)$$

gives a series with alternate signs. Taking only a certain number of terms in the expansion will introduce certain errors that can be neglected under certain conditions. For example, if we consider only the first term in the binomial expansion, then

$$\frac{1}{\sqrt{z^2+r^2}} = \frac{1}{z} \qquad (9\text{-}14)$$

we can see that this is approximately true if $r^2/2z^3 \ll 1/z$ in (9-13)

$$z^{-1}\left(1+\frac{r^2}{z^2}\right)^{-1/2} = \left(\frac{1}{z} - \frac{r^2}{2z^3} + \frac{3r^4}{8z^5} - \cdots\right) \qquad (9\text{-}13)$$

The maximum value of r is R and therefore $R^2/2z^3 \ll 1/z$ which can be written

$$\frac{R^2}{2z^2} \ll 1 \qquad (9\text{-}15)$$

restriction

for

Amplitude

Approx.

to be accurate

We will now make an approximation for $\sqrt{z^2+r^2}$ in the exponent of (9-8) by using the binomial expansion

$$(1+x)^n = 1+nx+\frac{n(n-1)x^2}{2!}+\frac{n(n-1)(n-2)x^3}{3!}+\cdots \quad (9\text{-}16)$$

where $x^2 < 1$. Since

$$\sqrt{z^2+r^2} = (z^2+r^2)^{1/2} = z\left(1+\frac{r^2}{z^2}\right)^{1/2}$$

then expanding with the binomial expansion gives

$$z\left(1+\frac{r^2}{z^2}\right)^{1/2} = z\left(1+\frac{r^2}{2z^2}-\frac{r^4}{8z^4}+\cdots\right)$$

$$= \left(z+\frac{r^2}{2z}-\frac{r^4}{8z^3}+\cdots\right)$$

The exponential term can therefore be approximated by the first two terms of the binomial expansion as follows

$$\exp\left[-jk\sqrt{z^2+r^2}\right] = \exp\left[-jk\left(z+\frac{r^2}{2z}\right)\right] = \exp\left[-jkz\right]\exp\left[-jk\frac{r^2}{2z}\right] \quad (9\text{-}17)$$

provided the third term $\exp\left[jk(r^4/8z^3)\right]$ is equal or close to one, that is, we desire to make the term

$$\exp\left[jk\frac{r^4}{8z^3}\right] = 1$$

which means the exponent $\left[jk(r^4/8z^3)\right]$ should be near or equal to zero. We can consider the exponential to be of the form $e^{j\theta}$ which can be written as a magnitude of one at an angle of θ, that is,

$$e^{j\theta} = 1\underline{/\theta}$$

Since θ is in radians, to make it near or equal to zero we can impose the inequality $\theta \ll \pi$ as shown in Figure 9.5 and therefore

$$\frac{kr^4}{8z^3} \ll \pi \quad\text{or}\quad \frac{2\pi}{\lambda}\frac{R^4}{8z^3} \ll \pi$$

then

$$\frac{R^4}{4\lambda z^3} \ll 1 \quad (9\text{-}18)$$

restriction for Phase Approx to be accurate (handwritten annotation)

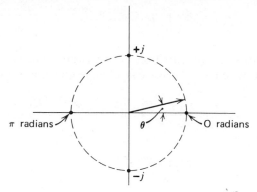

Figure 9.5   Plot of $1\underline{|\theta} = e^{j\theta}$. $e^{j\theta}$

Let us consider the example we previously used where $\lambda = 0.6$ microns and $R = 1$ cm; then by (9-15) we have

$$\frac{R^2}{2z^2} \ll 1 \tag{9-15}$$

or $z^2 \gg R^2/2$ then

$$z \gg \sqrt{\tfrac{1}{2}} = 0.707 \text{ cm}$$

and from equation (9-18)

$$\frac{R^4}{4\lambda z^3} \ll 1 \tag{9-18}$$

$$z^3 \gg \frac{R^4}{4\lambda} = \frac{R^4}{4 \times 0.6 \times 10^{-4}} = \frac{10^4}{2.4} = 4200$$

$$z \gg 16.2 \text{ cm}$$

From these results we can see that the phase effects given in (9-18) are more restrictive in (9-8) than the amplitude factor of (9-15). If we can assume that the conditions imposed by equations (9-15) and (9-18) are met (that is, $z \gg R$ so that the approximations hold), then we can make the following substitutions

$$\frac{1}{z} = \frac{1}{\sqrt{z^2 + r^2}} \tag{9-14}$$

$$\exp\left[-jk\sqrt{z^2 + r^2}\right] = \exp\left[-jkz\right] \exp\left[-jk\frac{r^2}{2z}\right] \tag{9-17}$$

into (9-8)

$$U(P) = \frac{2\pi A_0}{j\lambda} \int_0^R \frac{t(r)r}{\sqrt{z^2 + r^2}} \exp\left[-jk\sqrt{z^2 + r^2}\right] dr \tag{9-18}$$

giving
$$= \frac{2\pi A_0}{j\lambda} \int_0^R t(r) \frac{e^{-jkz} \, e^{-jk(r^2/2z)}}{z} r \, dr \qquad (9\text{-}19)$$

Because the transmission function of the zone plate is zero $(t(r) = 0)$, when $r > R$ or when $r < 0$, we can change the limits of integration as follows

$$U(P) = \frac{2\pi A_0}{j\lambda} \frac{e^{-jkz}}{z} \int_{-\infty}^{\infty} t(r) \frac{e^{-jk(r^2/2z)}}{z} r \, dr \qquad (9\text{-}20)$$

and since $k = 2\pi/\lambda$

$$U(P) = \frac{2\pi A_0 e^{-jkz}}{j\lambda z} \int_{-\infty}^{\infty} t(r) \, e^{-j\pi r^2/\lambda z} r \, dr \qquad (9\text{-}21)$$

Let us now make a substitution by setting $r^2 = v$ then

$$2r \, dr = dv$$

and

$$U(P) = \frac{2\pi A_0}{j\lambda z} e^{-j(2\pi z/\lambda)} \int_{-\infty}^{\infty} t(v) \, e^{-j(\pi v/\lambda z)} \, dv/2$$

$$= \frac{\pi A_0}{j\lambda z} e^{-j(2\pi z/\lambda)} \int_{-\infty}^{\infty} t(v) \, e^{-j(\pi v/\lambda z)} \, dv \qquad (9\text{-}22)$$

From (9-22) we can see that the complex amplitude $U(P)$ is directly related to the Fourier transform of $t(v)$; that is, (9-22) is in the form

$$F(\omega) = \int_{-\infty}^{\infty} f(\tau) \, e^{-j\omega\tau} \, d\tau \qquad \omega = \pi/\lambda z$$

where ω is the spatial frequency. The light intensity at P is

$$I(P) = U(P) U^*(P)$$

where

$$U^*(P) = \frac{\pi A_0^*}{-j\lambda z} e^{j(2\pi z/\lambda)} \int_{-\infty}^{\infty} t(v) \, e^{j(\pi v/\lambda z)} \, dv$$

$$I(P) = U(P) U^*(P) = \frac{\pi^2 A_0 A_0^*}{(\lambda z)^2} \left[\int_{-\infty}^{\infty} t(v) \, e^{-j(\pi v/\lambda z)} \, dv \right]$$

$$\times \left[\int_{-\infty}^{\infty} t(v) \, e^{j(\pi v/\lambda z)} \, dv \right]$$

We can let $A_0 A_0^* = I_0 =$ incident light intensity. Then

$$I_p = I_0 \left[\frac{\pi^2}{(\lambda z)^2} \right] \left[\begin{array}{c} \text{component of the power spectrum of } t(v) \\ \text{for the spatial frequency} \end{array} \right] \qquad (9\text{-}23)$$

if we write the Fourier transform of $t(v)$ as

$$F\left(\frac{\pi}{\lambda z}\right) = \int_{-\infty}^{\infty} t(v)\, e^{-j(\pi v/\lambda z)}\, dv \qquad (9\text{-}24)$$

The power spectrum is then

$$\text{power spectrum} = P\left(\frac{\pi}{\lambda z}\right) = F\left(\frac{\pi}{\lambda z}\right) F^*\left(\frac{\pi}{\lambda z}\right) \qquad (9\text{-}25)$$

and then (9-23) can be written

$$I(\lambda, z) = I_0(\lambda)\left(\frac{\pi}{\lambda z}\right)^2 P\left(\frac{\pi}{\lambda z}\right) \qquad (9\text{-}26)$$

Equation (9-26) gives the distribution of light intensity for any illuminating plane wave $I_0(\lambda)$ and any zone plate as defined by its power spectrum $P(\pi/\lambda z)$. The $(\pi/\lambda z)^2$ term is an attenuation factor that implies that the intensity decreases as $1/z^2$ from the zone plate—that is, the usual inverse square law. There are two important points indicated by (9-26) as to the light distribution: (1) The light intensity is directly related to the power spectrum of the transmission function of the zone plate. (2) The light intensity is, for a given plane wave $I_0(\lambda)$ illuminating the zone plate, only a function of $(\pi/\lambda z)$; that is, it is only a function of (λz).

Let us examine (9-26) in more detail with the aid of a graph (Figure 9.6). If we assume that the light is monochromatic (λ_0), then the curve in Figure 9.6 can represent a light distribution along the z axis plotted as a function of $\pi/\lambda z$; for example, the light intensity at z_1 is equal to the light intensity at z_2 and the intensity of the light at z_3 will be a maximum. If the light is not monochromatic (i.e., λ is not a constant), different curves can be found depending on the factor λ in $\pi/\lambda z$. Another interesting point is

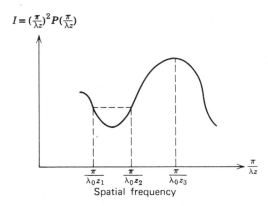

Figure 9.6   Plot of intensity versus spatial frequency.

that if a certain light distribution on the z axis is desired, equation (9-26) can be used to determine if a transmission law can be found to produce the desired light distribution. Certainly not all light distributions can be achieved. The only distribution along the z axis that is possible will be those of the form of (9-26)

$$I(\lambda, z) = I_0(\lambda) \left(\frac{\pi}{\lambda z}\right)^2 P\left(\frac{\pi}{\lambda z}\right) \tag{9-26}$$

Equation (9-26) states that the intensity along the z axis is the product of the spatial frequency squared, times the power spectrum.

Because the product λz determines the intensity distribution (see Figure 9.6), this also shows why it is not possible to make zone plates free of chromatic aberrations. If it were possible to produce such zone plates, all wavelengths would be focused at a given point z_0 and there would be no light at any other point. From (9-26) we can see that this is not allowed; that is, (9-26) shows the conditions

$$I(\lambda, z) = 0 \quad \text{for} \quad z \neq z_0 \tag{9-27}$$

and

$$I(\lambda, z) \neq 0 \quad \text{for} \quad z = z_0 \tag{9-28}$$

cannot be met for different values of λ. Since $I(\lambda, z)$ is determined by the product λz, $I(\lambda, z)$ will be different at $z = z_0$ for different λ. A simple example can show these conditions. Let us consider two wavelengths of light, each incident on a zone plate with the same intensity—that is, λ_0 and $m\lambda_0$ and two points z_0 and z_0/m. Then by (9-26) the intensity of the light $m\lambda_0$ at z_0/m is the same as the light intensity λ_0 in z_0; that is,

$$m\lambda_0 \frac{z_0}{m} = \lambda_0 z_0$$

In Figure 9.6 two such points are z_1 and z_2 if we let $z_0/m = z_2$.

Let us now examine how we might use the facts concerning zone plates for some practical purpose. Suppose, for example, that we desired to construct an instrument for determining the wavelength of the light incident on a zone plate; that is, we wish to build a zone plate that can be used as an interferometer. Such a device could be built if light of each different frequency could be focused to different points. At points of focused light corresponding to wavelengths of interest a photocell would determine the intensity, if any, of the light at that particular wavelength.

In order to construct a zone plate that could be used as an interferometer it is necessary to determine the best transmission function that will achieve the desired results. For a zone-plate interferometer it is desirable that each wavelength have its own focal point and one, and only one, focal point corresponding to each wavelength; that is, each point on the axis (where a photocell can be placed) corresponds to one wavelength, and only one. The transmission function that will achieve these results should have the following properties.

$$I(\lambda, z) = 0 \quad \text{when} \quad \lambda = \lambda_0 \quad \text{and} \quad z \neq z_0 \quad (9\text{-}29)$$

and

$$I(\lambda, z) \neq 0 \quad \text{when} \quad \lambda = \lambda_0 \quad \text{and} \quad z = z_0$$

Then from (9-26), assuming that the incident light on the zone plate has a constant intensity regardless of its wavelength, we can write

$$I(\lambda, z) = \left(\frac{\pi}{\lambda z}\right)^2 P\left(\frac{\pi}{\lambda z}\right) \quad (9\text{-}30)$$

From (9-29) we see that

$$\left(\frac{\pi}{\lambda z}\right)^2 P\left(\frac{\pi}{\lambda z}\right) = 0 \quad \text{except for } \lambda z = \lambda_0 z_0 \quad (9\text{-}31)$$

From (9-31) we can deduce that $P(\pi/\lambda z)$ must be zero everywhere except in one point $(\pi/\lambda z) = (\pi/\lambda_0 z_0)$. We can let

$$\lambda_0 z_0 = a^2 \quad (9\text{-}32)$$

The transmission function that we desire to find must satisfy the conditions $0 \leq t(v) \leq 1$ (to permit it to be placed on film). In addition we can further assume

$$t(v) = \frac{1}{2}\left(1 - \sin\frac{\pi v}{a^2}\right) \quad (9\text{-}33)$$

and since $v = r^2$, then

$$t(r) = \frac{1}{2}\left(1 - \sin\frac{\pi r^2}{a^2}\right) \quad (9\text{-}34)$$

The usual Fresnel zone plate is a square-wave approximation of the transmission function described in (9-34). The sine-wave transmission function [(9-33) and (9-34)] for a zone plate is shown plotted in Figure 9.7. When the transmission function is plotted as a function of v, it varies sinusoidally with v (where v is equal to the radius of the zone plate squared). When plotted as a function of the radius r, the transmission function is an increasing frequency sinusoidal waveform as shown in Figure 9.7(b). The transmission function $t(r)$ in (9-34) was plotted by selecting various

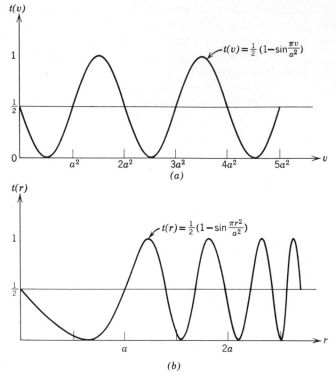

Figure 9.7 Plot of a sine-wave transmission function zone plate as a function of v and r. (a) As a function of v; (b) as a function of r.

values of r. The square-wave approximation can easily be determined by first observing the construction in Figure 9.8. Figure 9.8 shows a plot of a biased sine wave about the line $y = \frac{1}{2}$. The points of crossing on this bias line ($y = \frac{1}{2}$) can be found by noting.

$$y = \tfrac{1}{2}(1 - \sin \varphi) = \tfrac{1}{2} \tag{9-35}$$

when

$$\sin \varphi = 0 \rightarrow \varphi = N\pi$$

where $N = 0, 1, 2, 3. \ldots$ We can see therefore that a square-wave approximation for the sine wave would pass through these points; that is, each half cycle of the sinusoid is approximated by the square wave as shown in Figure 9.8. The square wave has a value of one in Figure 9.8 as follows:

$$f(\varphi) = 1 \qquad \text{for } (2N - 1)\pi \leqslant \varphi \leqslant 2N\pi \tag{9-36}$$

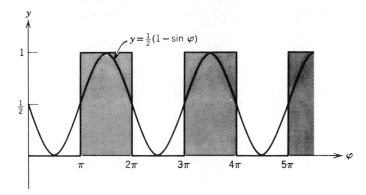

Figure 9.8 Sinusoid and its square-wave approximation.

where $N = 1, 2, 3, \ldots$. For all other values the square-wave is zero; that is, $f(\varphi) = 0$ for all other φ. Now we can let (9-35) correspond to (9-34) by letting $\varphi = (\pi r^2 / a^2)$. The square-wave approximation for (9-34) will have a value of one at points corresponding to the points shown in (9-36)—that is, when

$$(2N - 1)\pi \leqslant \varphi \leqslant 2N\pi$$

where $N = 1, 2, 3 \ldots$. Figure 9.9 shows a comparison of a sine-wave zone plate and its square-wave approximation.

The main difference between the two graphs of the transmission functions of the zone plates shown in Figure 9.9 can be seen from their power spectrums as introduced in (9-26),

$$I(\lambda, z) = I_0(\lambda) \left(\frac{\pi}{\lambda z}\right)^2 P\left(\frac{\pi}{\lambda z}\right) \tag{9-26}$$

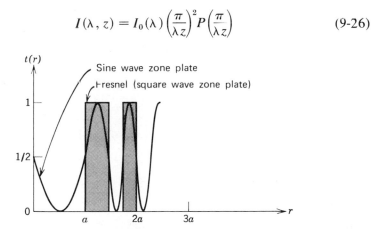

Figure 9.9 Comparison between a Fresnel zone plate and a sine-wave zone plate.

The power spectrum for the sine-wave zone plate can be calculated as follows:

$$P\left(\frac{\pi}{\lambda z}\right) = F[t(v)]F^*[t(v)] \qquad (9\text{-}37)$$

where $t(v)$ is the transmission function given in (9-33) and $F[t(v)]$ is given by

$$F[t(v)] = \int_{-\infty}^{\infty} t(v)\, e^{-j(\pi v/\lambda z)}\, dv = \int_{-\infty}^{\infty} \frac{1}{2}\left(1 - \sin\frac{\pi v}{a^2}\right) e^{-j(\pi v/\lambda z)}\, dv \qquad (9\text{-}38)$$

Since $t(v) = 0$ when $v > R^2$ and when $v < 0$, the limits of integration can be made from 0 to R^2 giving for (9-37)

$$P\left(\frac{\pi}{\lambda z}\right) = \int_0^{R^2} \frac{1}{2}\left(1 - \sin\frac{\pi v}{a^2}\right) e^{-j(\pi v/\lambda z)}\, dv \int_0^{R^2} \frac{1}{2}\left(1 - \sin\frac{\pi v}{a^2}\right) e^{j(\pi v/\lambda z)}\, dv \qquad (9\text{-}39)$$

Limiting the range of v to values less than R^2 makes $t(v)$ an aperiodic function that has an energy density spectrum rather than the power spectrum for periodic functions. In the case considered here we are only making use of the mathematical formulation of the conjugate product and the terminology of power spectrum will be used; that is, the energy density spectrum is more correct for aperiodic functions than for power spectrum; however, we are not interested in physical units here but only in the mathematical formulation, and reference to the power spectrum will be used in the remainder of this chapter with the meaning as indicated above.

Making use of the fact that

$$e^{\pm j\theta} = \cos\theta \pm j\sin\theta$$

we can write (9-39) as

$$P\left(\frac{\pi}{\lambda z}\right) = \int_0^{R^2} \left[\frac{1}{2}\left(1 - \sin\frac{\pi v}{a^2}\right)\cos\frac{\pi v}{\lambda z} - \frac{1}{2}\left(1 - \sin\frac{\pi v}{a^2}\right)j\left(\sin\frac{\pi v}{\lambda z}\right)\right] dv$$

$$\int_0^{R^2} \left[\frac{1}{2}\left(1 - \sin\frac{\pi v}{a^2}\right)\cos\frac{\pi v}{\lambda z} + \frac{1}{2}\left(1 - \sin\frac{\pi v}{a^2}\right)j\left(\sin\frac{\pi v}{\lambda z}\right)\right] dv \qquad (9\text{-}40)$$

Let

$$\alpha = \int_0^{R^2} \left(1 - \sin\frac{\pi v}{a^2}\right)\cos\frac{\pi v}{\lambda z}\, dv \qquad (9\text{-}41)$$

$$\beta = \int_0^{R^2} \left(1 - \sin\frac{\pi v}{a^2}\right)\sin\frac{\pi v}{\lambda z}\, dv \qquad (9\text{-}42)$$

and since we are looking for the real part of (9-39), we can find the power spectrum by evaluating (9-40) as follows:

$$P\left(\frac{\pi}{\lambda z}\right) = \frac{1}{4}\left[\int_0^{R^2}\left(1-\sin\frac{\pi v}{a^2}\right)\left(\cos\frac{\pi v}{\lambda z}\right)dv - \int_0^{R^2} j\left(1-\sin\frac{\pi v}{a^2}\right)\left(\sin\frac{\pi v}{\lambda z}\right)dv\right]$$

$$\left[\int_0^{R^2}\left(1-\sin\frac{\pi v}{a^2}\right)\left(\cos\frac{\pi v}{\lambda z}\right)dv + \int_0^{R^2} j\left(1-\sin\frac{\pi v}{a^2}\right)\left(\sin\frac{\pi v}{\lambda z}\right)dv\right]$$

$$= \frac{1}{4}[(\alpha-j\beta)(\alpha+j\beta)$$

$$= \frac{\alpha^2+\beta^2}{4} \tag{9-43}$$

To evaluate (9-43) let us first find α. Multiplying the terms in (9-41) we find α to be

$$\alpha = \int_0^{R^2}\left[\cos\frac{\pi v}{\lambda z} - \sin\frac{\pi v}{a^2}\cos\frac{\pi v}{\lambda z}\right]dv \tag{9-44}$$

Making use of the following trigonometric $\frac{1}{2}$ angle formula

$$\sin(A+B) = \sin A \cos B + \cos A \sin B \tag{9-45}$$
$$\sin(A-B) = \sin A \cos B - \cos A \sin B \tag{9-46}$$
$$\sin(A+B) + \sin(A-B) = 2\sin A \cos B$$

or

$$\sin A \cos B = \tfrac{1}{2}[\sin(A+B) + \sin(A-B)] \tag{9-47}$$

$$\alpha = \int_0^{R^2}\cos\frac{\pi v}{\lambda z}dv - \int_0^{R^2}\left[\sin\frac{\pi v}{a^2}\cos\frac{\pi v}{\lambda z}\right]dv \tag{9-48}$$

$$= \int_0^{R^2}\cos\frac{\pi v}{\lambda z}dv - \frac{1}{2}\int_0^{R^2}\sin\left(\frac{\pi v}{a^2}+\frac{\pi v}{\lambda z}\right)$$

$$-\frac{1}{2}\int_0^{R^2}\sin\left(\frac{\pi v}{a^2}-\frac{\pi v}{\lambda z}\right)dv \tag{9-49}$$

$$= \frac{\lambda z}{\pi}\sin\frac{\pi v}{\lambda z}\Big]_0^{R^2} + \frac{1}{2(\pi/a^2+\pi/\lambda z)}\cos\left(\frac{\pi v}{a^2}+\frac{\pi v}{\lambda z}\right)\Big]_0^{R^2}$$

$$+\frac{1}{2(\pi/a^2-\pi/\lambda z)}\cos\left(\frac{\pi v}{a^2}-\frac{\pi v}{\lambda z}\right)\Big]_0^{R^2} \tag{9-50}$$

$$\alpha = \frac{\lambda z}{\pi} \sin \frac{\pi R^2}{\lambda z} + \frac{1}{2(\pi/a^2 + \pi/\lambda z)}$$

$$\times \left[\left(\cos \left(\frac{\pi R^2}{a^2} + \frac{\pi R^2}{\lambda z} \right) \right) - 1 \right] + \frac{1}{2(\pi/a^2 - \pi/\lambda z)}$$

$$\times \left[\left(\cos \left(\frac{\pi R^2}{a^2} - \frac{\pi R^2}{\lambda z} \right) \right) - 1 \right] \tag{9-51}$$

To evaluate β, (9-42)

$$\beta = \int_0^{R^2} \left[\sin \frac{\pi v}{\lambda z} - \sin \frac{\pi v}{a^2} \sin \frac{\pi v}{\lambda z} \right] dv \tag{9-42}$$

we make use of the following trigonometric $\frac{1}{2}$ angle formula.

$$\cos (A + B) = \cos A \cos B - \sin A \sin B \tag{9-52}$$

$$\cos (A - B) = \cos A \cos B + \sin A \sin B \tag{9-53}$$

$$\cos (A + B) - \cos (A - B) = -2 \sin A \sin B$$

or

$$\sin A \sin B = \tfrac{1}{2} [\cos (A - B) - \cos (A + B)] \tag{9-54}$$

$$\beta = \int_0^{R^2} \sin \frac{\pi v}{\lambda z} \, dv + \frac{1}{2} \int_0^{R^2} \cos \left(\frac{\pi v}{a^2} - \frac{\pi v}{\lambda z} \right) dv$$

$$- \frac{1}{2} \int_0^{R^2} \cos \left(\frac{\pi v}{a^2} + \frac{\pi v}{\lambda z} \right) dv \tag{9-55}$$

$$= - \left(\frac{\lambda z}{\pi} \right) \cos \frac{\pi v}{\lambda z} \Big]_0^{R^2} + \frac{1}{2(\pi/a^2 - \pi/\lambda z)} \sin \left(\frac{\pi v}{a^2} - \frac{\pi v}{\lambda z} \right) \Big]_0^{R^2}$$

$$+ \frac{1}{2(\pi/a^2 + \pi/\lambda z)} \sin \left(\frac{\pi v}{a^2} + \frac{\pi v}{\lambda z} \right) \Big]_0^{R^2} \tag{9-56}$$

$$= - \frac{\lambda z}{\pi} \left[\left(\cos \left(\frac{\pi R^2}{\lambda z} \right) \right) - 1 \right]$$

$$+ \frac{1}{2(\pi/a^2 - \pi/\lambda z)} \sin \left(\frac{\pi R^2}{a^2} - \frac{\pi R^2}{\lambda z} \right)$$

$$- \frac{1}{2(\pi/a^2 + \pi/\lambda z)} \sin \left(\frac{\pi R^2}{a^2} + \frac{\pi R^2}{\lambda z} \right) \tag{9-57}$$

From (9-34) we see that the period of the sinusoidal transmission function is $2a^2$, that is, in the function

$$a^2 = \lambda_0 z_0$$

$$t(v) = \frac{1}{2}\left(1 - \sin\frac{\pi r^2}{a^2}\right) \tag{9-34}$$

when $r^2 = 2a^2$ the sine function starts to repeat itself because the value of the angle is 2π. Assuming that the zone plate consists of an integral number of cycles N, we find that the maximum limit on v (that is, r^2) is

$$v = R^2 = 2Na^2 \tag{9-58}$$

Equation (9-58) assures us that the zone plate consists of an integral number of cycles (or zones).
Substituting

$$R^2 = 2Na^2 \tag{9-58}$$

in (9-51) and (9-57) we find α and β to be

$$\alpha = \frac{\lambda z}{\pi}\sin\left(\frac{2N\pi a^2}{\lambda z}\right) + \frac{1}{2(\pi/a^2 + \pi/\lambda z)}\left[\left\{\cos 2\pi N\left(1+\frac{a^2}{\lambda z}\right)\right\} - 1\right]$$

indeterminate
at values
of $\lambda z = a^2$

$$+ \frac{1}{2(\pi/a^2 - \pi/\lambda z)}\left[\left\{\cos 2\pi N\left(1-\frac{a^2}{\lambda z}\right)\right\} - 1\right] \tag{9-59}$$

$$\beta = -\frac{\lambda z}{\pi}\left[\left\{\cos\left(\frac{2\pi Na^2}{\lambda z}\right)\right\} - 1\right] + \frac{1}{2(\pi/a^2 - \pi/\lambda z)}\sin 2\pi N\left(1-\frac{a^2}{\lambda z}\right)$$

$$- \frac{1}{2(\pi/a^2 + \pi/\lambda z)}\sin 2\pi N\left(1+\frac{a^2}{\lambda z}\right) \tag{9-60}$$

Let us consider value of λz near a^2. We can see that the first term in (9-59) tends towards zero with λz near a^2 since

$$\sin 2\pi = \sin 2\pi N = 0$$

The second term in (9-59) also tends towards zero with λz near a^2 since

$$\cos[2\pi N(2)] = 1$$

and

$$\left[\cos 2\pi N\left(1+\frac{a^2}{\lambda z}\right)\right] - 1 = [\cos 2\pi N(2)] - 1 = 0$$

The third term in (9-59) can be rewritten using the relationship

$$\sin\frac{A}{2} = \pm\sqrt{\frac{1-\cos A}{2}}$$

or

$$\cos A - 1 = -2\sin^2\frac{A}{2} \tag{9-61}$$

The third term in (9-59) then becomes

$$\alpha \approx \frac{1}{2(\pi/a^2 - \pi/\lambda z)}\left\{\left[\cos 2\pi N\left(1 - \frac{a^2}{\lambda z}\right)\right] - 1\right\} = -\frac{\sin^2\pi N(1 - a^2/\lambda z)}{(\pi/a^2 - \pi/\lambda z)}$$

We can now break $\sin^2$ up into sin times sin

$$\alpha \approx -\sin\pi N(1 - a^2/\lambda z)\frac{\sin\pi N(1 - a^2/\lambda z)}{(\pi/a^2 - \pi/\lambda z)}$$

$$= -\sin\pi N(1 - a^2/\lambda z)\frac{\sin\pi N(1 - a^2/\lambda z)}{\pi(1/a^2 - 1/\lambda z)}$$

$$= -\sin\pi N(1 - a^2/\lambda z)\frac{\sin\pi N(1 - a^2/\lambda z)}{\pi(1/a^2 - 1/\lambda z)}\frac{Na^2}{Na^2}$$

$$\alpha \approx \left[-\sin\pi N(1 - a^2/\lambda z)\right]\left[\frac{\sin\pi N(1 - a^2/\lambda z)}{\pi N(1 - a^2/\lambda z)}\right]Na^2$$

$$\alpha \approx \left[-Na^2\sin\pi N(1 - a^2/\lambda z)\right]\left[\frac{\sin\pi N(1 - a^2/\lambda z)}{\pi N(1 - a^2/\lambda z)}\right] \tag{9-62}$$

From (9-62) we can see that the first term is a sine term and the second term is a $(\sin x)/x$ term. When each term is examined separately as shown in (9-63) it can be seen that $\alpha \to 0$, that is,

$$\alpha \approx \left[-\overbrace{Na^2\sin\pi N(1 - a^2/\lambda z)}\right]\left[\overbrace{\frac{\sin\pi N(1 - a^2/\lambda z)}{\pi N(1 - a^2/\lambda z)}}\right] \to 0 \tag{9-63}$$

$$\text{as }\lambda z \to a^2$$

as $\lambda z \to a^2$
this term $\to 0$

as $\lambda z \to a^2$
this term $\to 1$

We can now rewrite (9-43) for the power spectrum by noting that

$$P\left(\frac{\pi}{\lambda z}\right) = \frac{\alpha^2 + \beta^2}{4} \tag{9-43}$$

by (9-63) as $\lambda z \to a^2$, $\alpha \to 0$ and (9-43) becomes

$$P\left(\frac{\pi}{\lambda z}\right) \to \frac{\beta^2}{4} \tag{9-64}$$

as $\lambda z \to a^2$. From (9-60) we can see that when $\lambda z \to a^2$

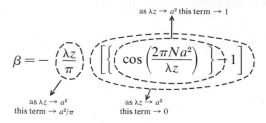

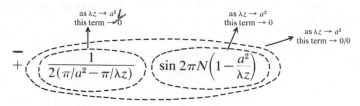

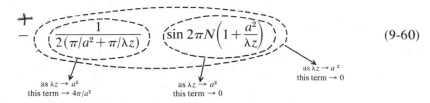

$$\tag{9-60}$$

Therefore as $\lambda z \to a^2$

$$\beta \approx +\frac{1}{2(\pi/a^2 - \pi/\lambda z)}\left[\sin 2\pi N\left(1 - \frac{a^2}{\lambda z}\right)\right] \tag{9-65}$$

Substituting the value of β given in (9-65) into (9-64) we get

$$P\left(\frac{\pi}{\lambda z}\right) = \frac{\beta^2}{4} \quad (\text{as } \lambda z \to a^2) \tag{9-64}$$

$$P\left(\frac{\pi}{\lambda z}\right) \to \frac{1}{4}\left[+\frac{\sin 2\pi N(1 - a^2/\lambda z)}{2(\pi/a^2 - \pi/\lambda z)}\right]^2 \tag{9-66}$$

$$= \frac{1}{4}\left[\frac{Na^2 \sin 2\pi N(1 - a^2/\lambda z)}{2\pi N(1 - a^2/\lambda z)}\right]^2 \tag{9-67}$$

and therefore as $\lambda z \rightarrow a^2$ the $(\sin x)/x$ term in (9-67) tends towards one. Therefore

$$P\left(\frac{\pi}{\lambda z}\right) = \frac{N^2 a^4}{4} \tag{9-68}$$

The intensity as $\pi/\lambda z \rightarrow \pi/a^2$ is found from (9-26) as follows

$$I(\lambda, z) = I_0(\lambda)\left(\frac{\pi}{\lambda z}\right)^2 P\left(\frac{\pi}{\lambda z}\right) \tag{9-26}$$

$$= I_0(\lambda)\left(\frac{\pi^2}{a^4}\right)\left[\frac{N^2 a^4}{4}\right] = \left[\frac{\pi^2 N^2}{4}\right] I_0(\lambda) \tag{9-69}$$

as $\pi/\lambda z \rightarrow \pi/a^2$.

Let us now consider $\pi/\lambda z \rightarrow 0$ and evaluate α and β as given in equations (9-59) and (9-60)

$$\alpha = \frac{\lambda z}{\pi} \sin\left(\frac{2N\pi a^2}{\lambda z}\right) + \frac{1}{2(\pi/a^2 + \pi/\lambda z)}\left[\left\{\cos 2\pi N\left(1 + \frac{a^2}{\lambda z}\right)\right\} - 1\right]$$

$$+ \frac{1}{2(\pi/a^2 - \pi/\lambda z)}\left[\left\{\cos 2\pi N\left(1 - \frac{a^2}{\lambda z}\right)\right\} - 1\right] \tag{9-59}$$

The first term of (9-59) can be written as a $(\sin x)/x$ as follows

$$\frac{2Na^2 \sin (2N\pi a^2)/\lambda z}{2Na^2 (\pi/\lambda z)} \text{ which when } \frac{\pi}{\lambda z} \rightarrow 0,$$

becomes $2Na^2$ since the $(\sin x)/x = 1$ when $x = 0$. The remaining two terms tend to zero as $\pi/\lambda z \rightarrow 0$ since

$$\cos 2Na^2\left(\frac{\pi}{a^2} + \frac{\pi}{\lambda z}\right) = \cos 2\pi N = 1$$

Therefore

$$\alpha \rightarrow 2Na^2 \tag{9-70}$$

as $\pi/\lambda z \rightarrow 0$. We can evaluate β in (9-60) by rewriting it as follows:

$$\beta = -\frac{\lambda z}{\pi}\left[\left\{\cos\left(\frac{2\pi Na^2}{\lambda z}\right)\right\} - 1\right] + \frac{1}{2(\pi/a^2 - \pi/\lambda z)} \sin 2Na^2\left(\frac{\pi}{a^2} - \frac{\pi}{\lambda z}\right)$$

$$+ \frac{1}{2(\pi/a^2 + \pi/\lambda z)} \sin 2Na^2\left(\frac{\pi}{a^2} + \frac{\pi}{\lambda z}\right) \tag{9-71}$$

Now as $\pi/\lambda z \to 0$ the first term leads to zero since

{ No!! the term is indet. but it does go to 0.}

$$\cos 0 = 1 \quad \text{and} \quad [(\cos 0) - 1] = 0$$

The second and third terms are also zero since

$$\sin 2\pi N = 0$$

and therefore

$$\boxed{\beta = 0} \quad as \quad \frac{\pi}{\lambda z} \to 0 \tag{9-72}$$

We can find the power spectrum (as $\pi/\lambda z \to 0$) by substituting in (9-43) the values of α and β as found in equations (9-70) and (9-72) giving

$$P\left(\frac{\pi}{\lambda z}\right) = \frac{\alpha^2 + \beta^2}{4} = \frac{[2Na^2]^2}{4} = N^2 a^4 \tag{9-73}$$

The intensity as $\pi/\lambda z \to 0$ is found from (9-26) as follows

$$I(\lambda, z) = I_0(\lambda)\left(\frac{\pi}{\lambda z}\right)^2 P\left(\frac{\pi}{\lambda z}\right) = I_0(\lambda)\left(\frac{\pi}{\lambda z}\right)^2 (N^2 a^4)$$

but since

$$\pi/\lambda z \to 0, \text{ then } I(\lambda, z) \to 0 \text{ and therefore } z \to \infty. \tag{9-74}$$

These results indicate that as $\pi/\lambda z \to 0$ we find that the DC term of the power spectrum will be $N^2 a^4$ and the corresponding intensity of the light at this focus ($z \to \infty$) will be zero. At the focus where $\pi/\lambda z \to \pi/a^2$ the power spectrum term is given by $(N^2 a^4)/4$ and the intensity is $(\pi^2 N^2)/4$. We can plot the power spectrum by normalizing as follows:

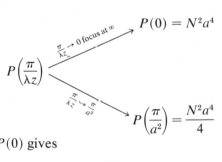

Multiplying by $1/P(0)$ gives

$$\frac{P(0)}{P(0)} = 1 \tag{9-75}$$

$$\frac{P(\pi/a^2)}{P(0)} = \frac{1}{4} \tag{9-76}$$

which can be plotted as shown in Figure 9.10.

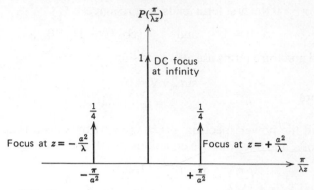

Figure 9.10 Power spectrum of a sine-wave zone plate as a function of $\pi/\lambda z$.

Figure 9.10 can be redrawn to show the location of the intensity peaks with respect to the distance z. This is done in Figure 9.11 by noting that in Figure 9.10 the focus on the $\pi/\lambda z$ axis is at π/a^2 or

$$\frac{\pi}{\lambda z} = \frac{\pi}{a^2}$$

$$z = \frac{a^2}{\lambda}$$

The focal point at a^2/λ on the z axis has an intensity of $\pi^2 N^2/4$. The point on the z axis at a^2/λ corresponds to a real focal point. The point at $-a^2/\lambda$ corresponds to a virtual focal point; that is, the focal points at $+a^2/\lambda$ and $-a^2/\lambda$ are the real and virtual focal points, respectively, which are similar to the real and virtual focal points of a lens. Note that the DC term of the zone plate's power spectrum does not produce a focal point since it corresponds to the intensity at $z = \infty$ which is zero.

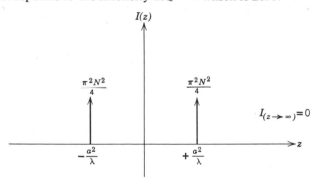

Figure 9.11 Intensity for a sine-wave zone plate as a function of the distance along the z axis.

CHARACTERISTICS OF A FRESNEL-TYPE ZONE PLATE

The Fresnel-type zone plate is essentially a square-wave approximation for the transmission function described in (9-34)

$$t(r) = \frac{1}{2}\left(1 - \sin\frac{\pi r^2}{a^2}\right) \qquad (9\text{-}34)$$

and illustrated in Figures 9.7 and 9.8. Let us assume the square wave has a transmission function $t(r)$ with spacings as shown in Figure 9.9. An expansion for a square wave such as shown in Figure 9.12 can be written as

$$t(v) = 0.5 - \frac{2}{\pi}\sin\frac{\pi v}{a^2} - \frac{2}{3\pi}\sin\frac{3\pi v}{a^2} - \frac{2}{5\pi}\sin\frac{5\pi v}{a^2}\cdots \qquad (9\text{-}77)$$

From (9-77) we can see that the average value of the square wave is 0.5. We can write the Fourier transform of a function as

$$F(\omega) = \int_{-\infty}^{\infty} t(v)\, e^{-j\omega v}\, dv$$

If we now let $\omega = \pi/\lambda z$, we find that the Fourier transform of (9-77) can be written

$$F\left(\frac{\pi}{\lambda z}\right) = \int_{-\infty}^{\infty} t(v)\, e^{-j\pi v/\lambda z}\, dv$$

This can be rewritten as a sum of integrals covering the range from $-\infty$ to $+\infty$

$$F\left(\frac{\pi}{\lambda z}\right) = \int_{-\infty}^{0} t(v)\, e^{-j\pi v/\lambda z}\, dv + \int_{0}^{R^2} t(v)\, e^{-j\pi v/\lambda z}\, dv + \int_{R^2}^{\infty} t(v)\, e^{-j\pi v/\lambda z}\, dv$$

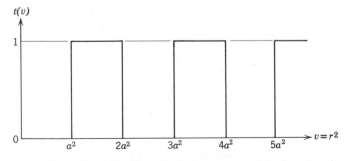

Figure 9.12 Transmission function $t(r)$ of a square wave plotted as a function of v.

Since v equals r^2, negative values of v are not possible. Therefore $t(v)$ is set equal to zero for negative v. For all values of v within the range of integration of the first and third integrals $t(v)$ is zero and therefore these integrals are zero. Therefore we can write the Fourier transform of $t(v)$ (9-77) as

$$F\left(\frac{\pi}{\lambda z}\right) = \int_0^{R2} t(v)\, e^{-j\pi v/\lambda z}\, dv$$

$$= \int_0^{R2} \left(0.5 - \frac{2}{\pi}\sin\frac{\pi v}{a^2} - \frac{2}{3\pi}\sin\frac{3\pi v}{a^2} - \cdots\right)(e^{-j\pi v/\lambda z})\, dv \qquad (9\text{-}78)$$

$$= \int_0^{R2} 0.5\, e^{-j\pi v/\lambda z}\, dv - \int_0^{R2} \frac{2}{\pi}\sin\frac{\pi v}{a^2}\, e^{-j\pi v/\lambda z}\, dv - \int_0^{R2} \cdots$$

$$= 0.5\int_0^{R2} e^{-j\pi v/\lambda z}\, dv - \frac{2}{\pi}\int_0^{R2} \sin\frac{\pi v}{a^2}\, e^{-j\pi v/\lambda z}\, dv - \int_0^{R2} \cdots$$

$$= 0.5\int_0^{R2} e^{-j\pi v/\lambda z}\, dv - \frac{2}{\pi}\int_0^{R2} \frac{e^{j(\pi v/a^2 - \pi v/\lambda z)} - e^{-j(\pi v/a^2 + \pi v/\lambda z)}}{2j}\, dv - \int_0^{R2} \cdots$$

$$= 0.5\int_0^{R2} e^{-j\pi v/\lambda z}\, dv - \frac{1}{\pi j}\int_0^{R2} e^{j(\pi v/a^2 - \pi v/\lambda z)}\, dv$$

$$+ \frac{1}{\pi j}\int_0^{R2} e^{-j(\pi v/a^2 + \pi v/\lambda z)}\, dv - \int_0^{R2} \cdots \quad \cdots \quad all\ terms$$

$$= 0.5\left[-\frac{\lambda z}{j\pi} e^{-j\pi v/\lambda z}\right]_0^{R2} - \frac{1}{\pi j}\left[\frac{1}{j(\pi/a^2 - \pi/\lambda z)} e^{j(\pi v/a^2 - \pi v/\lambda z)}\right]_0^{R2}$$

$$+ \frac{1}{\pi j}\left[\frac{1}{-j(\pi/a^2 + \pi/\lambda z)} e^{-j(\pi v/a^2 - \pi v/\lambda z)}\right]_0^{R2}$$

$$= 0.5\left[-\frac{\lambda z}{j\pi}\{e^{-j\pi R2/\lambda z} - 1\}\right] + \frac{1}{\pi(\pi/a^2 - \pi/\lambda z)}\left[e^{j(\pi R2/a^2 - \pi R2/\lambda z)} - 1\right]$$

$$+ \frac{1}{\pi(\pi/a^2 + \pi/\lambda z)}\left[e^{-j(\pi R2/a^2 + \pi R2/\lambda z)} - 1\right] \qquad (9\text{-}79)$$

$$F\left(\frac{\pi}{\lambda z}\right) = \frac{1 - e^{-j\pi R2/\lambda z}}{2j\pi/\lambda z} + \frac{1}{\pi(\pi/a^2 - \pi/\lambda z)}\left[e^{j(\pi R2/a^2 - \pi R2/\lambda z)} - 1\right]$$

$$- \frac{1}{\pi(\pi/a^2 + \pi/\lambda z)}\left[1 - e^{-j(\pi R2/a^2 + \pi R2/\lambda z)}\right] \qquad (9\text{-}80)$$

Writing (9-80) using the exponential form of the sine function, that i

$$\sin \theta = \frac{e^{j\theta} - e^{-j\theta}}{2j}$$

$$= \frac{\exp\left[-j(\pi R^2/2\lambda z)\right]}{\pi/\lambda z}\left[\frac{\exp\left[j(\pi R^2/2\lambda z)\right] - \exp\left[-j(\pi R^2/2\lambda z)\right]}{2j}\right]$$

$$+ \frac{2j\exp\left[j(R^2/2)(\pi/a^2 - \pi/\lambda z)\right]}{\pi(\pi/a^2 - \pi/\lambda z)}$$

$$\times\left[\frac{\exp\left[j(R^2/2)(\pi/a^2 - \pi/\lambda z)\right] - \exp\left[-j(R^2/2)(\pi/a^2 + \pi/\lambda z)\right]}{2j}\right]$$

$$- \frac{2j\exp\left[-j(R^2/2)(\pi/a^2 + \pi/\lambda z)\right]}{\pi(\pi/a^2 + \pi/\lambda z)}$$

$$\times\left[\frac{\exp\left[j(R^2/2)(\pi/a^2 + \pi/\lambda z)\right] - \exp\left[-j(R^2/2)(\pi/a^2 + \pi/\lambda z)\right]}{2j}\right]$$

$$= \exp\left[-j(\pi R^2/2\lambda z)(R^2/2)\right]\frac{\sin(\pi R^2/2\lambda z)}{\pi R^2/2\lambda z}$$

$$+ j\exp\left[j\frac{R^2}{2}\left(\frac{\pi}{a^2} - \frac{\pi}{\lambda z}\right)\right]\left(\frac{2}{\pi}\frac{R^2}{2}\right)\frac{\sin(R^2/2)(\pi/a^2 - \pi/\lambda z)}{(R^2/2)(\pi/a^2 - \pi/\lambda z)}$$

$$- j\exp\left[-j\frac{R^2}{2}\left(\frac{\pi}{a^2} + \frac{\pi}{\lambda z}\right)\right]\left(\frac{R^2}{2}\frac{2}{\pi}\right)\frac{\sin(R^2/2)(\pi/a^2 + \pi/\lambda z)}{(R^2/2)(\pi/a^2 + \pi/\lambda z)} \qquad (9\text{-}81)$$

DC & 1st Harmonic

Each of the terms is a $(\sin x)/x$ form and each is negligible far away from its center peak. Therefore each term in (9-81) can be evaluated as follows

$$\frac{\pi}{\lambda z} \to 0 \qquad F\left(\frac{\pi}{\lambda z}\right) \approx e^{-j(\pi R^2/2\lambda z)}\frac{R^2}{2}\frac{\sin(\pi/R^2/2\lambda z)}{(\pi R^2/2\lambda z)} \qquad (9\text{-}82)$$

$$\frac{\pi}{\lambda z} \to \frac{\pi}{a^2} \qquad F\left(\frac{\pi}{\lambda z}\right) \approx j\exp\left[j\frac{R^2}{2}\left(\frac{\pi}{a^2} - \frac{\pi}{\lambda z}\right)\right]\frac{R^2}{\pi}\frac{\sin(R^2/2)(\pi/a^2 - \pi/\lambda z)}{(R^2/2)(\pi/a^2 - \pi/\lambda z)}$$

$$(9\text{-}83)$$

$$\frac{\pi}{\lambda z} \to -\frac{\pi}{a^2} \qquad P\left(\frac{\pi}{\lambda z}\right) \approx -j\exp\left[-j\frac{R^2}{2}\left(\frac{\pi}{a^2} + \frac{\pi}{\lambda z}\right)\right]\left(\frac{R^2}{\pi}\right)$$

$$\frac{\sin(R^2/2)(\pi/a^2 + \pi/\lambda z)}{(R^2/2)(\pi/a^2 + \pi/\lambda z)} \qquad (9\text{-}84)$$

The power spectrum is defined as

$$P\left(\frac{\pi}{\lambda z}\right) = F\left(\frac{\pi}{\lambda z}\right)F^*\left(\frac{\pi}{\lambda z}\right)$$

then

$$\frac{\pi}{\lambda z} \to 0 \qquad P\left(\frac{\pi}{\lambda z}\right) \approx \frac{R^4}{4}\left[\frac{\sin\,(\pi R^2/2\lambda z)}{(\pi R^2/2\lambda z)}\right]^2 \to \frac{R^4}{4} \tag{9-85}$$

$$\frac{\pi}{\lambda z} \to \frac{\pi}{a^2} \qquad P\left(\frac{\pi}{\lambda z}\right) \approx \frac{R^4}{\pi^2}\left[\frac{\sin\,(R^2/2)\,(\pi/a^2-\pi/\lambda z)}{(R^2/2)\,(\pi/a^2-\pi/\lambda z)}\right]^2 \to \frac{R^4}{\pi^2} \tag{9-86}$$

$$\frac{\pi}{\lambda z} \to -\frac{\pi}{a^2} \qquad P\left(\frac{\pi}{\lambda z}\right) \approx \frac{R^4}{\pi^2}\left[\frac{\sin\,(R^4/2)\,(\pi/a^2+\pi/\lambda z)}{(R^4/2)\,(\pi/a^2+\pi/\lambda z)}\right]^2 \to \frac{R^4}{\pi^2} \tag{9-87}$$

From (9-26) we have the intensity

$$I = I_0\left(\frac{\pi}{\lambda z}\right)^2 P\left(\frac{\pi}{\lambda z}\right)$$

$$\frac{\pi}{\lambda z} \to 0 \qquad I \to 0$$

$$\frac{\pi}{\lambda z} \to \frac{\pi}{a^2} \qquad I \to I_0\left(\frac{\pi}{a^2}\right)^2\frac{R^4}{\pi^2} = I_0\frac{R^4}{a^4} \to I_0\,4N^2$$

$$\frac{\pi}{\lambda z} \to -\frac{\pi}{a^2} \qquad I \to I_0\left(\frac{\pi}{a^2}\right)^2\frac{R^4}{\pi^2} = I_0\frac{R^4}{a^4} \to I_0\,4N^2$$

since $R^2 = 2Na^2$

Plots of these results for the Fresnel-type zone plate are shown in Figure 9.13. It is interesting to compare the sine-wave zone plate with the Fresnel-type zone plate. From the solution of the sine-wave zone plate (summarized in Figure 9.11) we can see that there are only three focal points, one at infinity (intensity = 0), one virtual, and one real. The Fresnel-type zone plate has an infinite number of focal points in addition to those of the sine-wave plate. Figure 9.13 shows the corresponding points for the Fresnel-type zone plate as shown in Figures 9.10 and 9.11 for the sine-wave zone plate. If we compare the real focal points at π/a^2 for the two types of zone plates, we see that

$$\frac{I_\text{sine wave}}{I_\text{Fresnel}} = \frac{(\pi^2 N^2 I_0)/4}{4N^2 I_0} = \frac{\pi^2}{16} = \left(\frac{\pi}{4}\right)^2$$

that is, the intensity of the light at the point π/a^2 will be $(\pi/4)^2$ greater for the sine-wave zone plate than for the Fresnel-type zone plate. Figure 9.14 shows additional focal points for the Fresnel-type zone plate that do not exist for the sine-wave zone plate.

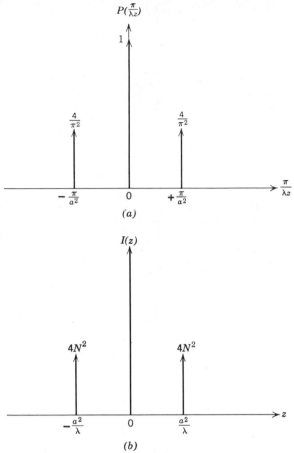

Figure 9.13 Power spectrum and intensity plots for a Fresnel-type zone plate (DC and first harmonic). (a) Power spectrum plotted as a function of $\pi/\lambda z$. (b) Intensity plotted as a function of z.

Since the power spectrum is defined as

$$P\left(\frac{\pi}{\lambda z}\right) = F\left(\frac{\pi}{\lambda z}\right)F^*\left(\frac{\pi}{\lambda z}\right)$$

we can see by inspection of (9-77) where each focal point occurs. We have solved for the DC focal point and the focal point corresponding to π/a^2. The component of the power spectrum at π/a^2 was found to have a value of R^4/π^2 in (9-86). By normalizing the zero-order terms to one (divide by $R^4/4$) the normalized value at π/a^2 is $(2/\pi)^2 = 4/\pi^2$ as shown in Figure 9.13. By similar reasoning we can conclude that other com-

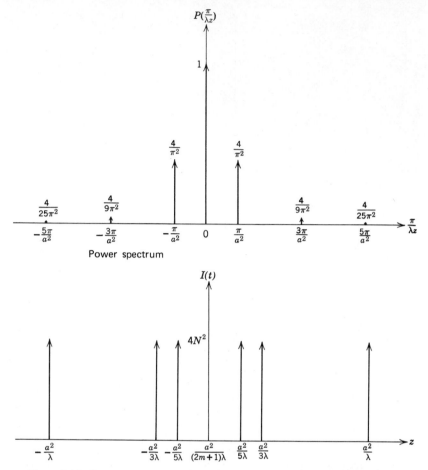

Figure 9.14   Power spectrum and intensity plots for the Fresnel-type zone plate.

ponents of the power spectrum for (9-77) occur at $3\pi/a^2, 5\pi/a^2, 7\pi/a^2 \cdots$
with corresponding (normalized) power spectrum components of $(2/3\pi)^2$,
$(2/5\pi)^2$, $(2/7\pi)^2 \cdots$. These points are shown plotted in Figure 9.14. The
intensities corresponding to each focal point is determined from (9-26).
Using (9-26) we can see that the intensity of each focal point will be the
same, that is,

$$I(\lambda, z) = I_0 \left(\frac{\pi}{\lambda z}\right)^2 P\left(\frac{\pi}{\lambda z}\right)$$

$$= I_0 \left(\frac{\pi}{\lambda z}\right)^2 \left[\frac{R^4}{4} P_N\left(\frac{\pi}{\lambda z}\right)\right]$$

where
$$P\left(\frac{\pi}{\lambda z}\right) = \frac{R^4}{4} P_N\left(\frac{\pi}{\lambda z}\right)$$

$P_N(\pi/\lambda z)$ is the normalized power spectrum and $R^4/4$ is the normalization factor from (9-85). Since $R^2 = 2Na^2$,

$$\frac{R^4}{4} = N^2 a^4$$

When $\dfrac{\pi}{\lambda z}$ equals	P_N equals	I equals
$\dfrac{\pi}{a^2}$	$\left(\dfrac{2}{\pi}\right)^2$	$I_0 \left(\dfrac{\pi}{a^2}\right)^2 N^2 a^4 \left(\dfrac{2}{\pi}\right)^2 = I_0 4 N^2$
$\dfrac{3\pi}{a^2}$	$\left(\dfrac{2}{3\pi}\right)^2$	$I_0 \left(\dfrac{3\pi}{a^2}\right)^2 N^2 a^4 \left(\dfrac{2}{3\pi}\right)^2 = I_0 4 N^2$
.	.	.
.	.	.
.	.	.
.	.	.
$\dfrac{(2m+1)\pi}{a^2}$	$\left(\dfrac{2}{(2m+1)\pi}\right)^2$	$I_0 \left[\dfrac{(2m+1)\pi}{a^2}\right]^2 N^2 a^4 \left[\dfrac{2}{(2m+1)\pi}\right]^2 = I_0 4 N^2$

With the approximation made the intensities at all focal points will be the same $(4N^2)$ but their locations will get closer together as $(2m+1)\lambda$ increases.

Let us now consider the exactness of the focus of a zone plate. We can consider the exactness of the focus by the distance that must be moved from the maximum intensity point (the focus) to the half power points on each side of the focus along the z axis. We shall first consider the solution for the Fresnel type of zone plate. The higher-order harmonics in (9-78) for the Fresnel-type zone plate can be considered in the same manner as the first harmonic was in (9-81). For the $(2m+1)$ harmonic an expression of the form

$$\frac{\sin \dfrac{\pi R^2}{2}\left(\dfrac{2m+1}{a^2} - \dfrac{1}{\lambda z}\right)}{\dfrac{\pi R^2}{2}\left(\dfrac{2m+1}{a^2} - \dfrac{1}{\lambda z}\right)} \tag{9-88}$$

is obtained which describes the amplitude distribution around the focal point at

$$\lambda z = \frac{a^2}{2m+1} \tag{9-89}$$

while the term

$$j \exp\left[j\frac{\pi R^2}{2}\left(\frac{2m+1}{a^2}-\frac{1}{\lambda z}\right)\right]$$

describes the corresponding phase distribution (where m is the order of the focus). From the $(\sin x)/x$ table we can see that when $x = 80°$, $(\sin x)/x \approx 0.707$ and therefore

$$\left(\frac{\sin x}{x}\right)^2 \approx \frac{1}{2}$$

If we let

$$x = \frac{\pi R^2}{2}\left(\frac{2m+1}{a^2}-\frac{1}{\lambda z}\right)$$

in (9-88), then when $x = 80°$, we have

$$x = \frac{\pi R^2}{2}\left(\frac{2m+1}{a^2}-\frac{1}{\lambda z}\right) = 80° \times \frac{\pi}{180°} = \frac{4\pi}{9}$$

$$\left(\frac{2m+1}{a^2}-\frac{1}{\lambda z}\right) = \left(\frac{4\pi}{9}\right)\left(\frac{2}{\pi R^2}\right) = \frac{8}{9R^2}$$

and since

$$R^2 = 2Na^2$$

then

$$\frac{2m+1}{a^2}-\frac{1}{\lambda z} = \frac{8}{9(2Na^2)} = \frac{4}{9Na^2} \tag{9-90}$$

If we now let z_f be the distance along the z axis to the mth focal point, then

$$\frac{2m+1}{a^2} = \frac{1}{\lambda z_f} \tag{9-91}$$

and

$$a^2 = (2m+1)\lambda z_f \tag{9-92}$$

and substitution (9-91) and (9-92) in (9-90) we get

$$\frac{1}{\lambda z_f}-\frac{1}{\lambda z} = \frac{4}{9N\left[(2m+1)\lambda z_f\right]} \tag{9-93}$$

Solving (9-86) for $z - z_f$ we get

$$\frac{z-z_f}{\lambda z z_f} = \frac{4}{9N\left[(2m+1)\lambda z_f\right]}$$

$$z - z_f = \frac{4(\lambda z z_f)}{9N(2m+1)\lambda z_f} = \frac{4z}{9N(2m+1)} \tag{9-94}$$

Solving (9-87) for z

$$z_f = z - \frac{4z}{9N(2m+1)} = z\left(1 - \frac{4}{9N(2m+1)}\right)$$

or

$$z = \frac{z_f}{1 - 4/[9N(2m+1)]} \tag{9-95}$$

Now substituting the value of z in (9-95) into (9-94) we get

$$z - z_f = \frac{4}{9N(2m+1)} = \frac{4}{9N(2m+1)}\left[\frac{z_f}{1 - 4/[9N(2m+1)]}\right]$$

$$= \frac{4z_f}{9N(2m+1) - 4} \tag{9-96}$$

Figure 9.15 shows the meaning of (9-96); z_f is a focal point as indicated by the peak value of the $(\sin x)/x$ curve. The difference in distance along the z axis of the half power point is the value $z - z_f$ given by (9-96).

Comparing equations (9-38) and (9-78) we can see that the first two terms of (9-78) are the same as those for a sine-wave zone plate. The only difference is an amplitude factor that does not affect the shape of the light intensity distribution near the focus (it just affects the amplitudes). Therefore the exactness of the focus of a sine-wave zone plate will be the same as that of the principal focal point ($m = 0$) of the Fresnel type of zone plate given by (9-96). From (9-96) we can see that the width of the focus of a zone plate is inversely proportional to the number of zones; that is, the greater the number of zones, the greater the resolution. Equation (9-96) indicates that for the approximations made, as the number of zones or the order of the focus increases so does the resolution.

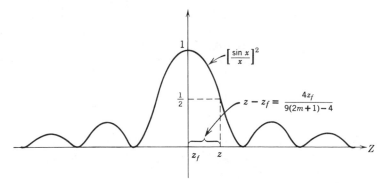

Figure 9.15 Resolution at a focal point.

As shown in Figure 9.14 the intensities of all focal points will be the same and their spacing along the z axis decreases (as N and m increase).

We have shown in Figure 9.2 that a Fresnel-type zone plate can be made by a combination of drafting and photography. A similar technique can be used for producing a sine-wave zone plate but this would not be as simple a procedure as for the Fresnel-type zone plate. The Fresnel-type zone plate can be considered to have a binary transmission function; that is, it is either opaque or transparent (a zero or a one). The sine-wave zone plate has a transmission function that varies sinusoidally. To draft a function that varies in density sinusoidally and then to photograph it so as to produce a sinusoidal transmission function is a relatively difficult task. There are other techniques that can produce a sine-wave zone plate with relative ease.

A technique for producing a sine-wave type of zone plate is by photographing Newton's rings. Fringes are produced when a flat plane glass plate M_0 and a spherical surface S are brought in contact with each other as shown in Figure 9.16. When a planoconvex lens with a radius of curvature R for its lower surface is placed in contact with the glass plate, Newton's rings can be formed. When the focal length of the lens is large enough, the rings can be viewed directly. It will be shown that the greater the focal length of the lens, the larger the radius of the Newton's rings.

Interference fringes are formed by reflection of the light from the lower surface S_1 of the lens interfering with the reflection of light from the glass plate M_1. Because the air space between the lower surface of the lens and

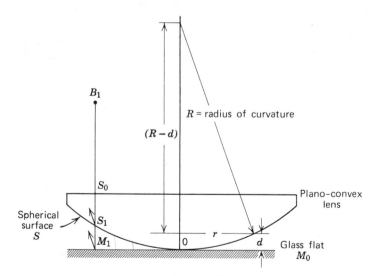

Figure 9.16 Formation of Newton's rings.

the glass plate is of variable thickness, d, the path length of the light from the lower surface of the lens to the glass plate is of variable length. When light strikes perpendicularly to the flat surfaces, the interference fringes appear circular with a common center around the point of contact. From the geometry of Figure 9.16 we can see

$$(R-d)^2+r^2 = R^2 \qquad (9\text{-}97)$$
$$R^2-2dR+d^2+r^2 = R^2 \qquad (9\text{-}98)$$

geom. only

$$-2dR+d^2 = -r^2 \qquad (9\text{-}99)$$
$$d(2R-d) = r^2$$

but since $d \ll R$

$$r^2 = 2Rd \qquad (9\text{-}100)$$

If we consider one incident ray, B_1, as shown in Figure 9.16, we can see that a portion of the light will be reflected at each surface. If the incident light is perpendicular to the flat surfaces, we can disregard the reflections from the top surface of the lens because these only tend to reduce uniformly the intensity to the lower reflecting surfaces. A portion of the incident ray, B_1, is reflected at S_1 and the remaining unreflected portion proceeds on to M_1 where it is also partially reflected. The portion of the ray striking M_1 will travel a distance, d, from S_1 to M_1, or a total of $2d$ for the light from S_1 to M_1 and back to S_1. This light will interfere with the light being reflected from S_1. When the path length, $2d$, is an integral number of wavelengths, we might expect to find constructive interference. Actually this is not the case. When the path length, $2d$, is an integral number of wavelengths, we find destructive interference. The reason for this is that when light is reflected so that one of the reflections takes place in a medium of high index of refraction at the boundary of a medium of low index of refraction (point S_1), and the other reflection takes place in a medium of low index of refraction at the boundary of a medium of high index of refraction (point M_1), there is always a 180-degree phase difference between the two reflections (plus any phase change due to path-length differences). In effect we can consider that the reflection at the lower surface M_1 introduces an additional phase shift of 180 degrees (or $\lambda/2$, where λ is the wavelength), so that when $d = \lambda/4$, $2d = \lambda/2$ and the light going from S_1 to M_1 and back to S_1 will experience a total phase shift of λ. Therefore when $d = \lambda/4$, we can expect constructive interference. This additional phase shift of $\lambda/2$ from the glass plate is indicated by the fact that the center ring is dark. This indicates that at the point of contact there must be destructive interference. The only way this can occur is if there is a phase shift of $\lambda/2$ upon reflection as stated.

Let us now examine the paths of an incident ray perpendicular to the plane top surface of the lens. In Figure 9.16 a ray B_1 is shown to strike the top surface of the lens at point S_0. At the upper surface S_0 a small portion of the light will be reflected back upon itself but the majority of the light will proceed through the lens to point S_1. At point S_1 a portion of the light is again reflected but the majority of the light will proceed on through the lens to be reflected by the glass plate at point M_1. The light reflected from point M_1 will interfere with the light being reflected from point S_1. The path length from point S_1 to M_1 and back to M_1 determines whether this interference will be constructive or destructive (or some intermediate value).

At the point where $r = 0$, that is, $d = 0$ a dark fringe is formed due to the 180 degree phase shift of reflection at the lower plane surface. The optical path determining the interference is therefore given by

$$\text{optical path} = 2d = \frac{r^2}{R} = n\lambda \qquad (9\text{-}101)$$

where n is the number of the dark ring.

$$r_n^2 = nR\lambda \qquad (9\text{-}102)$$

where r_n is the radius of the nth dark ring

$$r_n = \sqrt{nR\lambda} \qquad (9\text{-}103)$$

From this we can see that the radii of the rings are proportional to the square roots of natural numbers and the distances between successive rings decrease as n increases which is the definition we used for a zone plate. Therefore Newton's rings are in effect zone plates.

An interesting point to consider is the case where there is a slight separation h between the lens and the plane surface; that is, the lens and the glass plate are not in contact with each other as shown in Figure 9.17. In such a case interference fringes would still be produced but the center fringe can be adjusted to be any value from dark to light (by adjusting the separation between the lens and the glass plate). We have seen that there is a 180 degree phase shift between the reflection from the lower surface of the lens (corresponding to point C) and the reflection from the flat glass surface (corresponding to point D). This phase shift of 180 degrees can be considered to occur when the light is reflected at the plane glass surface or when reflections occur in a medium of low index of refraction from a surface of high index of refraction. This change of 180 degrees or $\lambda/2$ in the phase of the light must be added to any change in phase caused by the separation $(h + d)$. We can therefore deduce that the differ-

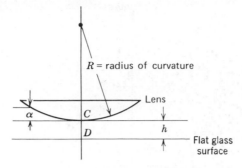

Figure 9.17   Formation of interference patterns by separating lens and mirror.

ence in path length between the light reflected from the lower lens surface and that coming from the plane glass surface is given by

$$\text{optical path} = \frac{\lambda}{2} + 2d + 2h \qquad (9\text{-}104)$$

where the $\lambda/2$ is added to account for the additional phase shift from the lower plane glass. From this we can conclude that when

$$\frac{\lambda}{2} + 2d + 2h = \begin{cases} \dfrac{2n+1}{2}\lambda \text{ gives destructive interference} \\ n\lambda \qquad \text{gives constructive interference} \end{cases} \qquad (9\text{-}105)$$

or

$$2d + 2h = 2h + \frac{r^2}{R} = \begin{cases} n\lambda \qquad \text{gives destructive interference} \\ \dfrac{2n-1}{2}\lambda \text{ gives constructive interference} \end{cases} \qquad (9\text{-}106)$$

depending on how we introduce the $\lambda/2$ phase change on reversal from the lower plane glass surface. We therefore can also use the equation

$$2h + \frac{r^2}{R} = n\lambda \qquad \text{gives destructive interference} \qquad (9\text{-}107)$$

or

$$r^2 + 2hR = n\lambda R \qquad (9\text{-}108)$$

We can determine the intensity of the light produced by the interference of the light being reflected from surface M_1 and S_1 by referring to Figure 9.18. From Figure 9.18 we can see that the separation between the surface S_1 (spherical surface of lens in Figures 9.16 and 9.17) and M_1 (glass flat of Figures 9.16 and 9.17) is

$$d = \frac{r^2}{2R} + h \qquad (9\text{-}109)$$

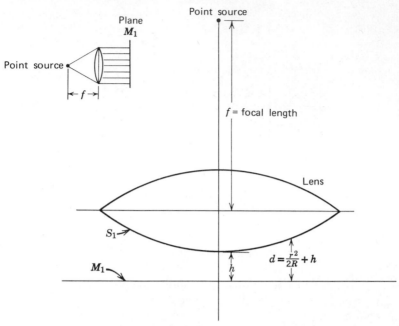

Figure 9.18   Newton's rings formation.

Let us consider that plane waves are incident perpendicular to the glass flat M_1. Figure 9.18 shows a point source of light at a focal distance from a convex lens producing a collimated light beam (plane waves) incident on M_1. In Figures 9.16 and 9.17 the light source had to be broad so that plane waves can be obtained.

From Figure 9.18 we see that plane waves can be made to be incident perpendicularly to plane M_1. Let us assume that the plane waves incident at M_1 have zero phase. The incident light on surface S_1 of Figure 9.18 can be written as $\exp\left[-jk(r^2/2R+h)\right]$. This is obtained by noting that the path length for the incident waves to the surface S_1 is shorter than the path length to the surface M_1 by a distance $r^2/2R+h$. Multiplying this distance by $k=2\pi/\lambda$ we obtain the difference in phase in radians; that is, dividing by λ gives the number of cycles and multiplying by 2π gives radians. The light reflected from M_1 has a phase at S_1 of $\exp\left[jk(r^2/2R+h)+j\pi\right]$. By similar reasoning we note that the path length for light incident to M_1 and reflected back to S_1 is longer than just the path length to M_1; we use a positive exponent whereas in the previous case the path length was shorter and we used a negative exponent. The additional phase term of $j\pi$ is present in the second expression because the light is shifted in phase 180 degrees or π radians upon reflection at M_1.

Let us assume that the light has an amplitude of $A/2$. Therefore the total light amplitude at any point will be given by the amplitude $A/2$ times the phase term (which is the sum of the incident light-phase term plus the phase term of the reflected light) that is,

total light amplitude

$$
= \frac{A}{2}\left\{\exp\left[-jk\left(\frac{r^2}{2R}+h\right)\right] + \exp\left[jk\left(\frac{r^2}{2R}+h\right)\right] + j\pi\right\}
$$

$$
= \frac{A}{2}\left\{\exp\left[-jk\left(\frac{r^2}{2R}+h\right)\right] + \exp\left[jk\left(\frac{r^2}{2R}+h\right)\right]e^{j\pi}\right\}
$$

$$
= \frac{A}{2}e^{j(\pi/2)}\left\{\exp\left[-jk\left(\frac{r^2}{2R}+h+\frac{\pi}{2k}\right)\right] + \exp\left[jk\left(\frac{r^2}{2R}+h+\frac{\pi}{2k}\right)\right]\right\}
$$

$$
= Ae^{j(\pi/2)}\left[\frac{\exp\left[-jk(r^2/2R+h+\pi/2k)\right] + \exp\left[jk(r^2/2R+h+\pi/2k)\right]}{2}\right]
$$

$$
= Ae^{j(\pi/2)}\cos k\left(\frac{r^2}{2R}+h+\frac{\pi}{2k}\right) \tag{9-110}
$$

The factor $e^{j(\pi/2)}$ in (9-110) contributes a constant phase that does not affect the results (e.g., it could have been eliminated if we assumed that that reference phase at M_1, Figure 9.18, was $-(\pi/2)$ instead of zero). We can therefore write (9-110) for the total light amplitude as

$$
\text{total light amplitude} = A\cos k\left(\frac{r^2}{2R}+h+\frac{\pi}{2k}\right)
$$

$$
= A\cos\frac{2\pi}{\lambda}\left(\frac{r^2}{2R}+h+\frac{\pi\lambda}{4\pi}\right)
$$

$$
= A\cos\left(\frac{2\pi r^2}{2R\lambda}+h\frac{2\pi}{\lambda}+\frac{2\pi^2\lambda}{4\pi\lambda}\right)
$$

$$
= A\cos\left(\frac{\pi r^2}{R\lambda}+\frac{2\pi h}{\lambda}+\frac{\pi}{2}\right) \tag{9-111}
$$

The intensity at any point (square of amplitude) is found by squaring (9-111) or

$$
\text{intensity} = A^2\cos^2\left(\frac{\pi r^2}{R\lambda}+\frac{2\pi h}{\lambda}+\frac{\pi}{2}\right)
$$

and since

$$
\cos\frac{1}{2}x = \sqrt{\frac{1+\cos x}{2}}
$$

therefore
$$\cos^2 \frac{1}{2}x = \frac{1}{2} + \frac{\cos x}{2}$$

$$= \frac{A^2}{2}\left\{1 + \cos\left(\frac{2\pi r^2}{R\lambda} + \frac{4\pi h}{\lambda} + \pi\right)\right\}$$

and since
$$\cos(x + \pi) = -\cos x$$

$$\text{intensity} = I = \frac{A^2}{2}\left\{1 - \cos\left(\frac{2\pi r^2}{R\lambda} + \frac{4\pi h}{\lambda}\right)\right\}$$

$$= \frac{A^2}{2}\left\{1 - \cos\frac{2\pi}{\lambda R}(r^2 + 2hR)\right\} \tag{9-112}$$

Plots of (9-112) are shown in Figure 9.19. From Figure 9.19 or equation (9-112) we can see that

$$I = 0$$

where
$$\cos\frac{2\pi}{\lambda R}(r^2 + 2hR) = 1$$

or
$$\frac{2\pi}{\lambda R}(r^2 + 2hR) = 2n\pi$$

where $n = 1, 2, 3, \ldots$. Then

$$2\pi(r^2 + 2hR) = 2\lambda R n\pi$$

$$r_n^2 + 2hR = \lambda R n \tag{9-113}$$

$$r_n^2 = \lambda R n - 2hR = R(\lambda n - 2h)$$

The radius of any zone for any n can be written as

$$r_n = \sqrt{R(\lambda n - 2h)} \tag{9-114}$$

and if $h = 0$,
$$r_n = \sqrt{\lambda R n} \tag{9-115}$$

Using (9-113) we can determine whether a particular pattern under consideration is a zone plate; and if it is a zone plate, we can determine the ring spacing and the intensity caused by the interference between the incident and reflected light (Figure 9.18). We can see from Figures 9.19(a), (b) and (c) that if we took a picture of the zone plate formed as shown in Figures 9.17 or 9.18, the intensity would be a sinusoidal function when plotted against r^2 or $r^2 + 2hR$. When we look at the actual interference pattern formed and plot the intensity as a function of the radius (not r^2 or $r^2 + 2hR$), the plot as shown in Figure 9.19(d) becomes more repre-

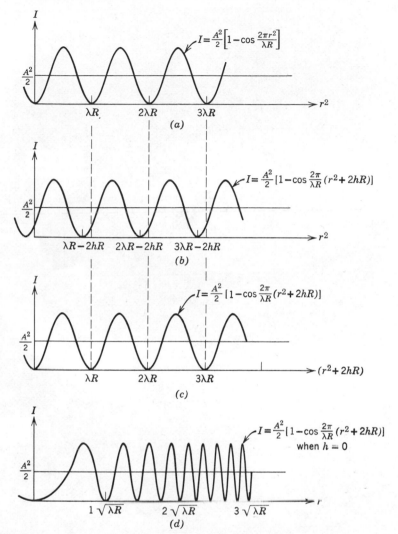

Figure 9.19 Plots of equation (9-112). (*a*) Plot of (9-112) [where $h=0$] versus r^2. (*b*) Plot of (9-112) versus r^2. (*c*) Plot of (9-112) versus (r^2+2hR). (*d*) Plot of (9-112) versus r for $h=n\lambda/2$.

sentative of what we see; that is, the location of intensity peaks producing the zone plate varies as the square root of natural numbers as indicated by equations (9-114) and (9-115).

Figure 9.20 shows a photograph of a 100-cm focal-length lens that was placed over an optical flat. A 100-cm lens is a rather long focal-length

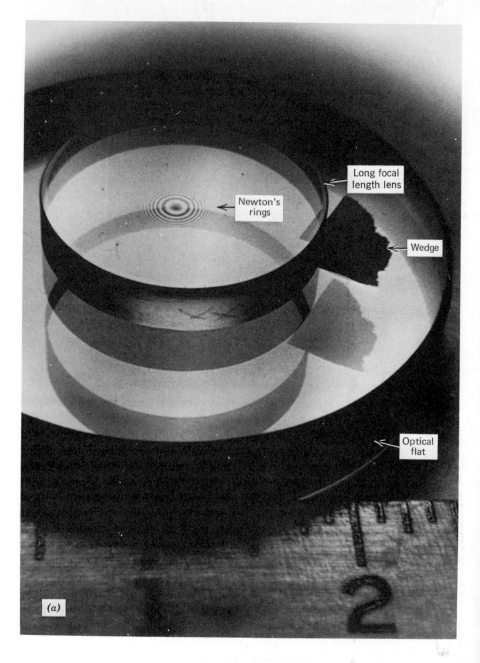

Figure 9.20a $\quad h = \dfrac{n\lambda}{2}$

430

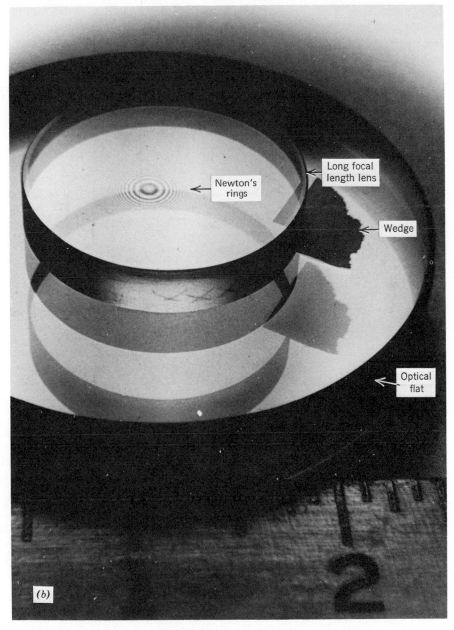

Figure 9.20b $\quad h = \dfrac{\lambda(2n-1)}{4}$

Lens over optical flat forming Newton's rings by reflected light.

lens and even so the sizes of the Newton's rings are rather small. With smaller focal length lenses the Newton's rings formed are smaller and may be easily overlooked. Figure 9.20 was photographed with light being reflected from the lens. Figure 9.20(a) shows the lens in contact with the optical flat or with spacing

$$h = \frac{n\lambda}{2}$$

between the lens and the optical flat; that is, with $d = 0$ and from (9-105) we get

$$\frac{\lambda}{2} + 2d + 2h = \frac{2n+1}{2}\lambda$$

$$2h = \frac{2n\lambda}{2} + \frac{\lambda}{2} - \frac{\lambda}{2}$$

$$h = \frac{n\lambda}{2} \qquad \text{for destructive interference}$$

Figure 9.20(b) shows the center of the Newton's rings clear; that is, there was constructive interference. This means that the spacing between the lens and the optical flat is again given by (9-105) or

$$\frac{\lambda}{2} + 2d + 2h = n\lambda$$

$$2h = n\lambda - \frac{\lambda}{2}$$

$$h = \frac{n\lambda}{2} - \frac{\lambda}{4} = \frac{4n\lambda - 2\lambda}{8} = \frac{2\lambda(2n-1)}{8}$$

$$h = \frac{\lambda(2n-1)}{4}$$

The wedges shown in Figure 9.20 were used to separate the lens from the optical flat to produce various values of h. By examination of Figure 9.19(d) and 9.20(a) it would appear that they correspond.

From Figure 9.19(d) we see the center intensity is zero. Figure 9.20(a) (which is a positive print, i.e., it appears the same as the original scene) shows that the center Newton ring is dark (zero intensity). We notice also that the spacing between the rings gets smaller as r increases and this is shown in Figure 9.19(d). Figure 9.20(b) shows the center ring to be light. This means that the spacing between the lens and the optical flat had to be such that constructive interference occurred at the center (with reflected light)

$$h = \frac{\lambda(2n-1)}{4}$$

Figure 9.21(a) shows a plot of the light intensity that exposed the Newton rings of Figure 9.20(b); that is, the intensity shown in Figure 9.21(a) exposed a negative, which was used to make the positive shown in Figure 9.20(b).

Up to this point we have been considering Newton's rings (zone plates) made by reflected light. It is possible to form Newton's rings by transmitted light. Figure 9.22 shows a convex lens in contact with an optical flat. The optical flat is illuminated perpendicularly with a portion of the light from ray AB reaching point C and is transmitted to point D. Another

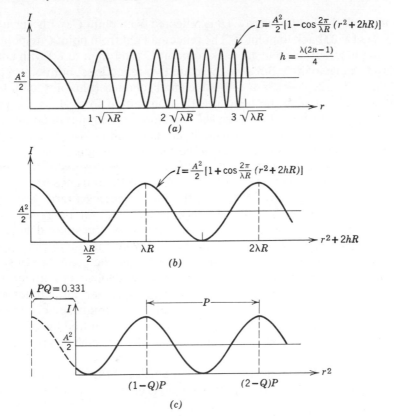

Figure 9.21   (a) Plot of equation (9-112) versus r for $h = \lambda(2n-1)/4$. (b) Plot of equation (9-116) versus $(r^2 + 2hR)$. (c) Plot of equation (9-123).

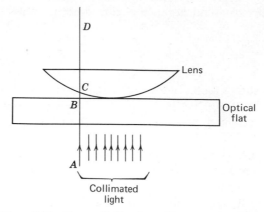

Figure 9.22 Newton's rings formed by transmitted light.

portion of the light from ray AB is reflected from point C (with a change of phase of $\lambda/2$) back to point B. The reflected light from point C to point B is again reflected (at point B) back to point C (and then to D) with another phase change of $\lambda/2$. Because there are two reflections, both with a phase change of $\lambda/2$, there is a full wavelength change (plus any path-length difference) between the light directly transmitted from A to D and that from A to C to B to D. We can see therefore at the point of contact, with only the two reflections (no path length between points B and C), there will be constructive interference at the center or a bright spot.

The interference phenomena discussed can be observed therefore by either reflected light or transmitted light. When transmitted light is used to view the Newton's rings, the interference patterns are caused by two reflections. The interference fringes viewed by reflection are therefore more apparent than when viewed by transmitted light. The reason for this is that in the reflection viewing the interfering rays are of approximately equal amplitude, whereas in the case of transmitted light the ray is reflected twice (causing greatly reduced amplitude). In the transmitted light viewing the difference between the amplitudes of the two rays is so great that complete cancellation cannot be accomplished, which makes viewing difficult. The rings with transmitted light will be the compliment of the reflected light pattern; that is, dark rings in one are bright rings in the other and vice versa.

Equation (9-112) indicated that the intensity of the interference pattern (Newton's rings) formed by reflected light is given by

$$I = \frac{A^2}{2}\left\{1 - \cos\frac{2\pi}{\lambda R}(r^2 + 2hR)\right\} \tag{9-112}$$

and is shown plotted in Figure 9.19(c). The corresponding equation for the intensity of the interference pattern (Newton's rings) formed by transmitted light is given by

$$I = \frac{A^2}{2}\left\{1 + \cos\frac{2\pi}{\lambda R}(r^2 + 2hR)\right\} \tag{9-116}$$

A plot of (9-116) is shown in Figure 9.23(b). Comparing the intensities of Figures 9.19(c) and 9.22(b) we see there is a phase difference of π between them. Let us now consider Figure 9.24(a) and determine if these rings are Newton's rings (a sinusoidal zone plate). Referring back to (9-113) we can write this equation for two successive rings on a zone plate (made with reflected light)

$$r_n^2 + 2hR = \lambda Rn \tag{9-113}$$

$$r_{(n+1)}^2 + 2hR = \lambda R(n+1) \tag{9-117}$$

Subtracting (9-117) from (9-113) we find

$$r_n^2 - r_{(n+1)}^2 = \lambda Rn - \lambda R(n+1) = \lambda R[n - (n+1)] = -\lambda R$$

$$r_{(n+1)}^2 - r_n^2 = \lambda R \tag{9-118}$$

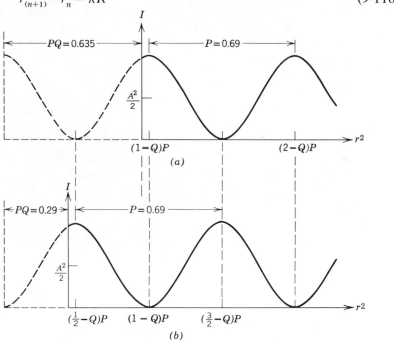

Figure 9.23 Comparison of phase shift for transmitted and reflected light Newton's rings. (*a*) Intensity for transmitted light. (*b*) Intensity for reflected light.

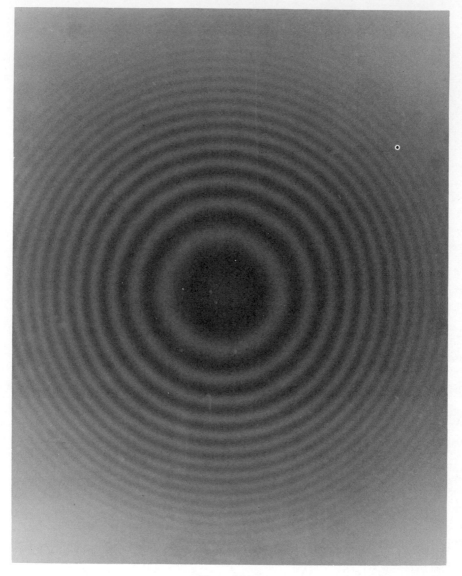

Figure 9.24*a***.**

From (9-118) we see that we can measure two successive radii (r_n and $r_{(n+1)}$) and from these measurements λR can be calculated. With λR known we can use (9-115)

$$r_n = \sqrt{\lambda R n} \qquad (9\text{-}115)$$

to find any radius and by use of (9-112) we can find the phase shift,

$$I = \frac{A^2}{2}\left[1 - \cos\frac{2\pi}{\lambda R}(r^2 + 2hR)\right] = \frac{A^2}{2}\left[1 - \cos 2\pi\left(\frac{r^2}{\lambda R} + \frac{2h}{\lambda}\right)\right]$$

and if we let $\lambda R = P$ and

$$\frac{2h}{\lambda} = Q$$

then (9-112) becomes

$$I = \frac{A^2}{2}\left[1 - \cos 2\pi\left(\frac{r^2}{P} + Q\right)\right] \tag{9-119}$$

The maximum intensity therefore occurs when

$$\cos 2\pi\left(\frac{r^2}{P} + Q\right) = -1$$

or

$$2\pi\left(\frac{r^2}{P} + Q\right) = (2n - 1)\pi$$

where $n = 1, 2, 3, 4, \ldots$.

$$\frac{r_n^2}{P} + Q = \frac{(2n - 1)}{2} \tag{9-120}$$

By substituting the value $P = \lambda R$ in (9-118) we obtain

$$r_{(n+1)}^2 - r_n^2 = P \tag{9-121}$$

We can now solve (9-120) for Q

$$Q = \frac{(2n - 1)}{2} - \frac{r_n^2}{P} \tag{9-122}$$

Equations (9-121) and (9-122) are for reflected light. Let us now find the corresponding equations for transmitted light. The corresponding equation to (9-119) for transmitted light is given by

$$I = \frac{A^2}{2}\left[1 + \cos 2\pi\left(\frac{r^2}{P} + Q\right)\right] \tag{9-123}$$

The maximum intensity therefore occurs when

$$\cos 2\pi\left(\frac{r^2}{P} + Q\right) = 1 \tag{9-124}$$

or
$$2\pi\left(\frac{r^2}{P}+Q\right) = 2n\pi \qquad (9\text{-}125)$$

Then
$$r_n^2 = (n-Q)P$$

$$r_{(n+1)}^2 = (n+1-Q)P$$

$$r_{(n+1)}^2 - r_n^2 = P \qquad (9\text{-}126)$$

and
$$Q = n - \frac{r_n^2}{P} \qquad (9\text{-}127)$$

From this we see that the frequency [equations (9-121) and (9-126)] of the function describing the Newton's rings is the same whether it is viewed by reflected or transmitted light. The phase, however, does change, [(9-122) and (9-127)] when viewed by reflected or transmitted light.

Let us now examine the Newton's rings of Figure 9.24(a) which were made by transmitted light as illustrated in Figure 9.22. We see immediately from Figure 9.24(a) that the center is dark (little or no light). We know that Newton's rings made by transmitted light will have a clear, or bright, center if the lens is in contact with the optical flat. We therefore know that there was a space between the lens and the optical flat causing a phase shift. By plane geometry techniques we can locate the center of the Newton's rings and measure the various radii. Careful measurement (a small measurement error can cause a large error) results in us finding the following values for each radius:

$$r_1 = 0.63$$
$$r_2 = 1.04$$
$$r_3 = 1.32$$
$$r_4 = 1.56$$
$$r_5 = 1.78$$
$$r_6 = 1.96$$
$$r_7 = 2.13$$
$$r_8 = 2.29$$
$$r_9 = 2.42$$

From (9-126) we see that we can take any two consecutive radii and determine the value of P (the frequency that the intensity function varies); for example, let us consider r_8 and r_9,

$$r_{n+1}^2 - r_n^2 = r_9^2 - r_8^2 = P$$

$$= 2.42^2 - 2.29^2 \approx 0.69$$

The approximation sign is used here because the value 0.69 was found to be a good fit to the radii measured. This was found by taking the difference of the squares or all the radii measured; that is, equation (9-126) was applied to all radii (r_1 through r_9) and the value 0.69 was found to be a good fit. Using this value for P the value of Q is found to be 0.48; for example, for the third bright ring

$$Q = n - \frac{r_n^2}{P} = 3 - \frac{1.32^2}{0.69} = 3 - \frac{1.74}{0.69} = 0.48$$

We notice that the value of Q (by definition) has to be positive. Figure 9.21(c) shows a plot of the intensity of the zone-plate rings shown in Figure 9.24(a). We see that the origin (as compared with the plot in Figure 9.21(b) is moved to a position corresponding to $0.48(2\pi)$ rad. The center of the zone plate of Figure 9.24(a) therefore has almost no light intensity and does not have a bright ring (maximum) until 1.04π rad. out from the center.

Let us now consider the rings shown in Figure 9.24(b). By plane geometry we can find the center and measure the radii to the various bright rings trying to measure to the peak brightness. Careful measurement shows the radii to be

$$r_1 = 0.25$$
$$r_2 = 0.87$$
$$r_3 = 1.20$$
$$r_4 = 1.46$$
$$r_5 = 1.67$$
$$r_6 = 1.87$$
$$r_7 = 2.04$$
$$r_8 = 2.20$$
$$r_9 = 2.35$$
$$r_{10} = 2.50$$
$$r_{11} = 2.62$$
$$r_{12} = 2.75$$
$$r_{13} = 2.86$$

Using either (9-121) or (9-126) we find P to be on the average

$$r_{n+1}^2 - r_n^2 = 0.69$$

From this we can make the assumption that these rings (at least up to the thirteenth ring) form a sinusoidal zone plate. Let us make the further assumption that these Newton rings (sinusoidal zone plate) were made by

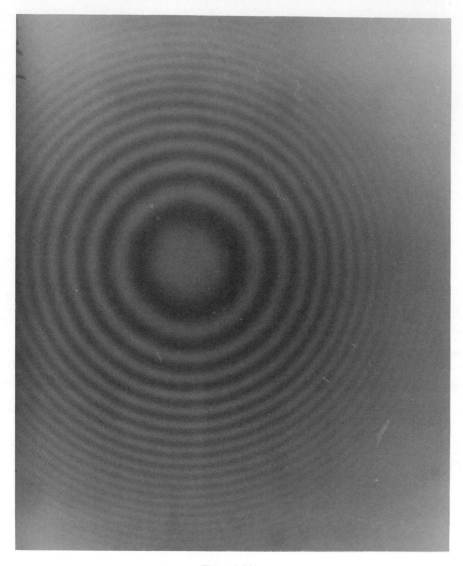

Figure 9.24b.

transmitted light. We can then find Q using (9-127) (and for illustration we can pick, say, the fourth ring); then

$$Q = n - \frac{r_n^2}{P} = 4 - \frac{2.13}{0.69} = 4 - 3.08 = 0.92$$

Figure 9.23(a) shows the intensity plotted against r^2 assuming that the Newton's rings of Figure 9.24(b) were made by transmitted light. Notice that the phase and amplitude of Figure 9.23(a) are in close agreement with Figure 9.24(b).

Let us now suppose that the Newton's rings of Figure 9.24(b) were made by reflected light. The value of P would still be the same (0.69), but the value of Q would be different. Q for reflected light is given by (9-122) and has a value of

$$Q = \frac{2n-1}{2} - \frac{r_n^2}{P} = \frac{7}{2} - \frac{2.13}{0.69} = 3.50 - 3.08 = 0.42$$

Figure 9.23(b) shows the intensity plotted against r^2 assuming that the Newton's rings of Figure 9.24(b) were made by reflected light. We notice that the phase and amplitude of Figure 9.23(b) are in close agreement with Figure 9.24(b).

When Newton's rings are constructed with transmitted light the interference pattern will be the complement of the reflected light pattern; that is, dark rings in one are bright rings in the other and vice versa. Newton's rings can also be viewed with white light. With white light there is a black center spot. As the radius increases from the center different monochromatic components of the source interfere causing colored rings to form around the black center. Further out from the center the appearance is that of uniform white-light illumination. When white light is used, the color of the reflected light from any point is complementary to the color of the transmitted light.

As the plane glass plate and the lens are moved apart the radius of the circle formed by the points (for a given value of d) will decrease. The interference rings will therefore appear to collapse toward the center (while new rings will appear to form at the periphery) as the distance h is increased. For each increase in separation of one quarter of a wavelength between the lens and glass plate the dark rings will be replaced by bright ones and vice versa. We can see therefore that we can make the center dark or any value, depending on whether transmitted or reflected light is used and the separation between the lens and optical flat.

Figures 9.24(a) through (m) show photographs of the Newton rings (taken by transmitted light) for the lens and optical flat shown in Figure 9.20. The interference pattern between the lens and optical flat changes when viewed by transmitted light as the separation between them is changed. Although the patterns appear different they are all zone plates with identical focal lengths. The difference between the various zone plates is the value of Q as calculated from (9-127). To be more specific; the periods P of each of these zone plates is the same (the same lens was used

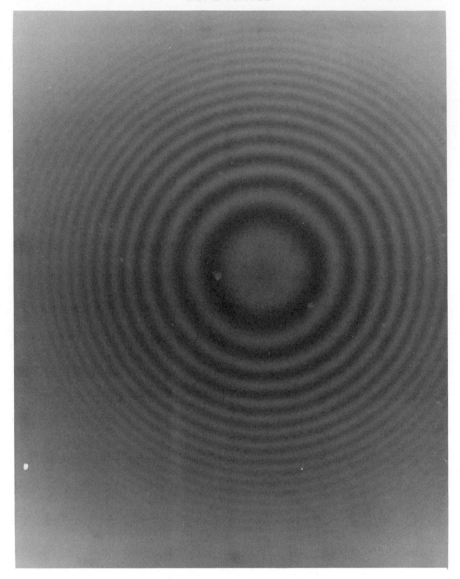

Figure 9.24c.

in all cases) and only the separation between the lens and optical flat was changed. As the separation changed so did the phase of the intensity pattern but not the respective periods.

It is important to understand Figure 9.24 when trying to analyze the

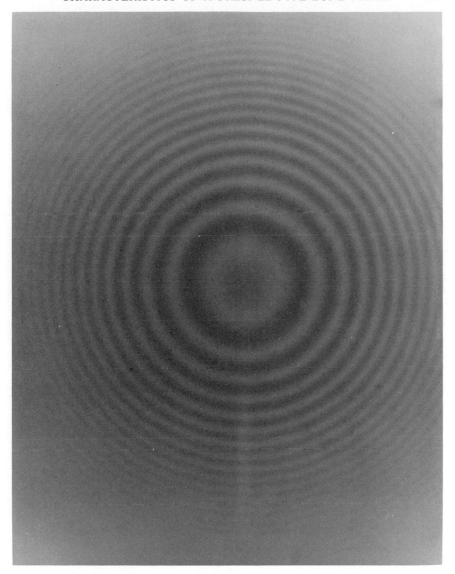

Figure 9.24*d.*

characteristics and formation of holograms in terms of the formation of multiple zone plates. A hologram can be considered as the superposition of many zone plates so positioned that an image (in 3D) can be formed by the individual focusing actions. We can consider a simple

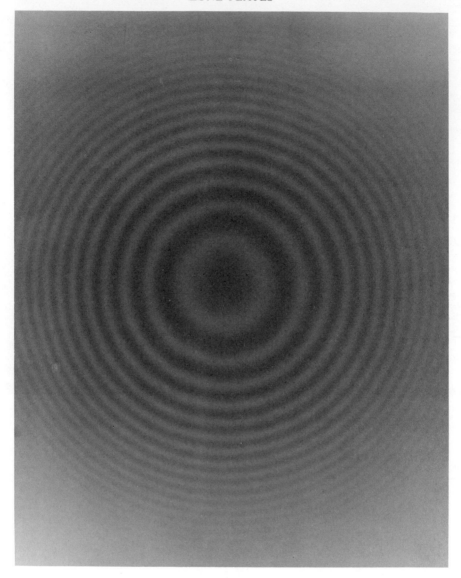

Figure 9.24*e*.

case of a hologram made of two point sources (that is, the hologram is the superposition to two zone plates). In order to superimpose two zone plates one on another the sum of their transmission functions should be no greater than one. To insure this we can let each of the amplitude

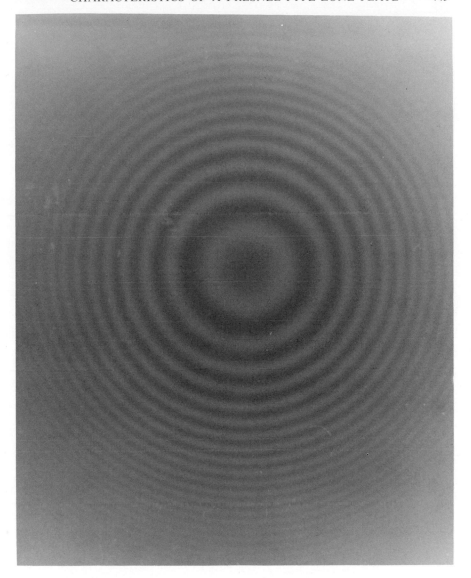

Figure 9.24f.

transmissions functions to be

$$0 < t_1(v) < 0.5 \quad \text{and} \quad 0 < t_2(v) < 0.5$$

This insures that at no point on the transmission function will exceed a

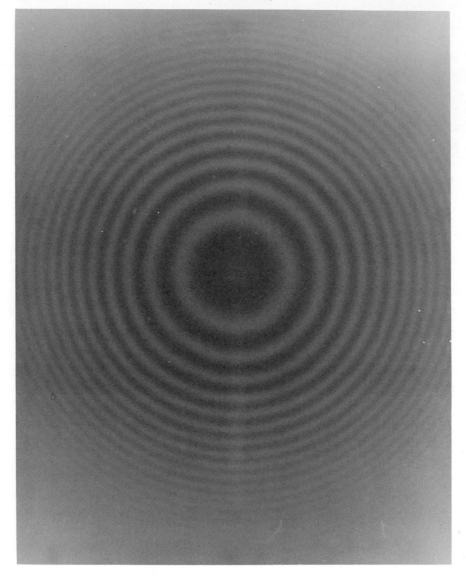

Figure 9.24*g***.**

transmission of one. When the amplitude of the transmission function is changed from (0 to 1) to (0 to 0.5), the intensity [intensity = (amplitude)^2] of each focus will be reduced to one quarter of its previous value. From

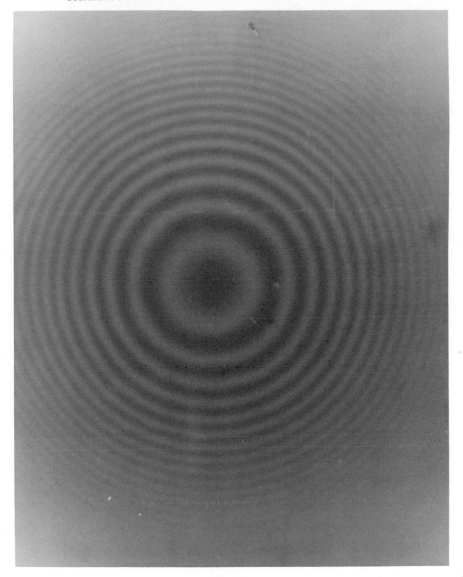

Figure 9.24h.

this we can see that a superposition of n zone plates (hologram of n point sources) will cause the intensity to fall to a value $1/n^2$ of the value of what a single zone plate can be on the same film.

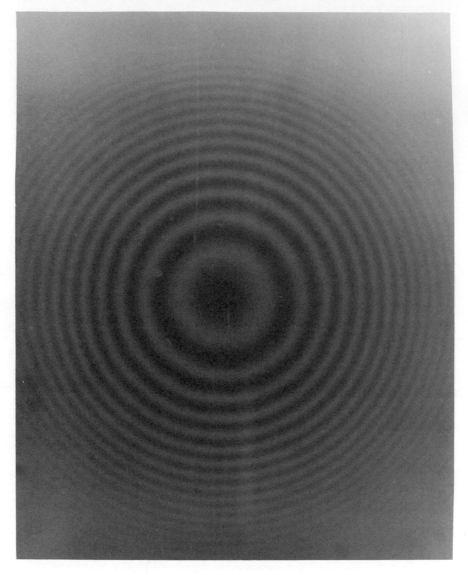

Figure 9.24*i*.

We have seen that a hologram can be considered as superimposed zone plates that are collectively capable of reconstructing an image. Let us consider a hologram consisting of two zone plates (i.e., the focal point of

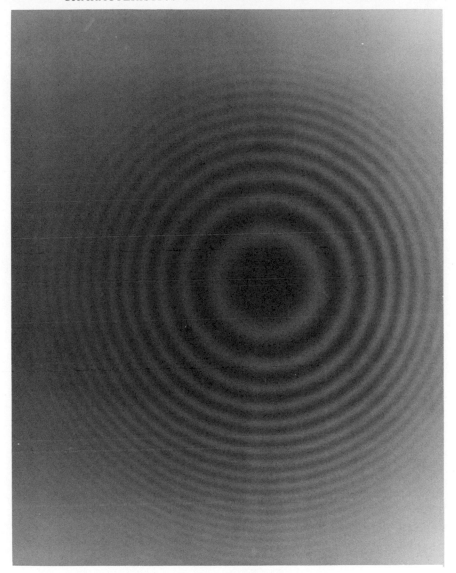

Figure 9.24*j*.

each zone plate produces one of the image points). We have seen that the focal point always lies on the axis of the zone plate. The distance that the focal point lies from the zone plate (on the axis of the zone plate) is a

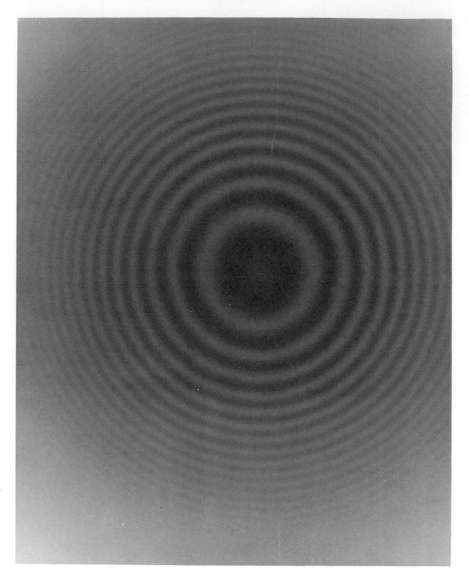

Figure 9.24*k*.

function of the wavelength of the light being used (see 9-3). We can see,
therefore, that the image produced by a hologram will not change its size
when viewed with different wavelengths of light. The location of the image

Figure 9.24*l*.

with respect to the hologram will vary depending on the wavelength of the light being used. In this sense therefore the hologram cannot be used for magnifying the image by changing the wavelength of light when viewing.

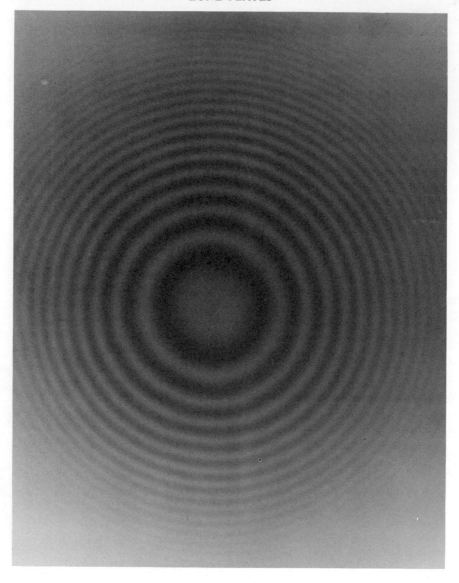

Figure 9.24*m***.**

10

Holography

GENERAL DISCUSSION OF PHOTOGRAPHY

Probably one of the main contributing factors that brought the photographic process into being was man's urge to be artistic. Louis Daguerre, who pioneered the photographic process, was originally an artist. William Talbot probably pioneered work in photography because he had difficulty drawing. Photography was compared initially with paintings or drawings. The ability of the photographic process to record minute detail was considered all that was necessary to satisfy every artistic need. As photographic techniques advanced one main drawback became apparent. The great length of time required to make an exposure did not permit moving objects to be recorded. Slow-moving objects were blurred and indistinct when photographed. The first photographic processes were not able to record colors but black and white drawings were common, so that photography was considered a simple method of producing artistic drawings.

The basic tools of photography are the camera and the photographic emulsion. It is usually erroneously believed that the camera can reproduce nature as the human eye sees it. The camera records objects as the unaided human eye can never see them. The human eye focuses on a small area at a time. An entire object or scene is seen only by letting the eye rove over the object in a series of short jumps. The brain sorts and integrates these small portions of the object so that the entire scene appears continuous. In this process the brain does something that the camera cannot. The brain automatically discards information that is considered unimportant while emphasizing others considered important according to individual preference. The camera records every detail within its field of view. The image is recorded with fine detail and with subtle tonal graduations. The images thus produced contain a great amount of information concerning the object. Photography over the years has been directed toward obtaining more information in the image concerning the object. There are some basic limitations which inherently limit the amount of information that a photographic image can contain concerning the object; for example, the maximum range of brightness in a photographic image is limited by the photographic emulsion, the resolution of the photographic image is limited by the lens capabilities as well as of the photographic emulsion, and the image of a three-dimensional object is reduced to a two-dimensional image. There are other problems associated with normal photography that limit its ultimate capabilities. These limitations are mainly caused by the method of viewing. Let us consider a colored three-dimensional scene to be photographed. The image will be produced on the two-dimensional photographic film emulsion. The finished two-dimensional image can be a transparency (and is viewed by transmitted light) or a print (and can be viewed by reflected light). In either case a replica of the original scene will be seen.

We will be able to discern that we are looking at a replica of the scene (and not the actual scene) for several reasons as follows:

(1) The film cannot accurately record colors and the color content of the light source used for viewing will further distort the color composition of the image.

(2) Distortion will be introduced unless the scene is viewed from the same relative perspective that the camera lens saw the scene. The image will not change in perspective relative to the original scene as the angle of viewing is changed. Changing the angle of viewing only distorts the image.

(3) The limitations of the depth of field (depth of the scene that is in focus) of the lens will not make the image of the scene truly representative; for example, if a scene is in focus from 20 ft to infinity, any objects from the film up to 20 ft will appear blurred. In viewing the image, if the eye is focused on an object that was 10 ft from the film, the blur will not be representative of the actual scene as it would be seen at the 10 ft distance.

(4) Any point in a scene that reflects light with high intensity (relative to other points in the object) will not be recorded with the same relative intensity (because of film limitations) as the remainder of the scene. A typical example of this is taking a picture of a person with the sun off to one side and behind the subject. This puts part of the person's face in a shadow. This type of exposure is usually quite difficult to make because of the limited contrast range of the film. If one exposes properly for the shadow detail, the bright sunlit area is overexposed. If the exposure is made for the bright sunlit area, the shadow detail will be underexposed and detail lost. Proper exposure in such a case is usually a compromise exposure.

(5) There is a limit to the amount of detail recorded on the photographic film; for example, if a picture is taken of a scene with a tree in the distance, there is a definite limit to how much the picture of the tree can be magnified. It is quite probable that details in the structure of any single leaf in the tree will not be discernible with magnification of the picture image, although if a telescope were placed where the camera took the picture such detail would be easily discernible. It is possible to take a picture of the telescopic magnified image but additional magnification of this photographic image would again be limited by the resolving power of the photographic film (and the lens system).

From the above it is apparent that there are certain basic limitations to the present photographic techniques. Let us suppose that we are in a position to order a photographic system without any of the above limitations. To be more specific let us determine what we would consider to be an ultimate photographic system (regardless of whether such a system could actually be achieved). The photographic system we shall now discuss (holography) is not intended as a replacement for present photography; it is really just another photographic technique.

INTRODUCTION TO HOLOGRAPHY

Probably the simplest way of visualizing an ultimate photographic recording system is by example. Let us imagine that we are in a room with one closed glass window through which we can see a scene. The closer we get to the window, the larger our view of the scene. By moving to the side of the window we can extend our field of view in a specific direction as is shown in Figure 10.1. In position A of Figure 10.1 we can see less of the scene than in position B which is closer to the window. In position C of Figure 10.1 we can see much further to the right than in either position A or B. From Figure 10.1 we can see that from position A, object 2 appears to the right of object 1. From positions B and C, object 2 appears to the left of object 1. We can also see that object 3 is visible from positions A and B but not from position C, while object 4 is only visible from position C.

In addition we can look at near or distant objects and see them in focus. By setting up a telescope in the room we can magnify either near or distant objects in the scene. The main point being brought out here is that by looking through the glass window we see the object scene. The scene's perspective and relative light intensities will change with the viewers relative position. An ultimate photographic system would be one that would allow us to remove the glass window, replacing it with a "picture" which would be such that no matter what tests we conducted from the

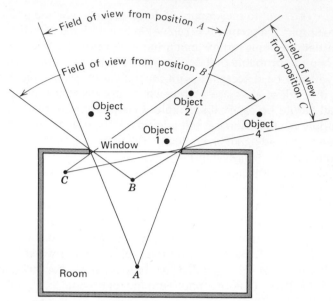

Figure 10.1 Field of view from several positions in a room.

room we would not be able to determine whether we were looking at the glass window or the "picture."

Let us redraw Figure 10.1 as shown in Figure 10.2 and examine it from a somewhat different viewpoint. If the object can be considered as a point

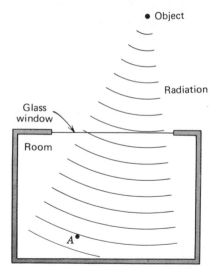

Figure 10.2 Coherent monochromatic radiation from a point source.

source of coherent monochromatic light, then spherical wavefronts will radiate from it as shown. From Figure 10.2 we can see that the radiation pattern through the glass window can easily be approximated.

From Figure 10.3 we can see that if we were able to produce the proper radiation pattern in the room, we would be able to "see" the object point source even when it is not present. Let us determine whether it is possible for us to replace the glass window with a "picture" that will permit us to "see" the point source object. We have seen in Chapter 9 on zone plates that a sine-wave zone plate will, when uniformly illuminated, diffract the light to a real and a virtual focal point at the same time. From Figure 10.3 we can see that if the window is a sine-wave zone plate with a virtual focus at the corresponding distance of the object from the window, no matter where we are in the room (radiation pattern) we will "see" the object. We can therefore, by substituting for our glass window a sine-wave zone plate and uniformly illuminating it as shown in Figure 10.4, obtain a three-dimensional type image of the object. That is, by changing our position in the room we would still see the object at the same relative position to the room but as we change our location in the room we would see the object at different relative positions to the apparent window (sine-wave zone plate). We have thus produced a simple three-dimensional image of a simple object. Let us consider how far we can extend this concept.

Let us suppose that we desire to make another three-dimensional "picture" of two point sources of monochromatic light (of the same

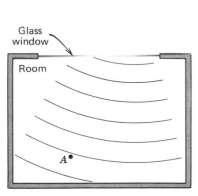

Figure 10.3 Desired radiation pattern through a glass window.

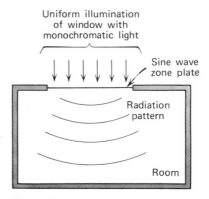

Figure 10.4 Virtual focus of a uniformly illuminated sine-wave zone plate.

frequency) as shown in Figure 10.5. We know from Young's experiment that two point sources which are spatially coherent (points on corresponding wavefronts can be made to interfere with each other) will produce an interference pattern of alternate bright and dark lines. If we consider the path length from point A to point source 2 as one-half wavelength longer than the path length from point A to point source 1, then point A will be dark. Under these conditions the amplitude of the light from point source 1 will cancel the amplitude of the light arriving from point source 2 at point A. This can only happen if the relative differences in the phases of the light leaving the point sources remain constant. The phase differences between the two point sources will remain constant if the two point sources are coherent. In our zone-plate analysis we saw that the various zones of the zone plate were constructed so that the diffracted light from all points on the zone plate arrive in phase (within 180°) at the focal point. We have also seen that Newton's rings can be made to produce sine-wave zone plates. These sine-wave zone plates were produced by the use of a planoconvex lens and an optical flat. The radius of curvature of the lens basically determines the focal length of the sine-wave zone plate produced. The phase of the light arriving at the focal point, with respect to that illuminating the sine-wave zone plate, can be changed by changing the spacing between the lens and the optical flat. We can therefore superimpose two sine-wave zone plates, each with the desired focal length, to produce the proper virtual point source images in Figure 10.5. We can also independently control the phase of the light from each of these two virtual point sources (by adjusting the spacing between the lens and the optical flat producing the zone plate).

We have now been able to construct two zone plates which, when superimposed on each other, will produce two virtual focal points at

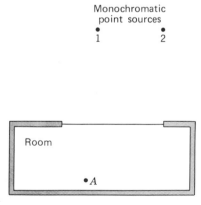

Figure 10.5   Formation of a radiation pattern between two monochromatic point sources.

desired locations with light being apparently emitted from them with the desired phase difference. We have not up to this point discussed the relative amplitudes of the two point sources. If we consider point source 1 to be twice the intensity of point source 2, then by simply adjusting the relative transmission functions of the two sine-wave zone plates we can control the relative intensities of the two virtual images. The maximum value of a transmission function is one. To allow for overlapping of the zone plates the sum of the transmission function of the two sine-wave zone plates must be no greater than one,

$$T_1 + T_2 = 1 \tag{10-1}$$

where T_1 is the maximum amplitude transmission of the zone plate corresponding to point source 1, and T_2 is the maximum amplitude transmission of the zone plate corresponding to point source 2.

Assuming the zone plates have the same number of zones the intensity of the image of T_1 will be twice the intensity of the image of point T_2 if

$$T_1^2 = 2T_2^2 \qquad \text{or} \qquad T_1 = \sqrt{2}T_2 \tag{10-2}$$

Substituting of T_1 in (10-1)

$$T_1 + T_2 = \sqrt{2}T_2 + T_2 = (1 + \sqrt{2})T_2 = 1$$

or

$$T_2 = \frac{1}{1 + \sqrt{2}} = 0.414 \tag{10-3}$$

and from (10-1)

$$T_1 + T_2 = 0.414 + T_1 = 1$$

$$T_1 - 0.586$$

We have now determined that we can superimpose sine-wave zone plates so as to place focal points at any desired location. We can independently control the amplitude of the light at each of these focal points as well as their phases. In addition there is no reason why we must limit ourselves to just two points. We can build up an image of any object by assuming it made up of many point sources. For each of the points (sources) in the object we can determine a corresponding zone plate which when placed in the window (and uniformly illuminated) would produce a virtual focus at the point (with the proper amplitude and phase). In this manner we can build up a virtual image of the object by superimposing the proper sine-wave zone plates in the window. We have thus constructed a picture (made up of sine-wave zone plates) which when uniformly illuminated will cause a virtual image to be formed. Let us now attempt to determine whether this virtual image formed by the sine-wave zone plates can be distinguished from the real object by measurements or observations made from inside the room.

Holographic Images — Real and Virtual

It was previously shown that a sine-wave zone plate has both a real and virtual focal point. We have seen that the virtual focus of many superimposed sine-wave zone plates can be formed to give rise to a virtual image. Likewise there is a real image formed by the same superimposed zone plates. From Figure 10.6 we can see that the real and virtual focal points are equally distant from the zone plate but on opposite sides of it.

Figure 10.7 shows the relative positions of the real and virtual images that might be formed from superimposed sine-wave zone plates. From position 1 in the room the real image would not be visible even though light is being "focused" to form an image because there is no light from the image reaching the eye. If a screen were placed where the real image is being formed, then by reflected light it would be possible to see the real image. Because the real image is being formed in three dimensions, the sharpest real image is formed only when the screen is shaped the same as the real image. If a screen (shown as a dotted line) were placed as shown in Figure 10.7, point A would appear sharp from position 1

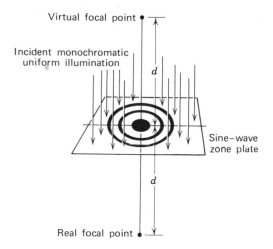

Figure 10.6   Real and virtual focal points of a uniformly illuminated zone plate.

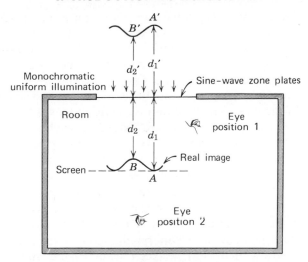

Figure 10.7   Real and virtual images from sine-wave zone plates.

while B, which is off the screen, would reflect out-of-focus light to the eye at position 1.

From position 2 in Figure 10.7 the real image would be visible (screen need not be used) because light from the image does reach the eye. This real image will be a three-dimensional type image; that is, the perspective will change with the observer's position in the room relative to the image.

Another interesting characteristic can be illustrated by Figure 10.7. Point A on the virtual image could be considered a low point on the object (it is further away from position 2 than point B'). Point B' on the virtual image can be considered a high point on the object (it is nearer to position 2 than point A'). The real image is just the opposite; that is, point A on the real image is a high point and point B is a low point. This means that if the virtual image is a true three-dimensional image, the real image is a pseudo-scopic image (a pseudo-scopic image can be seen in a three-dimensional viewer by reversing the two pictures; that is, the left picture is viewed by the right eye and vice versa).

Let us consider in more detail what we might observe from position 2 in Figure 10.7. Looking straight ahead we first see point B on the real image and behind it (as if superimposed) we see point B' on the imaginary image. This now introduces a condition that did not appear when we were looking out of the glass window at the object. Indeed this real image causes a condition that permits us to determine from inside the room whether we are looking at the glass window or the sine-wave zone plates;

for example, by use of a screen we can search for a real image and if it exists we will know the sine-wave zone plates are in the window.

The interference pattern formed by a plane wave and a ~~circular~~ *spherical* wave is a sine-wave type zone plate. Let us consider a special case so that we can analyze the formation of the real and virtual images in more detail. Figure 10.8 shows lens imaging monochromatic collimated light to a point (at the focus of the lens). The light diverges after it leaves the focal point. Two transparent screens A and A' are placed on each side of the focal point and a distance d from the focal point. Assuming there are no light losses as the light travels from transparent screen A to the focal point and from the focal point to transparent screen A', then light of equal but opposite phase illuminates both transparent screens. This can be seen by assuming the focal point as a point source of light. We know that a point source of light emits spherical waves. Two transparent screens equally distant from the point source will be illuminated identically by any particular wavefront. This can be seen from Figure 10.9 which is a representation of a portion of Figure 10.8. The portion of the spherical waves used in Figure 10.8 are darkened in Figure 10.9. The point P_1 is assumed to be equally distant from screens A and A'. At the instant shown a spherical wave radiating out from P_1 is just tangent to both screens (points 8 and 9). We can see in Figure 10.9 that if the wavefront starts with 0 degrees phase at point P_1 it will be 180 degrees out of phase at points 2 and 3. At points 4 and 5 it will be in phase having rotated 360 degrees. At points 6 and 7 it will again be in phase having rotated 720 degrees. Likewise at points 8 and 9 the wavefronts will be exactly in phase again (it is assumed the screens are spaced some integer multiple of wavelengths from point P_1) as this wavefront radiates outward further it will illuminate both screens identically; that is, its position on the screens at any instant relative to the axis will be the same. We can see, for example, that points 10 and 11 on the screens are equally distant from the axis and are being illuminated by different portions of the same wavefront. We know from Figure 10.8 that actually the light is propagating from transparent screen A to transparent screen A' and not as shown in

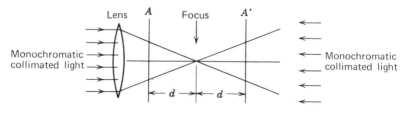

Figure 10.8   Zone-plate formation.

Figure 10.9 from point P_1 to screens A and A'. If we make the assumption that the phase at the focus in focus in Figure 10.8 is 0 degrees, we can assume that the phase of the light on screen A is $-\phi$; that is, if the light on screen A is $-\phi$, it will increase in phase to 0 degrees when traveling from the screen A to the focal point. There will be a further increase in phase of ϕ as the light travels from the focus to screen A'. This makes the phase on screen $A' = +\phi$ when the phase on screen $A = -\phi$. We have deliberately set up the screens to be a multiple of wavelengths from the focus so that the angle ϕ must be an integer multiple of 360 degrees so as to simplify the explanation. One additional simplification will be made and that is that the plane wavefronts arrive at the screens exactly when a spherical wavefront becomes tangent to the screen; for example, in Figure 10.9 plane wavefront **5** is at the screen just as a spherical wavefront is tangent to it at point 9. We will see later why these requirements simplify the explanation.

If we illuminate the screens A and A' in Figure 10.8 with both spherical waves (produced by the focusing action of the lens) and an additional collimated wave (obtained from the same monochromatic source), the interference pattern formed on both screens will be identical sine-wave zone plates — that is, provided the screens are an integer multiple of wavelength from the focus. Although both screens will have identical

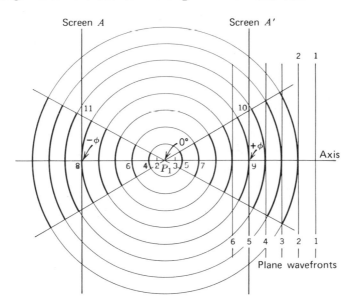

Figure 10.9   Determination of the phase of the spherical wavefronts illuminating screens A and A' of Fig. 10.8.

interference patterns (sine-wave zone plates), we have produced the one on screen A with converging spherical wavefronts while we have produced the identical sine-wave zone plate on screen A' with diverging spherical wavefronts.

Uniform illumination of the sine-wave zone plate on screen A' (Figure 10.10) with plane wavefronts will produce a real image (focus) at point F. When we uniformly illuminate screen A with plane wavefronts, we produce a virtual image (focus) at point F. Because the zone plates on both screens are identical, we can conclude that uniform illumination of sine-wave zone plate will produce both real and virtual images (focal points). Figure 10.11 shows one of the sine-wave zone plates of Figure 10.10 uniformly illuminated by plane wavefronts. The sine-wave zone plate diffracts this light in such a way that two spherical wavefronts are produced. One wavefront produced is a divergent spherical wavefront apparently originating at the virtual focus. The other wavefront produced is a convergent spherical wavefront that converges at the real focus.

Let us now consider the case in which the plane wavefronts do not arrive at the screens in phase with the spherical wavefronts (Figure 10.8). Figure 10.12 shows spherical wavefronts that are just tangent to the two screens A and A'. The plane wavefronts are shown to arrive at screen A' one quarter of a wavelength after the spherical wavefront was tangent to point 1. The plane wavefront will also arrive at screen A one quarter of a wavelength after the spherical wavefront was tangent at point 2. From this we can deduce that again the sine-wave zone plates produced on both screens will be identical. The appearance of the sine-wave zone plates will be different from those produced when the plane wavefronts arrive at the screen just as the spherical are tangent to them. In the previous chapter on zone plates we saw that sine-wave zone plates that have the same power spectrum will have the same focal points. We can see therefore that the sine-wave zone plates produced as in Figure 10.9 or 10.12 will have the same focal points. The phase of the light, however,

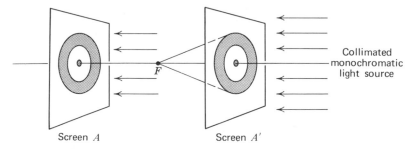

Figure 10.10 Focal points of zone plates produced as shown in Fig. 10.8.

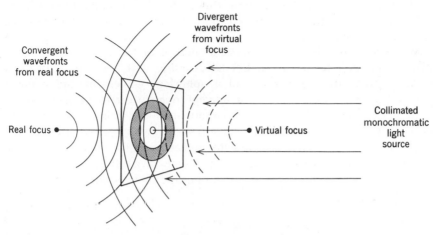

Figure 10.11 Convergent and divergent wavefronts produced by a uniformly illuminated zone plate.

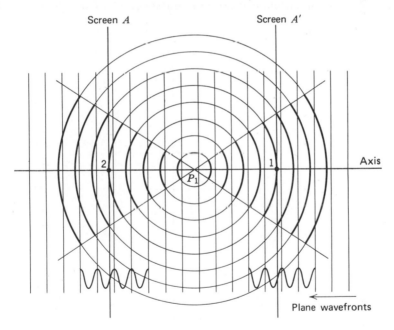

Figure 10.12 Plane wavefronts $\frac{1}{4}\lambda$ out of phase with spherical wavefronts at center of screens A and A'.

at the focal points will be different; that is, for this example there will be a $\frac{1}{4}$-cycle phase difference between the light at the focus produced by the respective zone plates.

We have thus obtained a physical picture of the formation of zone plates and their focal points. Let us examine with more mathematical rigor whether the interference of a spherical wavefront and a plane wavefront does indeed result in an interference pattern that is a sine-wave zone plate. Figure 10.13 shows a monochromatic point source of light producing spherical wavefronts, a portion of which strike the screen. Plane wavefronts from the same light source are also made to strike the screen. Because both the spherical and plane wavefronts are obtained from the same coherent source, the amplitude distribution on the screen will be the sum of the two; that is,

$$A = Se^{j\phi_s} + Re^{j\phi_R}$$

$$= S \exp [j\phi_s] + R \exp [j\phi_R]$$

where A is the amplitude distribution on the screen. The intensity on the

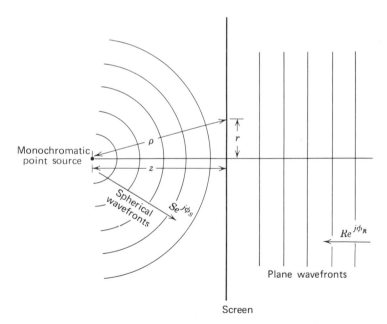

Figure 10.13 Interference between a spherical wavefront and a plane wavefront on a screen parallel to the plane wavefronts.

screen is given by the product of the amplitude and its complex conjugate or

$$I = AA^* = \{S \exp [j\phi_s] + R \exp [j\phi_R]\}\{S \exp (-j\phi_s) + R \exp [-j\phi_R]\}$$

$$= S^2 + R^2 + RS \exp [j(\phi_R - \phi_s)] + RS \exp [-j(\phi_R - \phi_s)]$$

$$= S^2 + R^2 + RS\left(\frac{2}{2}\right)\{\exp [j(\phi_R - \phi_s)] + \exp [-j(\phi_R - \phi_s)]\}$$

$$I = S^2 + R^2 + 2RS \cos (\phi_R - \phi_S) \qquad (10\text{-}5)$$

Referring to Figure 10.13 we can see that on the screen where $r = 0$, the phase difference between the spherical wavefront and the plane wavefront is given by

$$\theta_0 = \phi_z - \phi_R \qquad (10\text{-}6)$$

where ϕ_z is the phase of the spherical wavefront at the screen (at $r = 0$) and ϕ_R is the phase of the plane wavefront at the screen (at $r = 0$). At any point on the screen where $r \neq 0$ the phase difference between the spherical and plane wavefronts at any point is given by

$$\phi_S - \phi_R = \theta_0 + \phi_\rho - \phi_z \qquad (10\text{-}7)$$

From Figure 10.13 we can see that

$$\rho = \sqrt{r^2 + z^2} = z\sqrt{1 + \frac{r^2}{z^2}}$$

using the binomial expansion

$$(1 \pm x)^n = 1 \pm nx + \cdots$$

when $x^2 < 1$

$$\rho \approx z\left(1 + \frac{1}{2}\frac{r^2}{z^2}\right) \qquad (10\text{-}8)$$

when $r \ll z$. The difference $\rho - z$ is the difference in path length that the light must travel from the point source to the screen when $r \neq 0$. Substituting

$$\rho - z = z\left(1 + \frac{1}{2}\frac{r^2}{z^2}\right) - z = \frac{1}{2}\frac{r^2}{z} \qquad (10\text{-}9)$$

if $r \ll z$. The value $\frac{1}{2}(r^2/z)$ is the additional path length that the light must

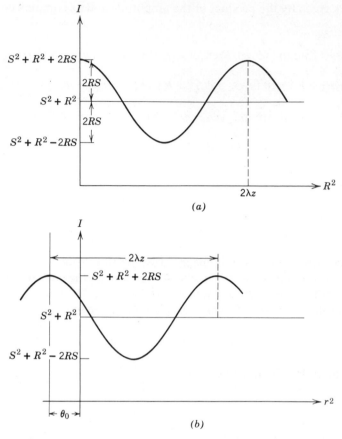

Figure 10.14 Comparison of interference patterns formed for different phases between plane and spherical wavefronts. (a) $I = S^2 + R^2 + 2RS \cos(\theta_0 + (\pi r^2/\lambda z^2))$ where $\theta_0 = 0$. (b) $I = S^2 + R^2 + 2RS \cos(\theta_0 + (\pi r^2/\lambda z^2))$.

travel from the point source to a point $r \neq 0$ on the screen. To convert this to a phase change we multiply by $2\pi/\lambda$ which gives for the phase difference

$$\frac{2\pi}{\lambda}\left[\frac{1}{2}\frac{r^2}{z}\right] = \frac{\pi r^2}{\lambda z}$$

Therefore the phase difference in radians at any point on the screen between the spherical and plane wave given by (10-7) is

$$\phi_S - \phi_R = \theta_0 + \phi_\rho - \phi_z = \theta_0 + \frac{\pi r^2}{\lambda z} \qquad (10\text{-}10)$$

Substituting (10-10) into (10-5) gives

$$I = S^2 + R^2 + 2SR \cos\left(\theta_0 + \frac{\pi r^2}{\lambda z}\right) \qquad (10\text{-}11)$$

We can see by plotting (10-11) as shown in Figure 10.14 that the intensity pattern on the screen will result in sine-wave zone plates. As the phase angle θ_0 changes, the appearance of the zone plate will also change. The period $2\lambda z$ of the intensity variation will remain the same regardless of the value of θ_0 and therefore the power spectrum of the zone plate will be the same. From the previous chapter we know that the location of the focal points of zone plates will be the same if their power spectra are identical (although the phase of the light at the focal points may be different).

GABOR HOLOGRAM

The hologram was first devised by Gabor to improve the resolving power of electron microscopes. Figure 10.15 illustrates how a Gabor-type hologram is made. An object is placed near a point source of light (focus in this case) and the film is positioned so as to be illuminated by the light directly from the point source and that light diffracted around, or through the object in the case of semi transparent objects, or both. The direct light forms a coherent background wave that interferes with the diffracted waves from the object. This interference pattern is recorded on photographic film and stores both the phase and amplitude of the diffracted light from the object. The reconstruction of the object is done by illuminating the exposed and developed photographic film (hologram) with coherent illumination similar to the coherent background illumination used in recording the hologram. This type of hologram has two major problems as follows:

(1) The object blocks most of the light to the film so that exposing the film properly is difficult (the diffracted light intensity is very low as compared to the direct light beam).

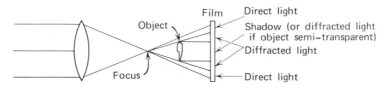

Figure 10.15   Gabor-type holograms.

(2) The real and virtual images are formed in line with each other; that is, there are twin or conjugate images formed with the same amplitude. When one of the reconstructed image is viewed, the other image (conjugate image) appears as a disturbing background.

FOURIER TRANSFORM HOLOGRAM

In the chapter dealing with the optical spectrum analyzer we saw that there is a Fourier transform relationship between the front and back focal planes of a lens. This relationship can also be made use of in holography. A Fourier transform hologram is an interference pattern between a Fourier transform of an object and the Fourier transform of a point source. The Fourier transform of a point source furnishes a plane reference beam for the Fourier transform hologram. The lens also takes the Fourier transform of the amplitude of the light diffracted by the object. The resultant interference pattern is a Fourier transform hologram.

Figure 10.16 shows how a Fourier transform hologram can be made. In this case we have shown a rectangular aperture that can contain a transparency and in the same plane there is a point source of light. Both the transparency and point source are uniformily illuminated from the same light source. The lens takes the Fourier transform of the amplitude of the light from the point source. This Fourier transform results in a constant light amplitude in the back focal plane of the lens.

This constant illumination (serving as a reference beam) interferes with the Fourier transform of the light amplitude being diffracted from the object. This results in a Fourier transform hologram.

The reconstruction of the object from the Fourier transform hologram is done by uniformly illuminating the hologram as shown in Figure 10.17.

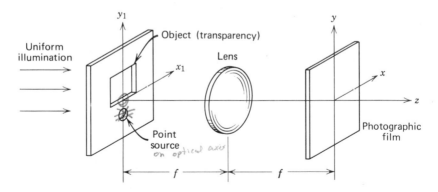

Figure 10.16   Formation of a Fourier transform hologram.

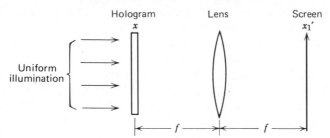

Figure 10.17 Reconstruction from a Fourier transform hologram.

The lens takes the Fourier transform of the light diffracted by the holo-gram. Because it was the spectrum of the object interfering with the reference beam that is recorded, the transform of the recorded object spectrum will produce an image of the object; that is, a Fourier transform of a Fourier transform is the original function.

This process can be analyzed to good advantage mathematically be-cause it can be used as a basis in the analysis of other types of holograms. Figure 10.18 shows an object or signal in the front focal plane of the lens. It was previously shown that a shift of ω_0 in the frequency domain is equi-valent to multiplication by $e^{-j\omega_0 t}$ in the time domain (or space domain).

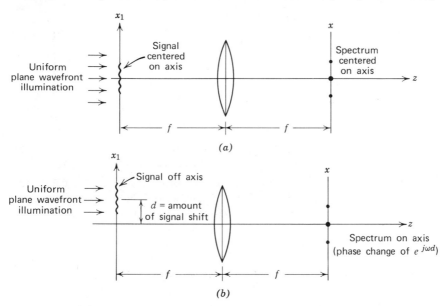

Figure 10.18 Effect of shifting signal in front focal plane of a lens on the spectrum. (a) Sig-nal centered on axis. (b) Signal off axis.

Because the Fourier transform process in reversible, we can also conclude that a shift, t_0, in position in the time domain (or space domain) is equivalent to multiplication by $e^{j\omega t_0}$ in the frequency domain. We can see therefore that an input signal to a spectrum analyzer resulting in a specific spectrum will, if this signal is displaced in the front focal plane of the lens, result in the same spectrum multiplied by a phase shift $e^{j\omega t_0}$. In order not to confuse this time domain with the space domain we shall use d as a displacement in distance where t_0 was a displacement in time.

The effect of displacing a signal can easily be visualized by assuming the signal in Figure 10.18 to be a Ronchi ruling of a given spatial frequency whose lines are horizontal. The spectrum will be vertically centered on axis no matter where the ruling is placed in the front focal plane of the lens (provided the lines remain horizontal). Because the eye only detects intensity, the spectra will look the same for all positions of the signal in the front focal plane but the phase of each spectrum for each displacement, d, will be different; that is, for a displacement d of the signal off axis the spectrum will be multiplied by a phase of $e^{j\omega d}$. Because ω has been defined for the optical Fourier transform as

$$\omega = \frac{2\pi x}{\lambda f}$$

where λ is the wavelength of the light, x is the distance along the x axis, and f is the focal length of lens, the phase shift in the spectrum for a given displacement d becomes

$$\exp\left[j\omega d\right] = \exp\left[j\frac{2\pi d}{\lambda}\frac{x}{f}\right]$$

Point Source and Object in Front Focal Plane

Figure 10.19 shows an object $O(x_1)$ in the front focal plane of a lens displaced off axis by an amount d. A point reference source is also in the front focal plane but is on axis. If we consider just the point reference source in the front focal plane of the lens, we know that the Fourier transform of it will appear in the back focal plane. We have previously seen that the Fourier transform of a point source is a constant. The point source in the front focal plane of the lens will therefore produce plane wavefronts in the back focal plane of the lens that can be represented by the constant R. Another way of considering this is that a lens collimates the light from a point source when it is placed a focal length from the lens. The object $O(x_1)$ which is centered at $x_1 = d$ can be expressed in the front focal plane as $O(x_1 - d)$. We can consider this object in the

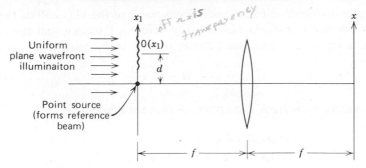

Figure 10.19 Signal plus a point source in front focal plane of lens.

front focal plane to be a photographic transparency. In the back focal plane of the lens we will obtain the Fourier transform of the object transparency in the front focal plane; that is, in the back focal plane we have

$$F(\omega) = \int_{-\infty}^{\infty} O(x_1 - d) e^{-j\omega x_1} \, dx_1 \qquad (10\text{-}12)$$

If the object transparency is centered on axis, its transform can be written

$$F(\omega) = \int_{-\infty}^{\infty} O(x_1) \, e^{-j\omega x_1} \, dx_1 \qquad (10\text{-}13)$$

A shift in position of d in the space domain (front focal plane of lens) is equivalent to multiplication by $e^{j\omega d}$ in the frequency domain (back focal plane of the lens). We can therefore see that a shift of the transparency in the front focal plane from a position centered on axis to a position centered about d off axis (but still in the front focal plane) will produce a Fourier transform in the back focal plane of

$$F(\omega) = e^{j\omega d} \int_{-\infty}^{\infty} O(x_1) \, e^{-j\omega x_1} \, dx_1 \qquad (10\text{-}14)$$

$$F(\omega) = F\left(\frac{2\pi x}{\lambda f}\right) = \int_{-\infty}^{\infty} O(x_1 - d) \, e^{-j\omega x_1} \, dx_1 = e^{j\omega d} \int_{-\infty}^{\infty} O(x_1) \, e^{-j\omega x_1} \, dx_1$$

$$(10\text{-}15)$$

that is, the Fourier transform of the displaced object $O(x_1 - d)$ is the same as that of the object $O(x_1)$ multiplied by a phase shift $e^{j\omega d}$. The total amplitude distribution in the back focal plane of the lens will be the sum of the Fourier transform of the point reference source and the object transparency in the front focal plane; that is, in the back focal plane we have

$$A(x) = R + F(\omega) \tag{10-16}$$

The intensity in the back focal plane of the lens is given by

$$
\begin{aligned}
I(x) &= A(x)A^*(x) \\
&= [R + F(\omega)][R + F^*(\omega)] \\
&= R^2 + RF(\omega) + RF^*(\omega) + F(\omega)F^*(\omega) \tag{10-17}
\end{aligned}
$$

Since

$$BB^* = |B|^2 \tag{10-18}$$

then

$$I(x) = R^2 + RF(\omega) + RF^*(\omega) + |F(\omega)|^2$$

or

$$I(x) = R^2\left[1 + \frac{F(\omega)}{R} + \frac{F^*(\omega)}{R} + \frac{|F(\omega)|^2}{R^2}\right] \tag{10-19}$$

We have previously seen that an exposed photographic plate has an amplitude transmission function of

$$T(x) = \left[\frac{E_1}{t_1 I(x)}\right]^{\gamma_1/2} = \left(\frac{E_1}{t_1}\right)^{\gamma_1/2} I^{-\gamma_1/2}(x) \tag{10-20}$$

where t_1 is the exposure time, γ_1 is the slope of linear portion of the film characteristic curve, and E_1 is the intersection of the linear portion of the characteristic curve extended to the $\log E$ axis.

The intensity distribution exposing the photographic plate for a Fourier transform hologram is

$$I(x) = R^2\left[1 + \frac{F(\omega)}{R} + \frac{F^*(\omega)}{R} + \frac{|F(\omega)|^2}{R^2}\right]. \tag{10-19}$$

The developed photographic plate has an amplitude transmission function of

$$T(x) = \left(\frac{E_1}{t_1}\right)^{\gamma_1/2} I^{-\gamma_1/2}(x) \tag{10-20}$$

when illuminated by plane wavefronts. By using the binomial expansion we can find $I^{-\gamma_1/2}(x)$; that is, the binomial expansion is given as

$$(1 \pm z)^{-n} = 1 \mp nz + \frac{n(n+1)z^2}{2!} \mp \cdots$$

where $z^2 < 1$ and if we let $n = \gamma_1/2$ and

$$z = \frac{F(\omega)}{R} + \frac{F^*(\omega)}{R} + \frac{|F(\omega)|^2}{R^2}$$

and assume $R \gg |F(\omega)|$ then the first two terms of the binomial expansion for $I^{-\gamma_1/2}(x)$ can be found as follows;

$$I^{-\gamma_1/2}(x) = \left[R^2 \left(1 + \frac{F(\omega)}{R} + \frac{F^*(\omega)}{R} + \frac{|F(\omega)|^2}{R^2} \right) \right]^{-\gamma_1/2}$$

$$= (R^2)^{-\gamma_1/2} \left(1 + \frac{F(\omega)}{R} + \frac{F^*(\omega)}{R} + \frac{|F(\omega)|^2}{R^2} \right)^{-\gamma_1/2} \qquad (10\text{-}21)$$

$$= (R^2)^{-\gamma_1/2}(1+z)^{-n} = (R^2)^{-\gamma_1/2}\left[1 - \left(\frac{\gamma_1}{2}\right)(z) \right]$$

$$I^{-\gamma_1/2}(x) = (R^2)^{-\gamma_1/2}\left[1 - \frac{\gamma_1}{2}\frac{F(\omega)}{R} - \frac{\gamma_1}{2}\frac{F^*(\omega)}{R} - \frac{\gamma_1}{2}\frac{|F(\omega)|^2}{R^2} \right]$$

then

$$T(x) = \left(\frac{E_1}{t_1}\right)^{\gamma_1/2} I^{-\gamma_1/2}(x) = \left(\frac{E_1}{t_1}\right)^{\gamma_1/2}\left[(R^2)^{-\gamma_1/2}\left(1 - \frac{\gamma_1}{2}\frac{F(\omega)}{R} \right.\right.$$

$$\left.\left. - \frac{\gamma_1}{2}\frac{F^*(\omega)}{R} - \frac{\gamma_1}{2}\frac{|F(\omega)|^2}{R^2} \right) \right]$$

$$= \left(\frac{E_1}{R^2 t_1}\right)^{\gamma_1/2}\left[1 - \frac{\gamma_1}{2}\frac{F(\omega)}{R} - \frac{\gamma_1}{2}\frac{F^*(\omega)}{R} - \frac{\gamma_1}{2}\frac{|F(\omega)|^2}{R^2} \right]$$

$$= \left(\frac{E_1}{R^2 t_1}\right)^{\gamma_1/2}\left[-\frac{\gamma_1}{2}\left(-\frac{2}{\gamma_1} + \frac{F(\omega)}{R} + \frac{F^*(\omega)}{R} + \frac{|F(\omega)|^2}{R^2} \right) \right]$$

If we let

$$K = -\frac{\gamma_1}{2}\left(\frac{E_1}{R^2 t_1}\right)^{\gamma_1/2}$$

then

$$T(x) = K\left[-\frac{2}{\gamma_1} + \frac{F(\omega)}{R} + \frac{F^*(\omega)}{R} + \frac{|F(\omega)|^2}{R^2} \right] = K\left[-\frac{2}{\gamma_1} + z \right] \qquad (10\text{-}22)$$

where K and the term $-2/\gamma_1$ are constants dependent on the photographic procedures, film developers, and techniques used. If we compare $I(x)$ with $T(x)$,

$$I(x) = R^2(1+z) = K'[1+z] \tag{10-23}$$

$$T(x) = K\left(-\frac{2}{\gamma_1}+z\right) = K(A+z) \tag{10-24}$$

where $A = -2/\gamma_1$, we can see that $I(x)$ and $T(x)$ have the same form except for the factor $(-2/\gamma_1)$ in the first term inside the brackets. The result is that we can produce an amplitude transmission function $T(x)$ proportional to the intensity $I(x)$. Since the intensity $I(x)$ exposes the photographic film to produce the hologram, we would like to illuminate the developed hologram and recreate the exact same light distribution that exposed the hologram; that is, the transmission function $T(x)$ of the hologram should produce the same light distribution that exposed it. We can easily understand that a given hologram with a transmission function $T(x)$ can be illuminated by light beams of different intensities to create images of different intensities; that is, this is similar to projecting a slide (transparency) on a screen—the greater the projector wattage, the greater the image brightness. The slide transmission function does not change as we illuminate it with beams of greater intensity but the projected image on the screen will become brighter.

It is therefore important to be able to determine the intensity distribution exposing a film and to be able to determine the transmission distribution $T(x)$ of the film (hologram) after it is developed.

The difference between the two terms, $T(x)$ and $I(x)$, equations (10-23) and (10-24), is the constant term that has no effect on the reconstructed image; that is, the constant term, it will be shown, gives rise to the DC or undiffracted beam that will not produce the desired reconstructed image. If R is not sufficiently greater than $|F(\omega)|$, additional terms which can act as noise terms must be considered in the binomial expansion for $I^{-\gamma_1/2}(x)$ in (10-20).

Illuminating the developed hologram with the reference beam R we obtain a light amplitude distribution at the plane of the hologram that is given by the product of $T(x)$

$$T(x) = K\left[-\frac{2}{\gamma_1}+\frac{F(\omega)}{R}+\frac{F^*(\omega)}{R}+\frac{|F(\omega)|^2}{R^2}\right]$$

$$= K\left[A+\frac{F(\omega)}{R}+\frac{F^*(\omega)}{R}+\frac{|F(\omega)|^2}{R^2}\right] \tag{10-25}$$

by R or

$$RT(x) = K\left[AR + F(\omega) + F^*(\omega) + \frac{|F(\omega)|^2}{R}\right]$$

The last term in this equation will vary because of its dependence on the transform $F(\omega)$ and will represent a noise term in the reconstruction. If the reference signal R is sufficiently greater than $|F(\omega)|$, this last term $|F(\omega)|^2/R$, will be negligible. We have been assuming that the original reference wave and reconstructing reference wave are equal, although this does not necessarily have to be true. If the reference wave used for reconstruction has an amplitude different from the original reference beam used to expose the photographic plate, the resultant reconstructed image will be more or less bright — that is, proportional to the reference beam used to reconstruct the image.

When $R \gg |F(\omega)|$, then $|F(\omega)|^2/R^2$ becomes negligible and

$$RT(x) = K[AR + F(\omega) + F^*(\omega)] \tag{10-26}$$

In order to reconstruct an image the hologram is placed in the front focal plane of a lens and then illuminated with the reference beam as shown in Figure 10.17. The light amplitude distribution in the back focal plane of the lens will be the Fourier transform of $RT(x)$ or

$$\int_{-\infty}^{\infty} RT(x)\, e^{-j\omega'x}\, dx = \int_{-\infty}^{\infty} K[AR + F(\omega) + F^*(\omega)]\, e^{-j\omega'x}\, dx \tag{10-27}$$

where

$$\omega' = \frac{2\pi x_1'}{\lambda f}$$

$$= KA \int_{-\infty}^{\infty} R e^{-j\omega'x}\, dx + K \int_{-\infty}^{\infty} F(\omega) e^{-j\omega'x}\, dx$$

$$+ K \int_{-\infty}^{\infty} F^*(\omega)\, e^{-j\omega'x}\, dx \tag{10-28}$$

To evaluate (10-28) we remember that the delta function was defined as

$$\delta(x) = \frac{1}{2\pi} \int_{-\infty}^{\infty} e^{-jkx}\, dk$$

$$\int_{-\infty}^{\infty} e^{-jkx}\, dk = 2\pi\delta(x)$$

therefore we can see that the first term of (10-28) becomes

$$\int_{-\infty}^{\infty} e^{-j\omega'x}\,dx = 2\pi\delta(\omega')$$

$$= 2\pi\delta\left(\frac{2\pi x_1'}{\lambda f}\right) \qquad (10\text{-}29)$$

and since

$$\delta(ax_1') = \frac{1}{|a|}\,\delta(x_1')$$

equation (10-29) becomes

$$\int_{-\infty}^{\infty} e^{-j\omega'x}\,dx = 2\pi\,\frac{\lambda f}{2\pi}\,\delta(x_1') = \lambda f\delta(x_1')$$

Then in (10-28) the first term can be written

$$KA\int_{-\infty}^{\infty} Re^{-j\omega'x}\,dx = KK_1AR\delta(x_1') \qquad (10\text{-}30)$$

where $K_1 = \lambda f$ and using (10-12) then (10-28) becomes

$$\int_{-\infty}^{\infty} RT(x)e^{-j\omega'x}\,dx = KK_1AR\delta(x_1') + K\int_{-\infty}^{\infty} e^{-j\omega'x}\,dx \int_{-\infty}^{\infty} O(x_1-d)$$

$$\times e^{-j\omega x_1}\,dx_1 + K\int_{-\infty}^{\infty} e^{-j\omega'x}\,dx \int_{-\infty}^{\infty} O(x_1-d)\,e^{j\omega x_1}\,dx_1 \qquad (10\text{-}31)$$

Equation (10-31) indicates that the focused light output of a plane-wave illuminated Fourier transform hologram (as shown in Figure 10.17) will consist of three terms. We will examine each term separately to determine the images it produces.

By changing the order of integration and substituting for ω' and ω we find the second term of (10-31) to be

$$K\int_{-\infty}^{\infty} e^{-j\omega'x}\,dx \int_{-\infty}^{\infty} O(x_1-d)\,e^{-j\omega x_1}\,dx_1 = K\int_{-\infty}^{\infty} O(x_1-d)\,dx_1$$

$$\int_{-\infty}^{\infty} e^{-j(\omega'x+\omega x_1)}\,dx$$

$$= K\int_{-\infty}^{\infty} O(x_1-d)\,dx_1 \int_{-\infty}^{\infty} \exp\left[-j\left(\frac{2\pi x_1'x}{\lambda f} + \frac{2\pi xx_1}{\lambda f}\right)\right]dx$$

$$= K\int_{-\infty}^{\infty} O(x_1-d)\,dx_1 \int_{-\infty}^{\infty} \exp\left[-j\left(\frac{2\pi x}{\lambda f}\right)(x_1'+x_1)\right]dx \qquad (10\text{-}32)$$

When integrating an exponential from $-\infty$ to $+\infty$, the integral is equal to a delta function as shown previously by the transform of a constant. Then

$$K \int_{-\infty}^{\infty} e^{-j\omega' x} dx \int_{-\infty}^{\infty} O(x_1-d) e^{-j\omega x_1} dx_1 = K \int_{-\infty}^{\infty} O(x_1-d) dx_1$$
$$\times [K_1 \delta(x_1+x_1')]$$

where $K_1 = \lambda f$

$$= KK_1 \int_{-\infty}^{\infty} O(x_1-d)\delta(x_1+x_1') dx_1 \qquad (10\text{-}33)$$

Applying the sifting property of the delta function the integration gives

$$K \int_{-\infty}^{\infty} e^{-j\omega' x} dx \int_{-\infty}^{\infty} O(x_1-d) e^{-j\omega x_1} dx_1 = KK_1 O(-x_1'-d) \qquad (10\text{-}34)$$

that is, the value of the delta function is zero except at the value where $x_1 + x_1' = 0$, therefore the product of $O(x_1-d)\delta(x_1+x_1')$ can have a value only where $x_1 = -x_1'$.

Referring back to (10-31) we can solve for the last term similarly to the way we solved for the middle term — that is, the last term of (10-31),

$$K \int_{-\infty}^{\infty} e^{-j\omega_1 x} dx \int_{-\infty}^{\infty} O(x_1-d) e^{j\omega x_1} dx_1$$

$$= K \int_{-\infty}^{\infty} O(x_1-d) dx_1 \int_{-\infty}^{\infty} e^{j\omega x_1} e^{-j\omega_1 x} dx$$

$$= K \int_{-\infty}^{\infty} O(x_1-d) dx_1 \int_{-\infty}^{\infty} e^{j\omega x_1} \exp\left[-j\frac{2\pi}{\lambda f}x_1' x\right] dx$$

$$= K \int_{-\infty}^{\infty} O(x_1-d) dx_1 \int_{-\infty}^{\infty} \exp\left[j\frac{2\pi x}{\lambda f}x_1\right] \exp\left[-j\frac{2\pi}{\lambda f}x_1' x\right] dx$$

$$= K \int_{-\infty}^{\infty} O(x_1-d) dx_1 \int_{-\infty}^{\infty} \exp\left[j\omega(x_1 \quad x_1')\right] \qquad (10\text{-}35)$$

$$= K \int_{-\infty}^{\infty} O(x_1-d) dx_1 \; K_1\delta(x_1-x_1') = KK_1 O(x_1'-d) \qquad (10\text{-}36)$$

From (10-31) we can see that the first term results in the plane wave being focused to a point by a lens; that is, the first term produces a focus on axis at $x_1' = 0$. The second term of (10-31) reduced to

$$KK_1 O(-x_1'-d) \qquad (10\text{-}34)$$

which indicates that this term represents an inverted image of the object $O(x_1)$ centered at $x_1' = -d$.

The third term of (10-31) reduces to

$$KK_1O(x_1'-d)$$ (10-36)

which indicates that this term represents an upright image $O(x_1')$ located at a distance $x_1' = d$. Figure 10.20(a) shows various images, their location on the x_1' axis, and their designation; for example, $O(x_1')$ represents an upright image where as $O(-x_1')$ represents an inverted image. $O(x_1'-d)$ re-

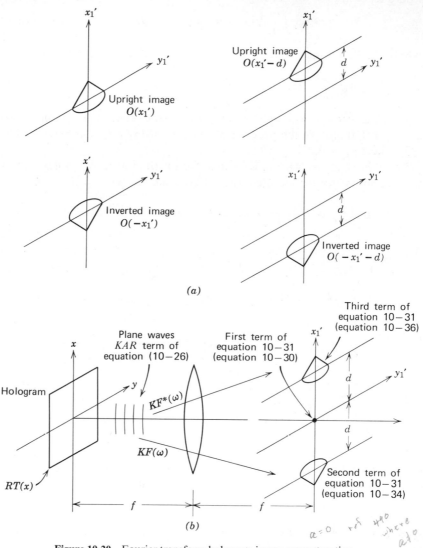

Figure 10.20   Fourier transform hologram image reconstruction.

presents an upright image displaced upward by an amount d; that is, the image is centered, when $x_1' = d$, whereas $O(-x_1' - d)$ represents an inverted image displace downward by an amount d. We see that the image is centered when $-x_1' = d$. Note that we displaced the object with respect to an x_1 axis and formed the Fourier transform on an x axis as shown in Figure 10.19. The reconstructed image we assumed to be located with respect to an x_1' axis as shown in Figure 10.20.

In the consideration above we assumed that the wavelength of the viewing light was the same as the wavelength of the recording light. We can generalize the analysis to show that magnification is possible in a Fourier transform hologram image if the viewing wavelength is not the same as the recording wavelength. Assume some focal length f for all wavelengths. From (10-27) we can define

$$\omega' = \frac{2\pi x_1'}{\lambda_v f}$$

where λ_v corresponds to viewing wavelength not necessarily equal to the recording wavelength λ. Then equation (10-29) becomes

$$\int_{-\infty}^{\infty} e^{-j\omega' x} \, dx = 2\pi\delta\left(\frac{2\pi x_1'}{\lambda_v f}\right)$$

Equation (10-32) can be written as

$$K \int_{-\infty}^{\infty} e^{-j\omega' x} \, dx \int_{-\infty}^{\infty} O(x_1 - d) \, e^{-j\omega x_1} \, dx_1$$

$$= K \int_{-\infty}^{\infty} O(x_1 - d) \, dx_1 \int_{-\infty}^{\infty} \exp\left[-j\left(\frac{2\pi x_1' x}{\lambda_v f} + \frac{2\pi x x_1}{\lambda f}\right)\right] dx$$

$$= K \int_{-\infty}^{\infty} O(x_1 - d) \, dx_1 \int_{-\infty}^{\infty} \exp\left[-j\frac{2\pi x}{\lambda f}\left(x_1 + \frac{\lambda}{\lambda_v} x_1'\right)\right] dx$$

Then Equation (10-33) becomes

$$K \int_{-\infty}^{\infty} e^{-j\omega' x} \, dx \int_{-\infty}^{\infty} O(x_1 - d) \, e^{-j\omega x_1} \, dx_1$$

$$= K \int_{-\infty}^{\infty} O(x_1 - d) \, dx_1 \left[K_1 \delta\left(x_1 + \frac{\lambda}{\lambda_v} x_1'\right)\right]$$

$$= K K_1 \int_{-\infty}^{\infty} O(x_1 - d) \delta\left(x_1 + \frac{\lambda}{\lambda_v} x_1'\right) dx_1$$

where $K_1 = \lambda_v f$
and (10-34) becomes

$$K \int_{-\infty}^{\infty} e^{-j\omega' x}\, dx \int_{-\infty}^{\infty} O(x_1 - d)\, e^{-j\omega x_1}\, dx_1 = KK_1 O\left(-\frac{\lambda x_1'}{\lambda_v} - d\right)$$

Similarly the equations leading to (10-36) change and become

$$K \int_{-\infty}^{\infty} e^{-j\omega^1 x}\, dx \int_{-\infty}^{\infty} O(x_1 - d)\, e^{j\omega x_1}\, dx = KK_1 O\left(\frac{\lambda}{\lambda_v} x_1' - d\right)$$

where the magnification is given by

$$\text{Magnification} = M = \frac{\lambda_v}{\lambda}.$$

in the Fourier Transform Hologram.

We can see, therefore, that unlike the holograms discussed in Chapter 9 the Fourier transform hologram does exhibit changes in magnification with changes in the wavelength of the viewing light.

To summarize, we have seen from Figure 10.19 that when an on axis point source and an off-axis object are on the front focal plane of a lens, the Fourier transform of the object in the back focal plane of the lens will have a phase shift of $e^{(j2\pi d/\lambda f)x}$ produced as a function of the shift of position d of the object. A photographic exposure of the Fourier transform in the back focal plane will result in three recorded terms if the amplitude of the reference beam is very much greater than the object spectrum. These three recorded terms reconstruct to a point focus and two images of the object as shown in Figure 10.20(b).

Point Source and Object not in Same Plane

Figure 10.21 shows an object placed so that it is not in the front focal plane of the lens. When the object is placed as shown in Figure 10.21, there is a phase shift introduced into the spectrum that is a function of the object position out of the front focal plane. When the object is displaced a distance a out of the front focal plane (away from the lens), the phase shift introduced into the spectrum is given by

$$\exp\left[-j\frac{\pi a}{\lambda}\left(\frac{x}{f}\right)^2\right] = \exp\left[-j\frac{\lambda a}{4\pi}\omega^2\right]$$

where

$$\omega = 2\pi x/\lambda f.$$

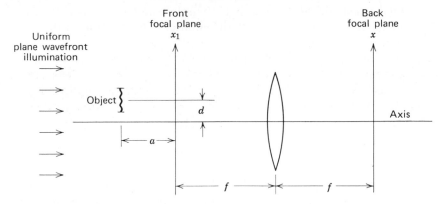

Figure 10.21 Fourier transform of an object not in the front focal plane of the lens.

Figure 10.21 shows an object displaced off axis by an amount d and out of the front focal plane of the lens by an amount a. If we consider the object only in one dimension along the x_1 axis, we can write the expression for the object as $O(x_1)$; that is, the object is defined by (or is a function of) the value of x_1. We know that if $O(x_1)$ is centered on axis and in the front focal plane of the lens, its Fourier transform will appear in the back focal plane of the lens; that is, if $T(\omega)$ is the Fourier transform of the object $O(x_1)$ in the front focal plane, it can be expressed as

$$T(\omega) = \int_{-\infty}^{\infty} O(x_1) \, e^{j(2\pi x/\lambda f)x_1} \, dx_1 \qquad (10\text{-}37)$$

or

$$T(\omega) = \int_{-\infty}^{\infty} O(x_1) \, e^{-j\omega x_1} \, dx_1 \qquad (10\text{-}38)$$

From the previous discussion we can see that the light amplitude distribution in the back focal plane produced by an object anywhere in or out of the front focal plane can be expressed as

$$T'(\omega) = e^{-j(\lambda a/4\pi)(\omega)^2} \int_{-\infty}^{\infty} O(x_1) \, e^{-j\omega x_1} \, dx_1 \qquad (10\text{-}39)$$

When the object is in the front focal plane of the lens ($a = O$), the equation reduces to

$$F(\omega) = e^{j\omega d} \int_{-\infty}^{\infty} O(x_1) \, e^{-j\omega x_1} \, dx_1 \qquad (10\text{-}40)$$

[see (10-14)]. From this we can see that when the object is out of the focal plane

$$T'(\omega) = e^{j\omega d} \, e^{-j(\lambda a/4\pi)(\omega)^2} \int_{-\infty}^{\infty} O(x_1) \, e^{-j\omega x_1} \, dx_1 = e^{-j(\lambda a/4\pi)(\omega)^2} \, F(\omega) \qquad (10\text{-}41)$$

By similar reasoning we can see that when the point source is in the front focal plane of the lens, the resultant amplitude distribution in the back focal plane of the lens can be represented by a constant, R. When the point source is out of the front focal plane (away from the lens), the resultant amplitude distribution in the back focal plane of the lens can be represented by $Re^{-j(\lambda a/4\pi)(\omega)^2}$; that is, constant amplitude multiplied by a phase shift term. We therefore can consider two cases with object out of the front focal plane of the lens (away from lens) as follows:

Case A: Point source out of front focal plane of (away from) lens $\rightarrow$ $Re^{-j(\lambda a/4\pi)(\omega)^2}$

Case B: Point source in front focal plane of lens $\rightarrow R$.

The amplitude distribution in the back focal plane of the lens for case A is

$$A(x) = T'(\omega) + Re^{-j(\lambda a/4\pi)(\omega)^2} \tag{10-42}$$

and

$$I(x) = A(x)A^*(x) = [T'(\omega) + Re^{-j(\lambda a/4\pi)(\omega)^2}][T'(\omega) + Re^{-j(\lambda a/4\pi)(\omega)^2}]*$$

$$= [e^{-j(\lambda a/4\pi)(\omega)^2}F(\omega) + Re^{-j(\lambda a/4\pi)(\omega)^2}][e^{j(\lambda a/4\pi)(\omega)^2}F^*(\omega) + Re^{j(\lambda a/4\pi)(\omega)^2}]$$

$$I(x) = |F(\omega)|^2 + R^2 + RF^*(\omega) + RF(\omega) \tag{10-43}$$

which is identical to (10-19) and produces images as shown in Figure 10.20(b).

The amplitude distribution in the back focal plane of the lens for case B is

$$A(x) = T'(\omega) + R \tag{10-44}$$

and the intensity is

$$I(x) = A(x)A^*(\omega) = [T'(\omega) + R][T'(\omega) + R]*$$

$$= [e^{-j(\lambda a/4\pi)(\omega)^2}F(\omega) + R][e^{j(\lambda a/4\pi)(\omega)^2}F^*(\omega) + R]$$

$$I(x) = |F(\omega)|^2 + R^2 + Re^{j(\lambda a/4\pi)(\omega)^2}F^*(\omega) + Re^{-j(\lambda s/4\pi)(\omega)^2}F(\omega) \tag{10-45}$$

$$I(x) = R^2\left[1 + \frac{e^{-j(\lambda a/4\pi)(\omega)^2}F(\omega)}{R} + \frac{e^{j(\lambda a/4\pi)(\omega)^2}F^*(\omega)}{R} + \frac{|F(\omega)|^2}{R^2}\right] \tag{10-46}$$

Equation (10-46) is in the same form as (10-19) except for the exponential, therefore the transmission function will be the same as (10-25) with the addition of the exponential,

$$T(x) = K\left[A + \frac{e^{-j(\lambda a/4\pi)(\omega)^2}F(\omega)}{R} + \frac{e^{j(\lambda a/4\pi)(\omega)^2}F^*(\omega)}{R} + \frac{|F(\omega)|^2}{R^2}\right] \tag{10-47}$$

where K and A have the same values as in (10-25); that is,

$$K = -\frac{\gamma_1}{2}\left[\frac{E_1}{R^2 t_1}\right]^{\gamma_1/2} \quad \text{and} \quad A = -\frac{2}{\gamma_1}$$

If we again assume that $R^2 \gg |F(\omega)|^2$ so that the term $|F(\omega)|^2/R^2$ can be neglected then

$$T(x) = K\left[A + \frac{e^{-j(\lambda a/4\pi)(\omega)^2}F(\omega)}{R} + \frac{e^{j(\lambda a/4\pi)(\omega)^2}F^*(\omega)}{R}\right] \quad (10\text{-}48)$$

Illuminating this hologram with plane waves R gives

$$RT(x) = KR\left[A + \frac{e^{-j(\lambda a/4\pi)(\omega)^2}F(\omega)}{R} + \frac{e^{j(\lambda a/4\pi)(\omega)^2}F^*(\omega)}{R}\right] \quad (10\text{-}49)$$

We can see from (10-48) that there are three terms recorded in the hologram. These three terms comprise the transmission function $T(x)$ of the hologram. When we illuminate the hologram with a reference plane wave R, the product $RT(x)$ is the resulting light leaving the hologram. From (10-49) we can see there is light corresponding to the three recorded terms of the transmission function $T(x)$ leaving the hologram. The greater the intensity of the illuminating reference beam, R, the greater will be the light corresponding to each term of the transmission function, $T(x)$.

The first term of (10-49), KRA, will be a background plane wave corresponding to the undiffracted part of the incident light. In order to see to what the light from the second term of (10-49) corresponds, let us refer to Figure 10.22. From Figure 10.22 we can see that an object $O(x_1)$ located a distance $(f+a)$ from the lens will produce a magnified image

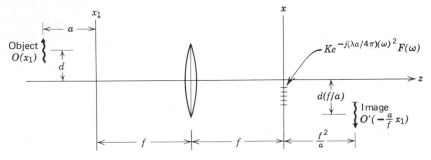

Figure 10.22 Fourier transform hologram reconstruction for the term $Ke^{-j(\lambda a/4\pi)(\omega)^2}F(\omega)$ in equation (10-49).

$O'[-(a/f)x_1]$ as shown; that is

$$\frac{1}{f} = \frac{1}{p} + \frac{1}{q} = \frac{1}{f+a} + \frac{1}{q}$$

$$\frac{1}{q} = \frac{1}{f} - \frac{1}{f+a} = \frac{(f+a)-f}{f(f+a)}$$

$$q = \frac{f(f+a)}{(f+a)-f} = \frac{f^2+fa}{a} = \frac{f^2}{a} + f$$

and

$$\text{magnification} = M = -\frac{q}{p} = \frac{-[f(f+a)]/a}{f+a} = -\frac{f}{a}$$

(the minus sign indicates the image is inverted).

Referring to Figure 10.22 we see that we have designated the object as $O(x_1)$ and the image as $O'[-(a/f)x_1]$ where a/f is 1/magnification. This can easily be understood by considering the object to have a sinusoidal transmission function as shown in Figure 10.23. In Figure 10.23(a) we start out with the object designated by

$$y = \sin x$$

where y is the transmission and x is location on the x axis. If we assume the lens magnifies this object by three ($M = 3$), the image should be three times larger than the object. If we multiply the function by M as shown in Figure 10.23(b), we change the amplitude of the sinusoid (transmission function) which would not represent a change of scale of the image by $M = 3$. Figure 10.23(c) shows the effect of multiplying the function variable by the magnification M and by the reciprocal of the magnification $1/M$. We can see that in order to have a magnification of $M = 3$ along the x axis we must divide the function variable by M. To summarize an object located $f+a$ from a lens and d off axis will have an image on the other side of the lens located at $f+f^2/a$ and $d(f/a)$ off axis (on the opposite side of the axis from the image). The image is therefore inverted and magnified by an amount $-f/a$. We can interpret the recording of the $F(\omega)$ term in (10-49) as representing the light passing through the back focal plane of the lens that focuses on the image $O'[-(a/f)x_1]$. With this interpretation, illuminating the hologram (placed on the x axis in Figure 10.22) with plane waves (corresponding to the Fourier transform of the point source) will reconstruct the image $O'[-(a/f)x_1]$ without any additional lenses because the $F(\omega)$ term in (10-49) will appear

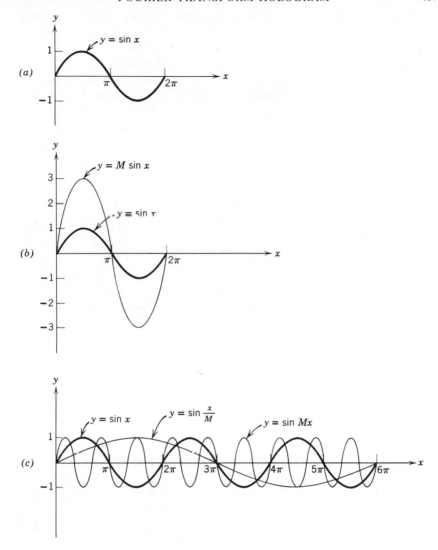

Figure 10.23 Scale changes for a sinusoidal function.

as one output of the illuminated hologram. To understand this better consider an object $O(x_1)$ in Figure 10.22 producing an image $O'[-(a/f)x_1]$ (the minus sign indicating that the image is inverted). The light passing through the x axis to form this image corresponds to the second term of (10-49), $Ke^{-j(\lambda a/4\pi)(\omega)^2}$, and is therefore identical to a portion of the light leaving a plane-wave illuminated hologram placed on the x axis. This means that the plane-wave illuminated hologram will reconstruct (without lenses) the image $O'[-(a/f)x_1]$ [just as did the object $O(x_1)$].

The light leaving the plane-wave illuminated hologram is made up of three terms as indicated by (10-49). We have already determined that the first term of (10-49) corresponds to plane waves leaving the hologram and the second term corresponds to light that will form an image $O'[-(a/f)x_1]$. Let us now consider the third term of (10-49). From the previous discussions the $F^*(\omega)$ term of (10-48) can be produced by an object $O(-x_1-d)$ as shown in Figure 10.24. [In Figure 10.24 the object is indicated as $O(-x_1)$ and shown displaced off axis by an amount d that is equivalent to the notation $O(-x_1-d)$.] The object $O(-x_1)$ produces a magnified virtual image $O''[-(a/f)x_1]$ since it is inside the focal length of the lens (the minus sign indicates that the images are inverted); that is,

$$\frac{1}{f} = \frac{1}{p} + \frac{1}{q} = \frac{1}{f-a} + \frac{1}{q}$$

$$\frac{1}{q} = \frac{1}{f} - \frac{1}{f-a} = \frac{(f-a)-f}{f(f-a)}$$

$$q = -\frac{f(f-a)}{a} = -\frac{f^2}{a} + f$$

and

$$\text{magnification} = M = -\frac{q}{p} = +\frac{[f(f-a)]/a}{f-a} = +\frac{f}{a}$$

We note that because M is positive, the lens will not invert the image of the object but because the object $O(-x_1)$ is negative, the image $O''[-(a/f)x_1]$ is negative; that is, because the object is inverted, it remains inverted since M is positive. The light from object $O(-x_1)$ in Figure 10.24 produces an image $O''[-(a/f)x_1]$ as shown. The light passing through the x axis forms this virtual image corresponding to the third term of (10-49) and is therefore identical to a portion of the light leaving a plane-wave illuminated hologram and will reconstruct (without lenses) the image $O''[-(a/f)x_1]$ [just as did the object $O(-x_1)$].

Figure 10.25 shows the combined results of illuminating the hologram with plane waves. The light from each term of (10-49) is indicated. We can see that two images are formed, each the same size and in line with each other but on opposite sides of the hologram; that is, one image is real and the other virtual. This hologram has the same apparent viewing

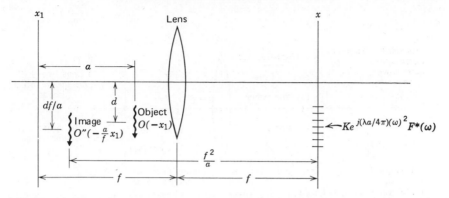

Figure 10.24   Fourier transform hologram reconstruction for the term $Ke^{j(\lambda a/4\pi)(\omega)^2}F^*(\omega)$ in equation (10-49).

difficulties as the Gabor hologram; that is, the real and virtual images are in line with each other and will tend to interfere with each other. The Fourier transform hologram can be viewed with a lens so that the real and virtual images are not in line. Let us examine Figure 10.26 to see the effect of a lens used for viewing when it is placed a focal length from the plane-wave illuminated hologram.

From Figure 10.26 we can see that the lens image $O''(x_1)$ of the hologram image $O''(-a/fx_1)$ will appear, because of the lens, at a

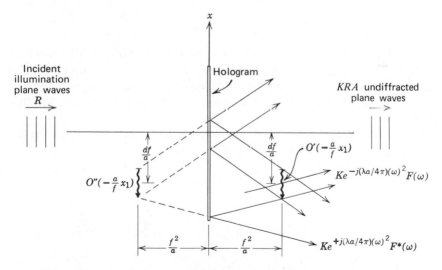

Figure 10.25   Fourier transform hologram reconstruction for equation (10-49).

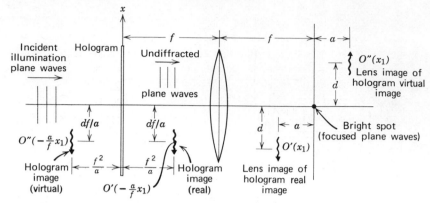

Figure 10.26 Viewing a Fourier transform hologram with a lens placed a focal length away from the hologram.

distance from the hologram, that is,

$$\frac{1}{f} = \frac{1}{p} + \frac{1}{q} = \frac{1}{f + (f^2/a)} + \frac{1}{q}$$

$$\frac{1}{q} = \frac{1}{f} - \frac{1}{f + (f^2/a)} = \frac{f + f^2/a - f}{(f + f^2/a)f} = \frac{f^2/a}{f(f + f^2/a)}$$

$$q = \frac{f(f + f^2/a)}{f^2/a} = f + a$$

$$M = -\frac{q}{p} = -\frac{f + a}{f + f^2/a} = -\frac{(f + a)a}{f(f + a)} = -\frac{a}{f}$$

Since $q = f + a$, the lens image $O''(x_1)$ of the hologram image $O''[-(a/f)x_1]$ is $2f + a$ from the hologram. This lens image $O''(x_1)$ is inverted and appears on the other side of the axis from the hologram image, $O''[-(a/f)x_1]$. Figure 10.27(a) shows a graphical construction for this image.

In a similar manner we can find where the lens forms the image for the hologram image $O'[-(a/f)x_1]$, that is,

$$\frac{1}{f} = \frac{1}{p} - \frac{1}{q} = \frac{1}{f - f^2/a} + \frac{1}{q}$$

$$\frac{1}{q} = \frac{1}{f} - \frac{1}{f - f^2/a} = \frac{f - f^2/a - f}{f(f - f^2/a)}$$

$$q = \frac{f(f - f^2/a)}{-f^2/a} = f - a$$

$$M = -\frac{q}{p} = -\frac{f-a}{f-f^2/a} = \frac{a-f}{f-f^2/a} = \frac{a-f}{f(1-f/a)} = \frac{a(a-f)}{f(a-f)} = \frac{a}{f}$$

Figure 10.27(b) shows that the lens image $O'(x_1)$ of the hologram image $O'[-(a/f)x_1]$ remains erect and on the same side of the axis. We can see therefore that the two images found by the lens are no longer in line but that one is formed on each side of the axis. This simplifies viewing of the Fourier transform hologram. We can also see that the hologram real image becomes a virtual lens image and the hologram virtual image becomes a lens real image. The lens images appear equally spaced on each side of the back focal plane of the lenses as shown in the composite Figure 10.27(c).

Figure 10.28 shows a point reference source located on the axis of the lens. When the point reference source is outside the front focal plane and away from the lens by a distance a, reference waves are produced in the back focal plane of the lens with a variable phase; that is, in the back focal plane we have plane waves that can be represented by $Re^{-j(\lambda a/4\pi)(\omega)^2}$ as shown in Figure 10.28(a). Figure 10.28(c) shows the point reference source closer to the lens by a distance a from the front focal plane. In the back focal plane of the lens we will again have reference waves with a variable phase that can be represented by $Re^{j(\lambda a/4\pi)(\omega)^2}$. When the point source is on axis and in the front focal plane, the plane waves in the back focal plane are represented by R as shown in Figure 10.28(b). From Figure 10.28 we can see that we can illuminate a hologram with plane waves by placing it in the back focal plane of a lens. We can further see that by positioning the point reference source with respect to the front focal plane we can choose the phase of the reference beam illuminating the hologram.

As summarized in Figure 10.25 we determined that the reconstruction of a Fourier transform hologram produces two images of the object plus undiffracted plane waves. The hologram of Figure 10.25 was produced with the point reference source on axis and in the front focal plane of the lens while the object was outside the front focal plane by a distance a. Let us consider illuminating this hologram by other than a reference source R as shown in Figure 10.28(b) and equation (10-49); for example, we can consider the case of the reference beam of Figure 10.28(a). The transmission function of the hologram is given by (10-48); that is,

$$T(x) = K\left[A + \frac{e^{-j(\lambda a/4\pi)(\omega)^2}F(\omega)}{R} + \frac{e^{j(\lambda a/4\pi)(\omega)^2}F^*(\omega)}{R}\right] \qquad (10\text{-}48)$$

Illuminating this hologram with reference waves of the form

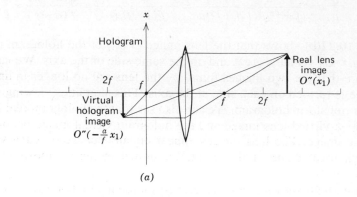

(a)

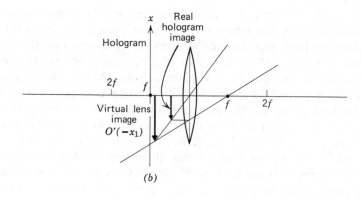

(b)

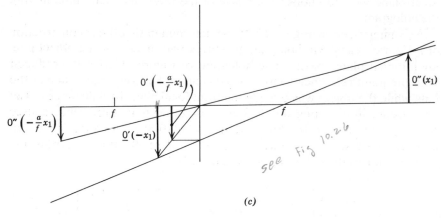

(c)

Figure 10.27   Graphical construction of lens image location for hologram images.

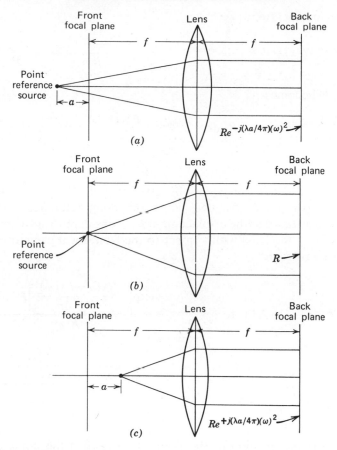

Figure 10.28 Plane waves produced by a point reference source on axis.

$Re^{-j(\lambda a/4\pi)(\omega)^2}$ produced as shown in Figure 10.28(a) we obtain

$$Re^{-j(\lambda a/4\pi)(\omega)^2}T(x) = KRe^{-j(\lambda a/4\pi)(\omega)^2}\left[A + \frac{e^{-j(\lambda a/4\pi)(\omega)^2}F(\omega)}{R}\right.$$

$$\left. + \frac{e^{j(\lambda a/4\pi)(\omega)^2}F^*(\omega)}{R}\right] \qquad (10\text{-}50)$$

We could have illuminated this hologram with reference waves of the form $Re^{j(\lambda a/4\pi)(\omega)^2}$ produced as shown in Figure 10.28(c), which gives

$$Re^{j(\lambda a/4\pi)(\omega)^2}T(x) = KRe^{j(\lambda a/4\pi)(\omega)^2}\left[A + \frac{e^{-j(\lambda a/4\pi)(\omega)^2}F(\omega)}{R} + \frac{e^{j(\lambda a/4\pi)(\omega)^2}F^*(\omega)}{R}\right]$$

$$(10\text{-}51)$$

Multiplying out (10-50) we have

$$Re^{-j(\lambda a/4\pi)(\omega)^2}T(x) = KR\left[Ae^{-j(\lambda a/4\pi)(\omega)^2} + \frac{e^{-j\lambda(2a/4\pi)(\omega)^2}F(\omega)}{R} + \frac{F^*(\omega)}{R}\right]$$

(10-52)

and (10-51) becomes

$$Re^{j(\lambda a/4\pi)(\omega)^2}T(x) = KR\left[Ae^{j(\lambda a/4\pi)(\omega)^2} + \frac{F(\omega)}{R} + \frac{e^{j(\lambda(2a)/4\pi)(\omega)^2}F^*(\omega)}{R}\right]$$

(10-53)

In general we can write equations (10-52) and (10-53) for the light amplitude distribution obtained when the point reference source for viewing is a distance b (the displacement b is considered $+$ away from lens and $-$ towards the lens) with respect to the front focal plane of the lens and can be written as

$$Re^{-j(\lambda b/4\pi)(\omega)^2}T(x) = KR\left[Ae^{-j(\lambda b/4\pi)(\omega)^2} + \frac{\exp\left[-j\dfrac{\lambda(a+b)(\omega)^2}{4\pi}\right]F(\omega)}{R}\right.$$

$$\left. + \frac{\exp\left[j\dfrac{\lambda(a-b)(\omega)^2}{4\pi}\right]F^*(\omega)}{R}\right]$$

(10-54)

where a was the location of the point source when the hologram was constructed. From (10-54) we can see that when $b = 0$, we have the case of the hologram being illuminated by a reference plane wave produced as shown in Figure 10.28(b). We have already determined the result of illuminating the hologram in this manner for viewing and the results are summarized in Figures 10.25 and 10.26.

If we now consider only the first term of (10-54), $KRAe^{-j(\lambda b/4\pi)(\omega)^2}$ we can see that this term has the form of the light distribution in the back focal plane of a lens caused by a point source a distance b out of the front focal plane of the lens (see Figure 10.28(a) and 10.28(c) where $b = +a$ and $b = -a$, respectively). The only effect of this term of the hologram is to attenuate the light from the point source by a factor KA. This means that the image of the point source produced by the hologram and lens will be attenuated when viewed.

Comparing the second and third terms of (10-54) with the second and third terms of (10-49)

$$\frac{e^{-j(\lambda(a-b)/4\pi)(\omega)^2}F(\omega)}{R} + \frac{e^{j(\lambda(a-b)/4\pi)(\omega)^2}F^*(\omega)}{R}$$ second and third terms of (10-54)

$$\frac{e^{-j(\lambda a/4\pi)(\omega)^2}F(\omega)}{R} + \frac{e^{j(\lambda a/4\pi)(\omega)^2}F^*(\omega)}{R}$$ second and third terms of (10-49)

we see that they have the same form. This means that the images formed by these two terms of (10-54) can be found similarly to the way they were found for the corresponding terms of (10-49) (summarized in Figures 10.25 and 10.26.

LENSLESS FOURIER HOLOGRAPHY

We have seen how the introduction of the Fourier transform can be used in the formation of a certain type of hologram. This type of hologram has certain advantages over other types in that the photographic film used to record the hologram does not have to have as high resolution as the films used to record other types of holograms; that is, recording of the Fourier transform can be done with lower-resolution films than required to record the Fresnel diffraction patterns. We have also seen how by means of a second Fourier transform reconstruction can be effected.

Figure 10.29 shows the interference pattern being produced between spherical waves originating at a point source P and those from the object

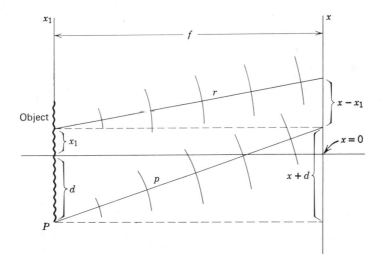

Figure 10.29 Formation of a Fourier transform hologram without lenses.

points (x_1). The amplitude variation of the object can be written as $A'(x_1)$. We can further assume that each point on the object acts as a point source radiating spherical waves of amplitude $A'(x_1)$. We can then write the expression for the spherical light waves from any point in the object as

$$A'(x_1)e^{jkr} = A'(x_1) \exp [jk(f^2 + (x - x_1)^2)^{1/2}] \quad (10\text{-}55)$$

where f is the distance from the object plane to the recording plane. If we assume $(x - x_1) \ll f$ and we use the first two terms of the binomial expansion for $[f^2 + (x - x_1)^2]^{1/2}$, that is,

$$(f^2)^{1/2} + \tfrac{1}{2}(f^2)^{-1/2}(x - x_1)^2 = f + \frac{(x - x_1)^2}{2f}$$

we can simplify (10-55) to

$$A'(x_1)e^{jkr} = A'(x_1)e^{jkf} \exp \left[jk \frac{(x - x_1)^2}{2f} \right] \quad (10\text{-}56)$$

To obtain the total light in the x plane from all points in the object we integrate over the object (sum of all infinite number of object points) and obtain what we can call the signal light $S(x)$. $S(x)$ is the light at any point x in the plane x or

$$S(x) = \int_O A'(x_1)e^{jkf} \exp \left[jk \frac{(x - x_1)^2}{2f} \right] dx_1 \quad (10\text{-}57)$$

where the integration is over the object O

$$= e^{jkf} \int A'(x_1) \exp \left[jk \frac{x^2}{2f} \right] \exp \left[-j \frac{kx}{f} x_1 \right] \exp \left[jk \frac{x_1^2}{2f} \right] dx_1$$

$$S(x) = e^{jkf} \exp \left[jk \frac{x^2}{2f} \right] \int A'(x_1) \exp \left[jk \frac{x_1^2}{2f} \right] \exp \left[-j \frac{kx}{f} x_1 \right] dx_1 \quad (10\text{-}58)$$

Equation (10-58) gives the light amplitude at each point x contributed by the object.

We can assume a point source reference P is radiating spherical waves of amplitude R_o. These reference spherical waves can be expressed as

$$R(x) = R_o e^{jkp} = R_o \exp [jk(f^2 + (x + d)^2)^{1/2}] \quad (10\text{-}59)$$

Again assuming $x + d \ll f$ and approximating the exponent by the first two terms of the binomial expansion we obtain

$$R(x) = R_o e^{jkf} \exp\left[jk \frac{(x+d)^2}{2f} \right] \qquad (10\text{-}60)$$

The total light amplitude due to both signal and reference is then the sum

$$A(x) = R(x) + S(x) \qquad (10\text{-}61)$$

$$A(x) = R_o e^{jkf} \exp\left[jk \frac{(x+d)^2}{2f} \right] + e^{jkf} e^{jk(x^2/2f)} \int A''(x_1) e^{-j(kx/f)x_1} \, dx_1 \qquad (10\text{-}62)$$

where we have substituted

$$A''(x_1) = A'(x_1) \exp\left[jk \frac{x_1^2}{2f} \right]$$

to simplify the notation. We can further define $F'(\omega)$ as the optical Fourier transform of $A''(x_1)$ where $\omega = kx/f$; i.e.,

$$F'(\omega) = \int A''(x_1) e^{-j\omega x_1} \, dx_1 \qquad (10\text{-}63)$$

and we can write

$$A(x) = R_o e^{jkf} e^{jk(x^2/2f)} e^{jk(xd/f)} e^{jk(d^2/2f)} + e^{jkf} e^{jk(x^2/2f)} F'(\omega) \qquad (10\text{-}64)$$

Factoring $R_o e^{jkf} e^{jk(x^2/2f)}$ in (10-64) we get

$$A(x) = R_o e^{jkf} e^{jk(x^2/2f)} \left[e^{jk(xd/f)} e^{jk(d^2/2f)} + \frac{1}{R_o} F'(\omega) \right]$$

$$= R_o e^{jkf} e^{jk(x^2/2f)} \left[e^{j\omega d} e^{jk(d^2/2f)} + \frac{1}{R_o} F'(\omega) \right] \qquad (10\text{-}65)$$

The intensity is

$$I(x) = A(x) A^*(x)$$

$$(10\text{-}66)$$

and from (10-65)

$$I(x) = R_o^2 \left[1 + \frac{e^{-j\omega d} e^{-jk(d^2/2f)}}{R_o} F'(\omega) + \frac{e^{j\omega d} e^{jk(d^2/2f)}}{R_o} F'^*(\omega) + \frac{|F'(\omega)|^2}{R_o^2} \right]$$

$$(10\text{-}67)$$

If R_o is made sufficiently large, the fourth term in (10-67) can be neglected giving

$$I(x) = R_o^2\left[1 + \frac{e^{-jk(d^2/2f)}}{R_o} e^{-j\omega d}F'(\omega) + \frac{e^{jk(d^2/2f)}}{R_o} e^{j\omega d}F'^*(\omega)\right] \quad (10\text{-}68)$$

If the photographic exposure were calculated as in the previous discussion, the terms inside the brackets in (10-68) would maintain the same form except for constant factors involving γ, exposure time, etc. We need not consider the constants here because we are mainly interested in the form of the variations rather than the absolute amplitude.

If the hologram is viewed in collimated light as shown in Figure 10.30 through a lens located a focal length f_L from the hologram, the transform of each of the terms in the brackets will be imaged in the back focal plane of the lens. The first term is a constant and as we have previously seen will be transformed to a DC spot on the optical axis. The second term will result in the transform of $e^{-j\omega d}F'(\omega)$ and the third term will give the transform of $e^{j\omega d}F'^*(\omega)$. The constant factors

$$\frac{e^{-jk(d^2/2f)}}{R_o} \quad \text{and} \quad \frac{e^{jk(d^2/2f)}}{R_o}$$

present in the second and third terms, respectively, do not affect the relative distribution of light; they determine the absolute phase and magnitude that is of no interest to us here.

We can now apply our previous results for the effect of a phase factor such as $e^{\pm j\omega d}$ which appears in a Fourier transform; that is, if the transform of $A''(x_1)$ is

$$F'(\omega) = \int A''(x_1)\, e^{-j\omega x_1}\, dx_1$$

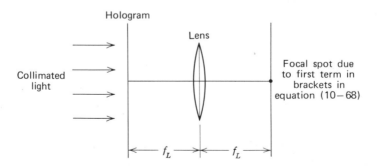

Figure 10.30   Fourier transform hologram reconstruction.

the transform of $A''(x_1 - d)$ is

$$e^{-j\omega d}F'(\omega) = \int A''(x-d)e^{-j\omega x_1}\,dx_1$$

and the transform of $A''(-x_1 - d)$ is

$$e^{j\omega d}F'^*(\omega) = \int A''(x-d)e^{-j\omega x}\,dx.$$

Thus the transform of the second term in (10-68) gives $A''(x-d)$ and the third term gives $A''(-x-d)$. If we refer back to our definition of

$$A''(x_1) = A'(x_1)\exp\left[jk\frac{x_1^2}{2f}\right]$$

we obtain

$$A''(x-d) = A'(x-d)\exp\left[jk\frac{(x-d)^2}{2f}\right] \qquad (10\text{-}69)$$

and

$$A''(-x-d) = A'(-x-d)\exp\left[jk\frac{(-x-d)^2}{2f}\right] \qquad (10\text{-}70)$$

Equation (10-69) is now the same as the original object $A'(x_1)$ except for a shift up a distance d from the optical axis and the phase term

$$\exp\left[jk\frac{(x-d)^2}{2f}\right]$$

The phase term does not affect the image because only intensity is seen or recorded and

$$I(x) = [A''(x-d)]\,[A''^*(x-d)]$$

$$= [A'(x-d)][A'^*(x-d)]$$

and the phase term drops out; that is, the conjugate of

$$\exp\left[jk\frac{(x-d)^2}{2f}\right] \quad \text{is} \quad \exp\left[-jk\frac{(x-d)^2}{2f}\right]$$

and their product is one.

Similarly (10-70) is the same as the original object except it is inverted due to the minus sign in front of x_1 and it is shifted down a distance d from the optical axis. These two terms are shown in Figure 10.31.

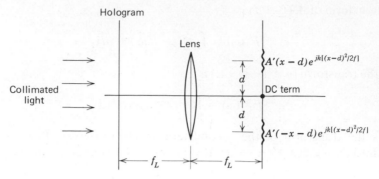

Figure 10.31 Reconstruction of the first, second, and third terms in equation (10-68).

In the discussion above we did not consider the effects of the relative magnitudes of the original object distance f and the viewing lens focal length f_L. We have previously defined $\omega = kx/f$, where f was the distance from the object to the recording plane. The ω in the optical Fourier transform is similarly defined, where f is the focal length. We will use the subscript L to denote the lens terms, that is,

$$\omega_L = \frac{kx'}{f_L}$$

Now let us consider the second term in (10-68), that is

$$e^{-j\omega d}F'(\omega) = e^{-j\omega d}\int A''(x_1)e^{-j\omega x_1}dx_1 \tag{10-71}$$

Taking a second transform through the lens we obtain

$$\int e^{-j\omega d}F'(\omega)e^{-j\omega_L x}\,dx = \int e^{-j\omega d}e^{-j\omega_L x}\,dx \int A''(x_1)e^{-j\omega x_1}\,dx_1 \tag{10-72}$$

Substituting for ω and ω_L equation (10-72) becomes

$$\int e^{-j\omega d}F'(\omega)e^{-j\omega_L x}\,dx = \int e^{-j(k/f)xd}e^{-j(k/f_L)x'x}\,dx \int A''(x_1)e^{-j(k/f)xx_1}\,dx_1 \tag{10-73}$$

Rearranging integrals gives

$$\int e^{-j\omega d}F'(\omega)e^{-j\omega_L x}\,dx = \int A''(x_1)\,dx_1 \int e^{-j(k/f)x[x_1 + d + (f/f_L)x']}\,dx. \tag{10-74}$$

Since

$$\int e^{-jkx}\,dx = 2\pi\delta(k)$$

the right hand side of (10-74) can be written

$$\int A''(x_1)\, dx_1 \left\{ 2\pi\delta\left[\frac{k}{f}\left(x_1 + d + \frac{f}{f_L} x' \right) \right] \right\}$$ (10-75)

Applying

$$\delta(az) = \frac{1}{|a|}\delta(z)$$ (10-76)

the term in the brackets can be rewritten to give

$$\int A''(x_1)\, dx_1 \left[\frac{2\pi f}{k} \delta\left(x_1 + d + \frac{f}{f_L} x' \right) \right]$$ (10-77)

The second integration is now given simply by the sifting property of the delta function, that is,

$$\int F(x_1)\delta(x_1 + a)\, dx_1 = F(-a)$$ (10-78)

Thus we obtain

$$\frac{2\pi f}{k} A''\left(-\frac{f}{f_L} x' - d \right)$$

The factor $2\pi f/k$ can be neglected here because it effects only the absolute magnitude. Thus we have found that the lens transforms $e^{-j\omega d}F'(\omega)$ into $A''(-(f/f_L)x' - d)$. The factor f/f_L in front of x_1 accounts for magnification due to relative lengths of f and f_L where magnification $M = f_L/f$. Thus we have shown that a lensless Fourier transform hologram can be made. This type of hologram does require a transform lens for reconstruction just as in the previously described hologram system.

11

Holographic Techniques

RECORDING A TRANSMISSION HOLOGRAM

The recorded interference pattern between a temporally coherent light wave from an object and a reference beam (from the same source) is referred to as a *hologram*. Holography is a type of interferometry and therefore interferometry techniques can be applied to holography. The hologram records information concerning the intensity, wavelength, and phase of the light. In conventional photography only intensity and wavelength (color) can be recorded. In holography the phase of the light reflected or transmitted by an object is recorded in addition to the parameters recorded by conventional photography.

We have previously seen that to record the phase of a light wave, the light source must emit temporally coherent light. It is for this reason that a laser is usually used as the light source when producing a hologram. Electrons at an excited, high-energy level are made to emit this energy only when stimulated thus producing coherent light. The light emitted from a laser is essentially monochromatic and therefore temporally coherent.

In order to record the phase content of a light wave it is necessary to be able to record an interference pattern of the light wave. An interference pattern is produced when two waves of the same wavelength arrive together at a set of points. The interference at a point can either be constructive, destructive, or anything in between. Methods of recording the interference pattern to produce a transmission hologram are shown in Figure 11.1.

A monochromatic, coherent beam of light from the laser is brought to a focus by a microscope objective lens. A pinhole is placed at the focal point of the lens to filter the light spatially so that any lens aberrations are removed and to provide a better point source of light. The light

502

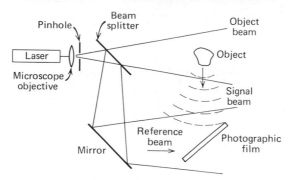

Figure 11.1a.

diverges past the focal point (pinhole) and strikes a plane glass that acts as a beam splitter as shown in Figure 11.1(a). The light beam is partially transmitted and partially reflected by the plane glass. It is usually desirable to have the reference beam intensity about ten times stronger than the signal beam. We can refer to the light beam reflected from the plane glass as the *reference beam* and the transmitted beam from the plane glass as the *object* or *signal beam*. The object beam is made to illuminate the object. The light from the object beam is reflected (by changing the configuration, transmitted light from the object can be used) from the object and made to illuminate the photographic plate. Each point on the object will reflect light in a different manner; that is, the phase of the reflected light from each point will vary depending on the path length from the plane glass to the object. Reflected light from each point on the object striking the photographic plate will interfere with the light from the reference beam. Because the light from the reference beam and

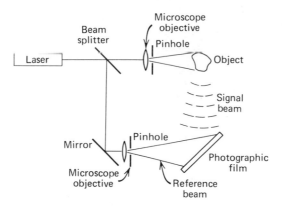

Figure 11.1b.

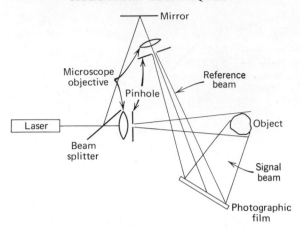

Figure 11.1c.

object beam come from the same coherent source, an interference pattern will be produced that will expose the photographic plate. The developed photograph is a transmission hologram. Alternate configurations for making transmission holograms are shown in Figures 11.1(b) and 11.1(c). We note that the requirement that the signal and reference beams have approximately equal path lengths from the beam splitter to the film is met in all configurations of Figure 11.1. In order to reconstruct (make visible) the object from a hologram a reference beam similar to that which was used to expose the hologram is used to illuminate the hologram as shown in Figure 11.2.

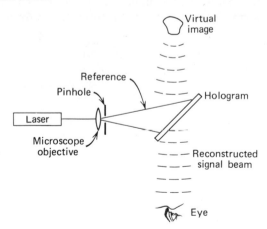

Figure 11.2.

It is possible to superimpose several holographic images on a single hologram. By using different incident angles of the reference beam, it is possible to produce on a single film many holographic images which can be reconstructed individually without appreciable crosstalk (one interfering with the viewing of another).

Viewing the Images of a Transmission Hologram

The general arrangement for viewing the images of a developed transmission hologram is shown in Figure 11.3. Monochromatic coherent light is used to illuminate the hologram and two wavefronts similar to that reflected from the original object are reconstructed. These wavefronts can be viewed separately (they travel in different directions) as shown by the two viewing angles in Figure 11.3. At one viewing angle the wavefronts appear to originate from an image behind the hologram plate. Since the light waves viewed do not actually pass through the image points, this image is called the *virtual* image. At the second viewing angle the wavefronts appear to originate from an image in front of the hologram plate. Since the light waves actually pass through the image points, this image is called the *real* image. The definition used for the real and virtual images are the same as those used in conventional optical imaging theory. When the same reference beam is used for exposing as well as viewing the hologram, an exact three-dimensional image of the original object will be seen as illustrated in Figure 11.3. This three-

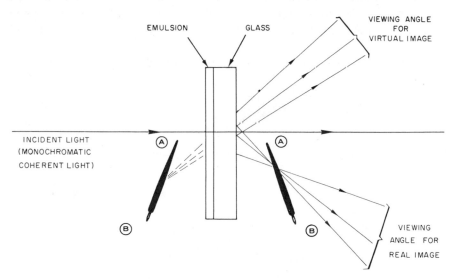

Figure 11.3 Viewing the images of a hologram.

dimensional characteristic of holograms is important. Just as in viewing an actual scene, the eye must refocus when looking at a near or a far portion of the reconstructed images. When the observer's view is obstructed by an object imaged in the foreground it is possible to look behind the obstruction by changing the angle of view (if possible within the viewing angle). As the observer changes his position within the viewing angle, a distinct change in the perspective of the image is perceived.

Variations of the viewing method discussed above are possible. In fact both theory and experiment indicates that the reconstruction process is sensitive to the angle of incidence used for the viewing light. The reconstructed images appear to be best when the viewing light is incident at the same angle as the original reference beam. In addition, it is found for some holograms (depending on the development process, and the signal and reference beam angle) that one of the images may be clearer than the other. That is, when the virtual image is very clear the real image is difficult to see and when the virtual image is clear the virtual image is difficult to see. It has been found for these holograms that if the virtual image is the clearest when the hologram is illuminated from one side, the real image is the clearest when the hologram is illuminated from the opposite side. This effect can be explained if the arrangement shown in Figure 11.3 is examined. If the direction of the incident light is reversed (i.e., imagine the light to travel from right to left), all the arrowheads indicating direction would be reversed in Figure 11.3. The images would then be viewed by looking toward the hologram from the left. Viewed from this new location, what had originally been a virtual image (when viewed from the right) now will appear as a real image (when viewed from the left.

Figure 11.4 shows the arrangement for viewing the virtual image of a

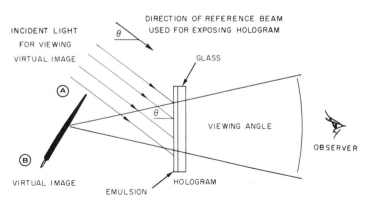

Figure 11.4 Viewing the virtual image.

hologram. As shown in Figure 11.4 the virtual image appears behind the hologram plane. In effect, the hologram acts as a window through which the virtual image is viewed. When the virtual image is viewed through the small section of the original hologram, the entire image may not be visible at one viewing angle. It is possible, however, to see all portions of the image by moving through the visual angle range. This truly gives the effect of looking through a window (e.g., to see into a room) since as the window is made smaller one is required to move through a larger angle to see all parts of a room.

It should be noted that whether the hologram is developed as a negative or positive the resulting image is the same. That is, the virtual image will always appear as a positive regardless of whether the hologram is a negative or a positive. This is easily understood if a negative is considered as the development of points of reinforcement and a positive is considered as the development of points of cancellation. Since the only difference will be a half wavelength shift in the position of the interference patterns, either hologram will contain the same pattern and therefore the same information. Thus the same positive virtual image (and real image) is produced in either case.

When the hologram is illuminated by a reference beam traveling in the direction opposite to that used during recordings, the hologram produces a wavefront pattern similar to the wavefront pattern reflected by the object used to make the hologram. A reconstructed real image appears suspended in mid-air between the observer and the hologram as shown in Figure 11.5. Although the real image is a precise replica of the illuminated surface of the object used to make the hologram, it appears to the viewer as though he were looking through the object from

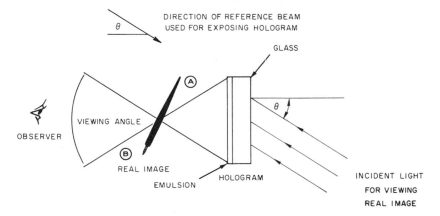

Figure 11.5   Viewing the real image.

the rear. This point can be shown by considering the configuration of Figure 11.3. In Figure 11.3 the original object is positioned so that point A is closer to the photographic film than point B and point A is on the top of the object. If we assume that the position of the virtual image is identical to that of the original object then in viewing the virtual image, it appears that the original object is still behind the hologram. When viewing the real image, the point B is still farther away from the hologram plate than point A. The real image however is now between the observer and the hologram plate, therefore point B is now closer to the observer than point A. This gives the effect of looking through the object from the rear. Point A is still the top of the object which means the real image is erect. Points further away from the observer in the virtual image are closer to the observer in the real image.

If a sheet of film were placed at point A of the real image, it would photograph a focused image of point A and a defocused image of point B. It is important to note that a focused image of point A is produced without any lenses. If point A and B are close together a true focused picture of the object can be obtained without lenses. If A and B are separated, individual pictures of the points A and B can be obtained by exposing separate films placed at A and B respectively. Therefore, the real image which is formed in front of the hologram can only be photographed piecewise since the reconstructed image is three-dimensional and is not in a single plane (the usual form of a photographic emulsion). Generally there is some initial difficulty in focusing one's eyes on the real image suspended in space between the hologram and the eye. This makes the real image more difficult to see than the virtual image.

Using the original transmission hologram it is possible to produce a second transmission hologram which will have a real image with true depth perception. Referring to Figure 11.3 let us assume that the hologram does not have a real image with true depth perception because the observer is effectively looking through the object as if from the rear. True depth perception in a real image can be accomplished by using the original hologram to produce another hologram as shown in Figure 11.6(a). The original hologram is illuminated by a reference beam which upon emerging from the hologram is reflected by a mirror onto a new photographic emulsion. The real image formed by the original hologram also illuminates the new photographic emulsion. The interference between the reference beam reflected from the mirror and the light from the real image interfere to form a second hologram. After development the second hologram is viewed as shown in Figure 11.6(b). The real image of this second hologram will have true depth perception when viewed within the visual angle.

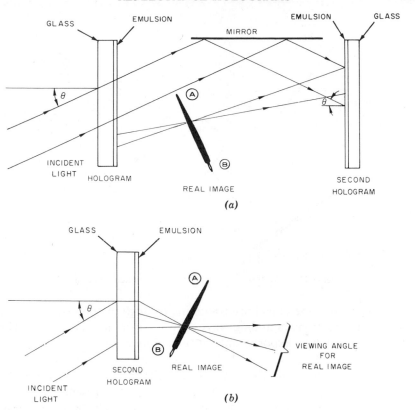

Figure 11.6   Hologram reproduced from another hologram.

REFLECTANCE HOLOGRAMS

In order to view a transmission hologram it is best to use a laser, (or other monochromatic light source which is spatially coherent) to illuminate the hologram. It is possible to produce a hologram type interference pattern which does not require a laser for viewing. Figure 11.7 shows a method for exposing a reflectance type hologram. The reflectance type hologram is made by illuminating the photographic plate from both sides, that is the signal beam illuminates one side of the plate while the reference beam exposes the other side of the plate. The laser beam is made to illuminate the object. The reflected light from the object (signal beam) illuminates the photographic plate (hologram). A portion of the laser beam is diverted by a beam splitter to a mirror which reflects the light onto the rear of the hologram. It is noted that the reference beam and the

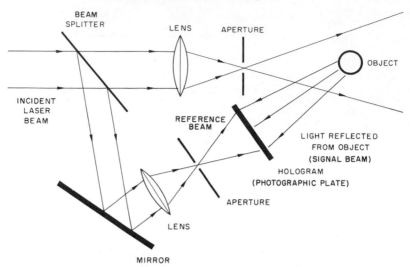

Figure 11.7   Method for producing a reflectance hologram.

signal beam from the object are travelling through the photographic plate in different directions. Standing waves are set up in depth in the emulsion. The emulsion will be exposed only at the points where the two waves add constructively. Transmission holograms form a slit type interference pattern in the emulsion. This pattern can either be on the surface or be one which is in depth in the emulsion (somewhat perpendicular to the plane of the emulsion, like venetian blinds). The reflectance holograms form an interference pattern in the emulsion which is in the form of planes which are essentially formed somewhat parallel to the plane of the emulsion. These planes formed in the emulsion can be considered to be an interference filter. The reflectance type of hologram can be viewed using a point white light source. Figure 11.8 shows a reflectance hologram being illuminated by a point white light source, (the flashlight bulb with its small filament can act as a point source, particularly at a large distance from the bulb). The reflectance type hologram is much more difficult to produce than the transmission type hologram. Some of the reasons for these difficulties are:

1. Exposing the reflectance type hologram is more difficult since sound vibrations tend to vibrate the emulsion. Considering the emulsion as a thin membrane held at its outer periphery, the similarity to the membrane of a drum is evident. As in a drum, any slight sound vibrations will tend to make the membrane vibrate. Since the exposure is being made in the emulsion in depth, (forming planes somewhat parallel to the hologram

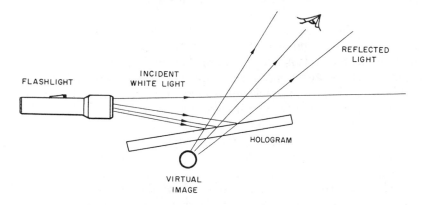

Figure 11.8   White light viewing of a standing wave type.

plane) any slight vibration will tend to blur it. In the transmission hologram a slight vibration (moving surface) will not be as evident as in the
reflectance hologram since the interference pattern is being recorded
along the direction of motion (somewhat perpendicular to the hologram
plane).

2. When the interference pattern is formed in depth (such as with a
reflectance hologram), any shrinkage of the film will change the spacing
between the exposed points of a hologram. This distortion is to be
expected since there is shrinkage in the normal processing of films. After
development the emulsion of the reflectance hologram must be expanded
in order to see the image. Not fixing the reflectance hologram plate
prevents much of the shrinkage since the basic content of the emulsion
after development will be the same as it was before development, that is
the unexposed silver halide will not be removed from the emulsion.
Shrinkage will not effect the transmission hologram as much as the reflection hologram since it will only tend to shorten the depth of the slats
(of the venetian blinds) but not appreciably change their spacing.

3. The absorption and dispersion will tend to weaken the signal beam
before it can penetrate deep enough to form the standing waves in depth.

4. When white light is used to view the reflectance hologram it acts as a
color filter (reflecting only the color used to form the hologram). If the
color filter formed by the hologram is not effective (because of fabrication
difficulties) then the color of the hologram image will change as the viewing angle is changed. The Bragg effect (see Appendix 5) predicts a downward shift in the frequency of the reflected light when the viewing white
light is incident at increased angles.

REQUIREMENTS FOR VIEWING HOLOGRAMS

Making a hologram requires temporal coherence because the light reflected from each point on the object must individually interfere with the reference beam. It is not necessary that the light from adjacent points on the object be able to produce an interference pattern. When viewing the hologram the opposite requirements are necessary. For viewing, adjacent points on the hologram plate must be able to produce an interference pattern. This requirement implies that spatial coherence is necessary for viewing. The interference pattern produced by the light from adjacent points on a hologram plate gives rise to a reconstructed wavefront similar to that which emanated from the original object. Temporal coherence is not required for viewing because each point in the light wave gives rise to a divergent wavefront. Some degree of temporal coherence is desired, however, since any variation in wavelength will tend to vary the focal length (change the divergence of the wavefronts). A point source of white light (as from a flashlight) has spatial coherence at large distances. The size of the point source as compared to the distance from the source determines the degree of spatial coherence. Temporal coherence is determined basically by the bandwidth of the light source. The use of a temporally noncoherent source will produce variations in depth of the object, while a lack of spatial coherence tends to produce longitudinal blurring.

The combination of the human eye and brain is capable of detecting partially obscured items. Basically the eye-brain combination integrates any variations out of a picture so that any slight blurring (due to spatial noncoherence) or slight variations in depth (temporal noncoherence) are eliminated. Figure 11.9 demonstrates the ability of the eye-brain combination to interpret and distinguish patterns although they are partially obscured. This basic process occurs unconsciously when viewing a hologram illuminated by partial spatially or temporally coherent light.

For some types of holograms a simple method of viewing is to place a ground glass in the light beam directly from the laser. The ground glass tends to diffuse the light in all directions, which normally might be considered undesirable for hologram viewing because spatial coherence is lost. Actually since the laser beam striking the ground glass can be considered essentially a point source, all rays striking the hologram at the angle corresponding to the reference beam angle will reconstruct the image. The hologram tends to pass only the rays corresponding to the reference beam angle (attenuating all others) thus tending to reconstruct the image without the necessity for critical orientation of the hologram with the reference beam angle. Ground glass reconstruction has some

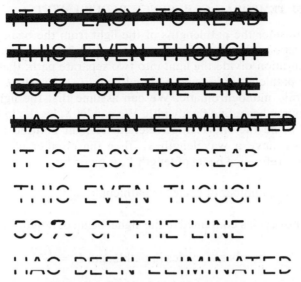

Figure 11.9 Eye-brain combination affects.

advantages for special cases, since the hologram does not have to be critically oriented for an image to be seen. In addition the beam-expanding lenses, or pinholes, or both, are not necessary, which makes for a simpler, cheaper viewing technique. This technique does, however, generally suffer from poorer (less sharp) images.

Each point on the object in reflecting (or transmitting) light acts as a point source. We saw previously how light from a point source is diffracted and tends to spread out. This means that the diffracted light from each point on the object illuminates the entire photographic plate (and not a single corresponding point as in conventional photography). This permits the hologram image to have a greater tonal range (since the recorded interference pattern does not easily become saturated) than the conventional photographic image. This also indicates that holograms can be made if objects which have shiney points without the use of a reference beam, (the point sources act as a reference beam).

Every small portion of the hologram contains information relating to the object (as viewed from that small portion of the hologram). This can easily be shown by illuminating for the real image a small spot on a hologram with an unexpanded laser beam. The real image can be seen from the particular perspective of the point being illuminated. The resolving power of the photographic emulsion is the basic limiting factor (before the diffraction limit) as to how small a hologram can be cut and still be capable of reconstructing the image. It is important therefore that the photographic emulsion used be one of high resolution.

HOLOGRAM TEMPORAL COHERENCE REQUIREMENTS

Let us consider the pathlengths of the light from the beam splitters in the configurations shown in Figures 11.1 and 11.7. The beam splitters in each configuration divide the light into two separate path; that is, a reference beam pathlength (L_R) and a signal beam pathlength (L_S). No light source is truly monochromatic. We can assume that the light radiation from any source has a fine bandwidth. We can let λ_2 be the longest wavelength and λ_1 be the shortest wavelength of light emitted from a light source. If we have a wavelength λ_1, then the number of wavelengths contained in a reference beam of length L_R is given by

$$\text{number of cycles in pathlength } L_R = \frac{L_R}{\lambda_1}$$

The number of cycles contained in the signal beam of length L_S is given by

$$\text{number of cycles in pathlength } L_S = \frac{L_S}{\lambda_1}$$

We can multiply by 2π to convert cycles to radians obtaining the number of radians represented by the difference in pathlength between L_R and L_S for wavelength λ_1

$$\varphi_1 = 2\pi \left(\frac{L_R}{\lambda_1} - \frac{L_S}{\lambda_1}\right)$$

The number of radians represented by these same pathlengths when we use light of wavelength λ_2 is given by

$$\varphi_2 = 2\pi \left(\frac{L_R}{\lambda_2} - \frac{L_S}{\lambda_2}\right)$$

The difference in the number of radians represented by these paths when we change from wavelength λ_1 to λ_2 is given by

$$\Delta\varphi = \varphi_1 - \varphi_2 = 2\pi \left(\frac{L_R}{\lambda_1} - \frac{L_S}{\lambda_1}\right) - 2\pi \left(\frac{L_R}{\lambda_2} - \frac{L_S}{\lambda_2}\right)$$

$$= \frac{2\pi}{\lambda_1}(\Delta L) - \frac{2\pi}{\lambda_2}(\Delta L) = 2\pi\Delta L \left(\frac{1}{\lambda_1} - \frac{1}{\lambda_2}\right) = 2\pi\Delta L \left(\frac{\lambda_2 - \lambda_1}{\lambda_1\lambda_2}\right)$$

A reasonably good assumption can be made here for most light sources, particularly lasers; that is, the wavelength λ_1 and λ_2 are close to the fundamental or center frequency λ_0, and therefore

$$\lambda_1\lambda_2 \approx \lambda_0^2$$

In order for the fringes to be recorded on the film $\Delta\varphi$ should change by less than $\pi/2$ radians. Therefore

$$\Delta\varphi = 2\pi\,\Delta L\left(\frac{\lambda_2 - \lambda_1}{\lambda_1\lambda_2}\right) = 2\pi\,\Delta L\left(\frac{\Delta\lambda}{\lambda_0^2}\right) < \frac{\pi}{2}$$

and

$$\Delta L < \frac{\lambda_0^2}{4\Delta\lambda}$$

For a given laser ($\Delta\lambda$ and λ_0 known) and reference beam pathlength L_R we can determine L_S which will give $\Delta\varphi < \pi/2$. A configuration where $\Delta\varphi < \pi/2$ will produce sharp fringes. Figure 11.10 shows the permissible change in L_S for a constant L_R, λ_0, and $\Delta\lambda$. An object which produces changes of L_S within this area will produce sharp fringes.

The pathlengths L_S and L_R are usually air pathlengths. Any change in index of refraction of either pathlength will cause a change in fringe pattern. Air pressure and temperature both effect air density (and therefore the index of refraction). A change in air density can effect both pathlengths. An increase in air density will increase the index of refraction and thus the optical pathlength. Since both pathlengths will effectively change proportionally to a density change, the result will be a change in a phase of the interference pattern but not a change in the coherence length. A change in density in only one path will result in a change in coherence length.

PHOTOGRAPHIC EMULSION RESOLUTION REQUIREMENTS FOR HOLOGRAPHY

The resolution of a hologram is not determined in the same manner as that of a normal photograph. The resolution of a normal photograph is related to the grain size of the silver halide crystals in the emulsion (it

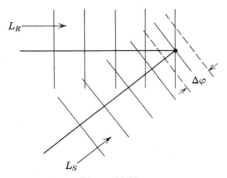

Figure 11.10.

is the blackening of the exposed silver halide crystals that forms the image). In a hologram the grain forms a type of diffraction grating that creates an image by diffracting light. Intuitively it can be seen that the resolution of a hologram is less dependent on grain size than a normal photograph. In holography the limitation is usually the size of the aperture. Here is another possible advantage of holography, since the size of the aperture can be considerably larger than most lenses.

Let us consider the interference pattern produced in a photographic emulsion when a plane-wave reference beam interferes with a plane-wave signal beam as shown in Figure 11.11. We shall consider that the reference and signal beams are in the plane of the paper as shown. The spacing between two dark points (fringes can be different when considered as a surface interference pattern or when considered as an interference pattern in depth. We shall consider the case with the film emulsion parallel to the stationary interference pattern formed between the signal and reference beams as shown in Figure 11.11 film position A. In this case the surface of the film will be uniformly exposed. There will be layers formed (in depth) within the emulsion parallel to the surface exposure. When the emulsion is in position B (surface perpendicular to the lines of the interference pattern), the spacing between fringes on the surface will be the same as the spacing within the emulsion. With the emulsion in position C, the spacing between fringes on the surface will be further apart than the spacing within the emulsion. In all three positions the spacing of the fringes within the emulsion will be the same. This means that for a given angle between the signal and reference beams the resolution requirements for an in-depth hologram will be constant regardless of the film orientation. The resolution requirements for a surface hologram will change with the orientation of the hologram in the interference pattern.

We shall consider the film resolution requirements only for a surface

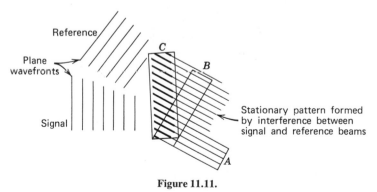

Figure 11.11.

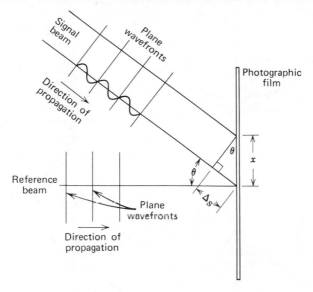

Figure 11.12.

hologram. On the surface of the hologram we select x to be the distance between two fringes. In order for x to be the distance between two dark points on the hologram the distance Δ_S in Figure 11.12 must be one wavelength. We can see from Figure 11.12 that

$$\lambda = \Delta_s = x \sin \theta$$

or

$$x = \frac{\lambda}{\sin \theta}$$

When the reference beam is perpendicular on the photographic plate, we can see that as $\theta \to 90$ degrees, the resolution requirements of the film goes up to a point where the film must be able to record the actual wavelength of the light being used. As the signal beam comes toward coincidence with the reference beam $\theta \to 0°$, the resolution requirements of the film go down.

We can consider another case where the interference pattern is produced by a reference beam and a signal beam at an angle θ to each other in the plane of the paper making equal angles with the normal to the film as shown in Figure 11.13(a). In order for point 1 and point 2 to be two recorded dark points on the hologram,

$$\Delta_S + \Delta_R = \lambda$$

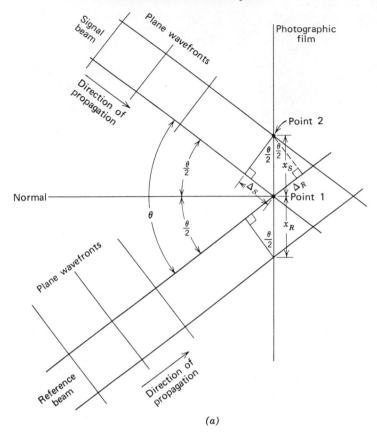

(a)

Figure 11.13a.

If we assume both signal and reference beams are in phase at point 1, then from point 1 the pathlength of the signal beam will change an amount Δ_S (negatively because we must go backward in time so that we can determine the phase of the signal wavefront when it intersected point 2) and the reference beam must advance from point 1 by an amount Δ_R (positively until the reference wavefront intersects point 2).

At point 1 both signal and reference waves are assumed in phase. The difference in pathlengths between the two beams is one wavelength when they both intersect point 2. The phase diagram Figure 11.13(b) shows the respective phase changes of the signal and reference beams from point 1.

From the diagram we can conclude that

$$\Delta_S = \Delta_R = \Delta \qquad \text{and} \qquad x_S = x_R = x$$

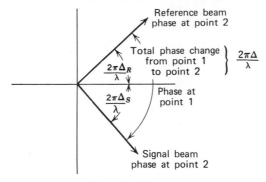

Figure 11.13b.

and therefore $\lambda = 2\Delta$. We can also see that

$$\sin\frac{\theta}{2} = \frac{\Delta_S}{x_S} = \frac{\Delta_R}{x_R}$$

or

$$\sin\frac{\theta}{2} = \frac{\Delta}{x}$$

$$\Delta = x\sin\frac{\theta}{2}$$

Therefore since $\lambda = 2\Delta$ then

$$\lambda = 2\left(x\sin\frac{\theta}{2}\right)$$

$$x = \frac{\lambda}{2\sin\theta/2}$$

From this we can see that the resolution requirements for the film are comparatively twice as great when the reference beam and signal beams make equal angles to the normal than when the reference beam is normal to the film.

We can consider a more general case where the signal beam makes an angle θ with the reference while the reference makes an angle of φ with the recording plate as shown in Figure 11.14. We will consider that both signal and reference beams are in the plane of the paper as shown in Figure 11.9(a). As before in order for point 1 and point 2 to be two dark points on the interference pattern we can consider that

$$\Delta_s + \Delta_R = \lambda$$

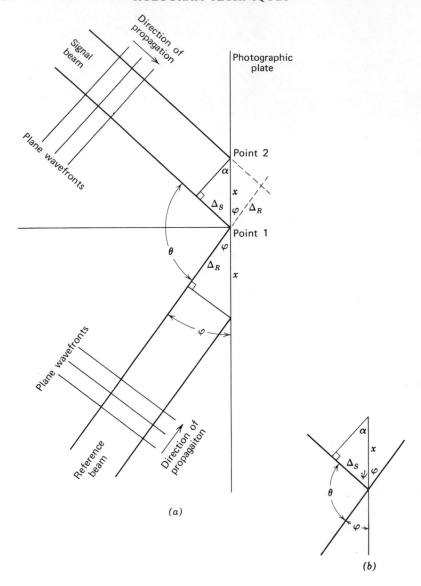

Figure 11.14.

Let us now consider only the portion of Figure 11.14(a) shown in Figure 11.14(b). We can see that the angles

$$\theta + \psi + \varphi = 180°$$

and therefore
$$\psi = 180° - (\theta + \varphi) = \pi - (\theta + \varphi)$$

In the triangle shown in Figure 11.14(b) we can see that
$$\alpha + \psi = 90° = \frac{\pi}{2}$$

and substituting of ψ
$$\alpha + \pi - (\theta + \varphi) = \frac{\pi}{2}$$
$$\alpha - \theta - \varphi = -\frac{\pi}{2}$$
$$\alpha = \theta + \varphi - \frac{\pi}{2}$$

From Figure 11.14(a) we can see that
$$\Delta_R = x \cos \varphi$$

and
$$\Delta_S = x \sin \alpha = x \sin \left(\theta + \varphi - \frac{\pi}{2}\right)$$

Since
$$\lambda = \Delta_R + \Delta_S$$
$$= x \cos \varphi + x \sin \left(\theta + \varphi - \frac{\pi}{2}\right)$$
$$= x \left[\cos \varphi - \cos (\theta + \varphi)\omega\right]$$
$$x = \frac{\lambda}{\cos \varphi - \cos (\theta + \varphi)}$$
$$= \frac{\lambda}{\cos \varphi - \cos \theta \cos \varphi + \sin \theta \sin \varphi}$$
$$= \frac{\lambda}{\cos \varphi (1 - \cos \theta) + \sin \varphi \sin \theta}$$
$$= \frac{\lambda}{2 \cos \varphi \sin^2 \theta/2 + \sin \varphi \sin \theta}$$
$$= \frac{\lambda}{2 \cos \varphi \sin^2 \theta/2 + 2 \sin \varphi \cos \theta/2 \sin \theta/2}$$
$$x = \frac{\lambda}{2 \sin \theta/2 (\cos \varphi \sin \varphi/2 + \sin \varphi \cos \theta/2)}$$
$$x = \frac{\lambda}{2 \sin \theta/2 \sin (\theta/2 + \varphi)}$$

Figure 11.15 is a nomograph of the equation

$$\frac{x}{\lambda} = \frac{1}{2 \sin \theta/2 \sin (\theta/2 + \varphi)}$$

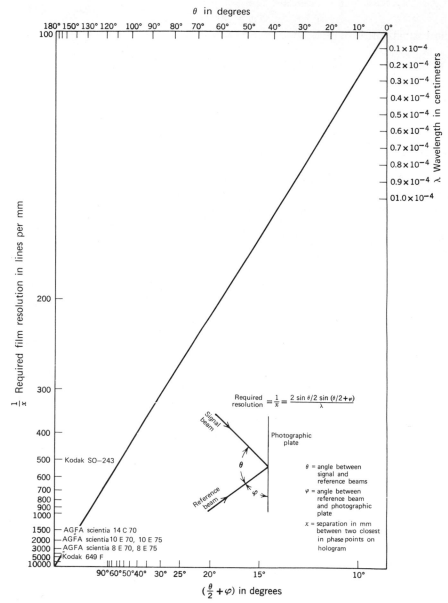

Figure 11.15.

This nomograph is a convenient way to determine the film resolution requirements for various angles between signal and reference beams (assuming they are plane waves) and for various angles between the photographic plate and the reference beam. Appendix 18 shows that the interference pattern between two-plane coherent collimated light beams varies sinusoidally in amplitude. It is easy to see that as the angle between the signal and reference beams gets smaller, the resolution requirements of the photographic film decreases. In addition as the angle between the reference beam and photographic plate gets smaller, the resolution requirements of the photographic film decreases. Let us assume that we want to make a hologram and we have a film that is capable of recording approximately 200 lines/mm but not more. For this example let us consider that we are using monochromatic light of wavelength 6×10^{-4} cm. We now connect these two points with a straight line. This straight line intersects the diagonal of the nomograph. Any line through this point that intersects the θ and $(\theta/2 + \varphi)$ scales will determine angles that permit the hologram to be recorded on photographic film having a resolution of 200 lines/mm (naturally any line passing to the right of the point of intersection on the diagonal intersecting the θ and $(\theta/2 + \varphi)$ scales will give angles that require lower film resolution for recording the hologram). As an example we can choose an angle of $\theta = 20°$ between the signal and reference beam. This choice permits us to draw a straight line from the 20-degree point on the θ scale through the intersection point of the first line drawn. Extending this line to the $(\theta/2 + \varphi)$ scale we see it intersects at the 20-degree points. This means that

$$\left(\frac{\theta}{2} + \varphi\right) = 20°$$

and since $\theta = 20°$ then $\theta/2 = 10°$ and $\varphi = 10°$. We can see therefore that when $\theta = 20°$, $\varphi = 10°$ we can produce a hologram (made with plane waves from a 0.6×10^{-4} wavelength light source) on a film that can record only 200 lines/mm. The nomograph also shows the limits of resolution for various representative types of films. It is possible therefore to select angles to permit holographic records to be made on relatively low resolution films.

SUBTRACTION OF SUPERIMPOSED IMAGES

In Chapter 7 we saw how to observe specific images from a multiple-image exposed photographic plate. For certain types of data processing it is desirable to have a technique to subtract a specific signal from a composite signal. Let us consider a signal A that is a darkened circle and a signal B that is a darkened rectangle as shown in Figure 11.16 to be posi-

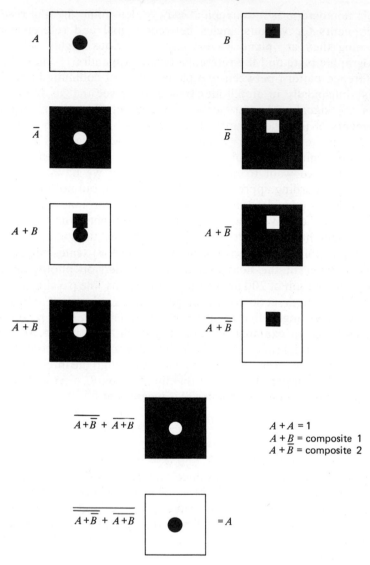

Figure 11.16.

tives. The negatives of these two signals are $\bar{A}$ (read "A not", i.e., everything that was A before is no longer A) and $\bar{B}$. From the sum of signals A plus B we would like to remove signal B. The "not" of $A + B$ is a clear rectangle and circle. If we add a darkened rectangle $B = \overline{A + \bar{B}}$ to the "not" of $A + B$; that is, $\overline{A + B} + \overline{A + \bar{B}}$ is a clear circle—the negative $\bar{A}$.

The positive $\bar{A}$ is A, a darkened circle. We have therefore removed the signal B from the signal $A + B$.

In order to accomplish this subtraction it was necessary to impose the following restrictions: (1) the signals had to be binary (i.e., either a black or a clear area on the film, no shades of gray were permitted); (2) the signal did not overlap; and (3) time was available to prepare the various negative and positive forms of the signals. A process has been outlined by Stroke*, et al, which indicates how to produce subtraction of signals by double exposing and eliminates some of the above restrictions. The procedure requires the use of coherent optics and illustrates the use of holography in data processing.

HOLOGRAPHIC SUBTRACTION BY DOUBLE EXPOSURE (STROKE ET AL)

Let us consider a signal $f_1(x)$ recorded on film. If this signal is the result of superimposing two signals $b(x)$ and $c(x)$, it will be given by the sum

$$f_1(x) = b(x) + c(x)$$

This signal is inserted in the front focal plane of a lens as shown in Figure 11.17.

A microscope objective is used to produce an effective point source at R_1 in the front focal plane of the lens. A point source can be represented by the mathematical expression $A\delta(x-a)$, where A is the amplitude and

*An Introduction to Coherent Optics and Holography in the Bibliography.

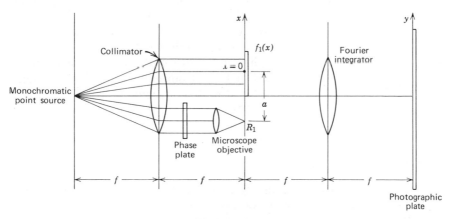

Figure 11.17.

$\delta(x-a)$ is a delta function that gives a point source displaced from $x = 0$ {of $f_1(x)$} by a. A delta function is defined by the following properties:

$$\delta(x) = 0 \quad \text{for } x \neq 0$$

for example,

$$\int \delta(x-a)f(x)\,dx = f(a)$$

$$\delta(x-a) = 0 \qquad \text{for } x \neq a$$

$$\int \delta(x-a)x^2\,dx = a^2$$

$$\int \delta(x-a)e^{j\pi x}\,dx = e^{j\pi a}$$

The phase plate can be adjusted to vary the phase ϕ of the point source (R_1). This can be done by putting a plane of optically flat glass in the light beam to the microscope objective. When the glass is exactly perpendicular to the light beam, it will have minimum thickness. The effective thickness can be increased by not having the glass perpendicular to the light beam. The phase change can therefore be selected by the orientation of the glass plate in the beams shown in Figure 11.18. Because the index of refraction of the glass is more than that of air, the effective pathlength (and thus the phase of the light) to the microscope objective can be changed. Phase plates can also be purchased commercially. These phase plates are optical flats which have a portion of the surface coated. The light leaving the coated portion will be phased differently from that leaving the uncoated portion of the glass. The thickness of the coating is made to change the phase of the light a specific amount for a specific wavelength of light.

The light amplitude at the photographic plate will be given by a Fourier transform of the light amplitude in the front focal plane of the lens, (Figure 11.17).

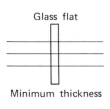

Glass flat

Minimum thickness

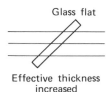

Glass flat

Effective thickness
increased

Figure 11.18.

$$F_1(y) = Ae^{-j(kay/f)} + B(y) + C(y)$$

Where

$F_1(y) = $ light amplitude as a function of y on the photographic plate

$Ae^{-jkay/f} = $ Fourier transform of $A\delta(x-a)$

$B(y) = $ Fourier transform of $b(x)$

$C(y) =$ Fourier transform of $c(x)$

$k = 2\pi/\lambda$ (λ is the wavelength of light source)

$f =$ focal length of the collimator and Fourier Integration lenses.

$a =$ displacement of R_1 from $x = 0$ coordinate of $f_1(x)$

The intensity is given by the relation

$$I_1(y) = F_1(y)F_1^*(y) = (Ae^{-j(kay/f)} + B + C)(Ae^{j(kay/f)} + B^* + C^*)$$

$$I_1(y) = A^2 + |B + C|^2 + A(B^* + C^*)e^{-j(kay/f)} + A(B + C)e^{j(kay/f)}$$

Before developing the photographic plate, a second exposure is made with the following changes. A new signal $f_2(x) = c(x)$ replaces $f_1(x)$ and the phase plate is adjusted for a phase shift of 180 degrees relative to that of the first exposure. The phase shift at the point source (R_1) is represented by the relation

$$A\delta(x - a)e^{\pm j\pi} = -A\delta(x - a)$$

since

$$e^{j\alpha} = \cos\alpha + j\sin\alpha$$
$$e^{\pm j\pi} = \cos\pm\pi + j\sin\pm\pi$$
$$e^{\pm j\pi} = -1$$

The amplitude of light at the photographic plate for this exposure with a 180 degree phase shift will be

$$F_2(y) = C(y) - Ae^{-j(kay/f)}$$

The intensity is given by

$$I_2(y) = F_2(y)F_2^*(y)$$
$$I_2(y) = A^2 + |C|^2 - AC^*e^{-j(kay/f)} - ACe^{j(kay/f)}$$

Exposure is defined as $E = tI$, where t is the time of exposure and I the exposing intensity. Multiple exposure is given by

$$E = \sum_n t_n I_n$$

In the case of double exposure that is being considered here the exposure is

$$E = t_1 I_1(y) + t_2 I_2(y)$$

Using the same exposure time in both cases $(t_1 = t_2 = t)$ and the same reference amplitude A the exposure equation becomes

$$E = t[I_1(y) + I_2(y)]$$

Substituting the expressions for $I_1(y)$ and $I_2(y)$,

$$E = t[2A^2 + |B+C|^2 + |C|^2 + AB^*e^{-j(ka y/f)} + ABe^{j(ka/f)y}]$$

$$E = 2tA^2\left[1 + \frac{|B+C|^2 + |C|^2}{2A^2} + \frac{B^*e^{-j(ka/f)y} + Be^{j(ka/f)y}}{2A}\right]$$

When processed in the linear region of the film characteristics the amplitude transmission function $H(y)$ of the film will be given by

$$H(y) = E^{-\gamma/2}$$

If we let

$$Y = \frac{|B+C|^2 + |C|^2}{2A^2} + \frac{B^*e^{-j(ka/f)y} + Be^{j(ka/f)y}}{2A}$$

then

$$H(y) = (2tA^2)^{-\gamma/2}(1+Y)^{-\gamma/2}$$

In practice the amplitude A of the reference light is made much greater than the amplitude of the signal light $|B+C|$. Examination of the equation defining Y shows that when $A \gg B+C$, then $A \gg B$ and $A \gg C$, for which Y will be small. Since Y is small, $H(y)$ can be approximated by the first two terms of a binomial expansion

$$H(y) = (2tA^2)^{-\gamma/2}\left(1 - \frac{\gamma}{2}Y\right)$$

Replacing Y gives

$$H(y) = (2tA^2)^{-\gamma/2}\left[1 - \underbrace{\frac{\gamma(|B+C|^2 + |C|^2)}{4A^2}}_{\text{zero-order term}} - \underbrace{\frac{\gamma B^*e^{-j(ka/f)y}}{4A}}_{\text{first-order down}} - \underbrace{\frac{\gamma Be^{+j(ka/f)y}}{4A}}_{\text{first-order up}}\right]$$

The first order up term is proportional to the transform of the signal $b(x)$. Viewing this first-order term through a lens (Fourier transform) produces the signal $b(x)$. Thus by inserting a 180 degree phase shift in the second exposure, the C term of the original sum $B+C$ is subtracted leaving B, which permits $b(x)$ to be reconstructed.

THREE-TERM BINOMIAL EXPANSION

We have seen that the expression for the exposure is given by

$$E = t[2A^2 + |B+C|^2 + |C|^2 + AB^* e^{-j(ka/f)y} + ABe^{j(ka/f)y}]$$

or

$$E = 2tA^2\left[1 + \frac{|B+C|^2 + |C|^2}{2A^2} + \frac{B^* e^{-j(2\pi a/f\lambda)y} + Be^{j(2\pi a/f\lambda)y}}{2A}\right]$$

When processed in the linear region of the characteristic curve, the amplitude transmission function $[H(y)]$ of the film will be given by

$$H(y) = E^{-\gamma/2}$$

Let

$$Y = \frac{|B+C|^2 + |C|^2}{2A^2} + \frac{B^* e^{-j(2\pi a/f\lambda)y} + Be^{j(2\pi a/f\lambda)y}}{2A} \tag{11-1}$$

then

$$H(y) = [2tA^2]^{-\gamma/2}[1+Y]^{-\gamma/2}$$

We shall again assume that the amplitude of the reference A is much greater than the amplitude of the signal light. Applying this condition to the expression for Y we have already shown that when Y is small and binomial expansion can be applied to the expression for $H(y)$, then

$$H(y) = [2tA^2]^{-\gamma/2}\left(1 - \frac{\gamma}{2}Y + \frac{(\gamma/2)((\gamma/2)+1)}{2}Y^2 - \cdots\right) \tag{11-2}$$

Let us consider (11-2) and substitute (11-1) for Y,

$$H(y) = (2tA^2)^{-\gamma/2}\left[1 - \frac{\gamma[|B+C|^2 + |C|^2]}{4A^2} - \frac{\gamma B^* e^{-j(2\pi a/\lambda f)y}}{4A}\right.$$

$$-\frac{\gamma Be^{j(2\pi a/\lambda f)y}}{4A} + \frac{\gamma}{2}\left(\frac{\gamma}{2}+1\right)\frac{[|B+C|^2 + |C|^2]^2}{8A^4}$$

$$+\frac{(\gamma/2)((\gamma/2)+1)}{2}\frac{[|B+C|^2 + |C|^2]B^* e^{-j(2\pi a/\lambda f)y}}{4A^3}$$

$$+\frac{(\gamma/2)((\gamma/2)+1)}{2}\frac{[|B+C|^2 + |C|^2]Be^{j(2\pi a/\lambda f)y}}{4A^3}$$

$$+\frac{(\gamma/2)((\gamma/2)+1)|B|^2}{4A^2}$$

$$+\underbrace{\frac{(\gamma/2)((\gamma/2)+1)}{8A^2}}_{H_2^*}B^{*2}e^{-j(4\pi a/\lambda f)y}$$

$$+\underbrace{\frac{(\gamma/2)((\gamma/2)+1)}{8A^2}}_{H_2}B^2 e^{j(4\pi a/\lambda f)y}\right]$$

The terms in this expansion can be grouped by exponential terms as

$$H(y) = H_0 + H_1^*(y)e^{-j(2\pi a/\lambda f)y} + H_1(y)e^{j(2\pi a/\lambda f)y}$$

$$+ H_2^*(y)e^{-j4\pi ay} + H_2(y)e^{j4\pi ay} \tag{11-3}$$

where

$$H_0(y) = [2tA^2]^{-\gamma/2}\Big\{1 - \frac{\gamma[|B+C|^2 + |C|^2]}{4A^2}$$

$$+ \frac{(\gamma/2)((\gamma/2)+1)[|B+C|^2 + |C|^2}{8A^4} + \frac{(\gamma/2)((\gamma/2)+1)|B|^2}{4A^2}\Big\}$$

$$H_1(y) = [2tA^2]^{-\gamma/2}\Big\{-\frac{\gamma B}{4A} + \frac{\gamma}{2}\Big(\frac{\gamma}{2}+1\Big)\Big[\frac{B(|B+C|^2 + |C|^2)}{8A^3}\Big]\Big\}$$

$$H_1^*(y) = [2tA^2]^{-\gamma/2}\Big\{-\frac{\gamma B^*}{4A} + \frac{\gamma}{2}\Big(\frac{\gamma}{2}+1\Big)\Big[\frac{B^*[|B+C|^2 + |C|^2]}{8A^3}\Big]\Big\}$$

Consider the first-order term $H_1(y)$

$$H_1(y) = [2tA^2]^{-\gamma/2}\Big\{-\frac{\gamma B}{4A} + \frac{\gamma}{2}\Big(\frac{\gamma}{2}+1\Big)\frac{B[|B+C|^2 + |C|^2]}{8A^3}\Big\} \tag{11-4}$$

The first term in parenthesis is seen to be proportional to the transform of the signal $b(x)$. Viewing this first-order term through a lens (Fourier transform) produces the signal $b(x)$. Thus by the insertion of a 180 degree phase shift in the second exposure, the C term of the original sum $B+C$ is subtracted leaving the remainder B, which permits $b(x)$ to be reconstructed. This is the result described earlier when terms such as the second term in the parenthesis of (11-4) are neglected.

The second term of $H_1(y)$ can be considered as noise. Defining the signal as

$$S = \frac{\gamma B}{4A}$$

and the noise as

$$N = \frac{\gamma}{2}\Big(\frac{\gamma}{2}+1\Big)B\Big[\frac{|B+C|^2 + |C|^2}{8A^3}\Big]$$

The signal-to-noise ratio will be given by

$$\frac{S}{N} = \frac{4A^2}{((\gamma/2)+1)[|B+C|^2 + |C|^2]} \gg 1 \tag{11-5}$$

since A is very large compared to signal amplitudes. This high signal-to-noise ratio is confirmed by the results of experiment.

HOLOGRAPHY WITH MOVING OBJECTS

Mechanical stability is extremely important in making a two-beam holo-gram. The interference pattern between the signal and reference beams can only form a hologram when the difference in the optical path length of the beams from the laser are within the coherent length of the laser and the interference pattern must be stationary in order for a hologram to form. In setting up the configuration to make a hologram the optical length from the laser to the hologram for the signal and reference beam must be equal (or within the coherence length of the laser). Any motion in the signal or reference beam greater than about one-quarter of the wavelength of the light used will prevent a sharp interference pattern (hologram) from being recorded. Figure 11.19(a) shows the usual two-beam method of making a hologram. In this case the illuminated portion of the object reflects light to form the signal beam. Any movement of the object will effect only the signal beam and therefore prevent the interference pattern at the film from being stationary and no hologram will be formed.

Figure 11.19(b) shows an alternate configuration for making a hologram which forms a silhouetted image of the object. The hologram is made from the diffused light from the ground glass interfering with the reference beam. The diffused light is therefore the signal beam. The object blocks a portion of this signal beam thus preventing the film from receiving light from certain portions of the ground glass. The silhouetted image has the usual three-dimensional characteristics of a hologram. Because the object is back lighted, of course no detail will be seen in the object and only its edges will be sharp.

For certain applications the silhouetted image hologram can offer advantages over the normal hologram. Let us consider the effect of the

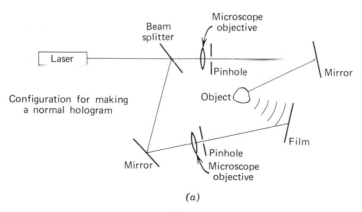

(a)

Figure 11.19*a*.

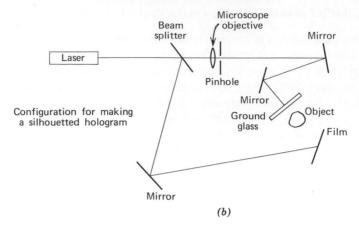

Figure 11.19b.

object moving (in any direction) while the silhouetted hologram is being exposed. The moving object will act to control the amount of light from specific points on the ground glass from reaching the film. Movement of the object will change the density of some of the recorded interference fringes (dependent of the object size, direction of motion, and speed of movement), and produce some new fringes (from portions of the ground glass that were previously blocked). The effect of changing the interference pattern because of the motion of the object is to make the hologram image appear blurred. The reconstructed hologram image will appear to have blurred edges because the interference patterns corresponding to the locations of the moving object will have been recorded. By determining the maximum density of the reconstructed blurred edge the probable path of the edge can be determined.

When light from the ground glass strikes the object so as to be reflected, a normal hologram will be produced; that is, those portions of the object will appear illuminated on reconstruction. If the object moves during exposure, no interference fringes will be produced and no reconstructed image can be obtained. If the object moves during the exposure of a silhouetted hologram, those portions of the object that are illuminated will appear black on the reconstruction because no interference patterns from those points will be recorded. For the same reason any portion of the object that is transparent or translucent will also appear black on reconstruction if the object moves during the exposure of the silhouetted hologram, that is, only the hologram of the unobstructed portions of the ground glass will be recorded and appear bright. In the silhouetted holograms the mechanical stability of the configuration is still required except for the object; that is, a normal hologram is taken of the ground glass which does require the usual mechanical stability. We can see therefore that we have traded detail in the reconstructed image for relaxed mechanical stability of the object during exposure.

ONE-STEP HOLOGRAPHIC PROCESS

We can determine the effect of development on the holographic images produced by first considering the exposure of the photographic film as shown in Figure 11.20. A signal beam exposes the photographic film with an amplitude $A(x, y)$ where x, y are the coordinates in the recording plane — that is, the film plane. The phase of the signal beam at each point in the recording plane is given by $e^{j\varphi_s(x, y)}$. In the recording plane the reference beam amplitude is given by $R(x, y)$ and its phase is given by $e^{j\varphi_R(x, y)}$. Since the hologram is being recorded with coherent light we can add the amplitudes of the signal and reference beams to determine the incident amplitude on the photographic film. The incident light amplitude on the photographic film is given by

$$A(x, y) \, e^{j\varphi_s(x, y)} + R(x, y) \, e^{j\varphi_R(x, y)}$$

The exposure of photographic film is proportional to the intensity of the light rather than its amplitude. To change the incident light amplitude to the equivalent incident light intensity we multiply the amplitude by its complex conjugate, that is,

$$I = K(Ae^{j\varphi_s} + Re^{j\varphi_R})(Ae^{-j\varphi_s} + Re^{-j\varphi_R}) \tag{11-6}$$

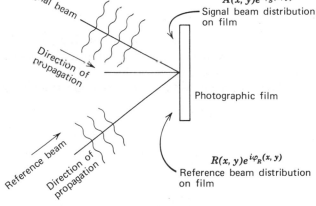

Figure 11.20 Effect of development on hologram image.

where K is the constant of proportionality between amplitude and intensity, that is,

$$I \propto A^2 \qquad I = KA^2$$

We also dropped the notation (x, y) for simplicity in expressing the equations. It is understood that the amplitude and phase of any specific point in the (x, y) plane of the recording medium is being determined. Multiplying out (11-6) we get

$$\begin{aligned} I &= K\left[A^2 + R^2 + ARe^{j\varphi_s - j\varphi_R} + ARe^{-j\varphi_s + j\varphi_R}\right] \\ &= K\left[A^2 + R^2 + ARe^{j(\varphi_s - \varphi_R)} + ARe^{-j(\varphi_s - \varphi_R)}\right] \\ &= K\left[A^2 + R^2 + AR\left\{e^{j(\varphi_s - \varphi_R)} + e^{-j(\varphi_s - \varphi_R)}\right\}\right] \\ &= K\left[A^2 + R^2 + 2AR\left\{\frac{e^{j(\varphi_s - \varphi_R)} + e^{-j(\varphi_s - \varphi_R)}}{2}\right\}\right] \end{aligned}$$

Since

$$\frac{e^{j\theta} + e^{-j\theta}}{2} = \cos\theta$$

we can write

$$I = K\left[A^2 + R^2 + 2AR\cos(\varphi_s - \varphi_R)\right] \qquad (11\text{-}7)$$

Photographic film exposure E is given by

$$E = It$$

where I is the intensity of the light, and t is the time of exposure. The density of the film after development is given by

$$D = \gamma\left(\log_{10}\frac{E}{E_f}\right)$$

where E_f is the point of intersection of the linear slope of the film characteristics curve with the exposure axis. The intensity transmission T of a film with a density D can be expressed by $T = 10^{-D}$. This is just transmission not transmission intensity. The amplitude transmission G can therefore be expressed as

$$G = T^{1/2} = 10^{-D/2} = 10^{-(\gamma/2)[\log_{10}(E/E_f)]}$$

$$G = 10^{\log_{10}\left(\frac{E}{E_f}\right)^{-\gamma/2}} = 10^{\log_{10}\left(\frac{E_f}{E}\right)^{\gamma/2}}$$

$$\log_{10}\left(\frac{E_f}{E}\right)^{\gamma/2} = \log_{10} G$$

$$G = \left(\frac{E_f}{E}\right)^{\gamma/2} = \left(\frac{E_f}{It}\right)^{\gamma/2}$$

$$G = \left[\frac{E_f}{K\left[A^2 + R^2 + 2AR\cos\left(\varphi_s - \varphi_R\right)\right]t}\right]^{\gamma/2}$$

$$= \left(\frac{E_f}{Kt}\right)^{\gamma/2}\left[A^2 + R^2 + 2AR\cos\left(\varphi_s - \varphi_R\right)\right]^{-\gamma/2}$$

$$= \left(\frac{E_f}{Kt}\right)^{\gamma/2}(R^2)^{-\gamma/2}\left[1 + \frac{A^2 + 2AR\cos\left(\varphi_s - \varphi_R\right)}{R^2}\right]^{-\gamma/2}$$

$$= \left[\frac{E_f}{KtR^2}\right]^{\gamma/2}\left[1 + \frac{A^2 + 2AR\cos\left(\varphi_s - \varphi_R\right)}{R^2}\right]^{-\gamma/2} \tag{11-8}$$

If $A < R$, we can use the binomial expansion giving

$$G = \left[\frac{E_f}{KtR^2}\right]^{\gamma/2}\left[1 - \frac{\gamma}{2}\left[\frac{A^2}{R^2} + \frac{2A}{R}\cos\left(\varphi_s - \varphi_R\right)\right]\right.$$

$$\left. + \frac{\gamma(\gamma+2)}{8}\left[\frac{A^2}{R^2} + \frac{2A}{R}\cos\left(\varphi_s - \varphi_R\right)\right]^2 - \cdots\right]$$

$$= \left[\frac{E_f}{KtR^2}\right]^{\gamma/2}\left[1 - \frac{\gamma A^2}{2R^2} + \frac{\gamma(\gamma+2)}{8}\frac{A^4}{R^4} + \cdots \left(\frac{A}{R}\right)^n\right.$$

$$- \frac{\gamma}{2}\frac{2A}{R}\cos\left(\varphi_s - \varphi_R\right) + \frac{\gamma(\gamma+2)}{8}\frac{4A^3}{R^3}\cos\left(\varphi_s - \varphi_R\right)$$

$$+ \cdots \left(\frac{A}{R}\right)^n \times \cos\left(\varphi_s - \varphi_R\right)$$

$$\left. + \frac{\gamma(\gamma+2)}{8}\left[\frac{4A^2}{R^2}\cos^2\left(\varphi_s - \varphi_R\right) + \cdots \cos^n\left(\varphi_s - \varphi_R\right)\right]\right] \tag{11-9}$$

If $A \ll R$ we can neglect the $(A/R)^n$ terms for $n > 1$; for example, when $A = R/10$, $A/R = 1/10$, $A^2/R^2 = 1/100$, $A^3/R^3 = 1/1000$, then

$$G \approx \left[\frac{E_f}{KtR^2}\right]^{\gamma/2}\left[1 - \frac{\gamma A}{R}\cos\left(\varphi_s - \varphi_R\right)\right] + \left[\text{neglected terms}\right]$$

If we illuminate this hologram with a reference beam $Re^{j\varphi_R}$, the amplitude transmission is the product of the amplitude transmission G of the hologram with the illuminating beam; that is,

$$GRe^{j\varphi_R} = \left[\frac{E_f}{KtR^2}\right]^{\gamma/2}\left[1-\frac{\gamma A}{R}\cos(\varphi_s-\varphi_R)\right]Re^{j\varphi_R}$$

using the relationship

$$\cos(\varphi_s-\varphi_R) = \frac{e^{j(\varphi_s-\varphi_R)}+e^{-j(\varphi_s-\varphi_R)}}{2}$$

and letting

$$K_1 = \left[\frac{E_f}{KtR^2}\right]^{\gamma/2}$$

Then

$$GRe^{j\varphi_R} \approx \underbrace{K_1Re^{j\varphi_R}}_{\text{DC}}-\underbrace{\frac{K_1\gamma A}{2}e^{j\varphi_s}}_{\text{image}}-\underbrace{\frac{K_1\gamma A}{2}e^{-j(\varphi_s-2\varphi_R)}}_{\substack{\text{Conjugate}\\\text{image}}}$$

$$(11\text{-}10)$$

TWO-STEP HOLOGRAPHIC PROCESS

We have seen in the one-step holographic process that the developed hologram (considered as a surface hologram) has an amplitude transmission given by

$$G_A = \left[\frac{E_f}{Kt}\right]^{\gamma_1/2}[A^2+R^2+2AR\cos(\varphi_s-\varphi_R)]^{-\gamma_1/2}$$

Putting the developed hologram G_A back at the same location that it was when exposed and removing the signal beam, we can expose a second photographic plate by illuminating it with a reference beam as shown in Figure 11.21. The second photographic plate will be exposed by the product of the incident light beam $(Ae^{j\varphi_R})$ and the amplitude transmission function G_A, that is $Ae^{j\varphi_R}G_A$. If we assume that the reference is a collimated (plane wavefronts) beam, then we can represent the illuminating beam by R'.

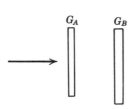

The developed second photographic plate will have an amplitude transmission given by

Figure 11.21 Reexposure

$$G_B = \left[\frac{E_f}{Kt_2}\right]^{\gamma_2/2} [I_2]^{-\gamma_2/2} = \left[\frac{E_f}{Kt_2}\right]^{\gamma_2/2} [[R'G_A]^2]^{-\gamma_2/2}$$

$$= \left[\frac{E_f}{Kt_2}\right]^{\gamma_2/2} [R'G_A]^{-\gamma_2}$$

$$= \left[\frac{E_f}{Kt_2}\right]^{\gamma_2/2} \left[R'\left\{\left(\frac{E_f}{Kt}\right)^{\gamma_1/2} (A^2+R^2+2AR \cos (\varphi_s-\varphi_R))^{-\gamma_1/2}\right\}\right]^{-\gamma_2}$$

$$= \left[\frac{E_f}{Kt_2}\right]^{\gamma_2/2} \left[\frac{E_f}{Kt}\right]^{-\gamma_1\gamma_2/2} [R']^{-\gamma_2} \left[\{A^2+R^2+2AR \cos (\varphi_s-\varphi_R))^{-\gamma_1/2}\right]^{-\gamma_2}$$

If we let

$$C = \left[\frac{E_f}{Kt_2}\right]^{\gamma_2/2} \left[\frac{E_f}{Kt}\right]^{-\gamma_1\gamma_2/2} [R']^{-\gamma_2}$$

then

$$G_B = C \{A^2+R^2+2AR \cos (\varphi_s-\varphi_R)\}^{\gamma_1\gamma_2/2}$$

If $\gamma_1\gamma_2 = 2$ then the transmission of the second hologram becomes

$$G_B = C[A^2+R^2+2AR \cos (\varphi_s-\varphi_R)]$$

$$= CR^2\left[\frac{A^2}{R^2}+1+2\frac{A}{R} \cos (\varphi_s-\varphi_R)\right] \qquad (11\text{-}10)$$

If $A \ll R$ then

$$G_B = CR^2\left[1+\underbrace{\frac{A^2}{R^2}}_{smaller}+\underbrace{2\frac{A}{R} \cos (\varphi_s-\varphi_R)}_{small}\right] \qquad (11\text{-}12)$$

Since A^2/R^2 is very small, it can be neglected, giving

$$G_B = CR^2\left[1+2\frac{A}{R} \cos (\varphi_s-\varphi_R)\right]$$

Illuminating this hologram with a reference beam $Re^{j\varphi_R}$ the amplitude transmission is the product of the amplitude transmission G_B of the hologram with the illuminating beam; that is

$$G_B Re^{j\varphi_R} = CR^2\left[1+2\frac{A}{R} \cos (\varphi_s-\varphi_R)\right]Re^{j\varphi_R}$$

Using the relationship

$$\cos(\varphi_s - \varphi_R) = \frac{e^{j(\varphi_s - \varphi_R)} + e^{-j(\varphi_s - \varphi_R)}}{2}$$

we obtain

$$G_B Re^{j\varphi_R} = \underbrace{CR^3 e^{j\varphi_R}}_{\text{D.C.}} + \underbrace{CR^2 Ae^{j\varphi_s}}_{\text{Image}} + \underbrace{CR^2 Ae^{-j(\varphi_s - 2\varphi_R)}}_{\substack{\text{Conjugate} \\ \text{image}}} \qquad (11\text{-}13)$$

Appendix 16 indicates the functions that must be recorded on film to eliminate various types of blurs from images.

STABILIZATION OF HOLOGRAPHIC FRINGES

Several techniques have been devised to stabilize the interference fringes produced in the hologram plane. These techniques which prevent the movement of the interference fringes significantly reduce the mechanical stability requirements required for making good holograms. One basic technique is to image one fringe from the holographic plane on a photomultiplier tube. The output is then used as an error signal to change the path length of either the signal or reference beam to maintain the fringe at a particular position.

The piezoelectric crystals used in some phonograph pickup arms have properties that can be useful for stabilizing the fringe location. When two faces of the piezoelectric crystal are compressed (as by vibrations caused by the phonograph needle), a voltage is produced across two opposite faces of the crystal. This voltage is amplified to produce the phonograph output. By applying a voltage to two opposite faces of the crystal it can be made to contract or expand across two opposite faces depending on the polarity of the applied voltage.

A mirror mounted on the piezoelectric crystal is made to reflect either the signal or the reference beam. The mirror can be moved so as to change the optical pathlength and keep the fringe at a constant location by applying the proper voltage (error signal from the photomultiplier). The mirror movement does not have to be more than one wavelength to maintain the fringe location constant (i.e., there will always be some position within one wavelength movement that will maintain the fringe location at a constant position). If the mirror is required to be moved more than one wavelength (in either direction), additional circuits can be used to step the mirror to its proper location (within its one wavelength of travel) should more than one wavelength of mirror travel be required (to maintain the fringe's constant position).

Figure 11.22 shows a signal and reference beam incident to a photographic emulsion. The signal beam is shown perpendicular to the photographic film plane and the reference beam is shown at an angle θ to the signal beam. After development the photographic emulsion (hologram) is illuminated by the reference beam that is incident to the film plane at the same angle as when the emulsion was exposed. A virtual image can be seen at the same location as the object during construction; that is, the virtual image will appear at an angle θ to the reference beam. A real image will also appear at an angle θ on the other side of the reference beam and on the other side of the hologram from the virtual image. If the geometry used during construction of the hologram was such that the angle θ is so large that the real image would appear on the same side of the hologram as the virtual image, then it would not be seen; that is,

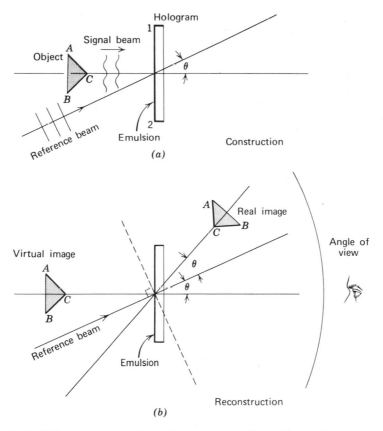

Figure 11.22 Hologram construction and reconstruction (the real image forms on opposite side of hologram from the virtual image). (*a*) Construction, (*b*) Reconstruction.

the real and virtual images can be seen only when they lie on different sides of the hologram. From Figure 11.22 we can see that the real and virtual images are located as mirror images to each other from a plane perpendicular to the reference beam at the hologram plane.

Figure 11.23 shows a hologram construction in which the geometry has been selected such that the angle θ between the signal and reference beams is so large that the real and virtual images would not appear on opposite sides of the hologram during reconstruction. Assuming the

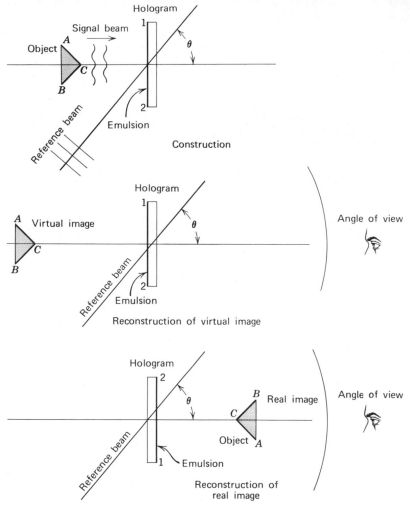

Figure 11.23 Location and relation of the real and virtual images when only one can appear at a time.

reference beam incident to the emulsion at the same angle as when constructed, the real or virtual image is reconstructed depending on how the reference beam is incident to the hologram emulsion.

An observer viewing the virtual image would see it as it existed when the hologram was constructed. When reviewing the real image, the object appears to have reversed relief; that is, the low portions of the original object now appear to be the high portions of the real image and vice versa. In addition the real image will appear to have a left to right reversal.

ACOUSTICAL HOLOGRAPHY

The same type of theoretical analysis that accounts for optical holograms can be applied to any system where energy is carried by similar types of coherent wave trains. One such coherent type of wave train can be produced by a high-frequency sound source. The coherent high-frequency sound source is then used in a manner similar to an optical coherent source to *illuminate* the object. The acoustical hologram is formed by the sound waves diffracted from the object, forming the signal beam, and a direct beam forming the reference beam. A microphone pickup can be moved in a plane at right angles to the object and the sound source to scan the sound energy. The microphone is connected to a CRT oscilloscope which converts the sound energy at any point in the scanned plane to a light output at a corresponding point on the CRT. The holograms thus produced on the CRT are "pictures" of the acoustical interference pattern. The interference pattern produced on the CRT can then be photographed to produce an optical hologram. This hologram permits the acoustically illuminated object to be seen in three dimensions by coherently illuminating the acoustical interference pattern (hologram). The images formed by acoustical methods do not have as much detail in them as optical holograms because of the difference in wavelengths used to form the holograms.

An alternate method for producing acoustical holograms makes use of the audio interference pattern formed at the surface of a liquid; that is, the two sound beams are made to interfere at the surface of a liquid to produce minute ripples. The ripple pattern now acts as a hologram; for example, if an object is placed in one of the sound beams, an image of the internal structure of the object can be seen by reflecting quasi-coherent light off the rippled surface. A photograph of the ripple pattern produces a permanent hologram-type record that can also be used to view the internal structure of the object.

The sensitivity of the scanned-type acoustical hologram is much greater than that of the surface ripple technique hologram. The scanned-type acoustical hologram, however, lacks the inherent speed and real time processing capabilities that are offered by the surface ripple hologram technique. Acoustical holograms are being considered for use in certain types of chemical diagnostic tests,—for example, as a replacement for x-ray monography in which details of soft tissues may be seen that would pass x-rays. The technique appears capable of producing images with detail approaching the resolution expected on the basis of the wavelengths and apertures used.

360-DEGREE HOLOGRAMS

Two different techniques can be used to obtain a 360-degree holographic image of an object. One 360-degree hologram technique developed at Lake Forest University is to place the object in a cylinder as shown in Figure 11.24. The photographic film is taped to the inner surface of the cylinder. The laser beam is expanded using a 97-power microscope objective. This high-power objective greatly expands the laser beam so that both the object and the photographic emulsion are simultaneously illuminated. The diffracted light from the object (in Figure 11. 24) interferes with the direct light (reference beam) incident to the photographic emulsion. The film is placed back in the cylinder after developing and illuminated by the high-divergent beam, and an image of the object in three dimensions can be seen. Because the object was illuminated on all sides (except the bottom), which could also have been illuminated with a mirror), as one moves around the cylinder looking through the glass walls at the hologram all sides of the object can be seen in three dimensions;

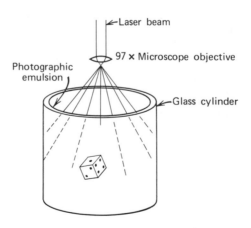

Figure 11.24.

that is, one can walk around the object and see not only the front of it (in three dimensions) but also its sides and rear.

Another technique developed by the Bell Telephone Laboratories (Murray Hill, New Jersey) is to make a composite of many narrow holograms of the object, each displaying a different three-dimensional view of the object. The first view is exposed with a slit at the end of the film plate. The object is rotated slightly and the slit moved a slit space for each subsequent exposure. The slit space and rotation of the entire object are such that film plate is exposed when the object is rotated 360 degrees. This hologram is viewed to see all sides of the object by moving ones head from side to side.

This is unlike conventional holograms (which present a three-dimensional image of only a limited angular view of an object); both methods above permit a 360-degree view of the object. The cylindrical hologram has disadvantages in that the film must surround and be bigger than the object and when viewed, the film must be oriented the same way so as to reconstruct a distortionless image. This cylindrical hologram is therefore somewhat awkward to use and handle. The rotated multiple-exposed hologram has disadvantages of multiple exposures being required and some disorientation because although all views appear, they are not in the same location.

COMPUTER GENERATION OF HOLOGRAMS

Holograms have been synthesized by computers. The interference pattern between two (or more) waves in a plane (corresponding to the plane of the hologram) can be determined by a computer. If the signal and reference beams can be described mathematically (as they usually can) the computer can mathematically determine the interference pattern (hologram) corresponding to a specific configuration. By photographing the computer printout (and usually reducing the scale photographically) an optical hologram can be produced. This synthesized hologram can produce an image of an object which only existed as a mathematical expression. Computer generated holograms can have applications in synthesizing filters for optical data processing systems.

MEASUREMENT OF SMALL DISPLACEMENTS

Small displacements of an object can be determined using holography. A hologram of an object is made prior to the object being displaced. The holographic image of the object can be made to coincide with the actual object. It is therefore not possible to distinguish between the actual object and its holographic image. Any slight displacement of the object with its holographic image will cause a Moiré pattern (see Appendix 6) to appear. The Moiré pattern can be used to measure the displacement. By making a double exposed hologram of the object (once in its original position and once in its displaced position) a permanent record of the Moiré pattern can be obtained.

FLY'S EYE LENS

The $\mathscr{F}$/number of a lens is given by the ratio of the focal length to the diameter; that is,

$$\mathscr{F}/\text{number} = \frac{F}{D}$$

where F is the focal length and D is the diameter. We can see therefore that if we keep the focal length constant, we can keep making the diameter of each of the lenses smaller and this will increase the $\mathscr{F}$/number of each of the respective lenses. By making the lenses smaller we can place more lenses in the space originally taken by the single camera lens. As we keep making each of the lenses smaller (so that we can fit a great many in the original space) we will find that the $\mathscr{F}$/number of each goes up, which will require a longer exposure for each lens image. We can reduce the focal length of each lens in an attempt to decrease the $\mathscr{F}$/number but there are mechanical limitations to this.

A fly's eye lens is usually made of molded plastic, which is flat on one side and has multiple convex lenses on the other side (see Figure 11.25). A typical fly's lens array is made of plastic material and can contain over 2000 lenses in a $2\frac{1}{4}$ by $2\frac{1}{4}$ in. square format. The lenses are made to focus on the same plane in air (in front or on the back surface of the plastic). A typical effective aperture of each lens is approximately $\mathscr{F}/3$ giving a resolution of over 300 lines/mm.

The plastic thickness and index of refraction provide some of the limitations on the focal length of each lens. The limitations on making the diameter of each lens smaller are the film speed and the diffraction limit;

Figure 11.25 Fly's eye lens.

that is, as the diameter is made smaller, the diffraction is increased in proportion to the energy in the image. When the aperture is so small it is diffraction limited, the aperture acts as a point source of light and not as an imaging device. For practical purposes the lens diameter is usually made approximately 0.043 in., which is a compromise between a small diameter lens and the diffraction limit. The lenses are also usually not made circular since not all the surface area would be used and there would be a reduced amount of total light energy striking the film. Figure 11.26 shows the fly's eye lenses when made circular showing the areas corresponding to a loss of light energy which does not occur when the lenses are made hexagonal.

The fly's eye lenses offer an alternate method to produce hologram-type images or can be used in conjunction with holography. The great advantage of the fly's eye three-dimensional hologram-type image is that the film can be exposed and the image viewed without the need for temporal coherent light. Let us consider three lenses of a fly's eye array and the image produced by a collimated incident beam. Each lens focuses the collimated beam and produces a point focus in the focal plane as shown in

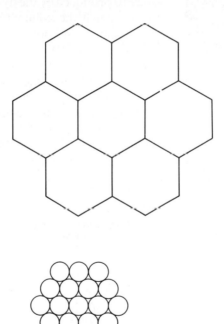

Figure 11.26.

Figure 11.27. We can see therefore that if an object is sufficiently far from the fly's eye lens (i.e., a distance much greater than the focal length of the lens), the image will be located approximately in the focal plane. (Usually the plastic is made thick so that the rear flat surface and the focal plane coincide. This permits the photographic film to be rested on the flat side of the plastic allowing simpler orientation and registration with the lenses.)

After exposures the photographic emulsion is developed as a positive (i.e., where light struck the emulsion, the final emulsion is clear and where light did not strike the film, the emulsion is dark) and placed back in registration with the fly's eye lens as it was when exposed. We can also see from Figure 11.27 that if we illuminate the three points in the focal plane, the lenses would produce the three collimated beams as shown. We can reason therefore that the image of any object (located sufficiently far from the fly's eye lens) will be produced by collimated beams produced by the fly's eye lens. We can re-establish the collimated light beams by illuminating the positive image formed by them on the film. The illuminated positive image will reconstruct the original collimated beams from the object in such directions that they will converge to the locations on the object and thus reconstruct the three-dimensional image of the object (without the need for coherent light). We can see from Figure 11.28 that a real image is formed which is pseudoscopic; that is, the highs appear as low and the lows as high just as in the real image of a hologram.

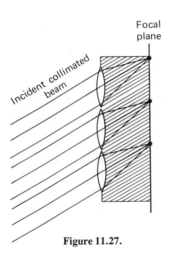

Focal plane

Incident collimated beam

Figure 11.27.

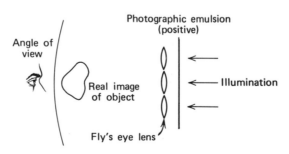

Angle of view

Photographic emulsion (positive)

Real image of object

Illumination

Fly's eye lens

Figure 11.28.

The real (pseudoscopic) image can be used as an object for a second fly's eye picture. The second fly's eye picture (of the first) will be an orthoscopic image that can be either real or virtual depending on the location of the second fly's eye lens and film during exposure. Figure 11.29 shows a pseudoscopic image produced by a fly's eye lens and two possible recording positions for a second fly's eye lens. Position *A* is in the converging beam and position *B* is in a divergent beam of the first fly's

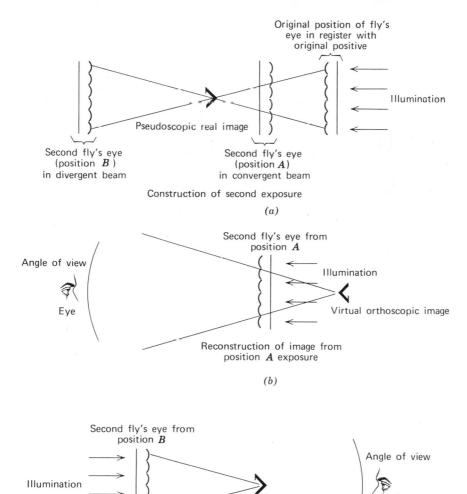

Figure 11.29   Reconstruction of image from position *B* exposure.

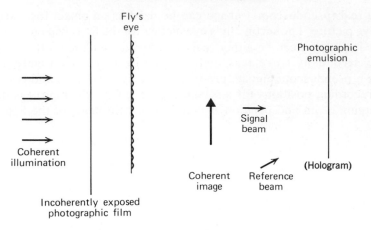

Figure 11.30.

eye lens. When the second recording is made with the fly's eye lens in position A (i.e. in a converging beam), the reconstructed image is virtual and orthoscopic as shown in Figure 11.29(b). The reconstructed image will be real and orthoscopic as shown in Figure 11.29(c) when the recording is made with the fly's eye in a divergent beam.

The fly's eye lens therefore permits hologram-type images to be taken with white light. If the incoherently recorded image made with the fly's eye lens is illuminated with coherent light, a coherently illuminated three-dimensional image is available for making a hologram as shown in Figure 11.30.

12

Properties and Techniques for Photographic Reproductions

ELEMENTS OF PHOTOGRAPHIC FILM PROCESSING

It was stated earlier that at present photographic film is probably the most versatile and simplest medium to use to produce transmission functions for coherent optical systems. The term *photographic film* refers to an emulsion that is a suspension of silver halide crystals in a protective colloid (usually gelatin). Photographic processes involve the interaction of light with certain of the silver halides. Generally for photographic purposes silver halides are considered to be either silver bromide, silver chloride, or silver iodide. When radiation falls on a photographic silver halide film, there is a tendency for a decomposing reaction to occur. This decomposing reaction is quite complex but can be naively represented for silver bromide by the expression,

$$\underbrace{A_g^+ + B_r^-}_{\substack{\text{silver} \\ \text{bromide}}} + \underbrace{\text{energy}}_{\text{light}} \;\rightarrow\; \underbrace{A_g + B_r}_{\text{silver} \;\; \text{bromide}}$$

This expression states that silver bromide crystals are made up of positive silver ions and negative bromide ions, and those silver halide crystals will decompose to metallic silver and bromide when light energy is added. This process can be demonstrated with almost any photographic film by exposing a portion of the film to bright sunlight for about five minutes. The exposed portion of the film will darken while the unexposed portion of the film will not darken. By placing a negative over the film while exposing to the sunlight it is possible to have good positive images produced on the film (these used to be called *sun pictures*. A package of negatives and out-dated photographic paper was usually sold as a kit for making sun pictures) which can be viewed for relatively long periods of time in somewhat subdued light without requiring any further processing. With time the image deteriorates.

Developing

For practical purposes it is desirable to obtain a permanent image with very little energy required to expose the film. The development process makes it possible to reduce the exposure time by over six orders of magnitude (as compared to that required to produce a sun picture). When light falls on the silver halide in a photographic emulsion, it is reduced to extremely small particles of silver that form a latent image which is initially invisible. Continued exposure to the light will make a latent image visible (sun pictures). The latent image can also be made visible by a photographic developer. A photographic developer is a reducing agent that acts on the extremely small particles of silver making up the latent image. Each of the extremely small particles of silver making up the latent image acts as a catalytic center for further reduction of the silver halide crystals into silver. A good developer is one that selectively reduces *exposed* silver bromide crystals more rapidly than it reduces the unexposed crystals. The advantages or disadvantages of any particular developer lies in its ability to *selectively* reduce the exposed silver bromide and not the unexposed silver bromide. For any particular developer this selectivity will depend primarily on its temperature and alkalinity when used. The term *alkali* is meant to apply to those soluble metallic hydroxides that react with acids to form salts. Metals that form such alkalis are called *alkali metals* while salt solutions with similar characteristics are called *alkaline*. The degree of alkalinity of a solution is determined by its pH. Values of pH above 7 are considered alkaline. The higher the pH number, the more alkaline a solution.

Some commonly used chemical compounds that have the required property of selective development of the exposed silver halide crystals are metol, phenidone, and hydroquinone. Combinations of these or other developing agents often improve the performance of the developer better than that of any developer used separately. Combining different developers together and varying the ratio of each usually will have a marked influence on the performance of the developer.

In reducing the exposed silver bromide in the film emulsion to metallic silver the developer becomes oxidized or depleted. It is important therefore that during the developing process a fresh supply of developer replace the worn-out developer. We can see that the amount of fresh developer necessary to accomplish this will be proportioned to the amount of silver halide reduced.

Accelerators

Most developing agents require an alkaline (high pH) solution before they can act to any great extent as a reducing agent. The presence and amount of an alkaline accelerator in the developing solution will determine the rate of development. Some commonly used accelerators are sodium hydroxide, borax, sodium bicarbonate, sodium carbonate, and sodium metaborate.

The type and concentrations of accelerators used with a specific developer are extremely important because generally the more alkali added, the higher the development rate and the higher the contrast obtained, the less selective the developer becomes to exposed versus unexposed silver halide (i.e., there is a tendency for the fog level to increase; fog in this case can be considered the development of unexposed silver halide crystals). Usually the alkalinity of a developer decreases as more film is processed, thus changing its characteristics. In order to maintain uniform development activity a large amount of alkali is usually added to the development solution so that as the developer tends to produce acid, this acid is neutralized by the alkali but sufficient alkali still remains in solution to maintain the pH at the desired level. This method sometimes increases the fog level because of too fast a development rate. The development time can also be too short to control easily with reasonable accuracy or the image produced may be too contrasty. The accelerator is chosen to provide the desired pH requirements of the developer; for example, when a low (or mild) alkalinity like pH = 9 is required, it is desirable to use an accelerator such as borax to achieve it, rather than a small amount of a strong alkali such as sodium hydroxide because of its instability. The accelerator, because of its high pH, tends to decompose the oxidized or used developer and thus maintains the developer fresh.

Buffering

A buffer is a solution composed of a combination of ingredients whose pH is unaffected by the addition of alkali or acids and acts as an accelerator to maintain uniform development activity. Some commonly used buffer combinations are; borax and boric acid, sodium hydroxide and boric acid, sodium carbonate and sodium bicarbonate, and sodium hydroxide and sodium phosphate.

Preservative

The developer being an oxidizer is itself oxidized not only when reducing the silver halide, but also when in contact with air. In order to slow down the oxidization of the developer when in contact with air a preservative (the most commonly used is sodium sulfite) is added. Sodium sulfite acts as a weak solvent for silver halide which in some types of developers tends to improve the grain and speed of the emulsion (these developers will act partially as a physical developer and will be discussed later). In addition sodium sulfite will react with the initial oxidation products formed in the developer inhibiting them from complete breakdown.

Restrainer

Restrainers are added to a developer to slow down the development process, usually to prevent the formation of fog. We have seen that the developer activity can be increased by increasing the alkalinity of the developer but this will also tend to reduce the ability of the developer to select only the exposed silver halide for reduction. A restrainer (most commonly used is potassium bromide) is usually added to prevent this fogging action from occurring.

Neutralizer (Short-Stop)

After development the film is placed in a solution to neutralize the alkalinity of the developer. Usually a solution of acetic acid is used rather than plain water to neutralize the alkali because the acid can rapidly neutralize any alkaline developer remaining in the gelatin emulsion. This rapid lowering of the pH to a point where development stops prevents aerial oxidation of the developing agent (which could form staining products). The acid bath also acts to dissolve or restrain the formation of calcium scum, preserve the acidity of the fixing bath, and decrease the swelling of the gelatin emulsion. (We have previously noted that swelling of the gelatin can be a major problem in optical processing and therefore minimizing this effect is desirable).

Fixing

The metallic silver image formed during development in the emulsion is intermixed with undeveloped silver halide crystals. The undeveloped silver halide is still light sensitive and can still be reduced to metallic silver. For this reason the image is not permanent. In order to make the image permanent it is necessary to remove these undeveloped silver halide crystals from the emulsion. In some cases a nonpermanent image is tolerated to avoid possible distortions to the gelatin during fixing. A developed, but not fixed image, can remain for several months without fading. Generally sodium thiosulfate (hypo) is used to remove the undeveloped silver halide crystals from the gelatin emulsion by making them water soluble. Ammonium thiosulfate also converts the undeveloped silver halide crystals to a water-soluble form faster than sodium thiosulfate but is more expensive and for this reason is usually not used. The usual time that the film is allowed to stay in the fixer solution is twice the time required to clear the film. The time required to clear the film is the time it takes to remove the milky color caused by the silver halides in the emulsion. Allowing the film to remain too long in the fixer will dissolve some silver from the image, which reduces the image quality.

Hardeners

Hardeners are usually added to the fixing solution to toughen the wet gelatin emulsion and make it resistant to abrasion. Hardening of gelatin is similar to the tanning of rawhide. Hardeners are used to reduce the swelling of the gelatin in water (and water solutions) as well as to increase the temperature required for solution in water.

Potassium alum, chrome alum, or ammonium alum is usually added to the fixer to harden photographic emulsions and to raise their melting point. Gelatin can also be hardened by heat, ultraviolet radiation, aluminum chloride, aluminum sulfate, or formaldehyde. Formaldehyde tends to result in *after hardening*, which is a gradual increase in hardening of the coated emulsion for months after processing. Although hardening is often taken for granted it is extremely important for coherent optical data processing because the dimensional stability of the film is vital (dimensional changes can result in a change of the transmission function).

Rinsing

Rinsing of the processed film is done to remove residual chemicals remaining in the emulsion that can cause staining; for example, residual hypo can cause silver sulfite to form causing a brownish staining to form. Without rinsing it is also possible for the processing chemicals to crystallize on the surface of the film as the film is dried.

Rinsing of an emulsion is generally considered to be an exponential function; that is, if 50 percent of the residual chemicals are removed in 5 min, then 50 percent of the remaining residual chemical will be removed in the next 5 min. A brief rinse will prevent the processing chemicals from crystallizing during drying and act to prevent staining over relatively long periods of time (the longer the rinse, the longer the film can be stored without staining).

Films are usually placed in a water-softening liquid to reduce the surface tension (such as calgon or photoflo) after the rinsing is completed which prevents spotting caused by the rinse water collecting on the film in drops when drying.

Drying

Drying is relative since gelatins usually contain some water even when they are considered dry. Drying is usually accomplished by normal evaporation. It is also done sometimes by circulating clean dry air at room temperature (or at elevated temperatures to speed drying). Sometimes for laboratory work drying can be speeded up by using an 80-percent alcohol rinse prior to drying.

FILM-PROCESSING TECHNIQUES

The silver halide photographic film processes described above can be combined in different ways. The desired end results dictate the procedure to be followed. Fundamental photographic film-processing procedures follow.

Negative Images

The usual steps involved in silver halide photographic film developing to produce a negative image are as follows:

1. Expose the light-sensitive material.
2. Develop the exposed light-sensitive material to produce a *negative* image.
3. Fix the negative by removing any remaining light-sensitive materials to assure its permanence.
4. Wash to remove any harmful residue remaining from the fixing process.
5. Dry the negative.

Positive Images

Method I. (Either contact printing or projection printing). The steps involved in the normal development process for making positive silver halide images are as follows:

1. Expose the light sensitive material.
2. Develop the exposed light-sensitive material to produce a *negative.*
3. Fix the negative by removing any remaining light-sensitive materials to assure its permanence.
4. Wash to remove any harmful residue remaining from the fixing process.
5. Dry.
6. Expose a second photographic sensitive material through the negative.
7. Develop the newly exposed material to produce a positive.

8. Continue through steps 3, 4, and 5 to complete the preparation of the positive.

Method 2. *Reversal Printing.* The reversal technique is another silver halide film-development method. In this process the exposed film is developed and then bleached. The bleaching removes the exposed metallic silver from the emulsion, leaving the unexposed silver halide crystals. This emulsion is then exposed to light and developed through the normal development process. The resulting image is now a positive instead of a negative. The following steps indicate the procedure for the reversal process.

Step 1: *First Development.* The first development is the most critical operation in the reversal process and therefore must be carefully controlled. It is in this step that the negative silver image is formed.

Step 2: *Stop Bath.* In this step either a stop bath or running water is used to stop the development action immediately and to remove any first developer from the film.

Step 3: *Bleach.* In this step the negative silver image is dissolved leaving only unexposed silver halide in the emulsion. (NOTE: The remaining steps can be accomplished in subdued light without affecting the results. Also timing is not critical as long as the times are long enough to carry out the processes.)

Step 4: *Rinse.* The film is rinsed in water to remove the bleach.

Step 5: *Clear.* The film is placed in a solution that removes stains which may later be caused by the bleach.

Step 6: *Rinse.* The film is again rinsed to remove the clearing solution.

Step 7: *Re-exposure.* The film is now exposed to light to expose the remaining silver halide.

Step 8: *Second Development.* In the second development the balance of the silver halide is converted to silver producing a positive silver image. (It has been found advantageous to develop slightly beyond the point when the image appears to be fully brought out.)

Step 9: *Rinse.* A stop bath or rinsing in water stops the second development.

Step 10: *Fix.* The film is now fixed and hardened. There should be little or no silver halide remaining for removal by the fixing bath; however, the hardening of the emulsion seems to be advantageous at this point.

Step 11: *Wash*. Washing in running water to remove the fixing solution from the emulsion reduces the possibility of stains forming.

Step 12: *Photoflow*. The film is rinsed in a photoflow solution to prevent spotting during drying. Photoflow is a solution that reduces the surface tension of water thus preventing drops from forming.

Step 13: *Drying*. The film is allowed to dry without any wiping.

In review, the first developer brings out the negative image in the usual manner by forming a silver image where the silver halide had been exposed to the light. This silver image can be removed by bleaching, leaving only unexposed and undeveloped silver halide. It was noted above that steps 3 through 13 are not critical; that is, the processes can be continued (within reason) longer than the normal times prescribed without appreciably affecting the results. When the bleached emulsion is exposed and developed, a positive image results. In other words wherever there had been a dense deposit of silver in the negative, there will be the least amount of silver in the positive. Likewise where there had been no silver in the negative image, there will be dense silver in the positive. The action of the bleach can be considered exactly opposite to that of the fixer in the normal developing process. The bleach removes the silver image without disturbing the undeveloped silver halide in the emulsion, while the fixer removes the undeveloped silver halides without affecting the silver image.

RELIEF IMAGES

A relief image is an image that appears to be carved or molded so as to appear to project from the surface.

Simulated Relief Images

A method of making an accentuated relief photographic print of a relief object (such as a coin) is to make a relatively low-contrast contact print (positive) from a relatively low-contrast negative of the coin. The negative and positive are then placed in contact with each other with a slight displacement (in the direction of the lighting on the subject) so as to accentuate the normal shadows and highlights without causing a double image. The resulting picture has distorted tonal gradation but appears to be an accentuated relief image.

Bas-Relief

Another method that produces an actual relief image is accomplished by making use of the swelling characteristics of gelatin. Gelatin can be selectively hardened which will permit a bas-relief image to be formed on the surface of the gelatin. A bas-relief image is a sculptural type of relief image in which the projection from the surrounding surface is slight and no part is undercut. Gelatin itself can be made light sensitive. Light sensitive gelatin can have some advantages over photographic (silver halide type) film for holographic applications (although silver halide emulsions form relief images during normal development). In holography the efficiency with which the images are reconstructed is important. Birchromated gelatin is light sentive and therefore the emulsion grain would (or should) be at the molecular level, which should be superior to silver halide-gelatin emulsions. There are then two basic reasons for pursuing the bichromatic-gelatin sensitized film for holographic application: namely, (a) high efficiency in image reconstruction, that is, the light is not attenuated by varying film densities but the phase of it is changed by varying thicknesses of the film; and (b) high resolution. Using a bichromated gelatin will therefore result in brighter reconstructed images which would require lower power lasers for reconstructing a given image brightness.

Let us assume that we desire to make a relief image of a negative. If we project ultraviolet rays (such as from a mercury arc) through the negative to expose sensitized bichromated gelatin, the shadow areas will become proportionately hardened by the exposure of the ultraviolet light. Just as ultraviolet light can cause sunburn it can also cause a chemical action to take place in the gelatin that causes it to harden. Hardened gelatin is somewhat waterproof and tends not to absorb water, while unhardened gelatin absorbs proportionately large quantities of water and swells. An image results in relief in the surface of the gelatin. The image produced has highlights that stand out and shadows that recede. This relief image can be cast in plaster while wet. In order for the final relief image in the gelatin to be representative of the object the density of the negative must be such that it varies according to the relief of the object being photographed and therefore the normal type of subject illumination is usually undesirable. A high-contrast negative is usually desired for this work.

Bichromated gelatin can be made by soaking 50 grams of gelatin for several hours (to a day) in 175 cc of water. The mixture is then heated with a double boiler until it is thoroughly dissolved. While stirring, 6.5 cc of glycerin is added to the gelatin. The solution is then strained and coated on glass plates while still warm. If it is found that the gelatin gets too hard on drying (the drying gelatin tends to curl and can get so hard it can actually crack the glass plate or rip the surface of the glass out), the amount of glycerin added should be changed. Any time after drying the plates can be sensitized by placing them in a 6-percent solution of ammonium bichromate for 15 min and then dried. The sensitizing and drying should be done in subdued light or preferably in total darkness. The plate is then exposed to the ultraviolet image. A 3-min exposure with a 100-watt mercury arc is a good starting point to determine the correct exposure. The plate is developed by placing in water until the gelatin swells to form the relief image. It can now be rinsed in relatively warm water to wash away the unhardened gelatin (so as to form a permanent relief image). This technique is suitable for making relief images for coins. With proper quality controls it may be possible to form a master for a plastic fly's eye lens. The multiple imaging required to expose the fly's eye lens on the bichromate gelatin can be accomplished with relative ease by using the light tunnel described in Chapter 1.

Films prepared as above exposed to a Helium-Neon laser (6328 Å) were found to be too slow to be practical. When exposed with an Argon laser, however, development could be obtained if the exposure time was increased a few orders (2 or 3) of magnitude over that required for 649F-type film. The development technique used to form holograms was to wash the emulsion in warm water about 100°F for a few minutes and then to air dry. It appears that dipping the emulsion in isopropanol (acetone, ether, or any similar substance to increase the drying speed) after washing improves the efficiency of the reconstructed image. Because gelatin will tend to absorb moisture from the atmosphere, this will tend to cause the relief image to distort with time. Coating the dried gelatin with lacquer will prevent some of this absorption from taking place without any significant change in the transmission properties of the gelatin. This indicates that bichromated gelatin (properly sensitized) can be useful in holography.

Emulsion Shrinkage Correction

One of the major problems in making reflection type holograms on photographic emulsions is caused by the shrinkage in thickness of the emulsion during the normal development processing. Shrinkage of the emulsion will cause the reconstructed reflected image's color to shift towards shorter wavelengths. Elimination of the fixing process will reduce the amount of shrinking. The unfixed hologram is unstable (in a relative short time, like one month, it can become too dark to be usable) and because of the unremoved silver halide in the emulsion the reconstructed image tends to be noisy. The gelatin emulsion can be swelled after fixing by soaking the plate in a 2–8% solution of triethanolamine, $(CH_2OHCH_2)_3N$, and then dried as usual. The triethanolamine can be washed out of the emulsion which will cause the emulsion to shrink after it is dried. Resoaking in triethanolamine will cause it to swell. It is therefore possible to swell the emulsion to a desired amount by controlling the concentration of the triethanolamine in the emulsion. Any error can be corrected by removal (or increasing concentration) of the triethanolamine. Different emulsions and optical densities require different concentrations of the triethanolamine to correct for the shrinkage of the emulsion during processing.

DEVELOPING SILVER HALIDE EMULSIONS

Each step outlined for silver halide film processing has a great many possible variations. The choice of the procedure to be followed in each step will depend on the final results desired. The principal step in silver halide film processing (and probably the most important step) is the development procedure. This step influences the final results more than any succeeding step. An understanding of what goes on during the development process as well as how the choice of developers can effect results is important.

The developing solution chemically reduces the silver halide crystals in the emulsion to metallic silver. When one looks at light-exposed silver halide crystals in an emulsion during development, they will appear to get blacker and blacker. The development process appears (within limits) to be a gradual process. The silver halide crystals that have been exposed to more light appear to get blacker faster during development. Portions of an emulsion that have not been exposed to light will not (or should not) turn black.

During the developing process the developer reacts with the light-exposed silver halide crystals making them blacker and blacker. The normal emulsion can be visualized as a gelatin medium with silver halide crystals suspended in it. This emulsion is usually placed on a transparent base for support. When light falls on the silver halide crystals in the emulsion, they absorb energy which make them developable. It appears that each crystal has a sensitive spot that absorbs the energy making the entire crystal developable. The developer is capable of penetrating the gelatin medium and during the development process gradually converts the silver halide crystals to metallic silver which makes them appear to turn blacker and blacker. When developer comes in contact with the silver halide crystal, not only the sensitive spot (which is very small compared to the entire crystal) but the entire crystal is developed (turned black).

Another important phenomenon during the developing process is that there is a definite amount of time required for the developer to penetrate the emulsion. It is logical to assume that the silver halide crystals near the surface of the emulsion will be developed faster and to a greater extent than those lower in the emulsion. It can be visualized that in a thick emulsion the developer will never penetrate all the way to the bottom of the emulsion. This will tend to make the crystals on the lower surface of the emulsion not be as fully developed as those on the upper surface. A portion of an emulsion that has been exposed to a greater amount of light will be blacker on development than portions that have not received as much light. Silver halide crystals are not uniformly distributed within the emulsion and therefore specific portions of the emulsion may not develop precisely to the corresponding proportion of light that exposed it. The silver halide crystals must be distributed uniformly, not only along the surface, but also in depth in order to obtain uniform exposure. The term *granularity* is used to indicate the distribution of the silver halide crystals in the emulsion.

DEVELOPERS

We have discussed how the developer converts the latent image formed during exposure into a visible image by changing the exposed silver halide crystals into black metallic silver. Because the action and type of developer used can change the results obtained with a given emulsion, it is reasonable to assume that for some purposes some experimentors will want to mix their own developer. A summary of the previous discussion will be given followed by specific details and formulas not previously given.

We have seen that a developer for black and white films usually consists of (a) the developing agent (which may be more than one type of developer), which converts the exposed silver halide to metallic silver, (b) a preservation, which reduces the developer deterioration with time, (c) an alkali, which determines the developer activity, and (d) a restrainer to reduce fog during development.

We have discussed how the properties of a developer are related to the relative proportions of the ingredients as well as to their chemical nature. The activity of the developer can usually be altered just as effectively by using more or less of a given alkali as by changing developers. We can see therefore that widely varying developing characteristics can be produced with relatively few chemicals.

Some fine-grain developers contain a solvent like hypo or sodium sulfite that dissolves some of the silver bromide in the emulsion so the solution can partially act like a physical developer (which will be discussed later). In addition water softeners and wetting agents may be added to the developer in order to obtain more uniform results and reduce the possibility of spotting the negative.

We shall discuss and illustrate basic types of negative developers as follows:

1. Normal-contrast developers.
2. High-contrast developers.
3. Low-contrast developers.
4. Special-purpose developers.
5. Fine-grain developers.

Normal-contrast developers

Normal-contrast developers are usually a combination of metol and hydroquinone developers. Metol-hydroquinone (MQ) developers are in very common use. The proportions of each in a developer will vary according to the type of film to be developed and the desired effect. Metol (also called Elon, Pictol, and Planetol) used alone is a high-speed (activity) developer which produces high film speed and a low-contrast image with some fog. Hydroquinone (also called *quinol*) used alone gives a high-contrast image with a high fog level. The more alkaline the solution of hydroquinone, the greater the development speed (activity) of the developer. Combination of these two developers at a specific pH with a restrainer to keep the fog level low results in a normal-contrast developer. Other normal-contast developers can be made by combining phenidone (which is a high-speed (activity) developer similar to metol but can be used in lower concentration) with hydroquinone. Another normal-contrast developer is a pyro developer. This developer is a high-activity developer that tends to produce a stained image. Glycin developer is a low-activity developer that tends to prevent or reduce aerial fog. Aerial fog is a chemical fogging of a negative caused by its being exposed to air when wet with developer. The use of glycin alone (or where slow development is required) is desirable because it produced almost no aerial fog.

Typical formulas for normal-contrast developers are represented by the following:

MQ-Type Developers. MQ developers combine the characteristics of metol (high film speed) and hydroquinone (high contrast and density) and are typified by Kodak D-11, D-19, DK-50, D-52, DK-60a, D-61a, D-72, and D-76 developers, Ansco 47, 48M, and 61.

The formula for Kodak Developer D-61a

Water about 50°C	500 cc
Kodak Elon developing agent	3.0 grams
Kodak sodium sulfite, desiccated	90.0 grams

Kodak sodium bisulfite	2.0 grams
Kodak hydroquinone	6 0 grams
Kodak sodium carbonate monohydrated	14.0 grams
Kodak potassium bromide	2.0 grams
Cold water to make	1 liter

For tank use dilute one part stock solution with three parts water.

Phenidone is similar in its action to metol but can be used in much lower concentrations. An alternate to the MQ developers therefore is the phenidone-hydroquinone developers. The main advantage of this type of developer is not in its photographic qualities but lies in the fact that it is not as apt to cause skin conditions in allergic people as can the MQ developers. Other developers such as para-aminophenol are in some ways similar to metol. Para-aminophenol can also be used as a metol substitute by those allergic to metol.

Kodak SD-1 is a pyro stain developer whose formula is as follows:

Water at 52°C	750 cc
Kodak sodium sulfite desiccated	1.4 grams
Kodak pyro	2.8 grams
Kodak sodium carbonate, monohydrated	6.2 grams
Cold water to make	1 liter

Dissolve chemicals in the order given and develop for about 6 min at 20°C, rinse, and fix in a large volume of plain hypo. The stain can be reduced by first removing entirely by bleaching in formula Kodak S-6 and redeveloping in a mildly staining pyro developer such as formula Kodak D-7. The developed stain image is formed along with the silver image during development. Photographic papers are usually sensitive to blue light only and therefore the yellow stain is photographically opaque because it absorbs blue light. Although the stained image looks as if it is low contrast to the eye, it will print with high contrast.

High-Contrast Developers

Hydroquinone is usually the most important ingredient of high-contrast developers, for example, by raising the proportions of hydroquinone in a metol-hydroquinone developer it becomes a high-contrast developer. It is logical to assume also that additional restrainer might be added to reduce fogging because of the increased hydroquinone. By increasing the pH the contrast can also be increased; for example, metol-hydroquinone developers usually use sodium carbonate as the alkali, but using sodium or potassium hydroxide can make a greater increase in pH. Using the hydroquinone alone without the metol in a high-alkaline solution can increase the contrast. Hydroquinone alone (without metol or other agents) tends to require a long development time before the image appears. This effect tends to develop an image simultaneous throughout the depth of the emulsion. This effect can be important for the reproduction of certain types of interference patterns (holograms) because it may be desired to produce them in depth and not just on or near the surface of the emulsion.

Ansco 70, and Kodak D-8, D-11, and D-178 are examples of high-contrast developers using hydroquinone, Kodak D-8 exemplifies a typical formula for a high-contrast developer.

Water about 32°C	750 cc
Kodak sodium sulfite desiccated	90.0 grams
Kodak hydroquinone	45.0 grams
Kodak sodium hydroxide	37.5 grams
Kodak potassium bromide	30.0 grams
Water to make	1 liter

Hydroquinone used with paraformaldehyde can produce an extremely high-contrast developer. This developer exhausts itself rapidly when developing and can therefore be used to produce certain special effects. During development the exhausted developer from a dark image area will diffuse across the boundary and thus hold back the density of the light area. Likewise the relatively fresh developer from the light image areas will diffuse across the boundary to the dark image areas. This effect tends to develop the edges of a boundary section to an extra density. This type of development is usually called a *Lith*-type development because it produces an extra sharp boundary line as required for a line or halftone dot image for graphic reproductions. A typical formula for an extremely high-contrast developer is as follows:

Sodium sulfite anhydrous	32 grams
Hydroquinone	23 grams
Paraformaldehyde	7 grams
Boric acid crystals	7 grams
Potassium metabisulfite	3 grams
Potassium bromide	2 grams
Water to make	1 liter

This solution should be allowed to stand several hours before use.

Low-Contrast Developers

Low-contrast developers are similar to MQ developers without hydroquinone but use only metol or para-aminophenol as the developing agent. The images produced are of comparatively low contrast and tend to be produced on the surface of the emulsion. This type of developer tends first to act on the surface before developing the image deeper in the emulsion. These characteristics can be desirable when making a surface-type hologram (i.e., the diffraction pattern is developed only on the surface of the emulsion and not in depth). This characteristic is also useful when developing the image of a spectrum because the halation effects will be minimized by surface development. A typical low-contrast formula is

Metol	2 grams
Sodium sulfite anhydrous	10 grams
Sodium carbonate anhydrous	10 grams
Potassium bromide	0.5 grams
Water to make	1 liter

Special-purpose Developers

POTA Developer. An article by Marilyn Levy* describes a developer called POTA made as follows:

1-phenyl-3-pyrazolidone	1.5 grams
Sodium sulfite	30 grams
Water to make	1.0 liter

Develop about 5 min at 95°F (or about 10 min at 68°F). This developer is low contrast and maintains a relatively linear exposure versus density curve. By reducing the gamma without loss of film speed (i.e., using the characteristics curves, the top speed was maintained) certain film qualities can be improved. This formula is important in optical data processing because it permits large variations in intensity to be recorded lineally; for example, a spectrum can easily be recorded lineally using this formula with a ratio of intensities of $10^4:1$ and maintain good resolving power. Photograph 1.14 in Chapter 1 is a positive made from a negative developed using a slight modification of this formula. Panatomic-X film was used. Figure 12.1(b) is a positive made from a similarly exposed Panatomic-X film but developed in a normal MQ developer. Figure 12.1(a) is the positive of a similarly exposed Panatomic-X film developed in a modified POTA developer.

*Marilyn Levy, "Wide latitude photography," *Photographic Science and Engineering*, **11**, Number 1, January–February 1967.

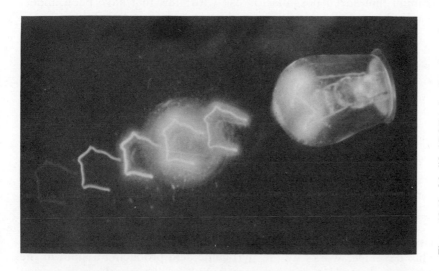

Figure 12.1a Normal MQ developed negative.

Figure 12.1b Modified POTA developed negative.

High-Energy Developers. High-energy developers are used to obtain the maximum film speed possible for use with films that have been underexposed. Pyro-metol is one such developer which is made as follows:

Solution I

Metol	2 grams
Potassium metabisulfite	6 grams
(or sodium bisulfite)	
Pyro	6 grams
Water to make	500 cc

Solution II

Sodium carbonate anhydrous	40 grams
Water to make	500 cc

Use equal parts of solution I and II and develop for about 6 min at 68°F.

This developer works by the pyrogallic acid producing a stain in the emulsion to provide a printing density and by the metol producing an image on the film even in a greatly underexposed area.

Another high-energy developer for underexposed negatives is caustic metol-hydroquinone developer as follows:

Water at 50°C	750 cc
Wood alcohol	48 cc
Metol	14 grams
Sodium sulfite desiccated	53 grams
Hydroquinone	14 grams
Sodium hydroxide (caustic soda)	9 grams
Potassium bromide	9 grams
Add cold water to make	1 liter

The sodium hydroxide should always be carefully dissolved in cold water to avoid spattering of alkali which can cause bad burns. Development time is about 5 min at 20°C. Kodak D-82 is an example of this developer.

High-Speed Developers. High-speed film development is important for real time optical data processors using photographic film as an input. Highly alkaline developers can be used to reduce the development time to less than a minute at 65°F. Some such developers are modified metol-hydroquinone developers that are normally highly alkaline. Strongly alkaline solutions require more than potassium bromide to keep fog down,

and a 0.1-percent solution of phenosafranine is used. One such developer
is

Sodium sulfite	3 grams
Hydroquinone	3 grams
Phenosafranine, 0.1 percent	2 cc
Water to make	50 cc

Add an equal volume of 12-percent sodium hydroxide solution just
before use.

Another high-speed developer using two solutions requires about 5
sec in the first solution, (which is made up of hydroquinone, 50 grams;
sulfurous acid, 8 percent, 250 cc; phenosafranine, 20 cc; and water to
make 1 liter), and 3 sec in the second solution (which consists of 30-per-
cent solution of sodium or potassium hydroxide with 1 part formalin solution added to every 40 parts of the 30-percent hydroxide solution). At
high temperatures (30°C) the development time can be reduced.

Tanning Developers. We have previously discussed that during development the gelatin hardens or tans and indicated how this effect might be
used in optical data processing. Tanning makes the gelatin insoluble and
tends to reduce swelling (and dimensional distortions). The amount of
tanning is usually proportional to the amount or metallic silver being
produced in the emulsion. A tanning developer can be made as follows:

Pyrocatechin	2 grams
Sodium sulfite, 1-percent	25 cc
Sodium hydroxide, 1-percent	50 cc
Water to make	1000 cc

The developer must be made immediately before use and at 68°F the
development time is about 15 min.

Another tanning developer is made as follows:

Solution I

Pyrogallic acid	4 grams
Sodium sulfite, anhydrous	10 grams
Water to make	1000 cc

Solution II

| Sodium carbonate, anhydrous | 28 grams |
| Water to make | 1000 cc |

Mix equal parts of solutions I and II immediately before use and develop film for about 6 min at 68°F. Alcohol added to tanning developers tends to increase their tanning ability. Tanning tends to water proof the gelatin. When tanning developers are used, areas that received the greatest exposure tend to have their development action reduced because the tanning will prevent fresh developer from reaching the emulsion. After development unhardened portions of the emulsion can be washed away in hot water leaving only the hardened gelatin in the form of a relief image. The developed metallic silver image can now bleach out if desired leaving only the relief image.

The relief image formed using a tanning developer requires the use of a silver halide emulsion. This technique differs therefore from that previously discussed for bichromated gelatin which requires no silver halide crystals; that is, the gelatin itself was made sensitive in one case while in the other it was used as a support for the light-sensitive silver halide crystals. Tanning developers seem appropriate when high film speeds are required but bichromated gelatin film will give the greatest possible resolution.

Fine-Grain Developers

Fine-grain developers reduce the tendency for the image-forming metallic silver in the emulsion to clump together to form relatively large grains during development. When enlargements are made, clumps can become obvious and are therefore to be avoided. Fine-grain developers are used to minimize the effects of clumping.

A typical normal fine-grain developer is Kodak Microdol-X. This fine-grain developer tends to produce a slightly brownish image. The brown image will print with higher contrast than it appears visually because it is opaque photographically; that is, photographic papers are usually sensitive to blue light which is absorbed by a brown stain. This developer thus maintains the effective film speed and produces a low-grain image with high acutance (sharpness).

The physical developers also produce fine-grain images. The fine-grain images are obtained by the addition of a solvent for silver halide. These developers dissolve some of the silver bromide crystals during development which minimizes clumping. Dissolved silver is then plated on portions of the emulsion that already have a metallic silver image formed.

PHYSICAL DEVELOPMENT

The normal photographic development process is termed *chemical development*. In chemical development the exposed photographic material is put into a reducing solution that causes the image to be reduced or oxidized. In the physical development process the silver halide is also initially reduced but after an initial slight chemical development, the remaining development is a silver-plating process—that is, taking silver out of the solution and depositing it on the slightly developed image. The differences between the two processes therefore is that in chemical development the silver halide within the emulsion is developed while in physical development the silver within the solution is plated on the emulsion to produce the image. In the physical development process the silver forming the image is supplied by the developer solution and not by the silver in the emulsion as in chemical development.

A physical developer solution usually consists of a source of silver, a silver solvent, and some normal chemical developing agent. The chemical developing agent initially produces a weak silver image required for physical development. The weak developed silver image in the emulsion is intensified essentially by plating silver on it from the developing solution. In the physical developing process very little of the silver halide in the emulsion is developed and therefore the grain of the final image is almost independent of the emulsion grain.

Reduced film speed cannot be avoided in physical development because chemical development is just an initial process and is stopped as soon as possible to allow silver plating from the solution. It is possible to achieve different ratios of chemical development to physical developing by slowing down the action of the sodium thiosulfate (hypo) on the silver halide. This is done by converting the silver bromide in the emulsion to silver iodide (which is less soluble in sodium thiosulfate solutions) by placing the film in a bath of potassium iodide before development. This permits the developer to act longer before its action is stopped by the sodium thiosulfate which removes the undeveloped silver halide from the emulsion.

The following formula can be used for physical development

Initial Bath

Sodium sulfite	25 grams
Potassium iodide	10 grams
Water to make	1000 cc

Solution A

Sodium sulfite	50 grams
Silver nitrate	30 grams
Sodium thiosulfate (hypo)	150 grams
Distilled H_2O to make	1000 cc

Step 1: Silver nitrate is dissolved in 200 cc H_2O.

Step 2: Sodium sulfite dissolved in 500 cc H_2O is slowly added to step 1 with stirring.

Step 3: A precipitate forms.

Step 4: Sodium thiosulfate is added and the precipitate redissolves.

Step 5: The remaining H_2O is added.

This solution will keep for long periods of time.

Solution B

Metol	2 grams
Sodium sulfite	10 grams
Hydroquinone	3 grams
Sodium hydroxide	3 grams
Distilled H_2O to make	100 cc

Procedure

Immerse film in initial bath for about $1\frac{1}{2}$ min. *Rinse well.* Place in a solution of 10 parts solution A, 1 part solution B, and 40 parts H_2O. This solution should be made immediately before using and should not be reused. Develop film for 25–30 min at 65°F.

Fix film in acid fixer containing 30 percent sodium thiosulfate for about 1 hr (silver iodide dissolves slowly). Wash for about 1 hr.

If desired a hardener of formalin (1 part to 80 parts H_2O) can be used any time in the process.

If film is milky (a fine silver deposit of film), it can be cleared by placing in 0.1 percent potassium bichromate and 0.2 percent sulphuric acid. The film should not remain in this solution too long or the image will be affected. (A 25-percent stronger solution will clean tanks and trays of the silver deposits.)

If film is underdeveloped it can be redeveloped (even after fixing because this process is essentially a silver-plating process. Rubber gloves should be worn to prevent staining hands.

MONOBATHS

A monobath is a single film-processing solution containing developers and fixers. Most developers are alkaline solutions. Fixers are about as efficient in alkaline developers as they are when used in acid fixing baths. Hardening agents (e.g., potassium alum) are only slightly effective as hardeners when added to alkaline developing solutions. Combining all three types of ingredients into one solution to develop, fix, and harden is therefore possible.

The combination of ingredients in a monobath solution is quite complex and must be carefully balanced because the fixer is dissolving the undeveloped silver halide out of the emulsion at the same time the developer is changing the exposed silver halide to metallic silver. If the developer is too vigorous, the resulting image will be of high contrast and effectively increase the emulsion speed. If the fixing action is too fast, the resulting image will be of low contrast and the emulsion speed will effectively be decreased. Proper balance between the developer and fixer occurs when the image forms before the fixer action removes the *developable* silver halide. Proper balance for one type of film or paper emulsion is not necessarily proper for another type of emulsion; that is, the amount of silver halide in the photographic emulsion primarily determines the amount of fixer required (a definite amount of fixer is required to clear a given emulsion in a given time). When determining a monobath solution for a particular emulsion, the usual starting point is the determination of the minimum amount of fixer to clear the emulsion in a given time. The next step is to determine the amount of developer to obtain a good image in this time.

The following chart indicates methods for changing the contrast, or emulsion speed, or both.

To Increase Contrast and for Emulsion Speed	To Decrease Contrast and/or Emulsion Speed
Make solution more alkaline	Make solution less alkaline
Increase amount of developer	Decrease amount of developer or increase amount of fixer.
Use more active developer	Use more active fixer.
Raise temperature	Lower temperature
For contrast use or increase amount of ascorbic acid or hydroquinone	Agitate more vigorously.
For emulsion speed use or increase amount of phenidone or metal	

A rise in monobath temperature increases the rate of development faster than the rate of fixing; that is, a rise in temperature tends to increase the contrast and emulsion speed. Making the monobath less alkaline will usually reduce the developer activity without affecting the fixer; that is, less alkaline solutions tend to decrease contrast and reduce emulsion speed. Vigorous agitation decreases the required development and fixing times but usually the required fixing time is decreased more than the required developing time; that is, more agitation tends to lower contrast and decrease emulsion speed. A typical monobath* is Monobath 438 made by H. S. Keelan then of the Boston University Research Laboratories in 1957 for Panatomic-X film.

Monobath 438		
		Purpose
Distilled H_2O (120°F)	700 mm	solvent
Hydroquinone	15 grams	developer
Sodium sulfite	50 grams	preservative
Phenidone	10 grams	developer
Potassium alum	18 grams	buffer
Sodium hydroxide	18 grams	alkali
Sodium thiosulfate	60 grams	fixer
Water to make	1000 mm	solvent

Expose Panatomic-X film at ASA-200 and process 4 minutes at 68°F.

It appears that for optical data processing the shrinkage of the emulsion might be reduced or eliminated by physical development techniques. If this shrinkage were eliminated, it would eliminate the need for the use of optical flats and oil of matching index of refraction (i.e., phase changes due to different thicknesses of film would not exist). Monobath techniques hold the promise of rapid film processing after exposure that would give a "real-time" optical capability to optical data processors.

REDUCERS

Reducing is a method for lowering the density of a negative by dissolving the silver out of the negative thus decreasing its density. There are three basic types of reducers as follows:

Type I: Subtractive Reducers

These reducers remove equal amounts of silver (reduce density) everywhere, thus reducing the density but not the contrast of the negative. The negative contrast may appear to increase but usually there is a slight decrease in gamma. Because the high, medium, and low densities are reduced, the effect appears to clear up shadow areas which makes them useful for treating fogged negatives, or overexposed negatives, or both. Kodak Reducer R-2 and Farmer's Reducer R-4a are examples of subtractive reducers.

*Monobaths are described in *Monobath Manual* by Grant Haist, Morgan and Morgan, Inc., New York, 1966.

Type II: Proportional Reducers

These reducers remove silver (reduce density) in proportion to the amount present in the negative. These reducers therefore decrease not only the density but also the image contrast (reduce the gamma). These reducers are useful for treating overdeveloped negatives. Kodak Farmer's Reducer R-4b, Reducer R-5, and R-8a are examples of proportional reducers. Farmer's Reducer R-4a which has been diluted will act proportionally. Using the standard two stock solutions A and B, a working solution is made up with only about one tenth the amount of potassium ferricyanide — that is, one tenth of stock solution A is used.

Type III: Superproportional Reducers

These reducers remove silver from the high-density areas and have little effect on the shadow (low) density areas. Because these reducers have a marked effect on the highlight (high) densities, they are useful for treating overdeveloped negatives of objects with high contrast. Kodak Persulfate Reducer R-15 is an example of a superproportional reducer.

PROCESSING PHOTOGRAPHIC (SILVER HALIDE) FILMS FOR COHERENT OPTICAL SYSTEMS

Phase Distortions

In our previous discussions we have shown the importance of being able to record accurately on photographic film a desired signal in the form of a transmission function. The transmission function should accurately represent the desired signal in both amplitude and phase. We have also discussed the possibility of making relief images in gelatin (containing no silver halide) that has been sensitized. In addition we have also seen that in the normal silver halide film development process a relief image is normally formed along with the darkened silver halide crystals (the relief image being a by-product of the development process that hardens the gelatin in the regions of greatest development). The relief image formed in silver halide photographic films is usually undesirable for coherent optical systems and steps are usually taken to eliminate it. In addition the permanence of a silver halide image depends on the processing technique used. The processing techniques for more permanent types of images are more costly and time consuming than processes producing less permanent images. The processing technique should be such that the transmission function would be only as permanent as the requirements of the system demanded.

When we expose a photographic emulsion with a Ronchi ruling, the latent image is dimensionally identical to the exposing light. When the exposed emulsion is placed in the developer, the gelatin starts to swell by absorbing the developer. This swelling will dimensionally distort the latent image. As the latent image is developed the gelatin becomes hardened in the developed regions. The swelled gelatin emulsion with its hardened developed image is placed in the short-stop bath (acetic acid) to neutralize the developer and then it is placed in the fixer. In the fixer the unexposed silver halide is dissolved out of the gelatin emulsion into the fixer solution. The removal of the unexposed silver halide crystals from the emulsion will leave voids in the emulsion. These voids in the emulsion will cause stresses and strains to develop in the gelatin emulsion as it dries. The end result can be that on drying the emulsion may shrink in the areas where the unexposed silver halide crystals have been removed. In the areas where the developed silver halide crystals remain, the emulsion can be higher than the areas that shrink. Developer, however, tends to harden the gelatin in areas where the silver halide has been turned to metallic silver. Hardened gelatin tends to be waterproof which would prevent swelling and therefore tend to prevent swelling in developed areas. From this we can conclude that the photographic processes must therefore be tailored to the requirements of the system in which it is to be used because in the final photographic product opposing reactions can be taking place — that is, shrinking and swelling.

Any differences in thickness in the emulsion will cause phase differences in the light leaving the transmission function. If the transmission function is of a Ronchi ruling, these phase differences can cause even harmonics to appear in the spectrum. We have previously shown how to sandwich the transmission function between two optical flats filled with an oil of matching index of refraction to maintain uniform thickness.

Removal of Amplitude Variations in Film (Phase Holograms)

We have seen how the stored data in the film can cause both the phase and amplitude of the light passing through the film to vary. We can further see that by filling the shrunken portions (surface relief image) of the gelatin thus making it flat (by using optical flats and an oil of matching index of refraction) we can eliminate unwanted phase changes and leave only the desired amplitude variations. If it were possible to control the phase variations and eliminate the amplitude variations for certain co-herent optical processing applications, more efficient use could be made of the available light; for example, a hologram can be viewed with less light and will appear brighter when made as a variable-phase recording than when made as a variable-density recording. Let us consider how we might develop a hologram (silver halide) as a phase variation rather than an amplitude variation. When a hologram is developed normally, there are density variations in the emulsion caused by the darkened metallic silver formed during development. The unexposed silver halide is removed during fixing. A relief image is formed on the surface of the emulsion because of the hardening of the gelatin during developing. Let us consider a hologram made with a Kodak 649F emulsion. The developed hologram can be treated with a solution of mercuric chloride ($HgCl_2$) (5 grams in 1000 cc water) which will change the metallic silver-to-silver mercurous chloride ($AgHgCl_2$, which is pink in color). Treatment of a Kodak 649F plate that has been exposed and developed in the normal way will result in the plate taking on an overall pinkish look. On drying this pinkish hologram the translucent pink crystals are relatively clear and have a different index of refraction from the gelatin which tends to change the phase of the light leaving it rather than affecting its amplitude.

The effect of mercuric chloride on a hologram can be illustrated by the following two examples. A Kodak 649F hologram that was overexposed was too dark after development for viewing as a hologram (although a very faint image could be seen through the very dark emulsion). This over-exposed hologram was placed in the mercuric chloride for about 2 min — that is, until all the black cast was gone and the overall cast of the holo-gram was pinkish. After drying, this hologram produced an excellent, bright, clear and sharp image when viewed, indicating that the phase hologram is more efficient than the amplitude hologram. Sandwiching this phase hologram between two glass flats with an oil of matching index of refraction did not cause the image to disappear. This indicated that there was no (appreciable) relief pattern on the surface and any phase changes were essentially caused by changes internal to the emulsion.

A second overexposed 649F hologram was placed in the mercuric chloride solution until it had an overall pinkish cast (about 2 min). When dry, this hologram produced a blurred image. The blurring appeared to be the result of more than the first-order diffraction pattern being bright. At least ten orders (i.e., separate identical images could be seen but each displaced one from another causing the blur) could now be seen whereas when originally viewed, only one order could barely be seen. Kodak Farmer's reducer (mixed as suggested on the packet) was then tried on this plate. The pink cast of the plate disappeared and the plate began to grey. Viewing (while wet from the Farmer's reducer) indicated that the higher-order images were being removed by the Farmer's reducer. By alternately viewing and returning the plate to the Farmer's reducer a final clear, bright, single image was obtained.

Figure 12.2 shows a photograph of a holographic image of some coins. The effects of the development distortions are evident. Each coin appears to produce a second image as if a double exposure was made. In the actual

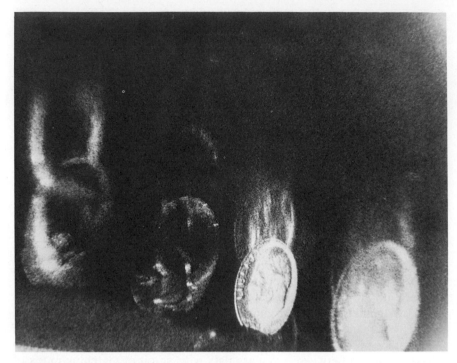

Figure 12.2 Example of hologram image distortions caused by improper development.

image more than one displaced image was evident but because of their low light levels, they were not recorded in the photograph. The camera was focused at the dime and the blurring of the other coins was caused by the shallow depth of field of the camera lens. The other coins are also sharp in the actual image.

Kostinsky Effect

We have seen that the choice of developer will affect the final results obtained using a specific development procedure. The development procedure used will itself affect the final results obtained. The Kostinsky effect illustrates the fact that not only will the type of developer and amount of agitation used during development affect the final results but also the actual type of image detail being developed will affect the final results.

Figure 12.3(a) shows a latent image of two closely spaced image points. Although the latent image is normally invisible, it is shown here for reference. Figure 12.3(b) shows the initial development of the latent

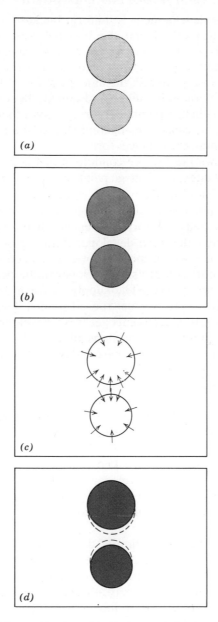

Figure 12.3 Images of four points on a clear background showing why they tend to move apart when developed. (*a*) Latent image. (*b*) Initial development. (*c*) Diffusion of fresh developer to replace used developer (used developer diffuses in the opposite direction to that shown). (*d*) Final developed image.

images. The initial development is uniform because all the developer is fresh. Figure 12.3(c) shows that the area between the exposed image points must supply fresh developer to the two image points, while the other areas must supply fresh developer to only one image point. Since the image points are closely spaced, the amount of fresh developer available between the image points is limited and quickly becomes depleted. The sections of the two image points bordering each other will therefore be underdeveloped. This causes an apparent shift of the image point centers because of the asymmetrical images formed.

Figure 12.4 shows the reverse conditions of Figure 12.3. In this case two closely spaced unexposed image points are surrounded by an exposed background. The latent image is again shown in Figure 12.4(a). Figure 12.4(b) shows the initial development. The initial development will tend to develop the latent image uniformly because all the developer is fresh. Figure 12.4(c) shows the fresh developer diffusing out of the two small image points to areas where the developer has been depleted. There is likewise a similar diffusion of used developer in the opposite direction to that shown in Figure 12.4(c). The supply of fresh developer is limited mainly by the area of the two unexposed points. The exposed borders around the unexposed image points get fresh developer mainly from the unexposed image points. The portions of the edges between the two unexposed image points will receive fresh developer from both unexposed image points and therefore will tend to be fully developed. The edges not between the two unexposed image points receive only one half the amount of developer and therefore do not develop fully. The image-point centers therefore appear to move apart because of the asymmetrical images formed being darker between the two points than elsewhere.

The effects described above are known as the *Kostinsky effect* and are related to the amount of fresh developer available and the uniformity of its distribution. The size, exposure, and spacing of the image points will affect the results as well as the film type used, the developer, the temperature and the amount of agitation.

We can see that the distribution of density within a developed image (especially when small) does not necessarily indicate the actual light intensity that exposed the emulsion. The most pronounced effects occur when there is a region of high density adjacent to a region of low intensity. In general the density will tend to increase as the size of an image point decreases (Eberhard effect). The reason for this is that fresh developer can be made to replace the exhausted developer more readily when the image points are small. Large image points will tend to have greater density at the edges than at the center.

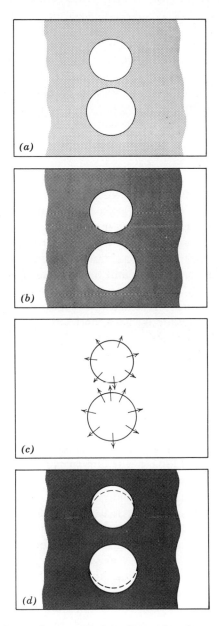

Figure 12.4 Images of two points on a black background tend to move apart when developed. (*a*) Latent image. (*b*) Initial development. (*c*) Diffusion of fresh developer to replace used developer (used developer diffuses in the opposite direction to that shown). (*d*) Final developed image.

PHOTOGRAPHIC FILM CHARACTERISTICS

We have discussed how to produce a developed image from an exposed photographic film and how the development procedure can affect the results. We have not discussed in any great detail how the choice of type of film can produce different results for a given exposure and development procedure. We will now discuss various film parameters and indicate the types and uses of the films most often used in optical processing.

High-Resolution Films

Many different types of film are commercially available. Each commercial film type is primarily made to meet specific photographic requirements. Black and white portrait film is made to meet the requirements of portrait studios — that is, thick base support for easy handling, emulsion surface matted to permit easy touchup and corrections, emulsion characteristics such that the toe of the characteristics curve (H&D) can be used (little or no flare light will exist) and the ratio of maximum-to-minimum light levels will be small. (Flare light is light entering the optical system off axis so as not to produce an image but to be reflected to the film as a general illumination.) Outdoor black and white films must be able to provide good contrast images even when high levels of flare light strike the film. Outdoors the ratio of maximum-to-minimum light levels are orders of magnitude greater than indoors in a studio. Outdoor films must be capable of recording light levels whose ratio of maximum to minimum are extremely high and with a high overall general illumination (flare light).

In optical data processing the choice of the film that is to be used must be guided by what is to be recorded. When large differences in light levels are to be recorded, outdoor type of film would seem desirable. For low light levels of recording, an indoor type film is desirable. There are, however, a great many factors other than light levels that must be considered before the choice of film to be used is made.

When high levels of light are being imaged on a photographic film, some of the light passes through the film. Some of this light will be reflected back into the emulsion from the rear surface(s) of the emulsion (and its support). Some of the light will therefore be refracted, some diffracted, and some reflected as it passes through the film. This will cause the light to spread out or diffuse beyond the image boundaries, which will cause reduced image sharpness. For example if a point source of light is imaged on the film (so that the light beam strikes the film normal to its surface), there will be a reflection of light from the rear surface of the film back into the emulsion. This reflected light will spread out beyond (even if the rear surface is parallel to the front surface) the boundary of the original image point and surround it. Actually it will appear that the original point is surrounded by a halo and the phenomenon is called *halation*. The thicker the emulsion, the smaller the halation effect (the emulsion absorbs or diffused most of the light so that there is little or no reflected light from the rear surface of the film). In thin emulsions halation is prevented by coating the base of the gelatin with a light-absorbing dye; that is, the light to which the emulsion is sensitive is absorbed by the dye. In processing this dye is removed. Certain films use a grey base emulsion support to reduce halation (this is usually accompanied by an increased exposure requirement during printing).

Certain black and white emulsions, such as the Lippmann type, can be used for reproducing color pictures by recording the interference pattern produced by an incident and reflected wave. The arrangement used in this color process is shown in Figure 12.5. A very high resolution emulsion is placed in contact with a reflecting surface such as mercury. Parallel incident light waves will pass through the glass plate and emulsion to the mercury from which it will be reflected back on itself. The incident waves would normally expose portions of the emulsion. In this case, however, the portion of the waves reflected back by the mercury will interfere with the incident waves in the emulsion. Now only the areas of the emulsion where the incident and reflected waves reinforce will be exposed. No exposure will take place where the incident and reflected waves cancel

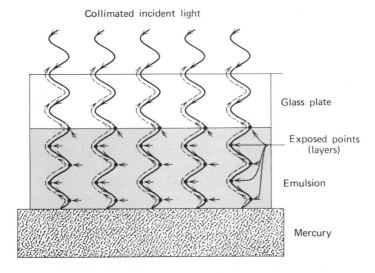

Collimated incident light

Glass plate

Exposed points
(layers)

Emulsion

Mercury

Figure 12.5   Lippmann color process.

each other. If the incident light is monochromatic, the location of exposed areas will be related to the wavelength of the incident light. In effect these exposed areas form a diffraction grating. When white light is incident on the developed emulsion at an angle, the color of the reflected light will correspond to the original color exposing the emulsion (refer to Appendix 5, Bragg's law). Appendix 17 is a discussion of how color is perceived.

Commercial films with emulsions similar to the Lippmann emulsion are available and are sold under the names Kodak 649F, Agfa-Gevaert 8E70, 8E75, 10E70, 14C70; these are very often used in coherent optical data processing (particularly for making holograms and filters).

A Lippmann-type emulsion can be made as described below. Prepare solution A as follows:

Solution A

Gelatin	20 grams
Water	390 cc
Filter and use at 35°C	

To 80 cc of solution A add

Water	10 cc
Silver Nitrate	4 grams

To balance of Solution A add 3.2 grams of potassium bromide and then slowly pour into gelatin the silver nitrate solution with no frothing. Both solutions should be at the same temperature. Stir gently and after $3\frac{1}{2}$ min add the following sensitizing dyes:

Pinacyanol (1 : 1000 alcohol solution)	4 cc
Orthochrom T (1 : 1000 alochol solution)	4 cc
Acridin orange (1 : 500 alcohol solution)	4 cc

The dyes must previously have been warmed to 35°C and added slowly with gentle stirring (they should be added at a uniform rate so that it takes about 45 sec to add). Apply to glass plate and place plates on a cold slab to set. Wash 10 min to remove the potassium nitrate in running water. Allow to dry horizontally. (NOTE: The process must be done in complete darkness when the sensitizing dyes are added to the gelatin solution.)

Techniques for Increasing Film Speed

One disadvantage of a Lippmann-type emulsion (or most other high-resolution emulsions) is that the film speed is low. Most development techniques that maintain the high resolution of the film will also tend to reduce the film speed. Any of the methods listed below will produce increased speed in photographic emulsions. An effective increase in film speed (as high as an effective increase of 10 times has been obtained) without any apparent adverse affects, such as increased grain size and lower resolution, has been obtained by these techniques. Any method of subjecting the exposed (or unexposed) film to additional uniform radiation will produce this effect. The method for adding the additional energy should be such that an extremely long exposure time is required to add the additional energy. At least one-half hour exposure should be required just to start to bring any noticeable effects to the underexposed areas. This means that the energy that is being added to the (normal or) underexposed film must be of extremely low level. These methods also work on normally exposed film. The increase in density of the underexposed areas (without affecting the areas that were exposed properly) gives improved contrast pictures for normally exposed films.

Most of the techniques work on the principle that a certain amount of energy is required to be absorbed by a silver halide crystal before it becomes developable. If a silver halide crystal absorbs some energy but not sufficient to make it developable, the additional energy to make the crystal developable can be added before or after the film exposure.

Method 1. Underexposed film is subjected to an additional exposure. This second exposure is a uniform controlled amount of light (for a predetermined length of time). This can be done by double exposing the film. The uniform exposure can easily be obtained by taking a picture of a piece of uniformly illuminated paper or wall. The film is then developed in the usual manner.

Method 2. Underexposed film is subjected to radioactive radiation for a predetermined length of time and then developed in the usual manner.

Method 3. Underexposed film is placed in a developing tank that has been coated with an electroluminescent matter (an electroluminescent night light is ideal for this). A voltage is then applied for a predetermined length of time permitting additional exposure of the previously exposed film. The film is then developed in the usual manner.

Method 4. Underexposed film is placed into a special developing tank. This tank has a coating of a phosphor material. The phosphor is energized by exposure to light and emits light at the proper rate and for the proper time to add the required energy to the film. The film is left in the tank for the required predetermined length of time and then developed in the usual manner.

Method 5. The underexposed film is placed in the normal development tank. Prior to development, however, the film is presoaked for 10–20 minutes in water. This soaking swells the gelatin and makes it more porous. After this presoaking the film is developed in the normal manner. Since the gelatin is more porous the developer penetrates the emulsion faster which permits the bottom layers of the emulsion to be developed to a much greater degree than if the emulsion were not presoaked. This tends to give the film increased density and greater contrast for a given development procedure.

Method 6. A few drops of 30 percent Hydrogen Peroxide (H_2O_2) is added to the water used for the presoak of Method 5. This procedure serves two purposes. The benefits of presoaking are obtained, and in addition the hydrogen peroxide tends to add energy to the film to make more of the slightly exposed silver halide crystals developable.

Preconditioning

It was previously pointed out that there are several difficult problem areas which must be overcome in order to make successful reflectance holograms. The dimensional stability of the emulsion is one of the most important considerations in making successful reflectance holograms. In fact the dimensional stability of the emulsion is extremely important in the entire area of optical data processing. It appears that placing the photographic plate in water for about five minutes and drying for an hour, (and repeating this procedure for five or six times) will improve the dimensional stability of the emulsion. The drying time can be reduced by using an alcohol rinse after the water soak. There appears to be an improvement in the resolution and signal-to-noise ratio of the holograms made from preconditioned plates. In addition it appears that the plates do not have to be used immediately after preconditioning but can be stored after preconditioning for later use.

Appendix 1

PROCEDURE FOR MAKING A REAR SURFACED MIRROR (MIRROR NOT FOR PRECISION USE)

Make two solutions as follows using only clean glass funnels, stirring rods, and containers.

Solution 1

Bring 10 oz of distilled (or demineralized water) to a boil in a glass container. Add 1 gram of silver nitrate and then 1 gram of sodium potassium tartrate (Rochelle salts). Boil for 5 to 6 min and then remove from the heat and add sufficient distilled (or demineralized) water to make 10 oz. Filter through No. 1 filter paper while still moderately hot and allow to cool to about 20°C (70°F).

Solution 2

In approximately 3 oz of distilled (or demineralized water in a glass beaker dissolve 1.5 grams of silver nitrate. Then carefully, while stirring, add *drops* of aqua ammonia (household ammonia may be used) to the solution. The solution will turn brown but continue stirring and adding ammonia until the solution becomes as clear as it was before the addition of any ammonia. Add another 1.5 grams of silver nitrate to the now clear solution. The brown color will again appear and will remain. Stir with glass rod until the silver nitrate is dissolved. Add enough water to make 10 oz and filter *twice* through two layers of diaper type cloth.

Procedure

The glass surface to be silvered must be chemically clean or the silver deposit will not be uniform. After washing the glass, clean it with ammonia using a cotton swab, rinse with water, and keep covered with water until ready for silvering. The glass surface that is to be silvered should be placed in a container which, when filled, will be deep enough to let the mixed solution cover the surface to a depth of approximately 1 in.

Carefully measure equal quantities of solution 1 and 2 which are at about 20°C. Add solution 1 to 2, mix thoroughly, and pour over surface to be silvered. Agitate the container enough to keep the liquid in a slight, even motion. It may require an hour to obtain a full deposit. A second coating may be applied if necessary. After the mirror is silvered, pour off the solution and rinse carefully with water. The silvered surface is very soft when wet so only the edges of the glass should be handled and not the mirrored surface.

When completely dry, the side of the glass that was face down in the solution will probably not be as uniformly silvered because the solution cannot flow evenly over it. The silver from this surface should be removed. Muriatic acid and a cotton swab can be used to remove the silver. Care should be exercised when using the acid so as not to let the acid run on the evenly silvered side of the glass or permit the fumes to damage it. After cleaning, rinse the mirror with water so that all smudges are removed.

The silvered surface, after it is dry, may be protected by spraying with lacquer. After the lacquer is dry, the mirror is ready for use.

Appendix 2A

$$\frac{\sin x}{x}$$

x	0°	10°	20°	30°	40°	50°	60°	70°	80°	90°
0°	+1.00000	+0.99493	+0.97982	+0.95493	+0.92073	+0.87782	+0.82699	+0.76915	+0.70532	+0.63662
100°	+0.56425	+0.48946	+0.41350	+0.33762	+0.26306	+0.19099	+0.12248	+0.05853	+0.00000	−0.05236
200°	−0.09798	−0.13642	−0.16740	−0.19083	−0.20675	−0.21536	−0.21702	−0.21221	−0.20152	−0.18566
300°	−0.16540	−0.14158	−0.11509	−0.08681	−0.05764	−0.02843	+0.00000	+0.02689	+0.05157	+0.07346
400°	+0.09207	+0.10705	+0.11814	+0.12521	+0.12824	+0.12732	+0.012266	+0.11455	+0.10337	+0.08957
500°	+0.07366	+0.05617	+0.03761	+0.01877	0.00000	−0.01809	−0.03499	−0.05026	−0.06350	−0.07439
600°	−0.08270	−0.08826	−0.09101	−0.09095	−0.08816	0.08283	0.07518	−0.06551	−0.05416	−0.04152
700°	−0.02799	−0.01401	0.00000	+0.01363	+0.02648	+0.03820	+0.04846	+0.05700	+0.06361	+0.06815
800°	+0.07053	+0.07074	+0.06881	+0.06487	+0.05907	+0.05164	+0.04282	+0.03293	+0.02227	+0.01118
900°	0.00000	−0.01093	−0.02130	−0.03080	−0.03918	−0.04620	−0.05169	−0.05551	−0.05758	−0.05787
1000°	−0.05643	−0.05331	−0.04865	−0.04261	−0.03541	−0.02728	−0.01849	−0.00930	0.00000	+0.00913
1100°	+0.01781	+0.02581	+0.03288	+0.03884	+0.04353	+0.04682	+0.04864	+0.04897	+0.04782	+0.04524
1200°	+0.04135	+0.03627	+0.03019	+0.02329	+0.01580	+0.00796	0.00000	−0.00783	−0.01531	−0.02221
1300°	−0.02833	−0.03350	−0.03759	−0.04048	−0.04211	−0.04244	−0.04149	−0.03930	−0.03596	−0.03158
1400°	−0.02631	−0.02032	−0.01380	−0.00696	0.00000	+0.00686	+0.01342	+0.01949	+0.02488	+0.02946
1500°	+0.03308	+0.03566	+0.03712	+0.03745	+0.03664	+0.03474	+0.03181	+0.02796	+0.02331	+0.01802
1600°	+0.01225	+0.00618	0.00000	−0.00610	−0.01195	−0.01736	−0.02219	−0.02628	−0.02954	−0.03186
1700°	−0.03319	−0.03351	−0.03281	−0.03112	−0.02852	−0.02508	−0.02093	−0.01619	−0.01101	−0.00556
1800°	0.00000	+0.00550	+0.01077	+0.01565	+0.02002	+0.02372	+0.02668	+0.02879	+0.03001	+0.03032
1900°	+0.02970	+0.02819	+0.02584	+0.02274	+0.01898	+0.01469	+0.01000	+0.00505	0.00000	−0.00500
2000°	−0.00980	−0.01425	−0.01823	−0.02162	−0.02432	−0.02626	−0.02739	−0.02768	−0.02713	−0.02576
2100°	−0.02363	−0.02080	−0.01737	−0.01345	−0.00916	−0.00463	0.00000	+0.00458	+0.00899	+0.01308
2200°	+0.01674	+0.01986	+0.02235	+0.02414	+0.02519	+0.02546	+0.02497	+0.02372	+0.02176	+0.01917
2300°	+0.01601	+0.01240	+0.00845	+0.00427	0.00000	−0.00423	−0.00830	−0.01209	−0.01547	−0.01836
2400°	−0.02067	−0.02234	−0.02332	−0.02358	−0.02313	−0.02198	−0.02017	−0.01777	−0.01485	−0.01151
2500°	−0.00784	−0.00396	0.00000	+0.00393	+0.00771	+0.01123	+0.01439	+0.01708	+0.01923	+0.0279
2600°	+0.02170	+0.02195	+0.02154	+0.02047	+0.01880	+0.01656	+0.01385	+0.01073	+0.00731	+0.00370
2700°	0.00000	−0.00367	−0.00720	−0.01049	−0.01344	−0.01596	−0.01798	−0.01944	−0.02030	−0.02054
2800°	−0.02015	−0.01916	−0.01760	−0.01551	−0.01297	−0.01005	−0.00685	−0.00347	0.00000	+0.00344
2900°	+0.00676	+0.00984	+0.01261	+0.01498	+0.01688	+0.01825	+0.01906	+0.01929	+0.01893	+0.01801
3000°	+0.01654	+0.01458	+0.01220	+0.00946	+0.00645	+0.00326	0.00000	−0.00324	−0.00636	−0.00927
3100°	−0.01188	−0.01411	−0.01590	−0.01720	−0.01797	−0.01819	−0.01786	−0.01698	−0.01560	−0.01376
3200°	−0.01151	−0.00893	−0.00609	−0.00308	0.00000	+0.00306	+0.00601	+0.00876	+0.01123	+0.01334
3300°	+0.01504	+0.01627	+0.01700	+0.01720	+0.01689	+0.01607	+0.01477	+0.01302	+0.01090	+0.00845
3400°	+0.00576	+0.00292	0.00000	−0.00290	−0.00570	−0.00830	−0.01064	−0.01265	−0.01426	−0.01543
3500°	−0.01612	−0.01632	−0.01603	−0.01525	−0.01402	−0.01236	−0.01035	−0.00803	−0.00547	−0.00277

587

APPENDIX 2A

x	0°	10°	20°	30°	40°	50°	60°	70°	80°	90°
3600°	0.00000	+0.00276	+0.00541	+0.00789	+0.01012	+0.01202	+0.01356	+0.01467	+0.01533	+0.01553
3700°	+0.01525	+0.01451	+0.01333	+0.01177	+0.00985	+0.00764	+0.00521	+0.00264	0.00000	−0.00262
3800°	−0.00516	−0.00752	−0.00964	−0.01146	−0.01292	−0.01398	−0.01462	−0.01481	−0.01454	−0.01384
3900°	−0.01272	−0.01123	−0.00940	−0.00729	−0.00497	−0.00252	0.00000	+0.00251	+0.00492	+0.00718
4000°	+0.00921	+0.01095	+0.01234	+0.01336	+0.01397	+0.01414	+0.01390	+0.01323	+0.01216	+0.01073
4100°	+0.00898	+0.00697	+0.00476	+0.00241	0.00000	−0.00240	−0.00471	−0.00687	−0.00881	−0.01048
4200°	−0.01181	−0.01279	−0.01337	−0.01355	−0.01331	−0.01267	−0.01165	−0.01028	−0.00861	−0.00668
4300°	−0.00456	−0.00231	0.00000	+0.00230	+0.00451	+0.00659	+0.00845	+0.01004	+0.01133	+0.01226
4400°	+0.01282	+0.01299	+0.01277	+0.01215	+0.01118	+0.00986	+0.00826	+0.00641	+0.00437	+0.00221
4500°	0.00000	−0.00220	−0.00433	−0.00632	−0.00811	−0.00965	−0.01088	−0.01178	−0.01232	−0.01248
4600°	−0.01227	−0.01168	−0.01074	−0.00948	−0.00794	−0.00616	−0.00421	−0.00213	0.00000	+0.00212
4700°	+0.00417	+0.00608	+0.00780	+0.00928	+0.01047	+0.01133	+0.01185	+0.01201	+0.01180	+0.01124
4800°	+0.01034	+0.00913	+0.00764	+0.00593	+0.00405	+0.00205	0.00000	−0.00204	−0.00402	−0.00586
4900°	−0.00752	−0.00894	−0.01008	−0.01092	−0.01142	−0.01157	−0.01138	−0.01083	−0.00996	−0.00880
5000°	−0.00737	−0.00577	−0.00390	−0.00198	0.00000	+0.00197	+0.00387	+0.00565	+0.00725	+0.00862
5100°	+0.00973	+0.01054	+0.01102	+0.01117	+0.01098	+0.01045	+0.00962	+0.00849	+0.00711	+0.00552
5200°	+0.00377	+0.00191	0.00000	−0.00190	−0.00374	−0.00546	−0.00700	−0.00833	−0.00940	−0.01018
5300°	−0.01065	−0.01079	−0.01060	−0.01010	−0.00929	−0.00820	−0.00687	−0.00533	−0.00364	−0.00185
5400°	0.00000	+0.00184	+0.00362	+0.06528	+0.00677	+0.00805	+0.00909	+0.00984	+0.01030	+0.01044
5500°	+0.01026	+0.00977	+0.00899	+0.00794	+0.00665	+0.00516	+0.00352	+0.00179	0.00000	

Appendix 2B

$$\left(\frac{\sin x}{x}\right)^2$$

x	0°	10°	20°	30°	40°	50°	60°	70°	80°	90°
0°	1.00000	0.98989	0.96004	0.91189	0.84774	0.77057	0.68392	0.59159	0.49747	0.40529
100°	0.31838	0.23957	0.17098	0.11399	0.06920	0.03648	0.01500	0.00343	0.00000	0.00274
200°	0.00960	0.01861	0.02802	0.03641	0.04274	0.04638	0.04710	0.04503	0.04061	0.03447
300°	0.02736	0.02005	0.01325	0.00754	0.00332	0.00081	0.00000	0.00072	0.00266	0.00540
400°	0.00848	0.01146	0.01396	0.01568	0.01645	0.01621	0.01505	0.01312	0.01069	0.00802
500°	0.00543	0.00316	0.00142	0.00035	0.00000	0.00033	0.00122	0.00253	0.00403	0.00553
600°	0.00684	0.00779	0.00828	0.00827	0.00777	0.00686	0.00565	0.00429	0.00293	0.00172
700°	0.00078	0.00020	0.00000	0.00019	0.00070	0.00146	0.00235	0.00325	0.00405	0.00464
800°	0.00497	0.00500	0.00474	0.00421	0.00349	0.00267	0.00183	0.00108	0.00050	0.00012
900°	0.00000	0.00012	0.00045	0.00095	0.00154	0.00213	0.00267	0.00308	0.00332	0.00335
1000°	0.00318	0.00284	0.00237	0.00182	0.00125	0.00074	0.00034	0.00009	0.00000	0.00008
1100°	0.00032	0.00067	0.00108	0.00151	0.00189	0.00219	0.00237	0.00240	0.00229	0.00205
1200°	0.00171	0.00132	0.00091	0.00054	0.00025	0.00006	0.00000	0.00006	0.00023	0.00049
1300°	0.00080	0.00112	0.00141	0.00164	0.00177	0.00180	0.00172	0.00154	0.00129	0.00100
1400°	0.00069	0.00041	0.00019	0.00005	0.00000	0.00005	0.00018	0.00038	0.00062	0.00087
1500°	0.00109	0.00127	0.00138	0.00140	0.00134	0.00121	0.00101	0.00078	0.00054	0.00032
1600°	0.00015	0.00004	0.00000	0.00004	0.00014	0.00030	0.00049	0.00069	0.00087	0.00101
1700°	0.00110	0.00112	0.00108	0.00097	0.00081	0.00063	0.00044	0.00026	0.00012	0.00003
1800°	0.00000	0.00003	0.00012	0.00025	0.00040	0.00056	0.00071	0.00083	0.00090	0.00092
1900°	0.00088	0.00079	0.00067	0.00052	0.00036	0.00022	0.00009	0.00003	0.00000	0.00003
2000°	0.00010	0.00020	0.00033	0.00047	0.00059	0.00069	0.00075	0.00077	0.00074	0.00066
2100°	0.00056	0.00043	0.00030	0.00018	0.00008	0.00002	0.00000	0.00002	0.00008	0.00017
2200°	0.00028	0.00039	0.00050	0.00058	0.00063	0.00065	0.00062	0.00056	0.00047	0.00037
2300°	0.00026	0.00015	0.00007	0.00002	0.00000	0.00002	0.00007	0.00015	0.00024	0.00034
2400°	0.00043	0.00050	0.00054	0.00056	0.00053	0.00048	0.00041	0.00032	0.00022	0.00013
2500°	0.00006	0.00002	0.00000	0.00002	0.00006	0.00013	0.00021	0.00029	0.00037	0.00043
2600°	0.00047	0.00048	0.00046	0.00042	0.00035	0.00027	0.00019	0.00012	0.00005	0.00001
2700°	0.00000	0.00001	0.00005	0.00011	0.00018	0.00025	0.00032	0.00038	0.00041	0.00042
2800°	0.00041	0.00037	0.00031	0.00024	0.00017	0.00010	0.00005	0.00001	0.00000	0.00001
2900°	0.00005	0.00010	0.00016	0.00022	0.00028	0.00033	0.00036	0.00037	0.00036	0.00032
3000°	0.00027	0.00021	0.00015	0.00009	0.00004	0.00001	0.00000	0.00001	0.00004	0.00009
3100°	0.00014	0.00020	0.00025	0.00030	0.00032	0.00033	0.00032	0.00029	0.00024	0.00019
3200°	0.00013	0.00008	0.00004	0.00001	0.00000	0.00001	0.00004	0.00008	0.00013	0.00018
3300°	0.00023	0.00026	0.00029	0.00030	0.00029	0.00026	0.00022	0.00017	0.00012	0.00007
3400°	0.00003	0.00001	0.00000	0.00001	0.00003	0.00007	0.00011	0.00016	0.00020	0.00024
3500°	0.00026	0.00027	0.00026	0.00023	0.00020	0.00015	0.00011	0.00006	0.00003	0.00001

x	0°	10°	20°	30°	40°	50°	60°	70°	80°	90°
3600°	0.00000	0.00001	0.00003	0.00006	0.00010	0.00014	0.00018	0.00022	0.00024	0.00034
3700°	0.00023	0.00021	0.00018	0.00014	0.00010	0.00006	0.00003	0.00001	0.00000	0.00001
3800°	0.00003	0.00006	0.00009	0.00013	0.00017	0.00020	0.00021	0.00022	0.00021	0.00019
3900°	0.00016	0.00013	0.00009	0.00005	0.00002	0.00001	0.00000	0.00001	0.00002	0.00005
4000°	0.00008	0.00012	0.00015	0.00018	0.00020	0.00020	0.00019	0.00018	0.00015	0.00012
4100°	0.00008	0.00005	0.00002	0.00001	0.00000	0.00001	0.00002	0.00005	0.00008	0.00011
4200°	0.00014	0.00016	0.00018	0.00018	0.00018	0.00016	0.00014	0.00011	0.00007	0.00004
4300°	0.00002	0.00001	0.00000	0.00001	0.00002	0.00004	0.00007	0.00010	0.00013	0.00015
4400°	0.00016	0.00017	0.00016	0.00015	0.00012	0.00010	0.00007	0.00004	0.00002	0.00000
4500°	0.00000	0.00000	0.00001	0.00004	0.00007	0.00010	0.00012	0.00014	0.00015	0.00016
4600°	0.00015	0.00014	0.00012	0.00009	0.00006	0.00004	0.00062	0.00000	0.00000	0.00000
4700°	0.00002	0.00004	0.00006	0.00009	0.00011	0.00013	0.00014	0.00014	0.00014	0.00013
4800°	0.00011	0.00008	0.00006	0.00004	0.00002	0.00000	0.00000			

Appendix 3

LASER OPERATION

Lasers (*L*ight *A*mplification by *S*timulated *E*mission of *R*adiation) are frequently used as light sources for optical data processing and therefore a basic understanding of the operation of a laser can be useful. The discussion will be limited to explaining only the most basic concepts of laser operation. Helium-neon (He-Ne) gas lasers are used quite extensively in optical processing because of their availability, stability, continuous visible output (versus pulsed output), and relatively low cost. Most often in optical data processing lasers are not used as light amplifiers but as oscillators (the feedback being accomplished optically with mirrors as will be described). Our discussion will therefore be confined to helium-neon gas lasers used as oscillators.

Figure A3.1 shows a quartz tube filled with helium-neon gas at low pressure. We know that neon signs emit red light when electrically excited. We see from Figure A3.1 that we have the same basic construction as used for neon signs. A high voltage can be applied between the two electrodes to excite the gas mixture. The gas mixture could also have

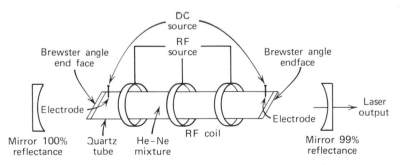

Figure A3.1   Basic configuration of He-Ne laser.

591

been excited by using radio frequency excitation, or a combination of high voltage and radio frequency excitation. The important point is that the gas mixture can be excited in some manner.

We know that atoms in a gas can be assumed to have discrete energy levels which are associated with the circular orbit of its electrons. For simplicity in explaining the operation of the helium-neon laser we shall assume that only two discrete energy levels above ground will be associated with each atom. Collisions between photons and atoms will give rise to three processes; that is,

1. Absorption or Stimulated Excitation. Stimulated excitation occurs when a photon of sufficient energy collides with an atom in energy state 1 above ground—it will raise the atom to energy state 2.

2. Stimulated Emission. Stimulated emission occurs when a photon of sufficient energy collides with an atom in energy state 2 above ground. This will lower the atom's energy level to energy state 1 by having *two in-phase* photons (coherent radiation) emitted from the atom after collision; that is, the original incident photon to the atom continues after the collision *plus* one additional photon in phase with it being released when the atom goes to energy state 1.

3. Spontaneous Emission. Spontaneous emission can occur when an atom in energy state 2 reverts back to energy state 1. When the atom spontaneously drops from energy state 2 to energy state 1, it will emit one photon (incoherent radiation) in the process.

It is to be noted that in stimulated excitation the incident photon raised the atom's energy level to energy state 2. The photon that raised the atom to energy level 2 is released in phase with an incident photon during stimulated emission. In stimulated emission the colliding photon reduces the energy of the atom and is not absorbed. We can see therefore that the photon has a disturbing influence on the atom. If the atom is in a low energy state, the photon will try to raise the energy state to the higher level. If the atom in a higher energy state, the photon will try to reduce its energy state to the lower level.

In any gas mixture stimulated excitation, stimulated emission, and spontaneous emission are occurring simultaneously. The net effect of the three interactions will be related to the number of atoms existing in energy state 1 and 2, and to the number of photons available (to produce stimulated excitation and emission). We can assume that the atoms in the gas mixture will have energy levels 1 and 2 only. It is reasonable to assume that more atoms will be in the energy level 1 than in energy level 2. More atoms in energy level 2 would indicate the availability of excess photons (by stimulated emission) to produce a light output. Such a condition is usually referred to as a *population inversion* and is a requirement for lasing. We shall now consider how a population inversion can be created in a gas mixture so that lasing can occur.

Excitation of the helium-neon mixture by electric discharge or by radio frequency excitation will create free electrons that will be accelerated by the applied field. Some of the free electrons will collide with helium atoms in the ground state and raise them to excited states.

Figure A3.2 shows that electron collision with helium will raise their level from the ground state to either of two excited states (indicated by level 1 or level 2). Level 2 for helium is called a *metastable state*. A metastable state is one that has a relatively long lifetime. Helium atoms therefore (which tend to remain) in this excited state are quite likely to collide with unexcited neon atoms before losing their energy. The energy level 2 for helium is nearly the same as the energy level 2 for neon. A collision of an excited helium atom with a ground level neon atom will therefore have a high probability of transferring its energy. A successful transfer of energy from the excited helium atom (at energy level 2) to an unexcited neon atom (at ground level) is represented by the helium atom returning to ground level and the neon atom being excited to the neon energy level 2 state. The excited neon atom (energy level 2) will emit red visible light (6328 Å) by stimulated emission when going to energy level 3 and will eventually return to the ground state permitting the process to repeat itself.

In order for lasing to be continuous it is necessary to excite (by resonant collisions with excited helium atoms) the neon atoms to energy level 2 at a faster rate than they are converting to energy level 3. This is

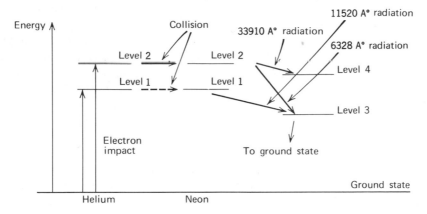

Figure A3.2   Energy levels of He-Ne.

necessary in order to maintain the population inversion. If the lifetime of the excited neon atom at energy level 2 is substantially longer (which it is) than the lifetime of the neon atom in energy level 3 the population inversion will be maintained.

Let us now consider an excited helium-neon gas mixture in the quartz tube mounted between two reflectors arranged as an interferometer. Let us assume a photon has been released in the tube towards the 100-percent reflectance mirror in Figure A3.1. If this photon encounters a neon atom in energy level 1, it will raise it to energy level 2. The neon atom raised to energy level 2 will tend to stay there because of its relatively long lifetime. If the photon encounters a neon atom in energy level 2, it will lower it to energy level 1 and one additional in-phase (with the incident) photon will be released in the process. The two in-phase photons will continue traveling toward the 100-percent reflectance mirror. These two photons will now continue the process of the original photon. We can see that either the number of photons will increase or the number of neon atoms in energy level 2 will increase by the time the process reaches the end of the tube. The light leaving the end of the tube will be reflected from the 100-percent reflectance mirror back into the tube where the process will continue. At the other end of the tube 99-percent of the output light will be reflected back into the tube (1 percent is transmitted as an output). We can see that at a certain point the photons traveling through the glass tube will upon being reflected from one mirror to the other encounter primarily excited neon atoms in energy level 2. At this point there will be essentially only photons being added in phase to the photon stream as collision with neon atoms at energy level 2 converts them to energy level 3. Red light (6328 Å) will be produced as long as the population inversion can be maintained.

The helium-neon gas is excited by an external source to invert the population of a pair of energy levels (level 2 and level 1) of energy difference ΔE. An electromagnetic wave of frequency $f = \Delta E/h$ where h is Planck's constant will then produce stimulated emission of radiation at the frequency f. The two mirrors and the laser tube act as interferometers tuned to the frequency f; that is, the laser acts as a resonant cavity and allows energy to build up. When the energy in the interferometer exceeds the energy absorbed or transmitted, lasing will take place. The amount of power transmitted is related to the rate at which the gas can be excited to maintain the population inversion. The lasing tends to take place at the interferometer frequency nearest the center of the emission line. Two other near infrared outputs can be obtained from a helium-neon laser (11,520 Å and 33,910 Å). These outputs are produced in a similar manner to that of the 6328 Å output. From Figure A3.2 we can see that the

33,910 Å output starts at the neon energy level but goes to a different energy level than when 6328 Å is radiated.

In Figure A3.1 we can see that the end faces of the quartz tube are placed at the Brewster angle. We have previously seen that 4 percent of light perpendicularly incident to a glass-air (or air-glass) boundary surface will be reflected back on itself. In order to select a particular laser oscillation mirrors are used that are highly reflective at the desired wavelengths.

Let us consider Figure A3.3 which shows perpendicularly incident light to surface A. At surface A approximately 4 percent of the light will be reflected back on itself. At surface B approximately 4 percent of the perpendicularly incident light will also be reflected back on itself. The perpendicularly incident light leaving the tube will therefore be approximately 8 percent less than the perpendicularly incident light to surface A. On reflection from the mirror back to the tube there will be another 4-percent loss on surface B and another 4-percent loss on surface A before it will enter the tube. There is a total loss of approximately 16 percent in the light leaving the tube before it will again enter the tube.

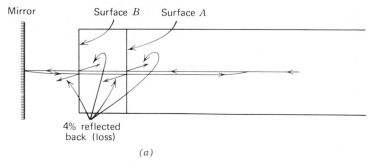

(a)

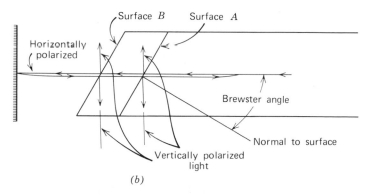

(b)

Figure A3.3.

This large loss at each end of the tube is sufficient to prevent lasing action. By mounting the end face at the Brewster angle these large losses can in part be prevented. Incident light at the Brewster angle to the end face surface A as shown in Figure A3.3B is broken into two parts. The horizontally polarized light is transmitted essentially without loss through the surface A interface. About 15 percent of the vertically polarized light to surface A is reflected outside the cavity. This means the light incident at the Brewster angle to surface B will be mainly horizontally polarized. The light incident to surface B will be at the Brewster angle (after refraction) and again 15 percent of the vertically polarized light will be reflected outside the cavity. Similar reflections will occur to the light reflected from the mirrors back to the tube. We can see therefore that the horizontally polarized light will pass the end faces with essentially no reflection losses. The vertically polarized light will eventually be reflected out of the cavity and therefore cannot cause lasing action. The light output from the laser is therefore nearly totally horizontally polarized. The Brewster angle is usually selected for the single-frequency axial mode. Other modes will also be transmitted with some attenuation.

The end mirrors of the laser can be plane or spherical or a combination of both. Two flat mirrors are generally not used because of the extremely high requirement for flatness (1/100 of a wavelength) and the critical alignment required. On the helium–neon laser the helium is used to enhance the excitation of the neon atom, although lasing can be obtained (with lower power output) using pure neon. For the laser we have described there will be several different discrete frequencies of oscillation. For optical data processing and holography we desire only one axial or longitudinal mode of oscillation. The discrete frequencies of oscillation are given by

$$\Delta V = \frac{c}{2L}$$

where c is the velocity of light in vacuum, L is the mirror separation, and ΔV is the frequency separation between discrete frequency output. For example,

$$L = 60 \, \text{cm}$$

$$\Delta V = \frac{3 \times 10^{10}}{120} = 25 \times 10^7 \, \text{cycles}$$

The addition of a Fabry-Perot etalon into the cavity can tune it to a single-mode frequency. Naturally the output will be lower since only one axial-mode output will be obtained and the other mode outputs will be lost.

A multimode laser can be made to be a single-mode laser by reducing the power output—that is, by reducing the power into the laser. The reason for this is that the cavity is essentially tuned to the center or axial frequency. Any other output frequency will be reduced. By reducing the power into the tube lasing is just maintained for the axial frequency without enough power into the tube to maintain lasing at other frequencies. A photocell connected to an electronic spectrum analyzer can be used to check the frequency content of the laser beam.

Appendix 4

THE MICHELSON INTERFEROMETER

The Michelson interferometer is a device that can be used to determine the degree of temporal coherence of essentially monochromatic light sources. The Michelson interferometer (Figure A4.1) causes light from a source S to be split into two beams by the action of the lower surface of glass plate O (which is half silver). One of these light beams is formed by light passing through the silvered back surface of plate O. This beam also

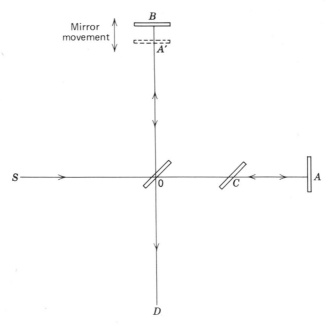

Figure A4.1   Michelson interferometer.

passes through plate C and falls on the front surface of mirror A. The portion of light reflected by the back surface of plate O forms the second beam which falls on the front surface of mirror B. The reflected beams from the two mirrors (A and B) recombine at the half silvered surface of O and enter a detector at point D.

The beam that goes to mirror B passes through plate O three times while the beam going to mirror A passes through plate O only once. The compensating plate C is usually inserted in a light beam going to mirror A in order to make the arrangements symmetrical. Mirror B is mounted on a slide that can be moved by a micrometer screw in the direction shown by the arrow.

An observer at point D sees the image of mirror A in mirror O at A'; that is, light from mirror A appears to the observer to be originating at A'. By moving mirror B with the micrometer screw, variations in intensity can be produced by interfering the light from mirrors A and B which is going to the observer at D. The difference in path lengths AO and BO permit different wavefronts from the original light source to interfere with each other.

The contrast of the intensity variations was defined by Michelson as $(I_{max} - I_{min})/(I_{max} + I_{min})$, where I_{max} is the maximum intensity produced when the light from mirrors A and B interfere constructively and I_{min} is the minimum intensity produced when the light from mirrors A and B interfere destructively.

Appendix 5

BRAGG EFFECT

The amount that a wave is diffracted is dependent on the wavelength. Light which has very short wavelengths is diffracted only slightly. X-rays which have extremely short wavelengths are therefore diffracted even less than light waves. An experiment to diffract x-rays based on the principle that crystals should be natural diffraction gratings (because of the close and symmetrical spacing of their atoms) was proposed by Max von Laue in 1912. This experiment was successful and provided understanding of both the nature of x-rays and the structure of crystals. Although the original experiment was performed by directing a narrow beam of x-rays on a crystal, the same effects can help explain certain effects that occur with holograms.

By varying the direction of the beam it is found that at certain discrete values of the glancing angle θ there will be peaks in the intensity of the diffracted beam. The diffracted beam also makes an angle θ with the crystal and can be detected with photographic film. The results were used to substantiate the wavelike nature of x-rays and the orderly arrangement of the atoms in crystals since the diffracted beam will have peak intensity only when waves from the various crystal planes reinforce each other.

Sir William H. Bragg and his son Sir William L. Bragg developed the law for crystal diffraction. Figure A5.1 shows a representation of the atoms in a crystal. A monochromatic beam of x-rays from a source S is directed on the crystal at a small angle θ. Reflection occurs at an angle of reflection that is equal to the angle of incidence at the various crystal planes. Each reflected ray is retarded more than the ray on the plane above it by an amount

$$CB + BD = 2AB \sin \theta$$

where AB is the spacing between the atomic planes.

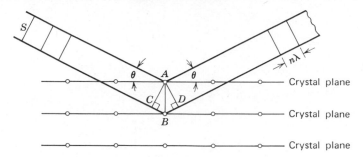

Figure A5.1   X-ray diffraction by a crystal.

This retardation must be an integral number of wavelengths if reinforcement is to occur and therefore

$$2AB \sin \theta = n\lambda$$

where n is an integer corresponding to the order of the diffracted beam.

From this we can see that the monochromatic x-rays will constructively interfere when

$$\lambda = 2AB \sin \theta$$

but there will also be reinforcement when the pathlength difference is $2\lambda, 3\lambda, \ldots$, etc. By referring to Figure A5.1 we can see that the x-rays will be reflected by all equally spaced atomic planes. Whenever the path difference between planes is one wavelength, the rays reflected from these layers will reinforce each other. This can be seen by noting that the distance between the first and second plane is λ, first and third plane is 2λ, first and fourth plane is 3λ, and therefore reflected rays from all these planes will reinforce each other for the direction in which the conditions of the Bragg equation are forfulled, that is,

$$2AB \sin \theta = n\lambda$$

From this relationship it is possible to determine either the atomic crystal spacing or the wavelength of the x-rays.

Appendix 6

MOIRÉ PATTERNS

Moiré patterns are figures produced when a family of curves are overlayed with another family of curves; that is, when a repetitive pattern is overlayed on another repetitive pattern, a moiré pattern is produced. Let us consider two series of Ronchi rulings as shown in Figure A6.1(a). This figure shows two Ronchi rulings whose lines are oriented perpendicular to each other. Each of the two Ronchi rulings can be seen at the edge of the figure, while the main central portion of the figure shows the intersection or overlap of the two rulings. It can be seen that the intersection of the black ruled lines of two perpendicularly crossed Ronchi rulings forms small squares as seen in Figure A6.1(a).

Figure A6.1(b) shows the effect of making the angle between the two Ronchi rulings less than 90 degrees. It can easily be seen from the figure that as the angle between the two Ronchi rulings becomes more than 90 degrees the squares become diamond shaped. It can also be deduced that the short diagonal of the diamond will bisect the angle between the two Ronchi rulings.

Figure A6.1(c) shows that as the two Ronchi rulings become parallel the diamond elongates. As the diamonds elongate, as shown in Figures A6.1(a) through A6.1(j), lines (moiré pattern) tend to appear which can also be considered as magnified images of the original Ronchi rulings. These magnified images can be used to measure very precisely the angle between the Ronchi rulings. In addition any irregularities in these magnified images would be caused by irregularities in the Ronchi rulings. The moiré pattern technique can therefore be used to determine the quality of Ronchi rulings.

Figure A6.2(a) through (f) shows the moiré patterns produced by two sets of intersecting lines. Essentially the same effects are produced with

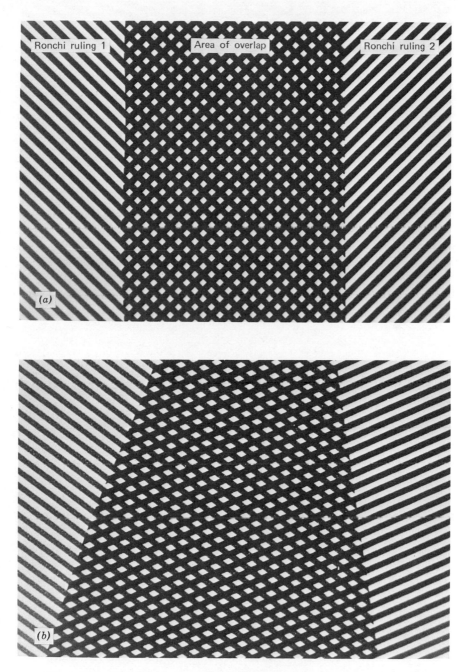

Inside image: Ronchi ruling 1   Area of overlap   Ronchi ruling 2

(a)

(b)

Figure A6.1   Moiré patterns produced by the intersection of two Ronchi rulings.

Figure A6.1c.

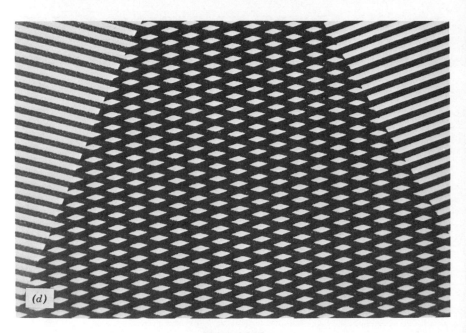

Figure A6.1d.

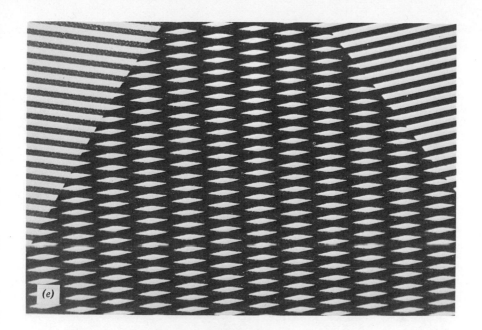

Figure A6.1e.

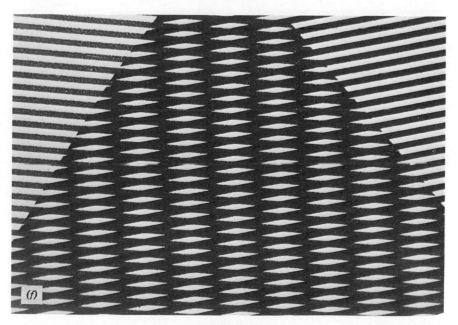

Figure A6.1f.

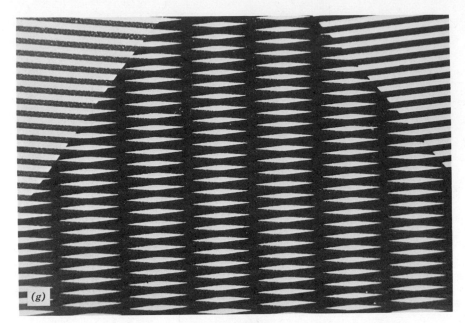

Figure A6.1g.

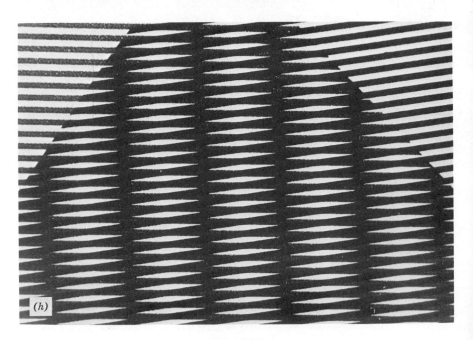

Figure A6.1h.

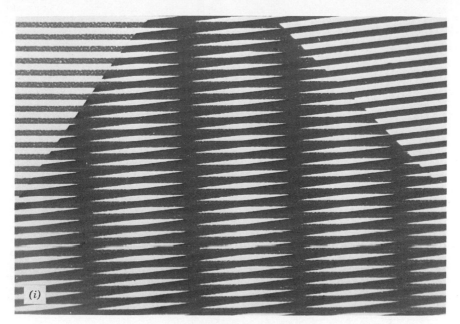

Figure A6.1*i*.

Figure A6.1*j*.

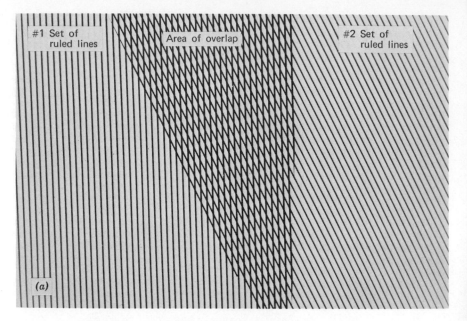

Figure A6.2   Moiré patterns produced by overlapping lines.

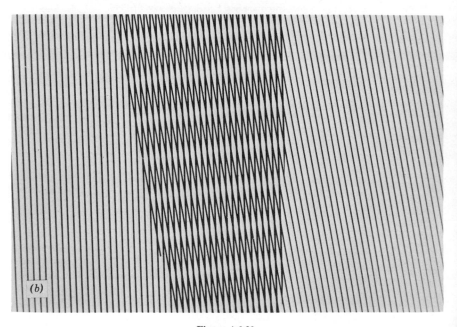

Figure A6.2*b*.

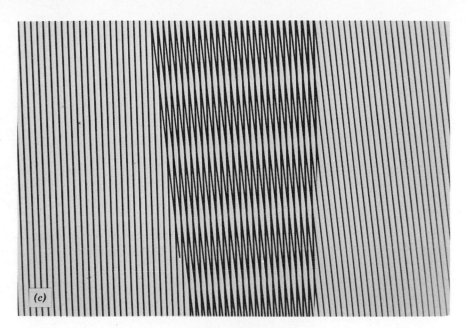

Figure A6.2c.

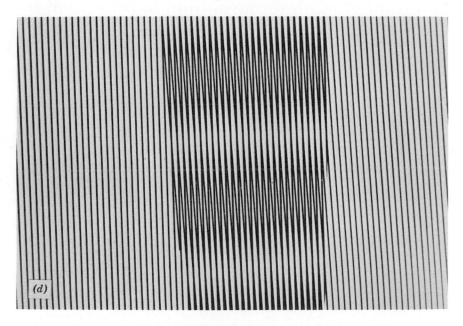

Figure A6.2d.

609

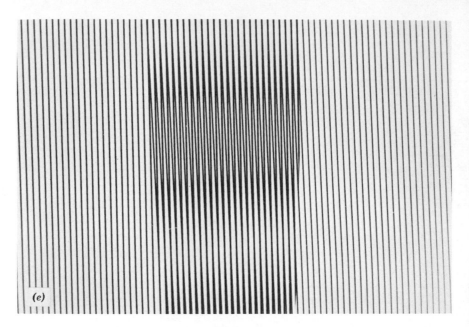

Figure A6.2e.

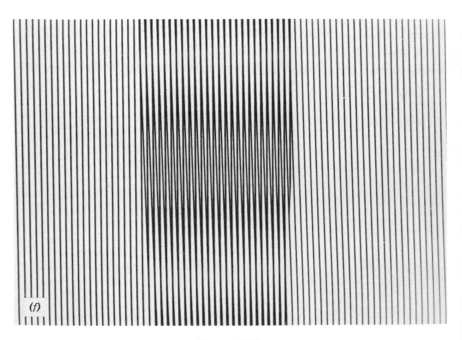

Figure A6.2f.

the ruled lines as are produced with the Ronchi rulings. As the lines become more parallel the moiré pattern line image becomes larger and any
irregularities become more apparent.

Figure A6.3(a) through (e) shows the moiré pattern produced by overlapping two identical sets of dot patterns. It can again be seen that as the
dot pattern orientations tend to coincide the moiré image pattern tends to
enlarge. Figure A6.4(a) through (e) shows that similar effects can be
produced by overlapping two identical sets of rectangular patterns.

Figure A6.5 shows the moiré pattern produced by overlapping sets of
evenly spaced concentric circles. Figure A6.5(a) shows one set of concentric circles that was used to produce the moiré patterns. Figure A6.5(b)
shows a double exposure of the concentric circles shown in Figure A6.5(a).
By moving the set of circles between exposures the two sets of concentric circles appear and their intersections produce the moiré patterns.
Figure A6.5(c) is an underexposed print of Figure A6.5(b) which shows
only the points of intersection (moirés pattern) between the two sets of
concentric circles. It clearly shows that the moiré pattern is the locus of
points of intersection of two overlapping families of curves. Figures
A6.5(d) and A6.5(e) show similar moiré patterns produced by varying the
spacing between the centers of the two sets of concentric circles.

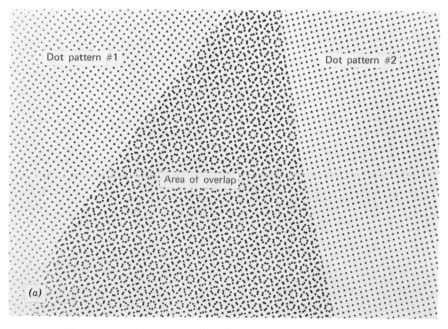

Figure A6.3   Moiré patterns produced by overlapping dot patterns.

Figure A6.3b.

Figure A6.3c.

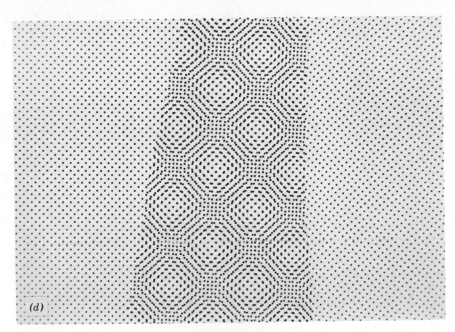

Figure A6.3*d*.

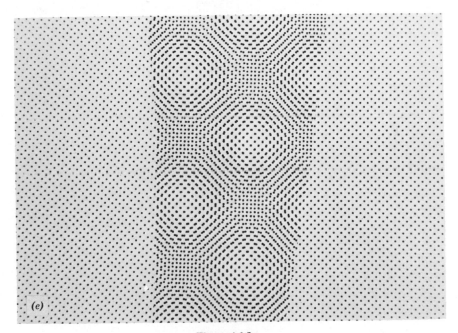

Figure A6.3*e*.

613

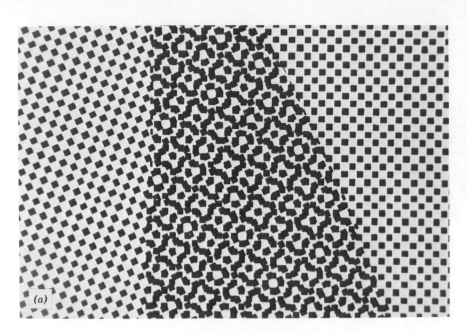

Figure A6.4a Moiré patterns produced by overlapping identical sets of rectangular patterns.

Figure A6.4b.

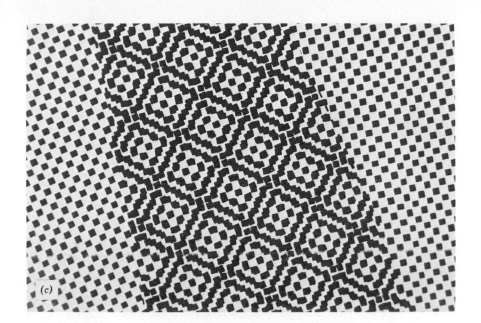

Figure A6.4c.

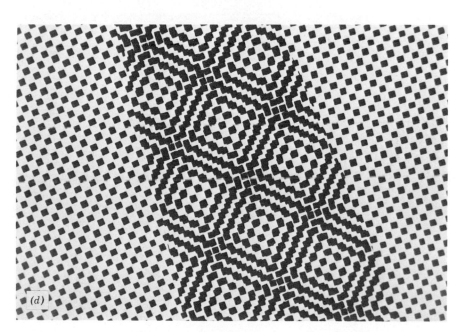

Figure A6.4d.

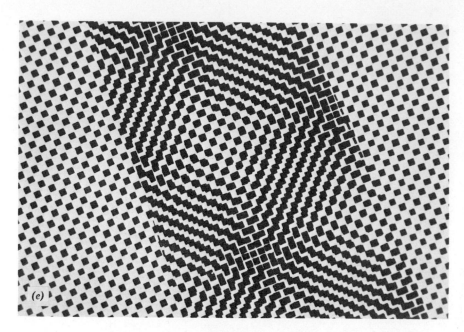

Figure A6.4e.

Figure A6.5a   Moiré patterns produced by overlapping evenly spaced concentric circles.

Figure A6.5*b*.

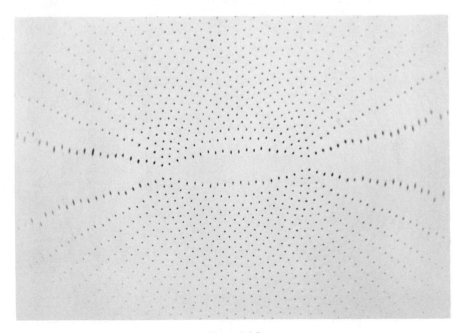

Figure A6.5*c*.

Figure A6.5*d***.**

Figure A6.5*e***.**

Figure A6.6(a) and (b) show the moiré patterns produced by over-lapping a set of evenly spaced lines with evenly spaced concentric circles. The photographs of Figure A6.6 were also made by double exposure.

There are a number of physical problems whose mathematical solution is identical in form to that describing a moiré pattern; for example, many types of diffraction patterns can be determined through the use of moiré patterns. The interference pattern between two coherent light beams can be determined through the use of moiré patterns. The interference pattern produced by the interference of two plane waves on a screen can be determined by scaling the spacing between equispaced parallel lines to the wavelength of the plane waves in the light beam.

Figure A6.2(a) can be used to determine the interference pattern produced by two plane waves on a screen. By assuming that each line is a maximum plane wavefront we can determine points of constructive or destructive interference in the area of overlap at any specific instant of time. If we consider the no. 1 and no. 2 set of ruled lines to be plane wavefronts in two different collimated beams of light, each traveling toward the area of overlap, we can easily determine the points of con-

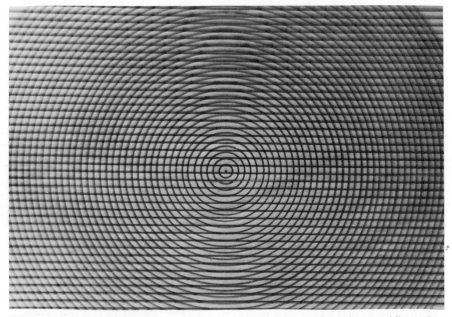

Figure A6.6a Moiré patterns produced by overlapping of a set of evenly spaced lines with evenly spaced concentric circles.

Figure A6.6b.

structive interference anywhere in the area of overlap. Let us consider Figure A6.2(a) to be a picture of the maximums of plane waves at a specific instant of time. Because in this case the plane wavefronts are evenly spaced (representing a temporally coherent collimated light beam), the locus of points of constructive interference in the area of overlap will remain fixed; that is, the next instant of time would show the wavefront maximums to have moved, producing points of intersection at different locations on the straight line locus through the points of intersection in Figure A6.2(a). Only when each wavefront has traveled a full wavelength from the positions shown will the wavefronts again intersect at the original points of intersection. The points of constructive interference on any screen placed in the area of overlap can be easily found. We can see therefore that the moiré patterns produced by placing the scaled equispaced parallel lines on equispaced parallel lines can represent the scaled interference pattern that can be expected by the interference of two collimated coherent plane light beams.

We can use Figure A6.5(a) to represent maximums of spherical waves (as a specific instant of time) radiating out from a temporally coherent point light source. Figure A6.5(b) shows the interference of two temporally coherent point light sources at a specific instant of time. It turns out that when the wavefronts of the two interfering light sources are identical to each other, the moiré pattern produced will be hyperbolas (when the displacement between the two identical patterns is small). When the displacement between the two identical patterns becomes large, the interference patterns becomes ellipses. Figure A6.6 can be used to show the interference pattern produced by temporally coherent spherical waves interfering with a collimated beam of temporally coherent light (plane waves).

Appendix 7

SOME COMMON DIFFRACTION PATTERNS

Appendix 7 shows photographs of diffraction patterns formed by uniformly illuminating varying types of apertures with monochromatic light. Figure A7.1 shows the far field pattern (i.e., the diffraction pattern formed a great distance away from the aperture) formed by uniformly illuminating with monochromatic light the rectangular aperture shown in Figure A7.2. The main diffraction pattern (Fraunhofer, or far field) of a rectangular aperture is a pattern of light points forming a cross. A weaker secondary pattern of points is formed perpendicular to each of the main points of light. It can be seen from Figure A7.2 that the rectangular aperture is quite poor since it has burrs and rounded corners. This accounts for the bright light points in the diffraction pattern lying diagonally to the main cross of points making up the diffraction pattern of a perfect rectangular aperture. The nonuniformity of the diminishing intensities starting from the center of the pattern is also caused by the poor rectangular aperture; that is, the intensity of the light should follow a $(\sin x/x)^2$ curve.

Figure A7.3 shows the far field (Fraunhofer) diffraction pattern of a circular aperture which is sometimes called an *Airy pattern*. A circular aperture (not perfectly circular which accounts for the rings in the Airy pattern not being truely circular) was uniformly illuminated by monochromatic light.

The Airy pattern is extremely important in optical data processing applications. It is important to realize that a small circular aperture does not transmit or act as a point source but acts as a source transmitting an Airy pattern. Only when using the central portion of the Airy pattern can the circular aperture be considered a good approximation to a point source.

622

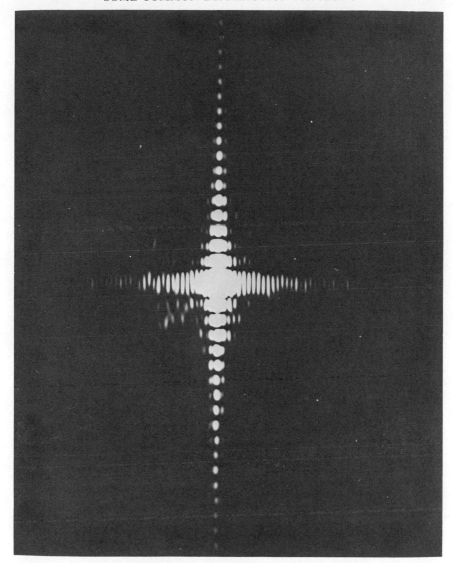

Figure A7.1 Diffraction pattern of rectangular aperture.

Figure A7.4 shows the diffraction pattern formed by uniformly illuminating a slit with monochromatic light. The slit acts as a source of cylindrical waves only when using the main central portion of the diffraction pattern. Just as the circular aperture cannot be considered as a true point source of spherical wave, so the slit cannot be considered as a true

Figure A7.2   Rectangular aperture.

source of cylindrical waves (because of the diffraction patterns formed by them).

Figure A7.5 shows the diffraction pattern formed by partially obscuring a monochromic point light source by a straight edge. Some light diffracts into the geometrical shadow and also causes the bands to form in the regions outside the geometric shadow.

Figure A7.3 Airy pattern of a circular aperture.

Figure A7.4 Diffraction pattern formed by uniformly illuminating a slit with monochromatic light.

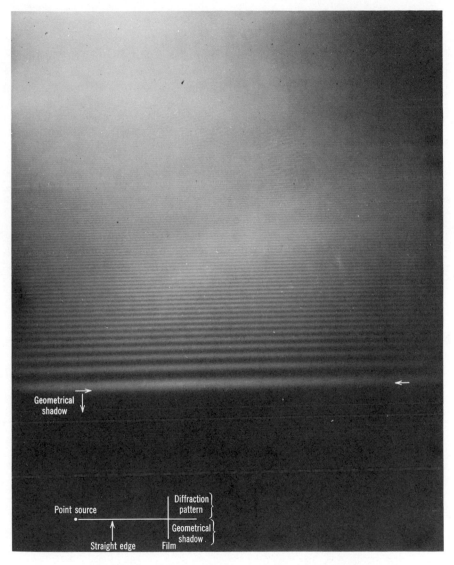

Figure A7.5   Diffraction pattern formed by a straight edge.

Appendix 8

Figure A8.1 shows the interference patterns produced by illuminating a plane sheet of glass from a point light source as shown in Figure A8.2. Figure A8.2 shows a point source of monochromatic light illuminating a plane sheet of thin glass. The light reflected from the far surface (from the point source of light) produces a virtual image of the point source. This virtual image will appear to be closer to the glass than the actual point source of light. The circular interference pattern produced in Figure A8.1 is therefore the result of the interference of two coherent point light sources.

Figure A8.1 shows the interference pattern produced by two monochromatic coherent point light sources. The difference in the interference patterns produced by the two pinholes as shown in Figure 2.15 and Figure A8.1 is the relative locations of the pinholes with respect to the screen. Figure A8.2 shows the configuration for producing the interference pattern of Figure A8.1 and Figure A8.3 shows the interference pattern produced by two similarly illuminated pinholes (with monochromatic light) each equally distant from the films.

Figure A8.4 shows the pattern produced by the interference between plane and spherical coherent light beams. To produce this photograph a source of monochromatic light was divided into two light beams by a beam splitter (plane glass); one beam was directed through a pinhole to the photographic film and the other beam was collimated (producing plane wavefronts). The Airy pattern of the spherical wave being produced by a small pinhole is evident in the photograph.

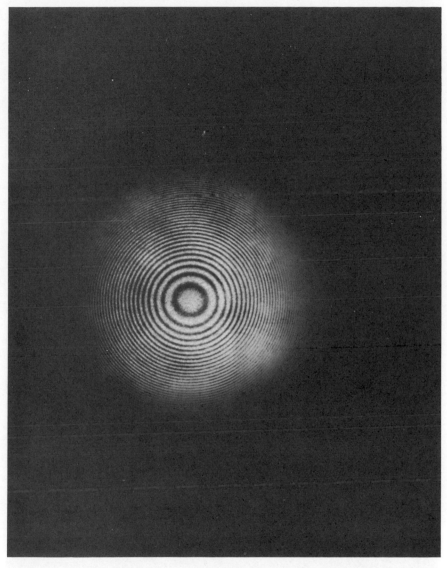

Figure A8.1 Circular interference patterns produced by two monochromatic point light sources.

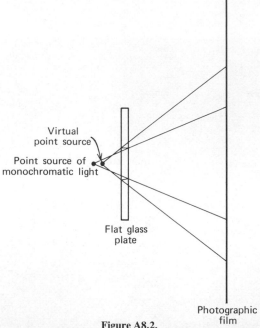

Figure A8.2.

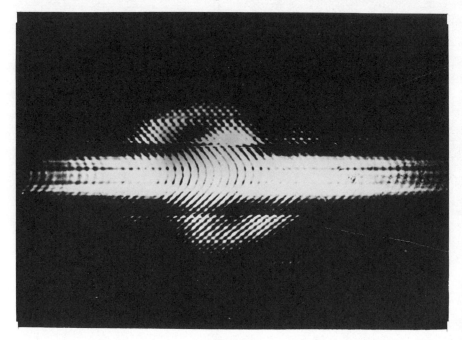

Figure A8.4.

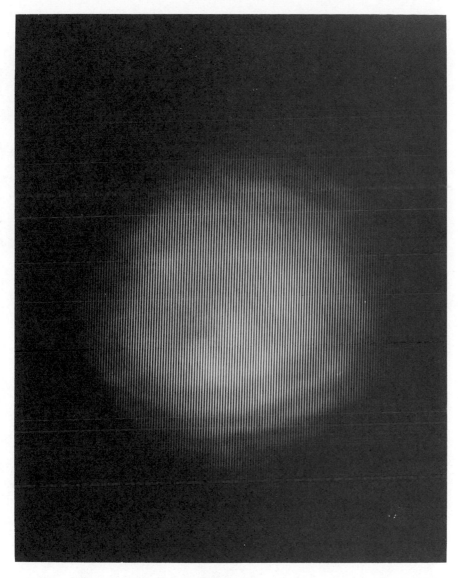

Figure A8.3.

Appendix 9

DERIVATION OF A_0 TERM IN EQUATION (3-1)

The equation expressing the components of a periodic wave is given by the Fourier series as

$$y = f(x) = A_0 + A_1 \cos x + B_1 \sin x + \cdots A_n \cos nx + B_n \sin nx$$

$$(A9\text{-}1)$$

To determine A_0 (a) both sides of (A9-1) can be multiplied by dx without changing the equality, and (b) both sides of (A9-1) can be integrated between the limits of 0 and 2π (one cycle) without changing the equality. Equation (A9-1) then takes the form

$$\int_0^{2\pi} y\, dx = \int_0^{2\pi} A_0\, dx + \int_0^{2\pi} A_1 \cos x\, dx + \int_0^{2\pi} B_1 \sin x\, dx$$

$$+ \int_0^{2\pi} A_2 \cos 2x\, dx + \int_0^{2\pi} B_2 \sin 2x\, dx$$

$$+ \int_0^{2\pi} A_3 \cos 3x\, dx + \int_0^{2\pi} B_3 \sin 3x\, dx$$

$$+ \cdots + \int_0^{2\pi} A_n \cos nx\, dx + \int_0^{2\pi} B_n \sin nx\, dx \qquad (A9\text{-}2)$$

Equation (A9-2) can be written as

$$\int_0^{2\pi} y\, dx = \int_0^{2\pi} A_0\, dx + \int_0^{2\pi} \sum_{n=1}^{n=\infty} (A_n \cos nx + B_n \sin nx)\, dx \quad (A9\text{-}3)$$

All terms involving sines and cosines in the last integral on the right will be zero; for example, the term

$$\int_0^{2\pi} B_n \sin nx\, dx = 0$$

The constant B_n can be brought outside the integral. This can be shown by the following:

$$\int_0^{2\pi} B_n \sin nx \, dx = B_n \int_0^{2\pi} \sin nx \, dx = B_n \left[\frac{(-\cos nx)}{n}\right]_0^{2\pi} = \frac{B_n}{n}[-(1-1)]$$

$$= \frac{B_n}{n}(0) = 0$$

Likewise the term

$$\int_0^{2\pi} A_n \cos nx \, dx = A_n \int_0^{2\pi} \cos nx \, dx = \frac{A_n}{n}[\sin nx]_0^{2\pi} = \frac{A_n}{n}(0-0) = 0$$

The area under the sinusoid over an integer number of cycles will always equal zero since the area under the sinusoid below the x-axis equals the area under the curve above the x-axis.

Therefore in (A9-3)

$$\int_0^{2\pi} \sum_{n=1}^{n=\infty} (A_n \cos nx + B_n \sin nx) \, dx = 0$$

and then

$$\int_0^{2\pi} y \, dx = A_0 \int_0^{2\pi} dx = 2\pi A_0 \qquad\qquad\text{(A9-4)}$$

Solving for A_0 gives

$$A_0 = \frac{1}{2\pi} \int_0^{2\pi} y \, dx = \frac{1}{2\pi} \int_0^{2\pi} f(x) \, dx \qquad\qquad\text{(A9-5)}$$

The term A_0 is the average value of the function $f(x)$ over one cycle.

For a periodic waveform the average of one cycle is the same as the average if found for two cycles, three cycles, or n cycles (where n is an integer).

DERIVATION OF THE A_n TERM IN EQUATION (3-1)

The equation expressing the components of a periodic wave is given by the Fourier series as

$$y = f(x) = A_0 + A_1 \cos x + B_1 \sin x + \cdots A_n \cos nx + B_n \sin nx$$
$$\text{(A9-1)}$$

To find A_n (a) both sides of (A9-1) are multiplied by $\cos mx \, dx$, where m is an integer and, (b) both sides of (A9-1) are integrated between the limits of zero and 2π (one cycle). Equation (A9-1) then becomes

$$\int_0^{2\pi} y \cos mx \, dx = \int_0^{2\pi} A_0 \cos mx \, dx + \int_0^{2\pi} A_1 \cos x \cos mx \, dx$$
$$+ \int_0^{2\pi} B_1 \sin x \cos mx \, dx + \cdots$$
$$+ \int_0^{2\pi} A_n \cos nx \cos mx \, dx + \int_0^{2\pi} B_n \sin nx \cos mx \, dx$$
$$\text{(A9-6)}$$

Equation (A9-6) can be written in the form

$$\int_0^{2\pi} y \cos mx \, dx = \int_0^{2\pi} A_0 \cos mx \, dx$$
$$+ \int_0^{2\pi} \cos mx \left[\sum_{n=1}^{n=\infty} A_n \cos nx + B_n \sin nx \right] dx \quad \text{(A9-7)}$$

In (A9-7) the term

$$\int_0^{2\pi} A_0 \cos mx \, dx = 0$$

since it represents the area under a cosine wave for a complete cycle (or integer multiple m thereof). The two remaining terms

$$\int_0^{2\pi} \cos mx \sum_{n=1}^{\infty} A_n \cos nx \, dx$$

and

$$\int_0^{2\pi} \cos mx \sum_{n=1}^{\infty} B_n \sin nx \, dx$$

can be evaluated by taking note of the following trigonometric relationship and adding them as follows:

$$\cos (A+B) = \cos A \cos B - \sin A \sin B$$
$$\underline{\cos (A-B) = \cos A \cos B + \sin A \sin B}$$
$$\cos (A+B) + \cos (A-B) = 2 \cos A \cos B \quad \text{(A9-8)}$$

$$\sin (B+A) = \sin B \cos A + \cos B \sin A$$
$$\underline{\sin (B-A) = \sin B \cos A - \cos B \sin A}$$
$$\sin (B+A) + \sin (B-A) = 2 \sin B \cos A \quad \text{(A9-9)}$$

From (A9-8) we have

$$\cos A \cos B = \tfrac{1}{2} \cos (A+B) + \tfrac{1}{2} \cos (A-B) \quad \text{(A9-10)}$$

and from (A9-9) we have

$$\sin B \cos A = \tfrac{1}{2} \sin (B+A) + \tfrac{1}{2} \sin (B-A) \quad \text{(A9-11)}$$

Returning now to the last two terms of (A9-7) and considering only the nth term; that is, n is a particular integer,

$$\int_0^{2\pi} \cos mx\, [A_n \cos nx + B_n \sin nx]\, dx = \int_0^{2\pi} A_n \cos mx \cos nx\, dx$$

$$+ \int_0^{2\pi} B_n \cos mx \sin nx\, dx$$

$$(A9\text{-}12)$$

We can let $A = mx$ and $B = nx$ in (A9-10) and (A9-11) and substitute these values in expression (A9-12) above giving

$$\int_0^{2\pi} A_n \cos mx \cos nx\, dx + \int_0^{2\pi} B_n \cos mx \sin nx\, dx$$

$$= A_n \int_0^{2\pi} \frac{\cos(mx+nx)}{2}\, dx + A_n \int_0^{2\pi} \frac{\cos(mx-nx)}{2}\, dx$$

$$+ B_n \int_0^{2\pi} \frac{\sin(mx+nx)}{2}\, dx + B_n \int_0^{2\pi} \frac{\sin(nx-mx)}{2}\, dx$$

$$(A9\text{-}13)$$

Now we can let $m = n$ or $m \neq n$. Looking at only the last two terms of expression (A9-13) and when
_m ≠ n,_

$$B_n \int_0^{2\pi} \frac{\sin(mx+nx)}{2}\, dx + B_n \int_0^{2\pi} \frac{\sin(nx-mx)}{2}\, dx$$

$$= \frac{B_n}{2} \int_0^{2\pi} \sin(m+n)\, x\, dx + \frac{B_n}{2} \int_0^{2\pi} \sin(n-m)\, x\, dx$$

$$= \frac{B_n}{2}\left[-\frac{\cos(m+n)x}{(m+n)} \right]_0^{2\pi} + \frac{B_n}{2}\left[-\frac{\cos(n-m)x}{(n-m)} \right]_0^{2\pi} \qquad (A9\text{-}14)$$

Substituting the upper limit (2π) makes the cosine equal to plus one for any integer values of m and n; that is,

$$\cos 2\pi = \cos(m+n)(2\pi) = 1$$

where m and n are integers.

Substituting the lower limit (0) makes the cosine equal to plus one for any integer value of m and n. Therefore,

$$\frac{B_n}{2(m+n)}[-\cos(m+n)x]_0^{2\pi} + \frac{B_n}{2(n-m)}[-\cos(n-m)x]_0^{2\pi}$$

$$= \frac{B_n}{2(m+n)}[-(1-1)] + \frac{B_n}{2(n-m)}[-(1-1)]$$

$$= 0$$

This means that when $m \neq n$, the B_n terms in expression (A9-13) are zero.

$\underline{m = n.}$ When $m = n$, the B_n terms in expression (A9-13) become

$$B_n \int_0^{2\pi} \frac{\sin (mx + nx)}{2} dx + B_n \int_0^{2\pi} \frac{\sin (nx - mx)}{2} dx$$

$$= \frac{B_n}{2} \int_0^{2\pi} \sin (nx + nx) \, dx + \frac{B_n}{2} \int_0^{2\pi} \sin (nx - nx) \, dx$$

$$= \frac{B_n}{2} \int_0^{2\pi} \sin 2nx \, dx + \frac{B_n}{2} \int_0^{2\pi} \sin 0 \, dx \qquad\qquad \text{(A9-15)}$$

Since $\sin (0) = 0$ then

$$\frac{B_n}{2} \int_0^{2\pi} \sin (0) \, dx = 0$$

The value of the term $[B_n/2 \int_0^{2\pi} \sin 2nx \, dx]$ is found from

$$\int \sin V \, dV = -\cos V$$

by letting $V = 2nx$, then $dV = 2n \, dx$. Making the substitutions we obtain

$$\int_0^{2\pi} (\sin 2nx) \, dx = \int_0^{4n\pi} \sin V \frac{dV}{2n} = \frac{1}{2n} \int_0^{4n\pi} \sin V \, dV$$

since

$$\frac{1}{2n} \int \sin V \, dV = -\frac{1}{2n} (\cos V)$$

then

$$\frac{1}{2n} \int_0^{2\pi} \sin 2nx \, 2n \, dx = \frac{1}{2n} [-\cos 2nx]_0^{2\pi} = \frac{1}{2n} [-(1-1)] = \frac{1}{2n} (0) = 0$$

We have thus proven that in expression (A9-12) the value of the B_n terms is zero at all times (when $m = n$ or $m \neq n$). The value of the A_n terms remain to be evaluated.

$\underline{m \neq n.}$ In expression A9-13 the A_n terms when $m \neq n$, are given by

$$A_n \int_0^{2\pi} \frac{\cos (mx + nx)}{2} dx + A_n \int_0^{2\pi} \frac{\cos (mx - nx)}{2} dx$$

$$= \frac{A_n}{2} \int_0^{2\pi} \cos (m + n) \, x \, dx + \frac{A_n}{2} \int_0^{2\pi} \cos (m - n) \, x \, dx$$

$$= \frac{A_n}{2}\left[\frac{\sin (m+n)x}{(m+n)}\right]_0^{2\pi} + \frac{A_n}{2}\left[\frac{\sin (m-n)x}{(m-n)}\right]_0^{2\pi}$$

$$= \frac{A_n}{2}[(0-0)] + \frac{A_n}{2}[(0-0)] = 0$$

When $m \neq n$, the A_n terms in expression (A9-13) are zero. We now have left the determination of the value of A_n when $m = n$.

$\underline{m = n}$. The A_n term in expression (A9-13) are evaluated as

$$\frac{A_n}{2}\int_0^{2\pi} \cos (m+n)\, x\, dx + \frac{A_n}{2}\int_0^{2\pi} \cos (m-n)\, x\, dx = 0 + \frac{A_n}{2}\int_0^{2\pi} dx = \pi A_n$$

The first integral is zero (i.e., $\sin 0 = \sin 2\pi = 0$) and the second term becomes a simple integral since $\cos 0 = 1$.

Now let us reexamine (A9-7),

$$\int_0^{2\pi} y \cos mx\, dx = \int_0^{2\pi} A_0 \cos mx\, dx$$

$$+ \int_0^{2\pi} \cos mx \left[\sum_{n=1}^{n=\infty} A_n \cos nx + B_n \sin nx \right] dx \qquad \text{(A9-7)}$$

or

$$\int_0^{2\pi} y \cos mx\, dx = \int_0^{2\pi} A_0 \cos mx\, dx + \int_0^{2\pi} \left[\sum_{n=1}^{n=\infty} A_n \cos mx \cos nx\, dx \right.$$

$$\left. + B_n \cos mx \sin nx \right] dx \qquad \text{(A9-16)}$$

In regard to (A9-16) we have found
1. The value of the A_0 term is zero for all values of m, except $m = 0$.
2. The value of the B_n term is zero for all values of m and n – that is, when $m \neq n$, and when $m = n$.
3. The value of the A_n term is zero when $m \neq n$.
4. When $m = n$ the value of the A_n term is $A_n \pi$.
We can now write (A9-16) as

$$\int_0^{2\pi} y \cos mx\, dx = A_m \pi \qquad \text{(A9-17)}$$

or

$$A_m = \frac{1}{\pi} \int_0^{2\pi} y \cos mx\, dx \qquad \text{(A9-18)}$$

This same result could have been reasoned out before making the integration by looking at expression (A9-13). We have previously seen that the area under a sinusoidal curve over one period (or integer multiple thereof) is zero. Using this reasoning we examine the four terms of expression (A9-13).

1. $A_n \int_0^{2\pi} \dfrac{\cos{(mx+nx)}}{2} dx$

When $m = n \neq 0$ and when $m \neq n$, this term must equal zero because we are integrating over an integral number of cycles of a cosine wave.

2. $A_n \int_0^{2\pi} \dfrac{\cos{(mx-nx)}}{2} dx$

When $m \neq n$, this term must be zero for the reason noted above. When $m = n$, we have $\cos 0° = 1$ and the integral becomes

$$\frac{A_n}{2} \int_0^{2\pi} dx = A_n \pi$$

3. $B_n \int_0^{2\pi} \dfrac{\sin{(mx+nx)}}{2} dx$

When $m = n$ and when $m \neq n$, this term must equal zero because we are integrating over an integral number of cycles of a sine wave.

4. $B_n \int_0^{2\pi} \dfrac{\sin{(nx-mx)}}{2} dx$

When $m \neq n$, this term must equal zero because we are integrating over an integral number of cycles of a sine wave. When $m = n$, the integral is zero because $\sin 0 = 0$ and the integral is therefore zero.

Using the same reasoning, the A_0 term in (A9-7) is zero which permits (A9-16) to be written directly into (A9-17),

$$\int_0^{2\pi} y \cos nx \, dx = A_n \pi$$

By following the same procedure but multiplying (A9-1) by $\sin mx$, the value of B_n can be found to be

$$B_n = \frac{1}{\pi} \int_0^{2\pi} y \sin nx \, dx \qquad \text{(A9-19)}$$

By using equations (A9-18) and (A9-19) we can find each of the coefficients in (A9-1) to be as follows:

$$A_1 = \frac{1}{\pi} \int_0^{2\pi} f(x) \cos x \, dx \qquad\qquad (A9\text{-}20)$$

$$B_1 = \frac{1}{\pi} \int_0^{2\pi} f(x) \sin x \, dx \qquad\qquad (A9\text{-}21)$$

$$A_2 = \frac{1}{\pi} \int_0^{2\pi} f(x) \cos 2x \, dx \qquad\qquad (A9\text{-}22)$$

$$B_2 - \frac{1}{\pi} \int_0^{2\pi} f(x) \sin 2x \, dx \qquad\qquad (A9\text{-}23)$$

$$A_3 = \frac{1}{\pi} \int_0^{2\pi} f(x) \cos 3x \, dx \qquad\qquad (A9\text{-}24)$$

$$B_3 = \frac{1}{\pi} \int_0^{2\pi} f(x) \sin 3x \, dx \qquad\qquad (A9\text{-}25)$$

$$A_n = \frac{1}{\pi} \int_0^{2\pi} f(x) \cos nx \, dx \qquad\qquad (A9\text{-}26)$$

$$B_n = \frac{1}{\pi} \int_0^{2\pi} f(x) \sin nx \, dx \qquad\qquad (A9\text{-}27)$$

and we previously found.

$$A_0 = \frac{1}{2\pi} \int_0^{2\pi} f(x) \, dx \qquad\qquad (A9\text{-}5)$$

The amplitudes of the harmonics present in the original function are
represented by the A_n and B_n terms and are known as *Fourier coefficients*.
The A_0 term represents the steady state value (or DC value).

In the sinusoidal waveform as shown in Figure A9.1 positioning the
origin at A makes the sinusoidal waveform a cosine function. Placing the
origin at B makes the sinusoidal waveform into a sine function. Obviously
the position of the origin does not change the actual sinusoidal waveform,
it only determines the reference point from which calculations are made.

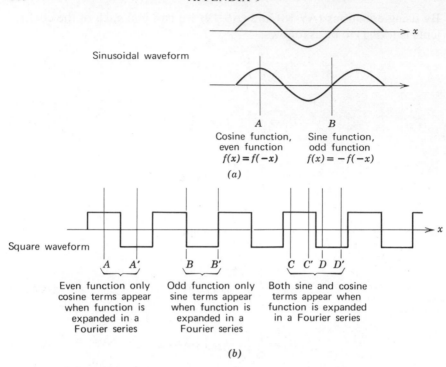

Figure A9.1 shows: Sinusoidal waveform

A
Cosine function,
even function
$f(x) = f(-x)$

B
Sine function,
odd function
$f(x) = -f(-x)$

(a)

Square waveform

A   A'   B   B'   C C' D D'

Even function only   Odd function only   Both sine and cosine
cosine terms appear   sine terms appear   terms appear when
when function is   when function is   function is expanded
expanded in a   expanded in a   in a Fourier series
Fourier series   Fourier series

(b)

Figure A9.1   Effect of varying the position of the origin on a Fourier.

The square waveform in Figure A9.1(b) shows that placing the origin at A or A' makes the square wave function into what is referred to as an *even function*. All even functions contain only the cosine terms, of (A9-1), when expanded in a Fourier series. Placing the origin at either B or B' makes the square wave function into what is referred to as an *odd function*. In all odd functions only sine terms appear when the function is expanded in a Fourier series (A9-1). When the origin is placed at positions such as C, C', D, or D', both sine and cosine terms appear when the function is expanded in a Fourier series (A9-1). A function is said to be an *odd function* if

$$f(t) = -f(-t) \qquad\qquad (A9\text{-}28)$$

and an even function if

$$f(t) = f(-t) \qquad\qquad (A9\text{-}29)$$

In practice functions are often encountered that are not odd but become odd after subtraction of the DC component. When physical measurements are taken, such as with a wave analyzer, the actual readings of the values of A_n and B_n can not be made. With a wave analyzer a value C_n is read for the coefficient of the nth harmonic that is related to A_n and B_n as follows:

$$C_n = \sqrt{A_n^2 + B_n^2} \tag{A9-30}$$

The value of C_n is independent of the position of the waveform with respect to the origin.

Appendix 10

Figure A10.1(a) shows the function $y = \sin \omega t$ plotted on a y and t axis and Figure A10.1(b) shows the same function plotted on a y and ωt axis.

When the same function $y = \sin \omega t$ is plotted on y and ωt as shown in Figure A10.1(b), we can see that the period $T = 2\pi$. The differences between the sine function graphs is the scale chosen for the abscissa.

In Figure A10.1(a) the abscissa chosen was t, and therefore the period is given by $T = 2\pi/\omega$. The value for y for a given value of ωt is the same in both graphs (Figures A10.1(a) and A10.1(b)). Choice of axes does affect the manner in which analysis of functions are treated.

Consider, for example, the term A_0 given in equation (3-8),

$$A_0 = \frac{1}{2\pi} \int_{x=0}^{x=2\pi} f(x) \, dx \tag{3-8}$$

If we again make the substitution $x = t$, then (3-8) becomes

$$A_0 = \frac{1}{2\pi} \int_{t=0}^{t=2\pi} f(t) \, dt \tag{A10-1}$$

Equation (A10-1) can be interpreted to mean that the average area under the curve over one period has a value A_0. We can see by inspection of Figures A10.1(a) and A10.1(b) that the value of A_0 is zero because the areas above and below the axis are equal. This could also have been determined analytically by integrating the function over one cycle and dividing by the length of the period. For Figure A10.1(a), the value of A_0 can be found from

$$A_0 = \frac{\omega}{2\pi} \int_{t=0}^{t=2\pi/\omega} \sin \omega t \, dt \tag{A10-2}$$

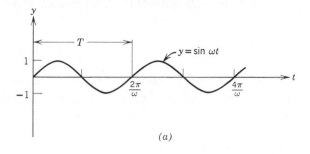

(a)

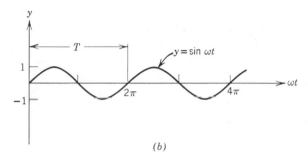

(b)

Figure A10.1   (a) Graph of the function of $y = \sin\omega t$ plotted on y and t axes. (b) Graph of the function $y = \sin\omega t$ and ωt axes.

where $\int_{t=0}^{t=2\pi/\omega} \sin\omega t \, dt$ is the area under the curve for one cycle and the length of a cycle is $2\pi/\omega$. The area under the curve is *divided* by the length of the period to determine the average value of the function over the one cycle. The values of A_0 for the sine function when plotted as shown in Figure A10.1(b) are found from

$$A_0 = \frac{1}{2\pi} \int_{\omega t=0}^{\omega t=2\pi} \sin\omega t \, d(\omega t) \qquad \text{(A10-3)}$$

We notice the limits of integration are changed to correspond to one cycle of the function being integrated. The integration is performed with respect to the way the function is plotted (i.e., for Figure A10.1(a) the integration is performed with respect to dt while in Figure A10.1(b) the function is integrated with respect to $d(\omega t)$. Dividing the integration by the length of one cycle (i.e., 2π for Figure A10.1(b)) gives the average value of A_0 for the function. Because (A10-3) is being integrated with respect to ωt and

ω is a constant, it can be brought outside the integral sign if we change the limits of integration correspondingly; that is,

$$A_0 = \frac{1}{2\pi} \int_{\omega t=0}^{\omega t=2\pi} \sin \omega t \, d(\omega t) = \frac{\omega}{2\pi} \int_{t=0}^{t=2\pi/\omega} \sin \omega t \, dt \qquad \text{(A10-4)}$$

We notice that equations (A10-4) and (A10-2) are identical even though they were determined in different manners. Figure A10.2 shows how each term in equation (A10-3) is related to the graph shown in Figure (A10.1(b)).

Figure A10.3 shows how the equation for A_0 will change depending on the function and how it is plotted.

The limits of integration are arbitrary except that the interval represented by them must be one cycle. Any of the following limits of integration would give equivalent results:

Variable of integration express in radians

$$\int_0^{2\pi} \quad \int_{-\pi}^{\pi} \quad \int_{-\pi/4}^{7\pi/4} \quad \int_{\pi/4}^{9\pi/4}$$

Variable of integration not expressed in radians

$$\int_0^T \quad \int_{-T/2}^{T/2} \quad \int_{-T/4}^{(3/4)T} \quad \int_{T/4}^{(5/4)T}$$

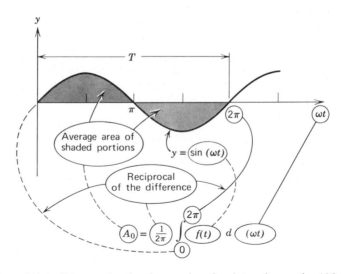

Figure A10.2 Diagram showing the meaning of each term in equation (A3-7).

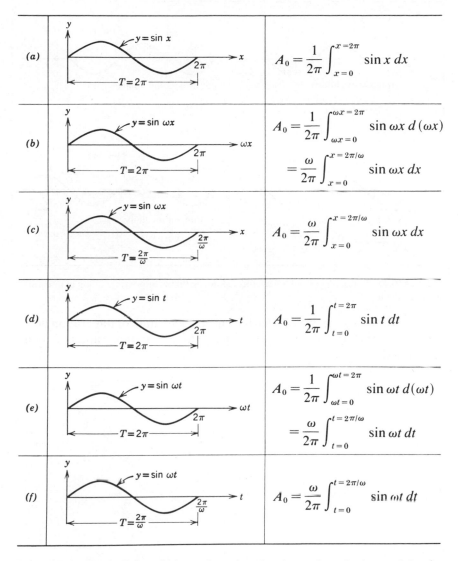

Figure A10.3 Graph of sinusoidal waveform plotted against various abscissas and showing how the corresponding value of A_0 is obtained.

Changing the limits of integration is equivalent to shifting the zero point. The graphs in Figure A10.4 show a sinusoidal waveform with different limits of integration imposed for one cycle. Integration over one

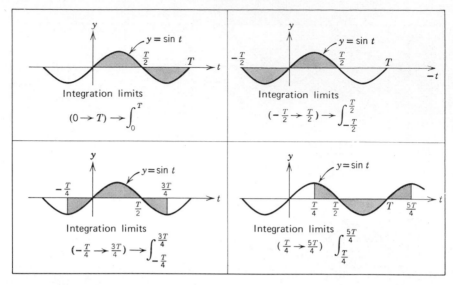

Figure A10.4 Plot of a sinusoidal waveform showing different possible limits of integration over one cycle.

cycle determines the area under the curve. The limits of integration determine the portion of the cycle over which the integration is being performed.

Appendix 11

As an example of other limit definitions of $\delta(t)$ we will consider a $(\sin kt)/kt$ function. Let us consider the integral,

$$\int_{-\infty}^{\infty} \frac{k}{\pi}\left[\frac{\sin kt}{kt}\right] dt = \frac{k}{\pi} \int_{-\infty}^{\infty}\left[\frac{\sin kt}{kt}\right] dt \qquad \text{(A11-1)}$$

The term under the integral $(\sin kt)/kt$ is an even function (i.e., it has the same values from 0 to ∞ as from 0 to $-\infty$.) Therefore we can rewrite the integral as

$$\int_{-\infty}^{\infty} \frac{k}{\pi}\left[\frac{\sin kt}{kt}\right] dt = \frac{2}{\pi} \int_{0}^{\infty}\left[\frac{\sin kt}{t}\right] dt \qquad \text{(A11-2)}$$

The integral $\int_0^{\infty} [\sin kt/t]\, dt$ is now in a standard form and can be found in a table of integrals to have a value of $\pi/2$. We now obtain the result from (A11-2) that

$$\int_{-\infty}^{\infty} \frac{k}{\pi}\left[\frac{\sin kt}{kt}\right] dt = 1 \qquad \text{(A11-3)}$$

If we take the limit as $k \to \infty$ we obtain

$$\lim_{k \to \infty} \int_{-\infty}^{\infty} \frac{k}{\pi}\left[\frac{\sin kt}{kt}\right] dt = 1 \qquad \text{(A11-4)}$$

and

$$\lim_{k \to \infty} \frac{k}{\pi}\left[\frac{\sin kt}{kt}\right] = 0 \qquad t \neq 0 \qquad \text{(A11-5)}$$

Equation (A11-5) becomes clear when we consider a $(\sin kt)/kt$ curve as k becomes larger. From Figures A11.1(a), (b), (c), and (d) we can see

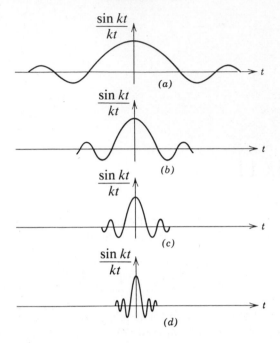

Figure A11.1.

that as k becomes larger the respective peaks move closer to the origin although the amplitudes of the respective peaks will be the same. In the limit as $k \to \infty$ the curve will be close to the origin although the individual peaks will still have the same respective amplitudes. Multiplying the $(\sin kt)/kt$ curve by k as $k \to \infty$ in effect makes the function go to infinity. Because the curve was clustered close about the origin (as $k \to \infty$), multiplying now by k drives the function to infinity. The function now fulfills the delta function requirement of being zero at all values except at $t = 0$. Then

$$\lim_{k \to \infty} \frac{k}{\pi} \left[\frac{\sin kt}{kt} \right] = \delta(t) \tag{A11-6}$$

The Fourier transform of a unit impulse function $\delta(t)$ is given by

$$F[\delta(t)] = \int_{-\infty}^{\infty} \delta(t)\, e^{-j\omega t}\, dt \tag{A11-7}$$

The integral on the right is readily evaluated by using the sifting property of a delta function which we will state here without proof.

$$\int_{-\infty}^{\infty} \delta(t - t_0) f(t)\, dt = f(t_0) \tag{A11-8}$$

Applying this property of a delta function to (A11-7), where $f(t) = e^{-j\omega t}$ and $t_0 = 0$, we obtain

$$F[\delta(t)] = [e^{-j\omega t}]_{t=0} = 1 \qquad\qquad \text{(A11-9)}$$

From this result we see that the spectral density of an impulse function is uniform; that is, all frequency components have the same relative amplitude.

Appendix 12

THE FOURIER TRANSFORM OF THE FUNCTION COS $\omega_0 t$ OVER A FINITE INTERVAL

Let us consider the function $y = \cos \omega_0 t$ as shown in Figure A12.1.

The function contains only three cycles as shown. The function is zero outside the region from $-3T/2$ to $+3T/2$. In order to find the Fourier transform we can use equation (3-81) from which we see that this finite function should be integrated from $-\infty$ to $+\infty$ as given in the following equation:

$$F(\underline{\cos \omega_0 t})_{3T/2} = \int_{-\infty}^{\infty} \underline{\cos \omega_0 t}\ e^{-j\omega t}\ dt \qquad (A12\text{-}1)$$

Where $\cos \omega_0 t$ has been underlined to indicate it existed only over the interval from $-3T/2$ to $+3T/2$. Since outside this interval the function is zero, we can find the Fourier transform by integrating over the finite interval from $-3T/2$ to $+3T/2$ or

$$F(\underline{\cos \omega_0 t}) = \int_{-3T/2}^{3T/2} \cos \omega_0 t\ e^{-j\omega t}\ dt \qquad (A12\text{-}2)$$

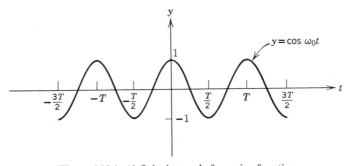

Figure A12.1 A finite interval of a cosine function.

Making use of the fact that

$$\cos \alpha = \frac{e^{j\alpha} + e^{-j\alpha}}{2} \qquad \text{(A12-3)}$$

and

$$\sin \alpha = \frac{e^{j\alpha} - e^{-j\alpha}}{2j} \qquad \text{(A12-4)}$$

we can write (A12-2) as

$$F(\cos \omega_0 t) = \int_{-3T/2}^{3T/2} \left[\frac{e^{j\omega_0 t} + e^{-j\omega_0 t}}{2} \right] e^{-j\omega t} \, dt \qquad \text{(A12-5)}$$

$$= \frac{1}{2} \int_{-3T/2}^{3T/2} \left[e^{(j\omega_0 t - j\omega t)} + e^{(-j\omega_0 t - j\omega t)} \right] dt \qquad \text{(A12-6)}$$

$$= \frac{1}{2} \int_{-3T/2}^{3T/2} \left[e^{jt(\omega_0 - \omega)} + e^{-jt(\omega_0 + \omega)} \right] dt \qquad \text{(A12-7)}$$

$$= \frac{1}{2} \int_{-3T/2}^{3T/2} e^{jt(\omega_0 - \omega)} \, dt + \int_{-3T/2}^{3T/2} e^{-jt(\omega_0 + \omega)} \, dt \qquad \text{(A12-8)}$$

$$= \frac{1}{2} \left\{ \left[\frac{e^{jt(\omega_0 - \omega)}}{j(\omega_0 - \omega)} \right]_{-3T/2}^{3T/2} + \left[\frac{e^{-jt(\omega_0 + \omega)}}{-j(\omega_0 + \omega)} \right]_{-3T/2}^{3T/2} \right\} \qquad \text{(A12-9)}$$

$$= \frac{1}{2} \left\{ \frac{e^{j(\omega_0 - \omega)(3T/2)}}{j(\omega_0 - \omega)} - \frac{e^{-j(\omega_0 - \omega)(3T/2)}}{j(\omega_0 - \omega)} - \frac{e^{-j(\omega_0 + \omega)(3T/2)}}{j(\omega_0 + \omega)} + \frac{e^{j(\omega_0 + \omega)(3T/2)}}{j(\omega_0 + \omega)} \right\}$$
$$\text{(A12-10)}$$

$$= \left[\frac{\sin (\omega_0 - \omega)(3T/2)}{(\omega_0 - \omega)} + \frac{\sin (\omega_0 + \omega)3T/2}{(\omega_0 + \omega)} \right] \qquad \text{(A12 11)}$$

$$= \frac{3T}{2} \left[\frac{\sin (\omega_0 - \omega)(3T/2)}{(\omega_0 - \omega)(3T/2)} + \frac{\sin (\omega_0 + \omega)(3T/2)}{(\omega_0 + \omega)(3T/2)} \right] \qquad \text{(A12-12)}$$

The two terms inside brackets of (A12-12) are plotted in Figure A12.2. From equation (A12-12) and Figure A12.2 it is apparent that each of these terms has the form of a sin x/x curve. Each of these terms can be plotted separately. A plot of the sum given by (A12-12) can be obtained by adding these two curves at each value of ω. From (A12-12) it is seen

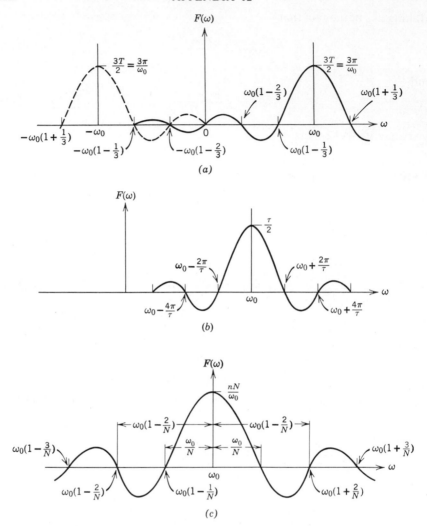

Figure A12.2   Plot of Fourier transform of cosine function of Fig. A12.1. (*a*) Plot of Fourier transform of cosine function of Fig. A12.1 (*b*) sin x/x curve centered about $+\omega_0$. (*c*) Plot of sin x/x in terms of cosine function of a finite interval of Fig. A12.1.

that a cosine waveform that is truncated (exists only within a finite interval) such as that shown in Figure A12.1 has a transform consisting of terms of the form (sin x)/x centered about the positive and negative value of the frequency of the waveform (ω_0).

At this point we will consider the effects of the interval length. If we had considered a general interval of length τ, the limits of integration in (A12-2) would have been $-\tau/2$ to $+\tau/2$ and the result corresponding to (A12-12) would have been

$$F(\underline{\cos \omega_0 t}) = \frac{\tau}{2}\left[\frac{\sin (\omega_0 - \omega)(\tau/2)}{(\omega_0 - \omega)(\tau/2)} + \frac{\sin (\omega_0 + \omega)(\tau/2)}{(\omega_0 + \omega)(\tau/2)}\right] \qquad \text{(A12-12a)}$$

Except for the peak location, the two terms inside the bracket have the same form and we will initially consider only the first term

$$\frac{\sin (\omega_0 - \omega)(\tau/2)}{(\omega_0 - \omega)(\tau/2)}$$

Figure A12.2(b) can be drawn immediately.

Now let us consider the interval length τ in terms of the number of cycles N in the interval and the period T of a cycle that gives

$$\tau = NT \qquad \text{(A12-13)}$$

The period of a cycle is given by

$$T = \frac{2\pi}{\omega_0} \qquad \text{(A12-14)}$$

Substituting in the equation for τ we obtain

$$\tau = \frac{2\pi N}{\omega_0} \qquad \text{(A12-15)}$$

Substituting the value of τ in (A12-15) in Figure A12.2(b) the location of the zero crossing (from ω_0) can be found in terms of the number of cycles N in the interval shown in Figure A12.1. For example, the first zero crossing (from ω_0) at $\omega_0 - 2\pi/\tau$ in Figure A12.2(b) becomes

$$\omega_0 - \frac{2\pi}{\tau} = \omega_0 - \frac{2\pi}{2\pi N/\omega_0} = \omega_0 - \frac{\omega_0}{N} = \omega_0\left(1 - \frac{1}{N}\right)$$

in Figure A12.2(c). The curves of Figure A12.2(a) are clearly the same as Figure A12.2(c) with $N = 3$.

As noted the distance from ω_0 to the first crossing (from ω_0) is

$$\omega_0 - \left[\omega_0\left(1 - \frac{1}{N}\right)\right] = \omega_0 - \omega_0 + \frac{\omega_0}{N} = \frac{\omega_0}{N}$$

The distance from ω_0 to the second crossing is

$$\omega_0 - \left[\omega_0\left(1 - \frac{2}{N}\right)\right] = \omega_0 - \omega_0 + \frac{2\omega_0}{N} = \frac{2\omega_0}{N}$$

The distance from ω_0 to the nth crossing is $n\omega/N$. Thus we find that as the interval (in Figure A12.1) is increased to indicate more cycles (that is, N increases) the distance between ω_0 (the center peak) and the first zero crossing (ω_0/N) gets smaller. All the zero crossings move closer to ω_0 as N becomes larger, which produces a sharper peak at ω_0. In the limit of an infinite number of cycles ($N \rightarrow \infty$) all the zero crossings approach ω_0 and the curve approaches the form of a delta function as previously shown analytically [see Appendix 11].

A significant fact that is indicated by the above discussion is that a compression in the time domain will show up as an expansion in the frequency domain and vice versa. We have seen that the time-domain signal $\cos \omega_0 t$ has frequency components at $\pm\omega_0$. The time-domain signal $\cos 2\omega_0 t$ represents a compression of $\cos \omega_0 t$ by a factor of two and will have frequency components at $\pm 2\omega_0$. From this we can see that the frequency spectrum has been expanded by a factor of two. This indicates that compression in the time domain is equivalent to expansion in the frequency domain and vice versa.

From (A12-12a) we see that the $(\sin x)/x$ terms are multiplied by $\tau/2$. The maximum amplitude of the function will occur when the $(\sin x)/x$ terms are equal to one. This occurs when $\omega = \pm\omega_0$ and the amplitude at these two points must be $\tau/2$ (or substituting $\tau = 2\pi N/\omega_0$ we have $\tau/2 = 2\pi N/2\omega_0 = \pi N/\omega_0$). For the special case of (A12-12a), where $N = 3, \tau/2 = 3T/2 = 3\pi/\omega_0$ as indicated in Figure A12.2(a).

THE GAUSSIAN FUNCTION

The Gaussian Function will transform into itself. Consider the function,

$$F(\omega) = \int_{-\infty}^{\infty} f(t) \, e^{-j\omega t} \, dt$$

and if

$$f(t) = e^{-t^2/2} \tag{A12-16}$$

then

$$F(\omega) = \int_{-\infty}^{\infty} e^{-t^2/2} \, e^{-j\omega t} \, dt \tag{A12-17}$$

$$= \int_{-\infty}^{\infty} e^{-(t^2/2 + j\omega t)} \, dt \tag{A12-18}$$

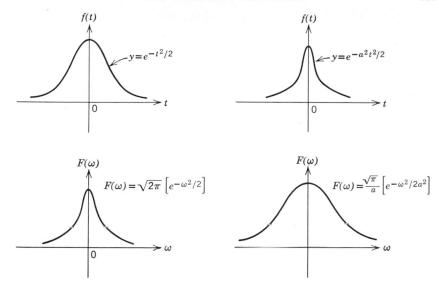

Figure A.12.3 Gaussian functions and their Fourier transforms.

completing the square in the exponent

$$F(\omega) = \int_{-\infty}^{\infty} e^{-1/2[t^2+2j\omega t+(j\omega)^2]} e^{(j\omega)^2/2} \, dt \qquad \text{(A12-19)}$$

$$= \int_{-\infty}^{\infty} e^{-1/2(t+j\omega)^2} e^{-\omega^2/2} \, dt \qquad \text{(A12-20)}$$

$\omega^2/2$ is a constant with respect to t and can be brought outside the integral

$$F(\omega) = e^{-\omega^2/2} \int_{-\infty}^{\infty} e^{-1/2(t+j\omega)^2} \, dt \qquad \text{(A12-21)}$$

letting $x = t + j\omega$, $dx = dt$

$$F(\omega) = e^{-\omega^2/2} \int_{-\infty}^{\infty} e^{-x^2/2} \, dx \qquad \text{(A12-22)}$$

The value of the integral can be found in integral tables to be,

$$\int_{-\infty}^{\infty} e^{-x^2/2} \, dx = \sqrt{2\pi} \qquad \text{(A12-23)}$$

and therefore

$$F(\omega) = [e^{-\omega^2/2}][\sqrt{2\pi}] \qquad \text{(A12-24)}$$

$e^{-\omega^2/2}$ is another Gaussian function and therefore the Fourier transform of a Gaussian function is another Gaussian function. More generally if

$$f(t) = e^{-a^2t^2/2} \tag{A12-25}$$

then

$$F(\omega) = \frac{\sqrt{\pi}}{a} e^{-\omega^2/2a^2} \tag{A12-26}$$

Appendix 13

CHANGE OF SCALE

When using Table A13.1, it is to be noted that the scale of the function can be changed to meet specific problem requirements. For example, the pulse function shown in the table as in Figures A13.1 might be used by changing the scale as in Figure A13.2. The scale change can be determined by the following considerations. We have previously shown that compression in the time domain is equivalent to expansion in the frequency domain and vice versa as shown above. Mathematically we can write a function $f(t)$ and its transform as $f(t) \leftrightarrow F(\omega)$. If we multiply the variable by a real constant a, then the Fourier transform of $f(at)$ is

$$F[f(at)] = \int_{-\infty}^{\infty} f(at)e^{-j\omega t}\, dt$$

We can now make a substitution letting $x = at$, then $dx = a\, dt$ and when $a > 0$

$$F[f(at)] = \int_{-\infty}^{\infty} f(x)e^{-j\omega(x/a)}\frac{1}{a}\, dx = \frac{1}{a}\int_{-\infty}^{\infty} f(x)e^{-j\omega(x/a)}\, dx$$

$$F[f(at)] = \frac{1}{a}F\left(\frac{\omega}{a}\right)$$

$$f(at) \leftrightarrow \frac{1}{a}F\left(\frac{\omega}{a}\right)$$

When $a < 0$, it can be shown that

$$f(at) \leftrightarrow \frac{-1}{a}F\left(\frac{\omega}{a}\right)$$

then

$$f(at) \leftrightarrow \frac{1}{|a|}F\left(\frac{\omega}{a}\right)$$

If $a = 2$ as in the example shown, then the pulse from $-T/2$ to $+T/2$ becomes a pulse from $-T$ to $+T$ as shown. The frequency plot changes correspondingly by a factor of $\frac{1}{2}$ as shown.

657

TABLE A13.1   FOURIER TRANSFORMS

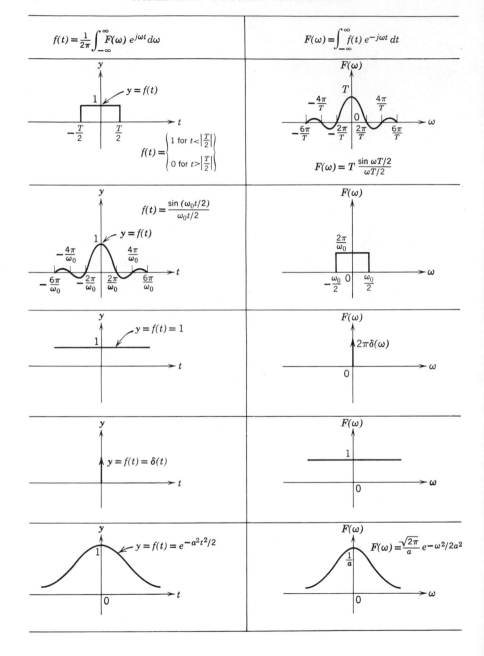

$$f(t) = \frac{1}{2\pi}\int_{-\infty}^{\infty} F(\omega)\, e^{j\omega t}\, d\omega \qquad\qquad F(\omega) = \int_{-\infty}^{\infty} f(t)\, e^{-j\omega t}\, dt$$

$$f(t) = \begin{cases} 1 \text{ for } t < \left|\dfrac{T}{2}\right| \\ 0 \text{ for } t > \left|\dfrac{T}{2}\right| \end{cases}$$

$$F(\omega) = T\,\frac{\sin \omega T/2}{\omega T/2}$$

$$f(t) = \frac{\sin (\omega_0 t/2)}{\omega_0 t/2}$$

$$y = f(t) = 1 \qquad\qquad 2\pi\delta(\omega)$$

$$y = f(t) = \delta(t)$$

$$y = f(t) = e^{-a^2 t^2/2} \qquad\qquad F(\omega) = \frac{\sqrt{2\pi}}{a}\, e^{-\omega^2/2a^2}$$

TABLE A13.1 FOURIER TRANSFORMS (continued)

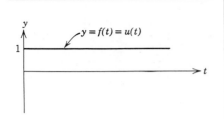

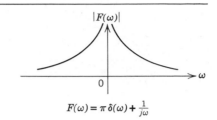

$$F(\omega) = \pi\,\delta(\omega) + \frac{1}{j\omega}$$

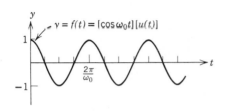

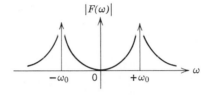

$$F(\omega) = \frac{\pi}{2}\,[\delta(\omega-\omega_0) + \delta(\omega+\omega_0)] + \frac{j\omega}{\omega_0{}^2 - \omega^2}$$

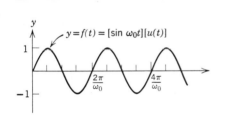

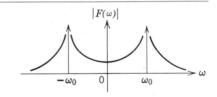

$$F(\omega) = \frac{\pi}{2j}\,[\delta(\omega-\omega_0) - \delta(\omega+\omega_0)] + \frac{\omega_0}{\omega_0{}^2 - \omega^2}$$

TABLE A13.1 FOURIER TRANSFORMS (continued)

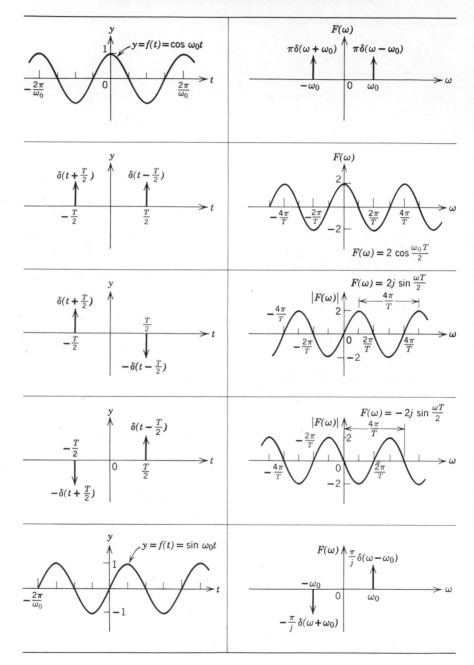

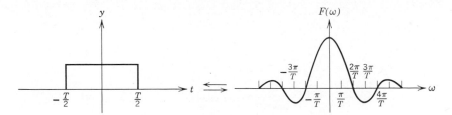

Figure A13.1.

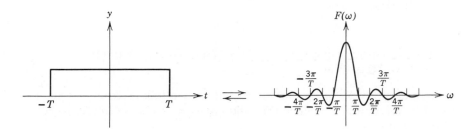

Figure A13.2.

Appendix 14

FOURIER TRANSFORM OF THE FUNCTION $\cos \omega_0 t$ AND THE SHIFTING THEOREM

Let us consider a cosine wave that is continuous from $-\infty$ to $+\infty$. Such a waveform is obtained by allowing the interval τ in (A12-12a) to go to infinity or

$$F(\cos \omega_0 t) = \lim_{\tau \to \infty} \frac{\tau}{2} \left[\frac{\sin (\omega_0 - \omega)(\tau/2)}{(\omega_0 - \omega)(\tau/2)} + \frac{\sin (\omega_0 + \omega)(\tau/2)}{(\omega_0 + \omega)(\tau/2)} \right]$$

$$(A14\text{-}1)$$

We have seen that in equation (A11-6) the limits are the same as for (A14-1) but we must put it in the same form. To do this we can rewrite (A14-1) as

$$F(\cos \omega_0 t) = \pi \lim_{\tau \to \infty} \frac{\tau}{2\pi} \left[\frac{\sin (\omega_0 - \omega)(\tau/2)}{(\omega_0 - \omega)(\tau/2)} + \frac{\sin (\omega_0 + \omega)(\tau/2)}{(\omega_0 + \omega)(\tau/2)} \right]$$

$$(A14\text{-}2)$$

Now the limit of (A14-2) is

$$F(\cos \omega_0 t) = \pi [\delta(\omega_0 - \omega) + \delta(\omega_0 + \omega)] \qquad (A14\text{-}3)$$

as indicated by equation (A11-6). This result is expected from Appendix 12 discussion where $N \to \infty$; that is, the only frequency components present will be at $\pm \omega_0$.

There is one more interesting fact that can be pointed out using the above example. We will show that a shift of ω_0 in the frequency domain is equivalent to multiplication by $e^{j\omega_0 t}$ in the time domain. To show this let us consider a function $f(t)$ and its Fourier transform,

$$f(t) \leftrightarrow F(\omega) \qquad (A14\text{-}4)$$

$$F(\omega) = F[f(t)] = \int_{-\infty}^{\infty} f(t) e^{-j\omega t} \, dt \qquad (A14\text{-}5)$$

Next we consider the Fourier transform of $f(t)e^{-j\omega_0 t}$,

$$F[f(t)e^{-j\omega_0 t}] = \int_{-\infty}^{\infty} f(t)e^{-j\omega_0 t}e^{-j\omega t}\, dt \tag{A14-6}$$

$$= \int_{-\infty}^{\infty} f(t)e^{-j(\omega-\omega_0)t}\, dt \tag{A14-7}$$

If we let $\omega' = \omega - \omega_0$, equation (A14-7) can be rewritten as

$$F[f(t)e^{j\omega_0 t}] = \int_{-\infty}^{\infty} f(t)\ e^{-j\omega' t}\, dt \tag{A14-8}$$

The right side of (A14-8) now has the same form as (A14-5) which means

$$F[f(t)e^{j\omega_0 t}] = F(\omega') \tag{A14-9}$$

Replacing ω' by $\omega - \omega_0$ as it was defined above, (A14-9) becomes

$$F[f(t)e^{j\omega_0 t}] = F(\omega - \omega_0)$$

where

$$F[f(t)] = F(\omega). \tag{A14-10}$$

We have thus shown that multiplication by a factor $e^{j\omega_0 t}$ in the time domain translates the whole frequency spectrum $F(\omega)$ by (ω_0). In communication theory this multiplication process is referred to as *modulation*. In communications work usually a signal $f(t)$ is multiplied by a sinusoid. Let us examine the modulation process in terms of the previously discussed examples. Let us assume we are multiplying a function $f(t)$ by the function $\cos \omega_0 t$. We have seen that the function $\cos \omega_0 t$ can be expressed as the sum of exponentials and therefore multiplication of $f(t)$ by $\cos \omega_0 t$ gives

$$f(t)\cos \omega_0 t = f(t)\frac{e^{j\omega_0 t} + e^{-j\omega_0 t}}{2} \tag{A14-11}$$

$$= \tfrac{1}{2}[f(t)e^{j\omega_0 t} + f(t)e^{-j\omega_0 t}] \tag{A14-12}$$

If we assume that $f(t)$ in (A14-12) is a rectangular pulse then by (A14-8) we can see that the $\sin x/x$ curve (produced by the rectangular pulse) will be translated by $+\omega_0$ and $-\omega_0$ giving the curve in Figure A12.2.

SHIFT IN TIME DOMAIN TRANSFORMS TO PHASE SHIFT IN FREQUENCY DOMAIN

The shifting theorem just shown indicated that a shift of ω_0 in the frequency domain is equivalent to multiplication by $e^{j\omega_0 t}$ in the time domain. We shall now show that a shift of τ in the time domain transforms to a multiplication by $e^{-j\omega\tau}$ in the frequency domain.

$$F(\omega) = \int f(t)e^{-j\omega t}\, dt$$

Consider a shift in time $f(t) \rightarrow f(t-\tau)$

$$F[f(t-\tau)] = \int f(t-\tau)\, e^{-j\omega t}\, dt$$

Similar to the method we used above we can let

$$x = t - \tau \qquad dx = dt$$

Substituting for t we obtain

$$\int f(x)e^{-j\omega(x+\tau)}\, dx = \int f(x)e^{-j\omega x}e^{-j\omega\tau}\, dx$$

$$= e^{-j\omega\tau}\int f(x)e^{-j\omega x}\, dx$$

$$= e^{-j\omega\tau}F(\omega)$$

Thus shift in time of $\pm\tau$ corresponds to phase term $e^{\mp j\omega\tau}$ in frequency domain.

Appendix 15

RONCHI RULING

A Ronchi ruling is used as an input to an optical filter as shown in Figure 7.1. Lens 1 forms the spectrum of the input Ronchi ruling. Lens 2 takes the Fourier transform of the spectrum formed by lens 1. Figure A15.1 is the image formed by the double Fourier transform of the input Ronchi ruling. Figure A15.1 is therefore an image of the input Ronchi ruling. Rug fibers were placed over the input Ronchi ruling to produce a noisy input. The imaged noisy ruling is shown in Figure A15.2. The spectrum of a Ronchi ruling was shown and discussed in Chapter 5. The spectrum of the noisy input is shown in Figure A15.3. A slit placed in the back focal plane of lens 1 will filter most of the noise from this spectrum. Figure A15.4 shows the filtered spectrum passed by the slit in the filter plane. Figure A15.5 shows the Fourier transform of the filtered spectrum. Figure A15.6 was underexposed to show the relative intensities of the lines formed.

Figure A15.1.

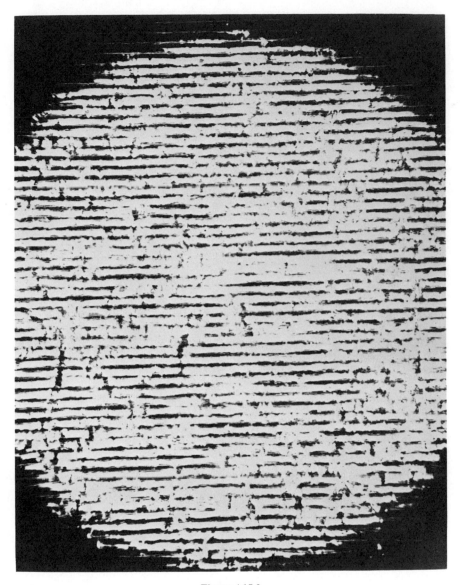

Figure A15.2.

Figure A15.3.

Figure A15.4.

Figure A15.5.

Figure A15.6.

671

Appendix 16

Certain images that have been degraded during the recording processes can be operated on to correct or compensate for the degradation. A degraded image that has been corrected to compensate for a specific degradation is called a *restored image*. Restoration of an image cannot be done without knowledge of the specific nature of the degradation. It is logical that no amount of image processing can increase the information content of an image. Processing any specific recorded image therefore usually means removal of undesirable noise or removal of extraneous portions of the image that have little or no bearing on the desired image.

A very simple illustration of image restoration was the example shown in Chapter 7 of removal of raster lines from a television image. Another simple example is that shown in Appendix 15. A Ronchi ruling is imaged through a two-lens system as shown in Figure 7.1. When the Ronchi ruling is buried in noise (rug fibers were placed over the Ronchi ruling and the random fibers created the noise image), a simple slit filter can eliminate most of the noise as shown in Figure A15.5.

Let us consider the feasibility of restoring images that have been blurred during exposure. This discussion is presented for the conceptual interest because the author is just starting work in this area. Since work is just starting in this area, results are not available; however, the reader will gain an appreciation for the difficulties that must be overcome in this type of work. The reader will gain understanding of how problems involving other types of motion can also be handled. Let us consider several different examples of blurred images created by relative motion between the camera and the entire scene. We shall assume that all the objects in the scene are in a fixed position relative to each other.

672

Case I: No Relative Motion ($v = 0$)

In the normal exposure in which there is no relative velocity between the scene and the camera the exposure (in one dimension for simplicity) is given by

$$E(x) = \int_0^T s(x)\, dt = Ts(x)$$

where $s(x)$ is the intensity being recorded at point x on the film, and 0 to T is the duration of the photographic film exposure. Such an exposure would produce an unblurred image because there is no relative motion between the camera and the scene.

Case II: Uniform Motion

Let us determine the exposure on the photographic film when the camera is in uniform motion during the entire exposure. In this case we can assume the camera to be travel-ing at a constant velocity v during the entire exposure. The camera can, however, be considered to be station-ary and the object to move so as to produce an image moving at constant velocity across the photographic film during the time interval from 0 to T (See Figure 16.1).

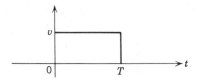

Figure A16.1.

The exposure along one dimension is then given by

$$E(x) = \frac{1}{v} \int_{x-vt}^{x} s(x')\, dx'$$

where x' is the coordinate of a point on the object (if there is no relative motion $x = x'$). We can rewrite this expression using two-step functions to represent the motion. A step function has a value of one only when the value in the parenthesis is positive (greater than zero). A step function $U(x-x')$ can be plotted as in Figures A16.2(a) and A16.2(b). The difference between these two pulses can be plotted as in Figure A16.2(c). We can therefore write the exposure

$$E(x) = \frac{1}{v} \int_{x-vt}^{x} s(x')\, dx'$$

$$= \frac{1}{v} \int_{-\infty}^{\infty} s(x')\left[U(x-x') - U(x-vT-x')\right] dx'.$$

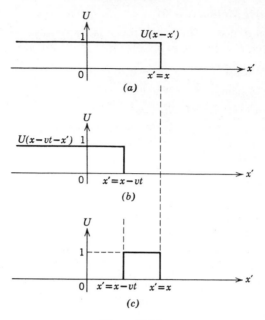

Figure A16.2.

Letting

$$R(x') = U(x') - U(x' - vT) = \begin{cases} 0 \begin{cases} x' < 0 \\ x' > vT \end{cases} \\ 1\{0 \leqslant x' \leqslant vT \end{cases}$$

then

$$E(x) = \frac{1}{v} \int_{-\infty}^{\infty} s(x') R(x - x') \, dx'.$$

This equation is in the form of a convolution and can be written as

$$E(x) = \frac{1}{v} s(x') \otimes R(x').$$

The Fourier transform of the convolution of two functions is the product of their respective spectrums, therefore

$$E(\omega) = \frac{1}{v} S(\omega) R(\omega).$$

From the Table A13.1, Fourier Transforms, and using the shifting theorem (A14-1) we obtain the Fourier transform of $R(x')$ as follows:

$$R(\omega) = (vT) [\exp(-j\omega vT/2)][\sin(\omega vT/2)/(\omega vT/2)].$$

Therefore

$$E(\omega) = (T)[\exp(-j\omega vT/2)][\sin(\omega vT/2)/(\omega vT/2)][S(\omega)]$$

Case III: Step Motion

Let us now consider the exposure obtained when the camera is held steady for an interval of time after the shutter has been opened and then the camera is moved at a constant velocity for an interval of time until the shutter is closed. A plot of the camera motion during the exposure time (from 0 to T) shows the camera motion to start at time T_1. Actually (Figure A16.3) the camera could be considered to be stationary and the object as moving so as to produce an image moving at constant velocity across the photographic film starting at time T_1. The exposure during the time interval from 0 to T_1 is similar to Case I and is given by

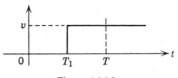

Figure A16.3.

$$E(x)_{0 \to T_1} = T_1 s(x).$$

The exposure during the time interval from T_1 to T is similar to Case II and is given by

$$E(x)_{T_1 \to T} = \frac{1}{v} s(x') \otimes R(x')$$

therefore the exposure during the time interval from 0 to T is the sum of the two exposures or

$$E(x)_{0 \to T} = E(x)_{0 \to T_1} + E(x)_{T_1 \to T} = T_1 s(x) + \frac{1}{v} s(x') \otimes R(x')$$

where

$$R(x') = \begin{cases} 0 \begin{cases} x' < 0 \\ x' > v(T - T_1) \end{cases} \\ 1\{0 < x' \leqslant v(T - T_1). \end{cases}$$

The Fourier transform of the exposure $E(x)_{0 \to T}$ is therefore

$$E(\omega) = \left\{ T_1 + (T - T_1) \exp\left[-j\frac{\omega v(T - T_1)}{2} \right] \left(\frac{\sin[\omega v(T - T_1)/2]}{[\omega v(T - T_1)/2]} \right) \right\} S(\omega)$$

Case IV: Uniform Pulsed Motion

Following similar reasoning we can see that camera motion given by Figure A16.4 can be expressed by

$$E(x)\big|_{0 \to T} = T_1 s(x) + (T - T_2) s[x - v(T_2 - T_1)] + \frac{1}{v} s(x') \otimes R(x')$$

where

$$R(x') = \frac{1}{0} \begin{cases} 0 \leqslant x' \leqslant v(T_2 - T_1) \\ \text{elsewhere} \end{cases}$$

The Fourier transform of the exposure therefore is given by

$$E(\omega) = \left[T_1 + (T - T_2)\, e^{-j\omega v(T_2 - T_1)} + (T_2 - T_1) \right.$$

$$\left. \exp\left[-j \frac{\omega v(T_2 - T_1)}{2} \right] \left\{ \frac{\sin\left[\omega v(T_2 - T_1)/2\right]}{[\omega v(T_2 - T_1)/2]} \right\} \right] S(\omega)$$

Let us now consider some of the photographic film characteristics discussed in Chapter 6. We have seen that the intensity transmission of a developed film is given by

$$\text{intensity transmission} = (E_1/E)^\gamma.$$

The amplitude transmission is given by

$$\text{amplitude transmission} = (E_1/E)^{\gamma/2} = g.$$

We can now write the amplitude transmission for each case above as follows:

Case I

$$g_1(x) = [E_1/E]^{\gamma/2} = [E_1/Ts(x)]^{\gamma/2} = K_1[1/s(x)]^{\gamma/2}$$

where $K_1 = [E_1/T]^{\gamma/2}$.

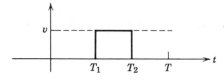

Figure A16.4.

Case II

$$g_{\text{II}}(x) = [E_1/E]^{\gamma/2} = [E_1]^{\gamma/2}[(1/v)(s(x') \otimes R(x'))]^{-\gamma/2}$$

$$= [E_1 v]^{\gamma/2}[S(x') \otimes R(x')]^{-\gamma/2} = K_{\text{II}}[s(x') \otimes R(x')]^{-\gamma/2}$$

where $K_{\text{II}} = [E_1 v]^{\gamma/2}$.

Case III

$$g_{\text{III}}(x) = [E_1/E]^{\gamma/2} = [E_1]^{\gamma/2}[T_1 s(x) + (1/v)(s(x') \otimes R(x'))]^{-\gamma/2}$$

$$= K_{\text{III}}[s(x)]^{-\gamma/2}[1 + (1/vT_1)\{[s(x') \otimes R(x')]/s(x)\}]^{-\gamma/2}$$

where $K_{\text{III}} = [E_1/T_1]^{\gamma/2}$.

Case IV

$$g_{\text{IV}}(x) = [E_1/E]^{\gamma/2} = [E_1]^{\gamma/2}[T_1 s(x) + (T - T_2)s'(x) + (1/v)\{s(x')$$
$$\otimes R(x')\}]^{-\gamma/2}$$

$$K_{\text{IV}}[s(x)]^{-\gamma/2}[1 + \{[(T - T_2)/T_1](s'(x)/s(x))\}$$

$$+ \{(1/vT_1)(\{s(x') \otimes R(x')\}/s(x))\}]^{-\gamma/2}$$

where $K_{\text{IV}} = [E_1/T_1]^{\gamma/2}$ and $s'(x) = s[x - v(T_2 - T_1)]$

If we now take this developed photographic film and place it in contact with an unexposed photographic film, this second exposed and developed photographic film will have an amplitude transmission function that we can write for each case as follows:

Case I. We have seen that the intensity transmission of an exposed and developed photographic film is given by

$$[E_1/E]^{\gamma_1} = [E_1]^{\gamma_1}[Ts(x)]^{-\gamma_1}.$$

Illuminating this photographic film uniformily with a light of intensity I and exposing a second photographic emulsion with the transmitted light produce a photographic film whose exposure is given by

$$E(x) = [E_1/Ts(x)]^{\gamma_1}It_2$$

where t_2 is the amount of time the second photographic film is exposed.

The amplitude transmission of the second developed photographic film is therefore given by

$$g'_{\mathrm{I}}(x) = [E_2/E]^{\gamma_2/2} = [E_2/(It_2)(E_1/Ts(x))^{\gamma_1}]^{\gamma_2/2}$$

$$= [(E_2 T^{\gamma_1})/(It_2 E_1^{\gamma_1})]^{\gamma_2/2}[s(x)]^{\gamma_1\gamma_2/2}$$

$$= K'_{\mathrm{I}}[s(x)]^{\gamma_1\gamma_2/2}$$

where

$$K'_{\mathrm{I}} = \mathrm{constant} = [(E_2 T^{\gamma_1})/(It_2 E_1^{\gamma_1})]^{\gamma_2/2}$$

Case II

$$g'_{\mathrm{II}}(x) = K'_{\mathrm{II}}[s(x') \otimes R(x')]^{\gamma_1\gamma_2/2}$$

where

$$K'_{\mathrm{II}} = \left[\frac{E_2}{t_2 I}\right]^{\gamma_2/2}[E_1 v]^{-\gamma_1\gamma_2/2}$$

Case III

$$g'_{\mathrm{III}}(x) = K'_{\mathrm{III}}[s(x)]^{\gamma_1\gamma_2/2}\left[1 + \frac{s(x') \otimes R(x')}{vT_1 s(x)}\right]^{\gamma_1\gamma_2/2}$$

where

$$K'_{\mathrm{III}} = [E_2/t_2 I]^{\gamma_2/2}[E_1/T_1]^{-\gamma_1\gamma_2/2}$$

Case IV

$$g'_{\mathrm{IV}}(x) = K'_{\mathrm{IV}}[s(x)]^{\gamma_1\gamma_2/2}\left[1 + \frac{(T-T_2)s'(x)}{T_1 s(x)} + \frac{s(x') \otimes R(x')}{vT_1 s(x)}\right]^{\gamma_1\gamma_1/2}$$

where

$$K'_{\mathrm{IV}} = [E_2/t_2 I]^{\gamma_2/2}[E_1/T_1]^{-\gamma_1\gamma_2/2}$$

and

$$s'(x) = s[x - v(T_2 - T_1)]$$

Setting the product $\gamma_1\gamma_2 = 2$, each of the above four cases reduces to

Case I

$$g'_{\mathrm{I}}(x) = K'_{\mathrm{I}}[s(x)]$$

Case II

$$g'_{\mathrm{II}}(x) = K'_{\mathrm{II}}[s(x') \otimes R(x')]$$

Case III

$$g'_{\mathrm{III}}(x) = K'_{\mathrm{III}}\left[s(x) + \frac{1}{vT_1}(s(x') \otimes R(x'))\right]$$

Case IV

$$g'_{IV}(x) = K'_{IV}\left[s(x) + \frac{(T-T_2)}{T_1}s'(x) + \frac{s(x') \otimes R(x')}{vT_1}\right]$$

The Fourier transform of each of the above cases is as follows:

Case I: $G'_I(\omega) = K'_I S(\omega)$

Case II: $G'_{II}(\omega) = K'_{II}S(\omega)R(\omega)$

Case III: $G'_{III}(\omega) = K'_{III}S(\omega)[1 + (R(\omega)/vT_1)]$

Case IV: $G'_{IV}(\omega) = K'_{IV}S(\omega)[1 + [(T-T_2)/T_1]\exp -j\omega v(T_2 - T_1)$
$$+ R(\omega)/vT_1\}$$

We have previously seen that a matched filter is one that eliminates any input phase terms in the output of the filter. In this particular application a filter is one that would eliminate any blurring terms from the output of the filter. We have seen that the blur terms have been represented by $R(\omega)$. In order to eliminate the $R(\omega)$ terms the filter should perform the function given by

$$G(\omega)H(\omega) = CS(\omega)$$

where C is a constant. The $H(\omega)$ term for each of the cases is as follows:

Case I: Since there is no motion, there is no filter required.

Case II:

$$G'_{II}(\omega) = K'_{II}S(\omega)R(\omega)$$

and

$$H(\omega) = K''_{II}/R(\omega)$$

where $R(\omega) \neq 0$.

Case III:

$$G'_{III}(\omega) = K'_{III}S(\omega)[1 + (R(\omega)/vT_1)]$$

$$H(\omega) = K''_{III}[1 + (R(\omega)/vT_1)]^{-1}$$

$$H(\omega) = K''_{III}\left[1 + \frac{T-T_1}{T_1}\left(\exp -j\omega v\frac{T-T_1}{2}\right)\frac{\sin \omega v(T-T_1)/2}{\omega v(T-T_1)/2}\right]^{-1}$$

Case IV:

$$G'_{IV}(\omega) = K'_{IV}S(\omega)[1 + \{(T-T_2)/T_1\}\{\exp(-j\omega v(T_2 - T_1))\}$$

$$+ R(\omega)/vT_1]$$

$$H(\omega) = K'_{IV}\left[1 + \frac{T-T_2}{T_1}(\exp - j\omega v(T_2 - T_1)) + \frac{R(\omega)}{vT_1}\right]^{-1}$$

$$H(\omega) =$$
$$K'_{IV}\left[1 + (\exp - j\omega v(T_2 - T_1))\left(\frac{T-T_2}{T_1} + \frac{T_2 - T_1}{T_1}\frac{\sin \omega v(T_2 - T_1)/2}{\omega v(T_2 - T_1)/2}\right)\right]^{-1}$$

For a look at some of the details involved in blur filters we will now restrict our discussion to Case II.

Case II: Since

$$R(\omega) = [vT][\exp(-j\omega vT/2)][\sin(\omega vT/2)/(\omega vT/2)]$$

then

$$H(\omega) = K'_{II}\frac{\exp j\omega vT/2}{vT\sin(\omega vT/2)/(\omega vT/2)}$$

where $\omega < 2\pi/vT$ otherwise the value of $H(\omega)$ becomes indeterminate.

Let us consider the action of this blur filter on various frequencies. We have seen that the curve for $R(\omega)$ follows a $(\sin x)/x$ curve as shown in the graph A of Figure A16.5. This means that at a value of

$$R(\omega) = 0$$

(i.e., when $\omega = +2\pi/vT$) the value of $H(\omega)$ becomes indeterminate because there is a zero in the denominator.

Let us consider graph B (Figure A16.5) which shows various image frequencies and their amplitudes (shown at constant levels). Since

$$G'(\omega) = K'S(\omega)R(\omega)$$

and

$$H(\omega) = K''/R(\omega)$$

then

$$G'(\omega)H(\omega) = CS(\omega)$$

The product of graph A and B (Figure A16.5(c)) (times a constant) shows how the various frequencies of the image are effected by the blurring. Graph C shows that every frequency other than the DC component will be attenuated by the blurring. It is possible to multiply graph C (Figure A16.5) by a curve which is the inverse of graph C as shown in graph D. This product will produce frequency components of constant amplitude. At frequencies of $\pm (2\pi n/vT)$ the filter output will be zero. Therefore any images containing frequencies equal to $\pm (2\pi n/vT)$ will

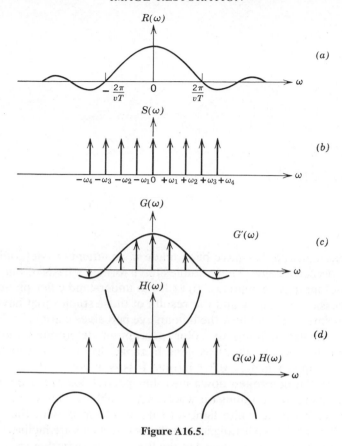

Figure A16.5.

be distorted (i.e., all frequency points on the $(\sin x/x$ curve that cross the ω axis will produce a zero output from the filter). Because most images do not contain discrete frequencies but rather a continuum of frequencies, the blur filter can only be made to function properly if the image frequencies are less than $\pm (2\pi/vT)$. This, unfortunately, means that the shortest period of an image component must be larger than vT. This makes this case trivial because the blur must be less than this smallest detail of interest in the image. If the frequencies at the $(\sin x)/x$ curve crossover points are known not to exist and the filter is made with more than the center loop of the curve, then for negative portions of the $(\sin x)/x$ curve, the product of graph A and B produces frequencies 180 degrees out of phase with the positive portions of the $(\sin x)/x$ curve. This makes the filter difficult to fabricate but then the elimination of a blur from an image has never been considered a simple task.

Appendix 17

COLOR

A great many studies have been made in an attempt to determine how humans perceive color. These studies have formed a basis of our present color photography techniques. In order to understand color photography it is necessary to understand the results of those studies that have determined the mechanics of how the human eye perceives color.

Initial studies relating to color perception determined the visible spectrum to include the wavelength range from approximately 0.4 microns to approximately 0.7 microns. It is well known that sunlight or white light can be broken down into the spectral colors it contains — for example, rainbows, oil films on water, soap bubbles. One of the familiar methods of breaking white light into its spectrum is with the use of a prism. The prism is so arranged that the two surfaces are inclined at some angle so the refraction produced at the first surface is further increased by the second surface. It is this process that increases the chromatic dispersion so the spectrum can be seen. Color is determined by the frequency of vibration — that is, the associated wavelength of the light. White light is composed of many wavelengths of light blended together. These wavelengths give rise to an infinite number of hues which make up the spectrum. These hues are usually grouped together broadly into six principle colors: red, orange, yellow, green, blue, violet. The angle of refraction of a prism varies with wavelength, shorter wavelengths being refracted more. No definite relationship covering all prisms exists between the wavelengths and the refraction angles. This means that prisms made of different substances will spread out the component colors of the spectrum to somewhat different extents. The normal eye can discern differences in wavelengths of 0.005 microns. Differences in hues do exist between colors whose wavelengths differ by less than 0.005 microns but it is not possible for the eye to distinguish this difference. Each hue is determined by a different frequency or wavelength of light. It would seem therefore that in order for the eye to determine color it would need a receptor in its sensitive area for each frequency of light that it can discern. This would mean that the eye would have in its sensitive area an infinite number of light-sensitive elements, each of which would be sensitive to light of but a single wavelength. This concept must be rejected as impracticable.

A more acceptable concept to explain how color is detected by the human eye is to accept a relationship between the eye and the brain. The sensitive receptors that make up the retina of the eye consist of two main types—rods and cones. Light falling on these receptors produces electric impulses that are transmitted along the optic nerves to the brain. It is the brain which interprets these many electric impulses into the sensation of sight.

One simple demonstration to show that different receptors are involved in sight makes use of a simple pendulum. A long pendulum (about 6 ft) is made to swing in a plane. An observer views the pendulum swinging from about 8 ft (from the plane in which the pendulum is swinging) standing symmetric with the pendulum swing. The observer can easily detect that the pendulum is swinging in a plane. The observer then puts a red filter over *one* eye (such as used in camera work for cloud accentuation) and views the pendulum with both eyes. The pendulum no longer appears to be swinging in a plane but now appears to be traveling in an ellipse. At the top of each swing (when the pendulum is motionless) it appears to be in the original plane. At the bottom of the swing (where it is traveling fastest) it appears to be furthest out of the actual plane of the swing. The direction of rotation that the pendulum appears to take will depend on which eye is being covered by the red filter (counter clockwise viewed from the pendulum pivot when the filter is over the right eye and clockwise when the filter is over the left eye). The explanation for this phenomena is that there are different receptor systems and the receptors for red light have a different response time than the white light receptors. The brain interprets this difference as a different location of the pendulum. At the top of the pendulum swing where it is motionless, it appears in its true location because the response of both optical receptor systems is fast enough to locate the pendulum in its true location. By placing a thin box in front of the plane of the pendulum swing, at the bottom of the pendulum swing the observer will see what appears to be the pendulum swinging through the box. Viewing horizontal motion on a television set using this red filter will also give an illusion of depth.

The retina of the eye corresponds to the film in a camera; that is, it is the sensitive receptor on which the light image is formed. We have stated that there are two main receptors in the retina — rods and cones. It is assumed that the purpose of the rods is primarily for the detection of weak light (as during twilight), being responsive to weak stimuli, but not being capable of distinguishing colors. The rods therefore probably are not involved in the above experiment. The cones function only in bright light and respond to various wavelengths of colors besides being more capable of distinguishing fine detail than are the night vision rods. Let us assume that the cone receptors can be considered to form three separate systems each responding to one third of the spectrum; that is, system one responds to violet and blue; system two responds to green and yellow; system three responds to orange and red. Each of these systems has a broad response; that is, the response of each system overlaps the adjacent system. Yellow light will excite the green-yellow system and also the orange-red system. It is the brain that interprets the electrical signals from each of the systems as a single response of yellow. It is possible to excite the green-yellow system with a green color and the orange-red system with a red so that the combination of the two colors is interpreted by the brain as yellow. The brain therefore can interpret true yellow impulses as yellow or can interpret a combination of green and red as yellow. These facts have been verified experimentally.

Experiment has shown that the simple mathematical equation shown below relates any four distinct colors as determined by the eye.

$$s(B) + t(C) + u(D) \equiv r(A)$$

where A, B, C, and D are four colors; r, s, t, and u are constants relating to units or amounts of each color; and "$\equiv$" is to be read "matches." The only restriction on the use of this equation is that no one of the four colors should be matchable by a mixture of any other two. This relation can be written as the algebraic equation

$$s + t + u = r$$

For example, if r units of color A are placed into one half of a chronometric field of a color-matching instrument, it can be matched by

s units of color B, plus
t units of color C, plus
u units of color D

Should any of the coefficients in the equations be negative, then these colors are added to color A and the mixture is then matched by the remaining colors.

When two colors are mixed together, the eye distinguishes this mixture as a third color. It cannot discern either of the two original colors in the third. The eye, therefore, cannot tell whether a given color is a sensation produced from a monochromatic light source (light of a single wavelength) or whether it is produced by a mixture of colors. Physical color-matching equipment operates differently from the eye in that the light is dispersed into a spectrum. Each of the component colors is then added together to determine the resulting hue. It is clear therefore that with regard to human perception it is necessary to take into account the physiology of what the eye sees when determining the color of a light. In color matching, as far as the eye is concerned, it is immaterial whether a monochromatic light or a color blend is used as a reference. The results will be the same since the eye cannot distinguish between them.

A better understanding of how the eye distinguishes between colors is obtained by an attempt to classify different colors into groups. Up to this point we have been separating the colors in accordance with their hue. Suppose, for example, we have a large number of colored cards which we wanted to classify into groups. We would normally place all the reds in one pile, the greens in another pile, and so on. This is straightforward and is usually the normal grouping that one thinks of when attempting to classify colors. However, it is not the only grouping that is possible. For example, it is possible to compare a red and a green. We could take a particular red card and look at its redness or intensity and then take a green card which closely approximates the intensity that appears on the red card; that is to say, the intensity of the red can be matched with the intensity of the green. We are thus classifying the intensities or brightness of different colors. Figure A17.1 shows a graph of the apparent brightness of colors (i.e. response of the eye to equal intensities) as a function of wavelength. Another type of color classification can be made in terms of saturation; that is, we can take a red which looks truly red and compare it with others which may not look so red. For example, a red mixed with white would still be red but would have a somewhat paler appearance.

As we previously pointed out a monochromatic light can be matched by a mixture of three independent colors. Three colors are independent when any one cannot be matched by any combination of the other two. It is possible therefore to make a chart that relates the quantity and proportions of three given independent colors required to match a given monochromatic light. When three independent colors are chosen, we consider these to be the primary colors of our system. It should be noted here that for any specific choice of real primaries there will be a few colors that cannot be matched. This fact was observed in the color equation as the possibility that one of the coefficients can be negative. The selection of

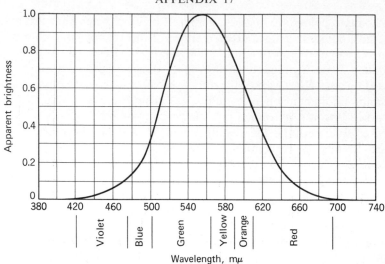

Figure A17.1   Apparent brightness (to the human eye) of light as a function of wavelength.

three colors as primaries is arbitrary (i.e., choosing the colors represented by B, C, D in the above equation to match a color A is arbitrary). Having chosen one set of primary colors it is possible to determine a second set of primaries from the first. The reason for this is that the specific values of the coefficients determined from the first set of primaries can be used to relate the second set of primaries. Once the relationship between the two sets of primaries is known it is possible to convert one into the other. The International Committee on Illumination has determined the coefficients of a hypothetical set of primaries to match the given wavelengths of monochromatic light. Selected values are tabulated below:

TABLE A17.1   Hypothetical
Color Coefficients

Wavelength (μ)	X	Y	Z
400	0.014	0.000	0.068
450	0.336	0.038	1.772
500	0.005	0.323	0.272
550	0.433	0.995	0.009
600	1.062	0.631	0.001
650	0.284	0.107	0.000
700	0.011	0.004	0.000

X, Y, and Z were chosen so that positive amounts of each could be mixed to match any given color regardless of hue or saturation. Thus they do not correspond to physical colors. They can be used in color equations, however, and the results can then be translated into a set of physical primaries in the manner described above. X, Y, and Z have the form of supersaturated colors (i.e., their saturation is greater than 100 percent). Their use offers two advantages. First, negative amounts of "color" are avoided in the equations. Second, Y has the property that it is also a direct quantative measure of the brightness of the resultant color of a mix of these three. Compare the values in the Y column with the values plotted in Figure A17.1.

The color equation implies the relation,

$$R = X + Y + Z$$

where R is the coefficient of the color being matched. For example 0.082 parts of a color with a wavelength of $400\,\mu$ can be matched by 0.014 parts of X, 0.000 parts of Y, and 0.068 parts of Z giving 0.082 parts of the color R.

Dividing the above equation by R gives

$$1 = \frac{X}{R} + \frac{Y}{R} + \frac{Z}{R}$$

Let $x = X/R$, $y = Y/R$, and $z = Z/R$. Then

$$x + y + z = 1$$

Here we see that only two of the coefficients x, y, z are independent; the knowledge of any two coefficients will determine the third. Let us take x and y as the independent coefficients. Using the values of X, Y, and Z are given in Table A17.1 the corresponding values for x and y can be determined by

$$x = \frac{X}{R} = \frac{X}{X+Y+Z} \qquad \text{and} \qquad y = \frac{Y}{X+Y+Z}$$

The values for x and y determined in this way are tabulated below.

TABLE A17.2

Wavelength (mμ)	x	y
400	0.17	0.00
450	0.16	0.02
500	0.01	0.54
550	0.30	0.69
600	0.63	0.37
650	0.73	0.27
700	0.74	0.26

The graph shown in Figure A17.2 is a plot of the values given in the above table. This plot is usually called a *chromaticity diagram* because it indicates the dominant hue and the saturation of any given color. Pure white, theoretically, should be an equal mixture of three primary colors; that is, $x = y = z = 0.33$. In Figure A17.2 we can see that $x = 0.33$ and $y = 0.33$, automatically determines that $z = 0.33$. This point has been labeled point A on the diagram.

The chromaticity diagram (Figure A17.2) is extremely valuable because it can graphically show all the known facts concerning mixing of colors. The outer perimeter of the curve depicts colors of complete purity or saturation. The wavelengths of these 100 percent pure colors are indicated on the perimeter of the curve. Colors that are not spectrally pure are located inside this curve. One such point is indicated by A. This point represents a mixture of equal parts of the three primary colors; that is, $x = y = z = 0.33$. The result of this mixture will appear to the eye as white light. Point B was selected such that its coordinates are $(0.2, 0.5)$. To obtain the dominant hue of point B a line is drawn between points A and B (point A corresponding to white light). This line is extended to intersect the spectrally pure curve at point C. Point C represents a wavelength of $510\,\text{m}\mu$. This monochromatic light (wavelengths, $510\,\text{m}\mu$) is the

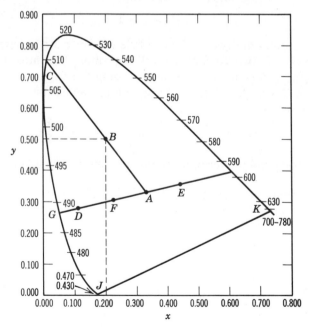

Figure A17.2   Chromaticity diagram.

dominant hue of the mixture represented by point B. The spectral purity of the color at point B will be represented by the ratio between the lines AB and AC or the ratio of 1 to 2.4. Therefore the spectral purity of the point B mixture is approximately 42 percent. Another way of stating this is to say that taking 2.4 parts of normal white light and one part of monochromatic light of wavelength 510 mμ will result in a color match of point B. The effects of mixing colors can also be demonstrated by the chromaticity diagram (Figure A17.2). Two colors D and E are to be mixed in the proportion of two parts D to one part of E. To determine the result connect a straight line between the points D and E and divide this line in the ratio of 2 to 1 parts. It is fairly obvious that the resulting mixture will be dominated by the color D, rather than E. Point F divides line DE in the ratio of two to one (i.e., $2DF = FE$) and since two parts of D were used, point F is situated on line DE closer to point D. The result of mixing colors D and E will be a color lying on a line joining D and E. In this particular case a line joining D and E also went through point A. Thus proportions of colors D and E could be mixed to give the appearance of white light. It is also clear from the chromaticity chart that an infinite number of pairs of points could have been selected such that a line joining them will pass through point A. The results of mixing any of these especially chosen colors (in the proper ratios) will appear to be white light. Colors which have this property are called *complimentary colors*. There are obviously an infinite number of complementary colors. The line joining D and E extended also intersects the 100 percent purity curve at 488 and 595 mμ. This means that when equal quantities of monochromatic light of these wavelengths are mixed, the resulting *sensation* will be white light (i.e., since $AH = AG$, equal parts of the two colors add to give the sensation of white light). It is possible to determine all the complementary pairs of colors by using the method above. Thus the sensation of white light can be produced by two monochromatic sources of proper wavelength, although white light from the sun normally contains most wavelengths.

Not all the colors to which the eye is sensitive appear in a natural spectrum of white light. Mixtures of red and blue are not present in a natural spectrum. Mixtures of these give rise to the pinks and violets. These pinks and violets which the eye can see as distinct colors have no counter part in the natural spectrum and no spectral matches of these colors are possible. Line JK on Figure A17.2 represents these pinks and violets on the chromaticity diagram.

Appendix 18

INTERFERENCE PATTERN BETWEEN TWO PLANE WAVES

We shall now use Figure A18.1 to indicate that the interference pattern between two coherent collimated light beams varies sinusoidally with intensity; that is, the intensity between two points of maximum intensity will vary sinusoidally. Figure A18.1(a) shows a representation of a monochromatic plane wave. The phase of the plane wave is constant in planes parallel to the x,y plane at any value of z at the instant shown. The amplitude of the plane curve is given by its y value. Figure A18.1(b) shows an instantaneous representation of the amplitude of two coherent, monochromatic-collimated, equal-amplitude light beams. If we consider the representation of the two waves as somehow solid, then as each wave moves in the direction of its travel the whole this wave will also move in the same direction. Because both waves travel in different directions (separate by an angle θ) but at the same speeds, the intersection of the crests of the waves will change position with respect to each other. Figure 18.1(c) shows the interference between two plane waves perpendicular to each other at an instant when the maximums of each wave are just arriving (and leaving) the area of interference. This can be seen more easily in Figure A18.2(a).

Figure A18.2(a) shows a top view of Figure A18.1(c) at an instant when the two waves are intersecting so as to produce four points of maximum constructive interference as shown. At these four points both waves are at their maximums and will produce points double the amplitude of either light beam. Let us follow the path of the point of constructive interference at the top of Figure A18.2(a). Figure A18.2(b) shows a representation of these two light beams an instant later than that shown in Figure A18.2(a). The wave peaks in both light beams have moved equal distances in their respective directions of travel. At this instant there will be only

690

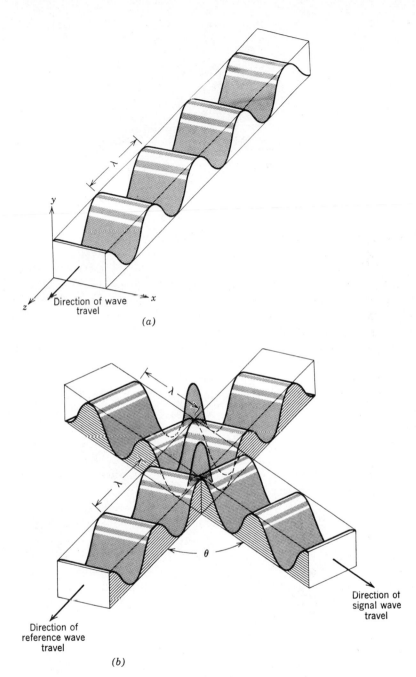

Direction of wave travel

(a)

Direction of reference wave travel

Direction of signal wave travel

θ

(b)

Figure A18.1.

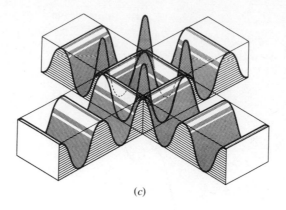

(c)

Figure A18.1(c).

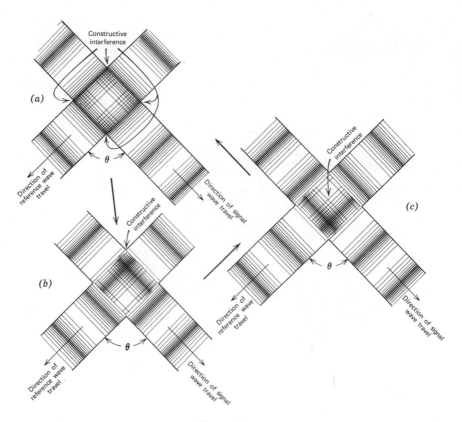

Figure A18.2.

one point of maximum constructive interference. The point at the top of Figure A18.2(a) has moved down to a position shown in Figure A18.2(b). Figure A18.2(c) shows a representation of these two light beams an instant later. The wave peaks in both light beams have moved equal distances in their respective directions of travel. At this instant the point of maximum constructive interference has moved down from the position shown in Figures A18.2(a) and A18.2(b). The point of maximum constructive interference is moving along a line that bisects the angle between the two light beams. Figure A18.2(a) now is a representation of an instant later than that shown in Figure A18.2(c). Points of maximum constructive interference are now at the bottom of Figure A18.2(a) and a new point appears at the top of Figure A18.2(a). The new point of maximum constructive interference at the top of Figure A18.2(a) is caused by the wave crests directly following the two that we have been following. Because the path the point of maximum constructive interference travels down is a line that bisects the angle between the two light beams, we can conclude that any photographic plate placed along this line will have a uniform (maximum) exposure and no information can be recorded.

Figure A18.3 shows a larger representation of the signal and reference beams with the photographic film placed in various positions. With the photographic film in position 1 it will be uniformally exposed to the maximum constructive interference value. In position 2 it is aligned with the plane wavefronts of the signal beam. In position 2 points a, b, c, and d will be exposed to the maximum constructive interference value. Between maximum points of exposure the exposure will be a smaller value. Actually the exposure will follow a sinusoidal curve between the points of maximum exposure; for example, between points a and b the exposure will follow a sinusoid, reaching a maximum at points a and b. At a point halfway between points a and b there will be a minimum. In position 3 only points a, e, f, g, and h will receive an exposure of a value equal to that of the maximum constructive interference value. Position 3 is perpendicular to the direction of travel of the points of maximum constructive interference (dotted lines) — that is, perpendicular to position 1. The highest photographic film resolution is required with the photographic plate in this position because the separation between the points of maximum interference (on the photographic plate) will be a minimum. In position 4 the photographic plate is aligned with the plane waves of the reference beam and points a, i, and j will be points of maximum constructive interference.

We will now determine the form of the intensity variations of the interference pattern between the two plane waves. In Figure A18.4 a signal and reference beam is shown making an angle $\theta/2$ with the y axis. If we assume that a photographic plate is on the x axis, that the signal beam

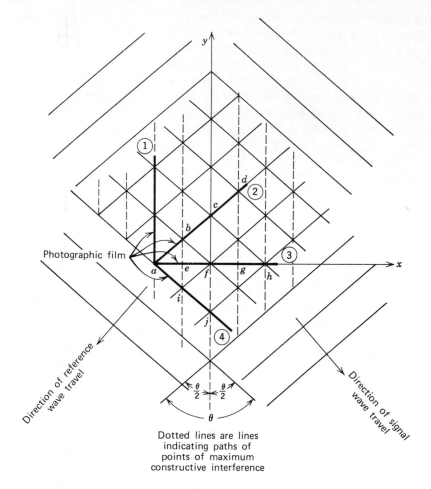

Photographic film

Dotted lines are lines
indicating paths of
points of maximum
constructive interference

Figure A18.3.

amplitude is S, and that the reference beam amplitude is R, then the signal
and reference beams can be represented as

$$\text{signal beam} = S \sin (k\alpha - \omega t)$$

and

$$\text{reference beam} = R \sin (k\beta - \omega t)$$

where $k = 2\pi/\lambda$. For example, $\sin (k\beta - \omega t)$ is a sine wave of wavelength
λ traveling in the positive β direction and having an angular frequency of

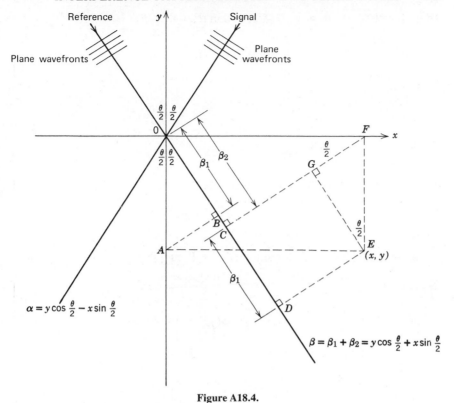

Figure A18.4.

ω, and t represents time. The sum of the amplitudes of the signal and reference beams can be expressed

$$\text{sum} = S \sin (k\alpha - \omega t) + R \sin (k\beta - \omega t)$$

From Figure A18.4 we can see that point E has coordinates (x,y) which will be intersected by a reference-beam plane wave when it is in position D. We would therefore like to find the length β in terms of the coordinates of point E. In triangle OAB we see that

$$\cos \frac{\theta}{2} = \frac{\overline{OB}}{\overline{AO}} = \frac{\beta_1}{y}$$

In triangle EFG we have

$$\cos \frac{\theta}{2} = \frac{\overline{EG}}{\overline{EF}} = \frac{\overline{EG}}{y}$$

then $\beta_1 = \overline{EG}$ and $\beta_1 = y \cos \theta/2$, and in triangle OFC

$$\sin \frac{\theta}{2} = \frac{\overline{OC}}{\overline{OF}} = \frac{\beta_2}{x}$$

or

$$\beta_2 = x \sin \frac{\theta}{2}$$

then

$$\overline{OD} = \beta_1 + \beta_2 = y \cos \frac{\theta}{2} + x \sin \frac{\theta}{2} = \beta$$

In a similar manner α can be found to be

$$\alpha = y \cos \frac{\theta}{2} - x \sin \frac{\theta}{2}$$

then

$$\text{sum} = S \sin \left(ky \cos \frac{\theta}{2} - kx \sin \frac{\theta}{2} - \omega t \right) + R \sin \left(ky \cos \frac{\theta}{2} + kx \sin \frac{\theta}{2} - \omega t \right)$$

If we let $S = R$ then

$$\text{sum} = S \sin \left(ky \cos \frac{\theta}{2} - kx \sin \frac{\theta}{2} - \omega t \right) + S \sin \left(ky \cos \frac{\theta}{2} + kx \sin \frac{\theta}{2} - \omega t \right)$$

$$= S \sin \left[\left(ky \cos \frac{\theta}{2} - \omega t \right) - \left(kx \sin \frac{\theta}{2} \right) \right]$$

$$+ S \sin \left[\left(ky \cos \frac{\theta}{2} - \omega t \right) + \left(kx \sin \frac{\theta}{2} \right) \right]$$

Making use of the relationships

$$\sin (B - C) = \sin B \cos C - \cos B \sin C$$

$$\sin (B + C) = \sin B \cos C + \cos B \sin C$$

we can write

$$\text{sum} = S \left[\sin \left(ky \cos \frac{\theta}{2} - \omega t \right) \cos \left(kx \sin \frac{\theta}{2} \right) \right.$$

$$- \cos \left(ky \cos \frac{\theta}{2} - \omega t \right) \sin \left(kx \sin \frac{\theta}{2} \right) \right]$$

$$+ S \left[\sin \left(ky \cos \frac{\theta}{2} - \omega t \right) \cos \left(kx \sin \frac{\theta}{2} \right) \right.$$

$$+ \cos \left(ky \cos \frac{\theta}{2} - \omega t \right) \sin \left(kx \sin \frac{\theta}{2} \right) \right]$$

$$= 2S \left[\sin \left(ky \cos \frac{\theta}{2} - \omega t \right) \cos \left(kx \sin \frac{\theta}{2} \right) \right]$$

If $S \neq R$ then

$$\text{sum} = S \sin \left(ky \cos \frac{\theta}{2} - kx \sin \frac{\theta}{2} - \omega t \right)$$

$$+ R \sin \left(ky \cos \frac{\theta}{2} + kx \sin \frac{\theta}{2} - \omega t \right)$$

$$= S \sin \left[\left(ky \cos \frac{\theta}{2} - \omega t \right) - \left(kx \sin \frac{\theta}{2} \right) \right]$$

$$+ R \sin \left[\left(ky \cos \frac{\theta}{2} - \omega t \right) + \left(kx \sin \frac{\theta}{2} \right) \right]$$

$$= S \sin \left(ky \cos \frac{\theta}{2} - \omega t \right) \cos \left(kx \sin \frac{\theta}{2} \right)$$

$$- S \cos \left(ky \cos \frac{\theta}{2} - \omega t \right) \sin \left(kx \sin \frac{\theta}{2} \right)$$

$$+ R \sin \left(ky \cos \frac{\theta}{2} - \omega t \right) \cos \left(kx \sin \frac{\theta}{2} \right)$$

$$+ R \cos \left(ky \cos \frac{\theta}{2} - \omega t \right) \sin \left(kx \sin \frac{\theta}{2} \right)$$

$$= (S + R) \sin \left(ky \cos \frac{\theta}{2} - \omega t \right) \cos \left(kx \sin \frac{\theta}{2} \right)$$

$$+ (R - S) \cos \left(ky \cos \frac{\theta}{2} - \omega t \right) \sin \left(kx \sin \frac{\theta}{2} \right)$$

The intensity is found by finding the average of the amplitude squared over a cycle or

$$\text{intensity} = \frac{1}{T} \int_T |A|^2 \, dt$$

When

$$S = R$$

$$I = \frac{1}{T} \int_T \left[2S \sin \left(ky \cos \frac{\theta}{2} - \omega t \right) \cos \left(kx \sin \frac{\theta}{2} \right) \right]^2 dt$$

$$I = \frac{4S^2 \cos^2 kx \sin (\theta/2)}{T} \int_T \sin^2 \left(ky \cos \frac{\theta}{2} - \omega t \right) dt$$

Making use of the relationship

$$\sin^2 \left(\frac{B}{2} \right) = \frac{1}{2} - \frac{\cos B}{2}$$

we have

$$I = \frac{4S^2 \cos^2 kx \sin (\theta/2)}{T} \int_T \left[\frac{1}{2} - \frac{\cos 2(ky \cos (\theta/2) - \omega t)}{2} \right] dt$$

$$= \frac{4S^2 \cos^2 kx \sin (\theta/2)}{2T} \int_T [1 - \cos 2(ky \cos (\theta/2) - \omega t)] \, dt$$

$$= \frac{2S^2 \cos^2 kx \sin (\theta/2)}{T} \left[\int_T dt - \int_T \cos 2(ky \cos (\theta/2) - \omega t) \, dt \right]$$

$$= \frac{2S^2 \cos^2 kx \sin (\theta/2)}{T} [T - 0] = 2S^2 \cos^2 kx \sin (\theta/2)$$

The integral of a cosine function over a cycle is zero and since

$$\cos^2 B = \tfrac{1}{2} + \tfrac{1}{2} \cos 2B$$

$$I = S^2 \left[1 + \cos \left(2kx \sin \frac{\theta}{2} \right) \right]$$

When $S \neq R$, we found the amplitude to be

$$\text{sum} = (S + R) \sin \left(ky \cos \frac{\theta}{2} - \omega t \right) \cos \left(kx \sin \frac{\theta}{2} \right)$$

$$+ (R - S) \cos \left(ky \cos \frac{\theta}{2} - \omega t \right) \sin \left(kx \sin \frac{\theta}{2} \right)$$

The intensity is found from

$$I = \frac{1}{T} \int_T |A|^2 \, dt$$

$$= \frac{1}{T} \int_T \left[(S + R) \sin \left(ky \cos \frac{\theta}{2} - \omega t \right) \cos \left(kx \sin \frac{\theta}{2} \right) \right.$$

$$\left. + (R - S) \cos \left(ky \cos \frac{\theta}{2} - \omega t \right) \sin \left(kx \sin \frac{\theta}{2} \right) \right]^2 \, dt$$

Multiplying this out we can eliminate the cross-product terms because the integral of a sine times a cosine term over a cycle is zero giving

$$I = \frac{1}{T} \int_T (S + R)^2 \sin^2 \left(ky \cos \frac{\theta}{2} - \omega t \right) \cos^2 \left(kx \sin \frac{\theta}{2} \right) dt$$

$$+ \frac{1}{T} \int_T (R - S)^2 \cos^2 \left(ky \cos \frac{\theta}{2} - \omega t \right) \sin^2 \left(kx \sin \frac{\theta}{2} \right) dt$$

$$= \frac{(S+R)^2}{T} \cos^2\left(kx \sin\frac{\theta}{2}\right) \int_T \frac{1}{2}\left(1 - \cos 2\left(ky \cos\frac{\theta}{2} - \omega t\right)\right) dt$$

$$+ \frac{(R-S)^2}{T} \sin^2\left(kx \sin\frac{\theta}{2}\right) \int_T \frac{1}{2}\left[1 + \cos 2\left(ky \cos\frac{\theta}{2} - \omega t\right)\right] dt$$

$$= \frac{(S+R)^2}{T} \cos^2\left(kx \sin\frac{\theta}{2}\right)\left[\int_T \frac{1}{2} dt - \int_T \frac{1}{2} \cos 2\left(ky \cos\frac{\theta}{2} - \omega t\right) dt\right]$$

$$+ \frac{(R-S)^2}{T} \sin^2\left(kx \sin\frac{\theta}{2}\right)\left[\int_T \frac{1}{2} dt + \int_T \frac{1}{2} \cos 2ky \cos\frac{\theta}{2} - \omega t\right] dt$$

Since the integral of a cosine function over one cycle is zero, we have

$$I = \frac{(S+R)^2}{2} \cos^2\left(kx \sin\frac{\theta}{2}\right) + \frac{(R-S)^2}{2} \sin^2\left(kx \sin\frac{\theta}{2}\right)$$

$$= \frac{S^2 + 2RS + R^2}{2} \cos^2\left(kx \sin\frac{\theta}{2}\right) + \frac{R^2 - 2RS + S^2}{2} \sin^2\left(kx \sin\frac{\theta}{2}\right)$$

$$= \left[\frac{S^2 + R^2}{2}\right] + RS\left[\cos^2\left(kx \sin\frac{\theta}{2}\right) - \sin^2\left(kx \sin\frac{\theta}{2}\right)\right]$$

$$= \frac{S^2 + R^2}{2} + RS \cos\left(2kx \sin\frac{\theta}{2}\right)$$

We shall now consider the actual intensity curve exposing the photographic plate for any given angle θ between the signal and reference beam and any angle φ between the reference beam and photographic plate. Figure A18.5 shows a general case where the photographic plate being exposed is on the line marked LL which makes an angle $\pi/2 - \theta/2 - \varphi$ with the x axis as shown. We have previously seen that when $S = R$ the intensity recorded on a photographic plate on the x axis is given by

$$I = S^2\left[1 + \cos\left(2kx \sin\frac{\theta}{2}\right)\right]$$

From Figure 18.5 we can see that

$$\sin\left(\frac{\theta}{2} + \varphi\right) = \frac{x}{l}$$

or

$$x = l \sin\left(\frac{\theta}{2} + \varphi\right)$$

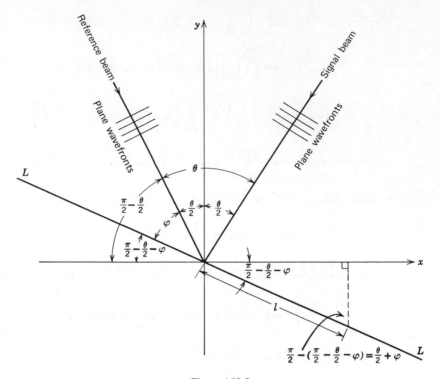

Figure A18.5.

then

$$I = S^2 \left[1 + \cos \left\{ 2kl \sin \left(\frac{\theta}{2} + \varphi \right) \sin \frac{\theta}{2} \right\} \right]$$

Since s, k, θ, and φ are constants for any given exposure, the intensity will follow a biased sinusoidal curve as given by this equation. The sinusoid will vary around a value (bias) of S^2 and will have a peak value of $2S^2$ and a minimum of zero. When $S \neq R$, we found the intensity to be given by

$$I = \frac{S^2 + R^2}{2} + RS \cos \left(2kx \sin \frac{\theta}{2} \right)$$

when the photographic plate was on the x axis. With the photographic plate on the line LL we have the intensity given by

$$I = \frac{S^2 + R^2}{2} + RS \cos \left[2kl \sin \left(\frac{\theta}{2} + \varphi \right) \sin \frac{\theta}{2} \right]$$

Since S, R, k, θ, and φ are constants for any given exposure of the photographic plate, it will be exposed to a biased sinusoidal curve as given by the equation. The bias value is

$$\frac{S^2 + R^2}{2}$$

and the peak value of the exposure is

$$\frac{S^2 + R^2}{2} + RS$$

The minimum value of exposure is $(S^2 + R^2)/2 - RS$.

References

ARTICLES

1. "Spatial Filtering Cleans Up Deep Space TV Noise," *EDN*, **26**, 28 (October 1968).
2. Lowell Rosen, "The Pseudoscopic Inversion of Holograms," *Proc. IEEE*, 118 (January 1967).
3. George W. Stroke et al., "Holographic Synthesis of Computer Generated Holograms," *Proc. IEEE,* 110, 111, (January 1967).
4. L. H. Lin, "Increase of Hologram-Image-Separation by Total Reflection," *Appl. Optics*, **6**, 2004 (November 1967).
5. Howard M. Smith, "Photographic Relief Images," *J. Opt. Soc. Am.*, **58**, No. 4, 533–539 (April 1968).
6. James Melcher, "Complex Waves," *IEEE Spectrum,* 86–103 (October 1968).
7. Adrianus Korpel, "Acoustic Imaging and Holography," *IEEE Spectrum*, 45–52 (October 1968).
8. Richard N. Einhorn, "Advances in Lasers," *Electronic Design*, **19**, 50–58, (September 12, 1968).
9. Richard N. Einhorn, "Harnessing the Laser to Electronic Systems," *Electronic Design*, **19**, 59–70 (September 12, 1968).
10. George W. Stroke, "A New Holographic Method for a Posteriori Image-Deblurring Restoration of Ordinary Photographic Using 'Extended-Source' Lens less Fourier-Transform, Holography Compensation," *Phys. Rev. Letters*, **27A**, No. 7, 405–406, (August 26, 1968).
11. George W. Stroke et al., "A Posteriori Holographic Sharp-Focus Image Restoration from Ordinary Blurred Photographs of Three-Dimensional Objects Photographed in Ordinary White Light," *Phys. Rev. Letters,* **26A**, No. 9, 443–444, (March 25, 1968).
12. George W. Stroke et al., "A Posteriori Image-Correcting 'Deconvolution' by Holographic Fourier-Transform Division," *Phys. Rev. Letters,* **25A**, No. 2, 89–90 (July 31, 1967).
13. George W. Stroke et al., "On the Absence of Phase-Recording or 'Twin-Image' Separation Problems in 'Garbor' (in-line) Holography," *Brit. J. Appl. Phys.,* **17**, 497–500, (1966).
14. Dr. Joseph L. Horner, "Do Computation at the Speed of Light," *Electronic Design*, **23**, 60–68 (November 7, 1968).
15. R. P. Chambers and J. S. Courtney-Pratt, "Bibliography on Holograms," *J. SMPTE*, **75**, 373–435 (April 1966).

702

16. E. N. Leith and J. Upatnieks, "Photography by Laser," *Sci. Am.*, **212**, 24–35 (June 1965).

17. Winston E. Koch, "Microwave Holography," *Microwaves*, 46–54 (November 1968).

18. Robert J. Collier, "Holography and Integral Photography," *Phys. Today*, 54–63 (July 1968).

19. Dr. D. J. DeBitetto, "A Holographic 3-D Movie with Constant-Velocity Film Transport," *Laser Focus*, 36–37 (September 1968).

20. M. B. Dobrin, "Optical Processing in the Earth Sciences," *IEEE Spectrum*, 59–66 (September 1968).

21. Robert Haawind, "Optical Computers Poised for Systems Role," *Electronic Design*, **8**, 25–32 (April 1968).

22. E. N. Leith and J. Upatnieks, "The Way-Out Wonderful World of Holography," *IEEE Student J.*, 2–9 (March 1966).

23. Kendall Preston, Jr., "Computing at the Speed of Light," *Electronics*, 72–83 (September 6, 1965).

24. L. J. Cutrona, E. N. Leith, and L. J. Porcello, "Optical Data Processing and Filtering Systems," *IRE Trans. Information Theory,* **IT-6**, No. 3, 386–400 (June 1960).

25. Edward L. O'Neill, "Spatial Filtering in Optics," *IRE Trans. Information Theory*, 56–65, (June 1956).

26. Peter Elias et al., "Fourier Treatment of Optical Processes," *J. Opt. Soc. Am.*, **42**, No. 2, 127–134 (February 1952).

27. J. Elmer Rhodes, Jr., "Analysis and Synthesis of Optical Images," *Am. J. Phys.*, **21**, 337–343 (May 1953).

28. Gerald Oster, "Theoretical Interpretation of Moiré Patterns," *J. Opt. Soc. of Am.*, **54**, No. 2, 169–175 (February 1964).

29. George C. Sherman, "Integral-Transform Formulation of Diffraction Theory," *J. Opt. Soc. of Am.*, **57**, No. 12, 1490–1498 (December 1967).

30. George C. Sherman, "Application of the Convolution Theorem to Rayleigh's Integral Formulas," *J. Opt. Soc. Am.*, **57**, No. 4, 546–547 (April 1967).

31. George C. Sherman, "Reconstruction Wave Forms with Large Diffraction Angles," *J. Opt. Soc. Am.*, **57**, No. 9, 1160–1168 (September 1967).

32. Undertake Research for Optical Memory Data Processing, July 1968. "Optical Memory Acts Like a Kaleidoscope." *Electronic Design*, **17**, 52–56 (August 1968).

33. Adam Kozma, "Photographic Recording of Spatially Modulated Coherent Light," *J. Opt. Soc. Am.*, **56**, No. 4, 428–432 (April 1966).

34. David G. Falconer, "Role of the Photographic Processing Holography," *Phot. Sci. Eng.*, **10**, No. 3, 133–139 (May–June 1966).

35. Gerald B. Brandt, "Techniques and Applications of Holography," *Electro-Technology*, 53–72 (April 1968).

36. "Holograms in the Round," *EDN*, 33 (May 1968).

37. Don B. Neumann, "Holography of Moving Scenes," *J. Opt. Soc. Am.*, **58**, No. 4, 447–454 (April 1968).

38. L. H. Lin and C. V. LoBranco, "Experimental Techniques in Making Multicolor White Light Reconstructed Holograms," *Appl. Opt.*, **6**, No. 7, 1255–1258 (July 1967).

39. D. H. R. Vilkomerson and D. Bostwick, "Some Effects of Emulsion Shrinkage on a Hologram's Image Space," *Appl. Opt.*, **6**, No. 7, 1270–1272 (July 1967).

40. Don B. Neumann and Harold W. Rose, "Improvement of Recorded Holographic Fringes by Feedback Control," *Appl. Opt.*, **6**, 1097–1104 (June 1967).

41. Viktor Met, "Working with Etalons," *Laser Technology*, 45–54 (September 1967).

42. George W. Stroke, "Holographic Image Deblurring Methods," *Laser Focus*, 42–43 (November 1968).

43. Karl Stetson, "Total Internal Reflection Holography," *Laser Focus*, 30–31, (November 1968).
44. A. Van der Lugt, "Signal Detection by Complex Spatial Filtering," *IEEE Trans. Information Theory*, 139–145 (April 1964).
45. Milton B. Dobrin et al., "Velocity and Frequency Filtering of Seismic Data Using Laser Light," *Geophysics*, **XXX**, No. 6, 1144–1178, (December 1965).
46. David C. Beste and Emmett N. Leith, "An Optical Technique for Simultaneous Beam Forming and Cross Correlation," *IEEE Trans. Aerospace Electronic Systems*, 376–384 (July 1966).
47. R. V. Pole, "3D Imagery and Holograms of Objects Illuminated in White Light," *Appl. Phys. Letters*, **10**, No. 1, 20–22, (January 1967).
48. Konrad Weniger, "A New Type of Display-Photochromic," *The Electronic Engineer*, 13–14 (June 1968).
49. Christopher Lucas, "A Four-Second Monobath," *Phot. Sci. Eng.*, **10**, No. 5, 259–262 (September–October 1966).
50. Margarete Ehrlich, "Characteristics of Dosmeter Films Processed in Phenidone-Thesulfate Mono-baths," *Phot. Sci. Eng.*, **9**, No. 1, 1–9 (January–February 1965).
51. A. R. Shulman, "Principles of Optical Data Processing for Engineers, *GSFC X-Document*, 521–66–434 (August 1966).
52. Gerald J. Grebowsky, "Fourier Transform Representation of an Ideal Lens in Coherent Optical Systems," *GSFC X-Document*, 521–68–292 (August 1968).

BOOKS

1. Max Born, Emil Wolf, *Principles of Optics*, 3rd ed. New York: Pergamon, 1965.
2. George W. Stroke, *An Introduction to Coherent Optics and Holography*. New York: Academic, 1966.
3. John B. DeVelis and George O. Reynolds, *Theory and Applications of Holography*. Reading, Mass.: Addison-Wesley, 1967.
4. Leo Levi, *Applied Optics, A Guide to Optical System Design*. New York: Wiley, 1968.
5. C. A. Taylor and H. Lipson, *Optical Transforms, Their Preparation and Application to X-Ray Diffraction Problems*. Ithaca, New York: Cornell University Press, 1964.
6. Grant Haist, *Monobath Manual*. Hastings-on-Hudson, New York: Morgan and Morgan, 1966.
7. C. B. Neblette, *Photography, Its Materials and Processes*. Princeton, New Jersey: Van Nostrand, 1962.
8. C. E. Kenneth Mees, *The Theory of the Photograph Process*. New York: MacMilliam, 1966.
9. Keith M. Hornsby, *Sensitometry in Practice*. London: Henry Greenwood, 1957.
10. Edward L. O'Neill, *Introduction to Statistical Optics*. Reading, Mass.: Addison-Wesley, 1963.
11. R. A. Houstoun, *Physical Optics*. Glasgow: Blackie, 1958.
12. R. Kingslake, *Applied Optics and Optical Engineering*. New York: Academic, 1965.
13. Francis A. Jenkins and Harvey E. White, *Fundamentals of Optics*. New York: McGraw-Hill, 1957.
14. Bruno Rossi, *Optics*. Reading, Mass.: Addison-Wesley, 1957.
15. R. W. Ditchburn, *Light*. Glasgow: Blackie, 1963.
16. Y. W. Lee, *Statistical Theory of Communication*. New York: Wiley, 1964.
17. J. Guild, *Interference Systems of Crossed Diffraction Gratings*. Oxford: Oxford University Press, 1956.

18. John Guild, *Diffraction Gratings as Measuring Scales*. Oxford: Oxford University Press, 1960.
19. John M. Stone, *Radiation and Optics*. New York: McGraw-Hill, 1963.
20. *The Focal Encyclopedia of Photography*. London: Focal Press, 1965.
21. H. Baines, *The Science of Photography*. New York: Wiley, Second Edition 1967.
22. Maurice Francon, *Diffraction*. Paris: Gauthier-Villars, 1964.
23. R. C. Jennison, *Fourier Transforms and Convolutions for the Experimentalist*. New York: Pergamon, 1961.
24. John A. Jamison, Raymond H. McFee, et al., *Infrared Physics and Engineering*. New York: McGraw-Hill, 1963.
25. William L. Wolfe, *Handbook of Military Infrared Technology*. Washington, D.C.: U.S. Government Printing Office, 1965.
26. Julius Adams Stratton, *Electromagnetic Theory*. New York: McGraw-Hill, 1941.
27. Charles A. Holt, *Introduction to Electromagnetic Fields and Waves*. New York: Wiley, 1963.
28. James T. Tippett, *Optical and Electro-Optical Information Processing*. The M.I.T. Press, 1965.
29. Harold V. Soule, *Electro-Optical Photography at Low Illumination Levels*. New York: Wiley, 1968.
30. Hugh Hildreth Skilling, *Fundamentals of Electric Waves*. New York: Wiley, 1942.
31. Joseph W. Goodman, *Introduction to Fourier Optics*. New York: McGraw-Hill, 1968.

Index

Aberrations, chromatic, 51
  lens, 51
  spherical, 51
Accelerators, 551
Acoustical holography, 541
Airy pattern, 241
Amplitude and intensity, 76
Angle, grazing, 25
  incidence, 7, 25
  polarizing, 92
  reflection, 7
  refraction, 15, 28
Angular frequency, 107
Aperture, relative, 53
Autocorrelation, 135, 335

Band pass filter, 326
Bas-relief, 556
Bichromated gelatin, 557
Block-diagram, 358
Bragg effect, 86, App. 5
Brewster, Sir David, 92
  Law of polarization, 93
Buffering, 551

Camera, pinhole, 52
  step and expose, 36
Characteristic curve, 310
Chromatic aberrations, 51
Circle of confusion, 57
Collimated light, 360
Coma, 51
Complex conjugate, 133
Coherent, fiber bundle, 41
  light beam, 73
Coherence, spatial, 73
  temporal, 74

Collimated light, 63, 360
Color, 89, App. 17
Convolution, 144
  and correlation, 147
  and Fourier Transforms, 148
  functions, 148
Convolved, 347
Complex amplitude, 389
Complex amplitude spectrum, 358
Complex notation, 115
Computer generated holograms, 543
Contact printing, 554
Correlation, 135
  autocorrelation, 135, 335
  crosscorrelation, 140, 334
  double side band, 384
  function, 340
Correlator, 223
Critical angle, 29
Crosstalk, 351
Curvilinear distortions, 51
Curvature, radius, 41

Day-nite mirrors, 23
D.C. spectral point, 97, 241
Delta function, 361
Density, 308
Depth, of field, 57
  of fucus, 57
Diffracted light, 389
Diffraction formula, 171, 389
Developing silver halide, 559
Developers, 560
Developing, 550
Diffraction, 78
  formula, 171, 389
  grating, 88

of light, 80
  patterns, 85, App. 7, 8
Directional filter, 326
Dispersion, 31, 88
Displacement, 135
Distance, hyperfocal, 58
  viewing, 57
Distortion, barrel, 51
  curvilinear, 51
  multiple color, 35
  pin cushion, 51
Double sideband correlation, 384
Drying film, 554

Electrochromic, 322
Emulsion, resolution, 515
  shrinkage correction, 558
  thickness changes, 86, 558
Equation, lens, 49
Euler's formula, 219
Eye resolution, 57

Fiber optics, 39
Field, depth of, 57
  of view, 55
Film processing, 554
Film resolution nomograph, 522
Fine grain developers, 568
First harmonic, 98, 241
Fixing, film, 552
Flow diagram, 359
Fly's eye lens, 544
  /number, 53, 195
Focal, length, 48
  plane, 48
  point, 43
  properties of lens, 170
Focus, depth of, 57
Focusing diffraction pattern, 87
Fountain, luminous, 32
Fourier, cosine transform, 123
  integrals, 106
  series, 95
  sine transform, 123
  transform, 120, 207, 358
  inverse transform, 122
Fourier transform, by diffraction, 156
  hologram, 470
  of a constant, 131
  of a Fourier transform, 208

of a rectangular pulse, 125
Fresnel, diffraction, 387
  zone plate, 413, 422
Fundamental, 98

Gabor hologram, 469
Gamma product, 320
Grating, diffraction, 88
Guides, light, 33

Hardened gelatin, 557
Hardeners, 553
High, contrast developers, 562
  energy developers, 566
  resolution films, 580
  speed developers, 566
High-pass filter, 325
Hologram, 444, 469
  360°, 542
Holographic, images, 460
  subtraction, 525
Holography, 453
  moving objects, 531
Huygen's principle, 79, 81, 89
Hyperfocal distance, 58

Image, multiple, 37
  object location, 44
  real, 44
  specular, 8
  virtual, 44, 460, 506
Immersion oil, 321
Impulse function, 362
Increasing film speed, 582
In-depth hologram, 516
Index of refraction, 12, 29, 31, 51
Intensity, and amplitude, 76
  distribution, 400
Interference, 64, 71
  patterns, 510, App. 7, 8
Interferometer, 400
Inverse Fourier transform, 122, 207

Kaleidoscope, 25
Kostinsky effect, 576

Laser operation, 91, App. 3
Length, focal, 48
Lens, abberations, 51
  aperture restriction, 193

coating, 55
equation, 49
plane, 47
thin, 48
Lenses, 46
Lensless Fourier holography, 495
Light, collimated, 63
diffraction, 80
guides, 33
interference, 71
polarized, 89, 91
velocity, 12, 14
white, 32
Lightbeams, coherent, 73
collimated, 63
non-collimated, 63
Lippmann color process, 581
Low contrast developers, 564
Low-pass filter, 326
Luminous fountain, 32

Magnification, spherical, 45
Matched filter, 134
Measurement, small displacements, 543
temporal coherence, 74, App. 4
Mirrors, concave, 45
convex, 45
day-night, 23
inclined reflection, 10
making, 20, App. 1
prismatic, 25
spherical, 41
Moire patterns, 81, App. 6
Monobaths, 570
M-Q developers, 561
Multiple, image recording, 346
slits diffraction, 81

nth harmonic, 101
Negative, 308
images, 554
Neutralizer, 552
Newton's rings, 422, 458
Normal contrast developers, 561
nth order wavefront, 85

Object, image location, 44
Obliquity factor, 174, 389
Oil matching index of refraction, 86
Opacity, 309

Optical, axis, 47
correlator, 223, 334, 376
filtering, 324, 370
memories, 39
path, 394, 424
tape printout, 39
tunnel, 36
orthoscopic, 548
path difference, 70

Pattern, diffraction, 85, App. 7
interference, 81
Moiré 81, App. 6
Phase, changes, 86
distortions, 573
holograms, 574
plates, 526
Phasors, 66
Photochromic, 322
Photographic, characteristics, 302, 580
film basics, 307
techniques, 308
Physical development, 568
Plane, focal, 48
lens, 47
wavefronts, 388, 466
Pinhole, 52
Point, focal, 43
source, 7, 360, 472, 482
white light source, 510
Polarization, angle, 92
Brewster's Law, 93
plane, 91
Positive, 308
images, 554
POTA developer, 564
Power spectrum, 132, 358, 399, 417
and autocorrelation, 143
Preservative, 551
Primary focal point, 48
Principal ray, 48
Printout, optical tape, 39
Prisms, 30
Projection printing, 554
Pseudoscopic, 547

Radius of curvature, 41
Rayleigh-Sommerfeld diffraction formula,
171
Ray, principal, 48

Real images, 44, 460, 507
Reconstruct a colored image, 351
Recording technique, 322
Reducers, 572
Reflectance holograms, 509
Reflections, air to glass, 17
  color distortion, 35
  inclined mirror, 10
  Law of, 7, 9
  multiple, 9, 18
  point source, 6
  polarizing by, 92
  specular, 8
  total, 28, 33
Refraction, 13
  angle, 15
  index, 12, 29, 31, 51
  ray displacement, 15
  oil matching index, 86
Relative aperture, 53
Relief images, 556
Resolution, 421
  eye, 57
  lens, 57
Restrainer, 552
Reversal printing, 555
Rinsing, 553
Ronchi ruling, 240

Schlieren photography, 58
Second harmonic, 101
Short stop, 552
Silver halide, 554, 573
Sine-wave zone plate, 422
Single side band correlator, 381
Snell's law, 15, 29

Solarization, 311
Special purpose developers, 564
Spectrum, analysis, 87
  analyzer, 238, 365
  energy density, 133
  power, 132
Spherical mirror magnification, 45
Square wave, 96, 402
Stabilization of fringes, 538
Standing waves, 510
Steady state, 97
Subtraction of images, 523
Sun pictures, 549

Tanning developers, 567
Temporal coherence, 514
Thermoplastics, 323
Transmission, 309
  function, 309, 401, 446
  hologram, 502
Triethanolamine, 558
Tunnel, optical, 36

Viewing, distance, 57
  holograms, 512
Virtual images, 44, 460, 506

Wavefront, 61
  first order, 82
  nth order, 85
  second order, 81
  zero order, 81
Wave trains, 62

Zero-order, 240
Zone plates, 387, 457